Mechanical E

BTEC National Engineering Specialis

Second Edition

Alan Darbyshire

AMSTERDAM • BOSTON • HEIDELBERG • LONDON • NEW YORK • OXFORD
PARIS • SAN DIEGO • SAN FRANCISCO • SINGAPORE • SYDNEY • TOKYO
Newnes is an imprint of Elsevier

Newnes is an imprint of Elsevier
Linacre House, Jordan Hill, Oxford OX2 8DP, UK
30 Corporate Drive, Suite 400, Burlington, MA 01803, USA

First edition 2003
Second edition 2008

British Library Cataloguing in Publication Data
A catalogue record for this book is available from the British Library

Library of Congress Cataloging-in-Publication Data
A catalog record for this book is available from the Library of Congress

ISBN: 978-0-7506-8657-0

For information on all Newnes publications visit our website at www.elsevierdirect.com

Typeset by Charon Tec Ltd., A Macmillan Company.
(www.macmillansolutions.com)

Printed and bound in Slovenia

08 09 10 10 9 8 7 6 5 4 3 2 1

Contents

Acknowledgements

The author and publisher would like to thank the following people for their contribution to the second edition:

W Bolton for writing chapter 6 "Electro, Pneumatic and Hydraulic Systems and Devices"

Mike Tooley for allowing us to adapt his chapters in *Engineering A Level* and *Higher National Engineering* for use as chapter 5 "Engineering Design"

Chapter opener and cover illustrations

Photo of train (chapter 1) courtesy of iStockphoto, Remus Eserblom, Image # 4619117

Photo of racing car (chapter 2) courtesy of iStockphoto, Jan Paul Schrage, Image # 4955692

Photo of turbine (chapter 3) courtesy of iStockphoto, Tomas Bercic, Image # 4056469

Photo of aircraft (chapter 4) courtesy of iStockphoto, Dan Barnes, Image # 4941515

Photo of CNC machine (chapter 5) courtesy of iStockphoto, Shawn Gearhart, Image # 1798391

Photo of robotic system (chapter 6) courtesy of iStockphoto, Paul Mckeown, Image # 5129181

Introduction

Welcome to the challenging and exciting world of engineering! This book has been written to help get you through six specialist units of the revised BTEC National Certificate and Diploma awards in Engineering. It provides the essential underpinning knowledge required of a student who wishes to pursue a career in engineering.

The book has been written by a highly experienced further education lecturer, who has over 30 years of practical teaching experience, with contributions from specialist lecturers in Engineering Design and Pneumatics and Hydraulics. Throughout the book I have adopted a common format and approach with numerous student activities, examples, end of chapter review questions and key points.

About the BTEC National Certificate and Diploma

The BTEC National Certificate and National Diploma qualifications have long been accepted by industry as appropriate qualifications for those who are about to enter industry or who are receiving training at the early stages of employment in industry. At the same time, these qualifications have become increasingly acceptable as a means of gaining entry into higher education.

BTEC National programmes in engineering attract a very large number of registrations per annum such that there are in excess of 35,000 students currently studying these qualifications in the UK by both part-time and full-time modes of study.

The BTEC National syllabus was recently reviewed and extensively updated and new programmes have been launched with effect from September 2007. The new scheme is likely to be adopted by all institutions that currently offer the programme as well as a number of others who will be offering BTEC qualifications for the first time.

Many organizations have contributed to the design of the new BTEC National Engineering programme including the Qualifications and Curriculum Authority (QCA), the Engineering Council and several Sector Skills Councils (SSC).

The Engineering Council continues to view the BTEC National Certificate/Diploma as a key qualification for the sector. They also recognize that BTEC National qualifications are frequently used as a means of entry to higher education courses, such as HNC/HND programmes and Foundation Degree courses.

How to use this book

This book covers six of the most popular specialist units that are common to many of the BTEC Engineering programmes. Each chapter covers one unit and contains *Text*, *Key points*, *Test your knowledge questions*, *Examples*, *Activities* and *Review questions*.

The *Test your knowledge questions* are interspersed with the text throughout the book. These questions allow you to check your understanding of the preceding text. They also provide you with an opportunity to reflect on what you have learned and consolidate this in manageable chunks.

Most *Test your knowledge questions* can be answered in only a few minutes and the necessary information, formulae, etc., can be gleaned from the surrounding text. *Activities*, on the other hand, make excellent vehicles for gathering the necessary evidence to demonstrate that you are competent in key skills. Consequently, they normally require a significantly greater amount of time to complete. They may also require additional library or resource area research time coupled with access to computing and other information technology resources.

Many tutors will use *Test your knowledge questions* as a means of reinforcing work done in class while *Activities* are more likely to be 'set work' for students to do outside the classroom. Whether or not this approach is taken, it is important to be aware that this student-centred work is designed to complement a programme of lectures and tutorials based on the BTEC syllabus. Independent learners (i.e. those not taking a formal course) will find complete syllabus coverage in the text.

The *Examples* not only show you how to solve simple problems but also help put the subject matter into context with typical illustrative examples. In order to successfully tackle this work you will need to have a good scientific calculator (and get to know how to use it).

Finally, here are some general points to help you with your studies:

- Allow regular time for reading – get into the habit of setting aside an hour, or two, at the weekend. Use this time to take a second look at the topics that you have covered during the week or that you may have not completely understood.
- Make notes and file these away neatly for future reference – lists of facts, definitions and formulae are particularly useful for revision!
- Look out for the inter-relationship between subjects and units – you will find many ideas and a number of themes that crop up in different places and in different units. These can often help to reinforce your understanding.
- Don't expect to find all subjects and topics within the course equally interesting. There may be parts that, for a whole variety of reasons, don't immediately fire your enthusiasm. There is nothing unusual in this; however, do remember that something that may not appear particularly useful now may become crucial at some point in the future!
- However difficult things seem to get – don't be tempted to give up! Engineering is not, in itself, a difficult subject, rather it is a subject

that demands logical thinking and an approach in which each new concept builds upon those that have gone before.

- Finally, don't be afraid to put your new ideas into practice. Engineering is about doing – get out there and do it!

Good luck with your BTEC Engineering studies!

Alan Darbyshire

Trains travelling at speeds in excess of 200 kmh^{-1} have greatly reduced the journey time to continental cities. Mechanical principles are applied in the calculation of the tractive effort required to maintain these speeds and the braking forces required to bring the trains to rest. Mechanical principles are also applied in the design of tilting mechanisms and the track infrastructure.

Photo courtesy of iStockphoto, Remus Eserblom, Image# 4619117

Chapter 1

Further Mechanical Principles and Applications

The design, manufacture and servicing of engineered products are important to the nation's economy and well-being. One has only to think of the information technology (IT) hardware, aircraft, motor vehicles and domestic appliances we use in everyday life to realise how reliant we have become on engineered products. A product must be fit for its purpose. It must do the job for which it was designed for a reasonable length of time and with a minimum of maintenance. The term 'mechatronics' is often used to describe products which contain mechanical, electrical, electronic and IT systems. It is the aim of this chapter to broaden your knowledge of the underpinning mechanical principles which are fundamental to engineering design, manufacturing and servicing.

Engineering Structures

Loading systems

Forces whose lines of action lie in a single plane are called *coplanar forces*. If the lines of action pass through a single point, the forces are said to be *concurrent* and the point through which they pass is called the *point of concurrence* (Figure 1.1).

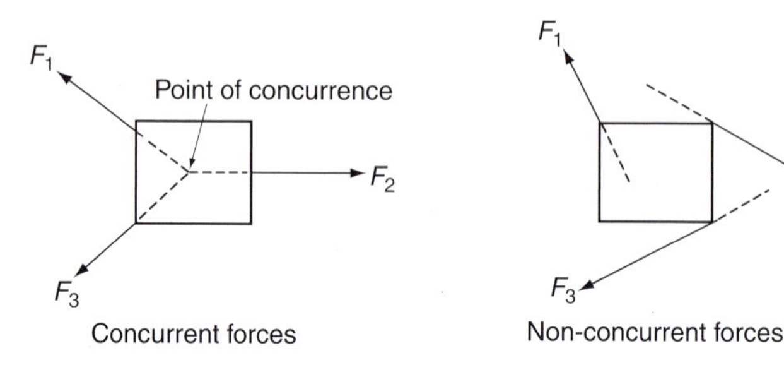

Figure 1.1 Coplanar force systems

> **KEY POINT**
> A system of concurrent coplanar forces can be reduced to a single resultant force.

A concurrent system of coplanar forces can be reduced to a single force acting at the point of concurrence. This is called the *resultant force*. If a body is subjected to a system of concurrent coplanar forces and is not constrained, it will move in the direction of the resultant force. To prevent this from happening, a force must be applied which is equal and opposite to the resultant. This balancing force, which will hold the body in a state of static equilibrium, is called the *equilibrant* of the system.

> **KEY POINT**
> A system of non-concurrent coplanar forces can be reduced to a single resultant force and a resultant couple or turning moment.

When a body is subjected to a system of non-concurrent coplanar forces, there is a tendency for the forces not only to make it move in a particular direction, but also to make it rotate. Such a non-concurrent system can be reduced to a single *resultant force* and a *resultant couple*.

If the body is to be held at rest, an equilibrant must again be applied which is equal and opposite to the resultant force. This alone however will not be sufficient. A balancing couple or turning moment must also be applied which is equal and opposite to the resultant couple.

If you have completed the BTEC First Diploma unit Mathematics and Science for Technicians and the BTEC National Certificate/Diploma unit Mechanical Principles and Applications, you will know how to find the resultant and equilibrant of coplanar force system graphically by means of a force vector diagram, and also by using mathematics. We will shortly be using both methods again but applied to more complex engineering structures. Here is a reminder of some of the main points of the mathematical or *analytical* method.

Sign convention

When you are using mathematics to solve coplanar force system problems you need to adopt a method of describing the action of the forces and couples. The following sign convention is that which is most often used:

(i) Upward forces are positive and downward forces are negative.
(ii) Horizontal forces acting to the right are positive and horizontal forces acting to the left are negative.
(iii) Clockwise acting moments and couples are positive and anticlockwise acting moments and couples are negative.

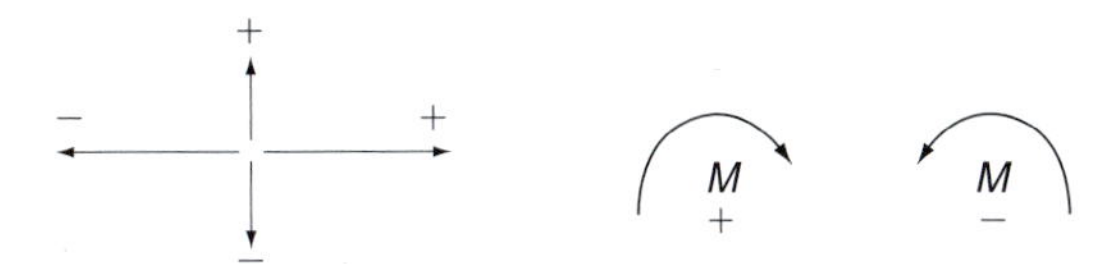

Figure 1.2 Sign convention

Resolution of forces

Forces which act at an angle exert a pull which is part horizontal and part vertical. They can be split into their horizontal and vertical parts or *components*, by the use of trigonometry. When you are doing this, it is a useful rule to always measure angles to the horizontal (Figure 1.2).

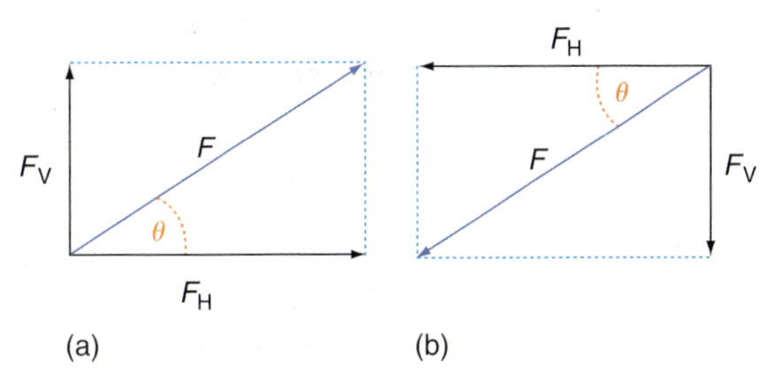

Figure 1.3 Resolution of forces

In Figure 1.3(a) the horizontal and vertical components are both acting in the positive directions and will be:

$$F_H = +F\cos\theta \quad \text{and} \quad F_v = +F\sin\theta$$

In Figure 1.3(b) the horizontal and vertical components are both acting in the negative directions and will be:

$$F_H = -F\cos\theta \quad \text{and} \quad F_v = -F\sin\theta$$

Forces which act upwards to the left or downwards to the right will have one component which is positive and one which is negative. Having resolved all of the forces in a coplanar system into their horizontal and vertical components, each set can then be added algebraically to determine the resultant horizontal pull, ΣF_H, and the resultant vertical pull, ΣF_V. The Greek letter Σ (sigma) means 'the sum or total' of the components. Pythagoras' theorem can then be used to find the single resultant force R of the system.

$$\text{i.e.}\quad R^2 = (\Sigma F_H)^2 + (\Sigma F_V)^2$$

$$\mathbf{R = \sqrt{(\Sigma F_H)^2 + (\Sigma F_V)^2}} \tag{1.1}$$

The angle θ which the resultant makes with the horizontal can also be found using:

$$\tan\theta = \frac{\Sigma F_V}{\Sigma F_H} \tag{1.2}$$

With non-concurrent force systems, the algebraic sum of the moments of the vertical and horizontal components of the forces, taken about some convenient point, gives the resultant couple or turning moment. Its sign, positive or negative, indicates whether its direction is clockwise or anticlockwise. This in turn can be used to find the perpendicular distance of the line of action of the resultant from the chosen point.

Example 1.1

Find the magnitude and direction of the resultant and equilibrant of the concurrent coplanar force system shown in Figure 1.4.

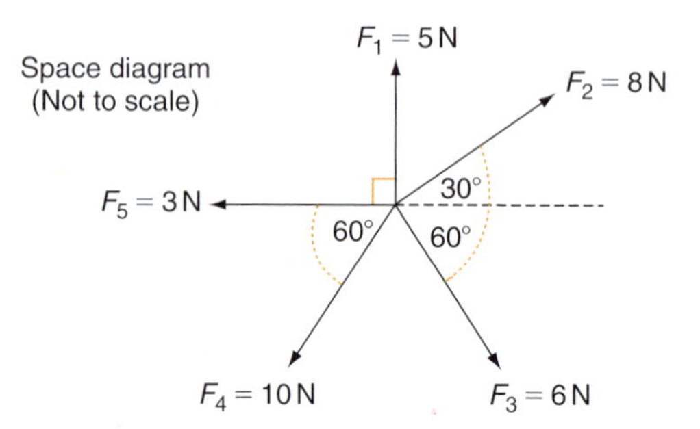

Figure 1.4

When you resolve the forces into their horizontal and vertical components it is essential to use the sign convention. A logical way is to draw up a table as follows with the forces and their horizontal and vertical components, set out in rows and columns.

Force	Horizontal component	Vertical component
$F_1 = 5\,\text{N}$	0	+5.0 N
$F_2 = 8\,\text{N}$	+8 cos 30 = +6.93 N	+8 sin 30 = +4.0 N
$F_3 = 6\,\text{N}$	+6 cos 30 = +3.0 N	−6 sin 60 = −5.2 N
$F_4 = 10\,\text{N}$	−10 cos 60 = −5.0 N	−10 sin 60 = −8.66 N
$F_5 = 3\,\text{N}$	= −3.0 N	0
Totals	$\Sigma F_H = +1.93\,\text{N}$	$\Sigma F_V = -4.86\,\text{N}$

The five forces have now been reduced to two forces, ΣF_H and ΣF_V. They are both negative and can be drawn as vectors (Figure 1.5).

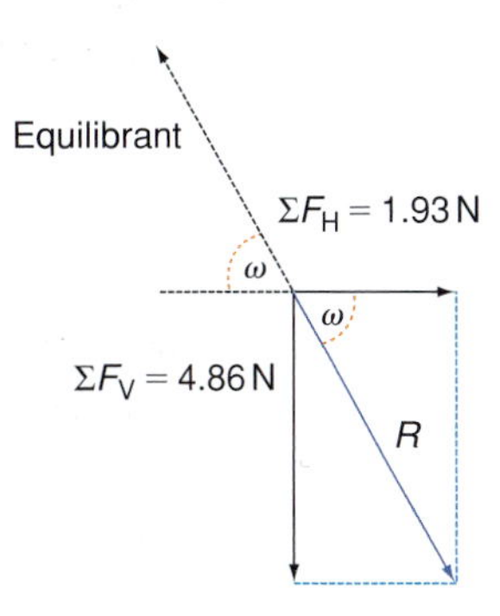

Figure 1.5

The resultant R is found by Pythagoras as follows:

$$R = \sqrt{1.93^2 + 4.86^2}$$

$$\mathbf{R = 5.23\ N}$$

The angle θ is found from:

$$\tan\theta = \frac{\Sigma F_V}{\Sigma F_H} = \frac{4.86}{1.93} = 2.52$$

$$\theta = \tan^{-1} 2.52$$

$$\boldsymbol{\theta = 68.3°}$$

The equilibrant, which is required to hold the system in a state of static equilibrium, is equal to the resultant but opposite in sense.

Bow's notation

Example 1.1 can be solved graphically by means of a force vector diagram drawn to a suitable scale. The process is known as *vector addition*. The force vectors are taken in order, preferably working clockwise around the system, and added nose to tail to produce *a polygon of forces*.

Should the final vector be found to end at the start of the first, there will be no resultant and the system will be in equilibrium. If however there is a gap between the two, this when measured from the start of the first vector to the end of the last represents the magnitude, direction and sense of the resultant. The equilibrant will of course be equal and opposite. It must be remembered that when you solve problems graphically, the accuracy of the answers will depend on the accuracy of your measurement and drawing.

Bow's notation is a useful method of identifying the forces on a vector diagram and also the sense in which they act. In the space diagram, which shows the forces acting at the point of concurrence, the spaces between the forces are each given a capital letter. Wherever possible, the letters should follow a clockwise sequence around the diagram. In the solution shown below, the force F_1 is between the spaces A and B and when drawn on the vector diagram it is identified by the lower case letters as force *ab*.

KEY POINT

Use capital letters and work in a clockwise direction when lettering space diagram using Bow's notation.

The clockwise sequence of letters on the space diagram, i.e. A to B, gives the direction of the force on the vector diagram. The letter a is at the start of the vector and the letter b is at its end. Although arrows have been drawn to show the directions of the vectors they are not really necessary and will be omitted on future graphical solutions.

Example 1.2

Alternative graphical solution (Figures 1.6 and 1.7)

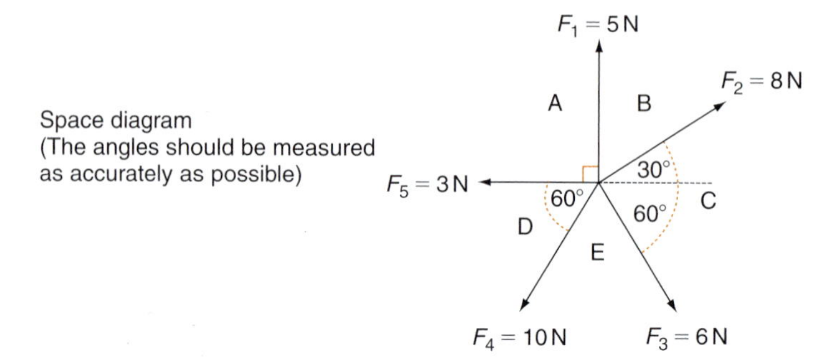

Figure 1.6

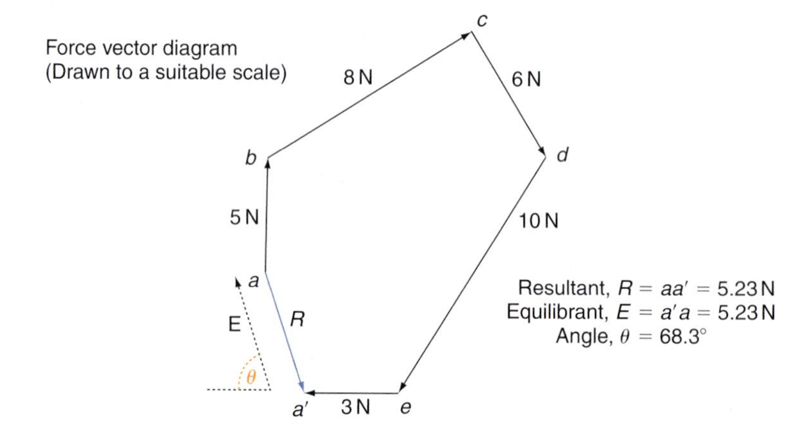

Figure 1.7

KEY POINT

When using the analytical method of solution it is a good idea to measure angles from the horizontal. All of the horizontal components of the forces will then be given by $F_H = F\cos\theta$ and all of the vertical components will be given by $F_V = F\sin\theta$.

Test your knowledge 1.1

1. What are coplanar forces?
2. What are concurrent forces?
3. What are the conditions necessary for a body to be in static equilibrium under the action of a coplanar force system?
4. What are the resultant and equilibrant of a coplanar force system?
5. What is Bow's notation?

To check your understanding of the preceding section, you can solve Review questions 1–3 at the end of this chapter.

Pin-jointed framed structures

Examples of framed structures which you see in everyday life are bicycles, roof trusses, electricity pylons and tower cranes. They are made up of members which are joined at their ends. Some of these are three-dimensional structures whose analysis is complex and it is only the two-dimensional or coplanar structures, which we will consider. There are three kinds of member in these structures (Figure 1.8):

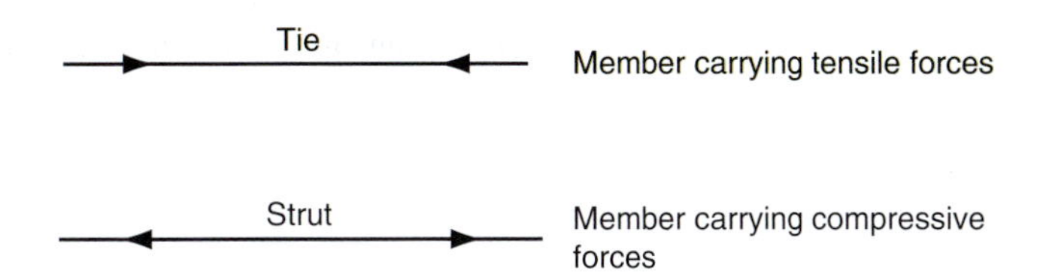

Figure 1.8 Representation of structural members

- *Ties*, which are in tension and shown diagramatically with arrows pointing inwards. You have to imagine yourself in the place of a tie. You would be pulling inwards to stop yourself from being stretched. The arrows describe the force which the tie exerts on its neighbours to keep the structure in position.
- *Struts*, which are in compression and shown diagramatically with arrows pointing outwards. Once again, you have to imagine the way you would be pushing if you were in the place of a strut. You would be pushing outwards to keep the structure in position and to stop yourself from being squashed.
- The third type is *redundant members*. A perfect framed structure is one which has just sufficient members to prevent it from becoming unstable. Any additional members, which may have been added to create a stiffer or stronger frame, are known as redundant members. Redundant members may be struts or ties or they may carry no load in normal circumstances. We shall avoid framed structures with redundant members as very often they cannot be solved by the ordinary methods of statics.

In reality the members are bolted, riveted or welded together at their ends but in our analysis we assume that they are pin-jointed or hinged at their ends, with frictionless pins. We further assume that because of this, the only forces present in the members are tensile and compressive forces. These are called *primary forces*. In practice there might also be bending and twisting forces present but we will leave these for study at a higher level.

When a structure is in a state of static equilibrium, the external active loads which it carries will be balanced by the reactions of its supports. The conditions for external equilibrium are:

1. The vector sum of the horizontal forces or horizontal components of the forces is zero.
2. The vector sum of the vertical forces or vertical components of the forces is zero.

KEY POINT

When a body or structure is in a static equilibrium under the action of *three* external forces, the forces must be concurrent.

3. The vector sum of the turning moments of the forces taken about any point in the plane of the structure is zero.

$$\text{i.e.}\quad \Sigma F_H = 0 \qquad \Sigma F_V = 0 \qquad \Sigma M = 0$$

We can also safely assume that if a structure is in a state of static equilibrium, each of its members will also be in equilibrium. It follows that the above three conditions can also be applied to individual members, groups of members and indeed to any internal part or section of a structure.

KEY POINT

Whichever method of solution you adopt, it is good practice to find all of the external forces acting on a structure before investigating the internal forces in the members.

A full analysis of the external and internal forces acting on and within a structure can be carried out using mathematics or graphically by drawing a force vector diagram. We will use the mathematical method first and begin by applying the conditions for equilibrium to the external forces. This will enable us to find the magnitude and direction of the support reactions.

KEY POINT

When you are finding the internal forces start with a joint which has only two unknown forces acting on it whose directions you know.

Next, we will apply the conditions for equilibrium to each joint in the structure, starting with one which has only two unknown forces acting on it. This will enable us to find the magnitude and direction of the force in each member.

Example 1.3

Determine the magnitude and direction of the support reactions at *X* and *Y* for the pin-jointed cantilever shown in Figure 1.9, together with the magnitude and nature of the force acting in each member.

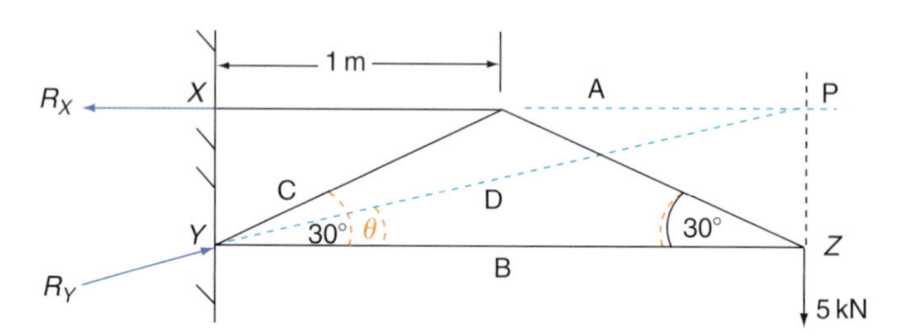

Figure 1.9

The wall support reaction R_X must be horizontal because it is equal and opposite to the force carried by the top member CA. Also, the three external forces must be concurrent with lines of action meeting at the point P.

Begin by locating the point P and drawing in the support reactions at *X* and *Y* in the sense which you think they are acting. If you guess wrongly, you will obtain a negative answer from your calculations, telling you that your arrow should be pointing in the opposite direction. Letter the diagram using Bow's notation, with capital letters in the spaces between the external forces and inside the structure. Now you can begin the calculations.

Finding distances *XY*, *YZ* and the angle θ:

$$XY = 1 \times \tan 30 = 0.577\ \text{m} \qquad YZ = 2.0\ \text{m}$$

$$\tan\theta = \frac{XY}{YZ} = \frac{0.577}{2.0} = 0.289$$

$$\boldsymbol{\theta = 16.1^\circ}$$

Take moments about Y to find R_X. For equilibrium, $\Sigma M_Y = 0$:

$$(5 \times 2) - (R_X \times 0.577) = 0$$
$$10 - 0.577\,R_x = 0$$
$$\mathbf{R_X = \frac{10}{0.577} = 17.3\,kN}$$

The force in member AC will also be **17.3 kN** because it is equal and opposite to R_X. It will be a tensile force, and this member will be a tie.

Take vertical components of external forces to find R_Y. For equilibrium, $\Sigma F_V = 0$:

$$R_y \sin 16.1^\circ - 5 = 0$$
$$0.277 R_Y - 5 = 0$$
$$\mathbf{R_Y = \frac{5}{0.277} = 18.0\,kN}$$

You can now turn your attention to the forces in the individual members. You already know the force F_{AC} acting in member AC because it is equal and opposite to the support reaction R_X. Choose a joint where you know the magnitude and direction of one of the forces and the directions of the other two, i.e. where there are only two unknown forces. Joint ABD will be ideal. It is good practice to assume that the unknown forces are tensile, with arrows pulling away from the joint. A negative answer will tell you that the force is compressive (Figure 1.10).

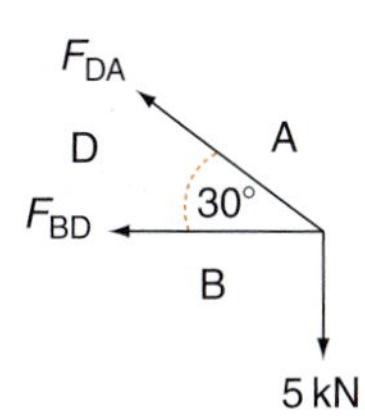

Figure 1.10

Take vertical components of the forces to find the force F_{DA} in member DA.

$$\text{For equilibrium, } \Sigma F_v = 0$$
$$F_{DA} \sin 30 - 5 = 0$$
$$0.5 F_{DA} - 5 = 0$$
$$\mathbf{F_{DA} = \frac{5}{0.5} = 10\ kN}$$

The positive answer denotes that the force F_{DA} in member DA is tensile and that the member is a tie. Now take horizontal components of the forces to find F_{BD} in member BD.

$$\text{For equilibrium, } \Sigma F_H = 0$$
$$-F_{DA} \cos 30 - F_{BD} = 0$$
$$-10 \cos 30 - F_{BD} = 0$$
$$\mathbf{F_{BD} = -10 \cos 30 = -8.66\ kN}$$

The negative sign denotes that the force F_{BD} in member BD is compressive and that the member is a strut. Now go to joint ADC and take vertical components of the forces to find F_{DC}, the force in member DC (Figure 1.11).

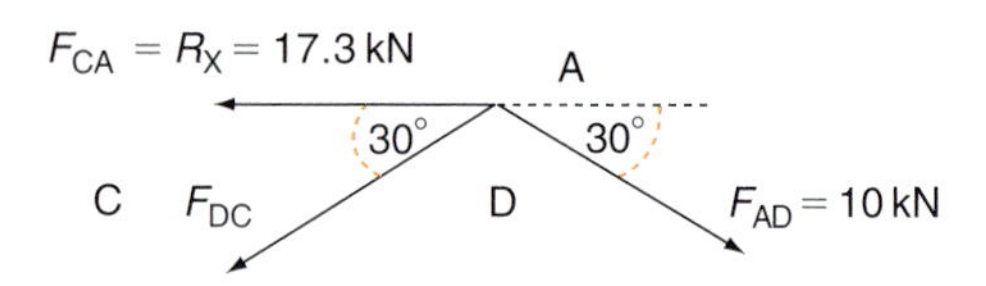

Figure 1.11

$$\text{For equilibrium, } \Sigma F_V = 0$$
$$-(10\sin 30) - (F_{DC}\sin 30) = 0$$
$$-10 - F_{DC} = 0$$
$$\mathbf{F_{DC} = -10\ kN}$$

The negative sign denotes that the force F_{DC} in member DC is compressive and that the member is a strut. You have now found the values of all the external and internal forces which can be tabulated and their directions indicated on the structure diagram as follows (Figure 1.12):

Reaction/member	Force (kN)	Nature
R_X	17.3	
R_Y	18.0	
AC	17.3	Tie
DA	10.0	Tie
BD	8.66	Strut
DC	10.0	Strut

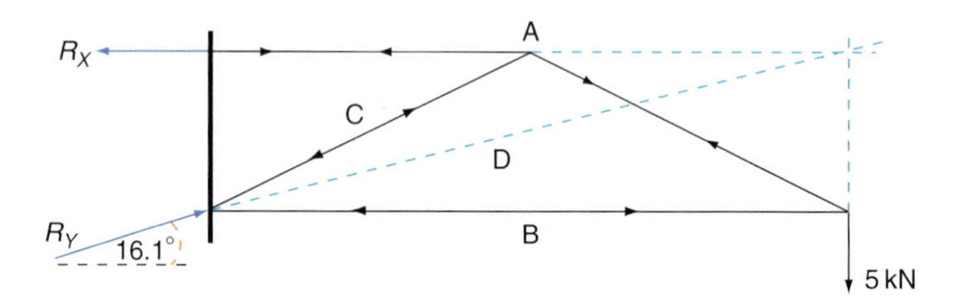

Figure 1.12

You may also solve framed structure problems graphically. Sometimes this is quicker but the results may not be so accurate. As before, it depends on the accuracy of your drawing and measurement. An alternative solution to the above framed structure is as follows. Once again, use is made of Bow's notation to identify the members.

Example 1.4

Alternative graphical solution.

Space diagram

Begin by drawing the space diagram shown in Figure 1.13 to some suitable scale. Take care to measure the angles as accurately as possible. The angle R_y can be measured, giving $\theta = 16.1°$.

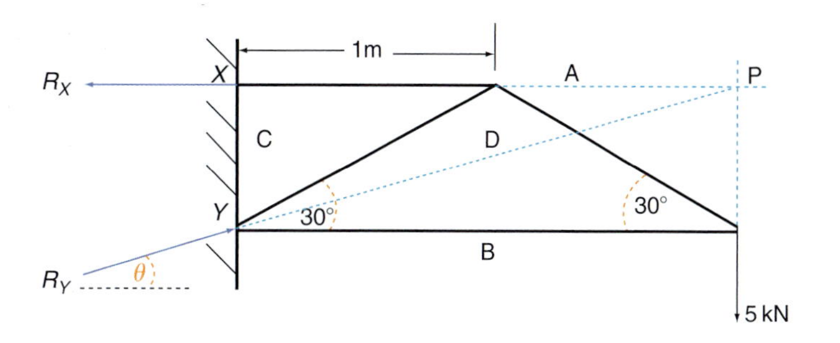

Figure 1.13

Now each of the joints is in equilibrium. Choose one where you know the magnitude and direction of one force and the directions of the other two. The joint ABD is in fact the only one where you know these conditions.

Draw the force vector diagram for this joint to a suitable scale, beginning with the 5 kN load. It will be a triangle of forces, which you can quite easily construct because you know the length of one side and the directions of the other two. The force vectors are added together nose to tail but do not put any arrows on the diagram. Bow's notation is sufficient to indicate the directions of the forces (Figure 1.14).

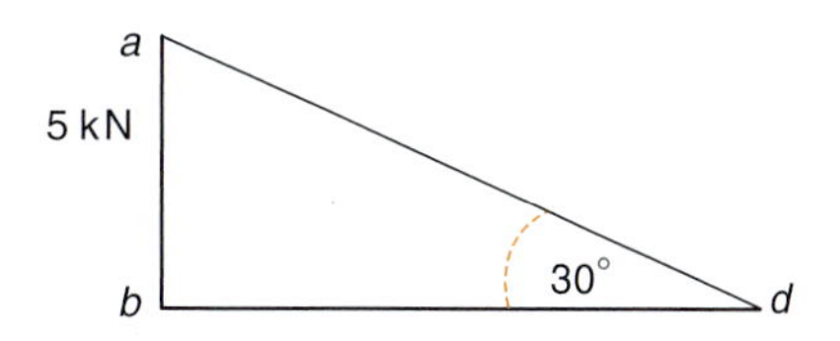

Figure 1.14

Accurate measurement gives,

Force in DA = da = 10.0 kN

Force in BD = bd = 8.66 kN

Remember, the clockwise sequence of letters around the joint on the space diagram ABD gives the directions in which the forces act on the joint in the vector diagram, i.e. the force ab acts from a to b, the force bd acts from b to d and the force da acts from d to a.

Now go to another joint where you know the magnitude and direction of one of the forces and the directions of the other two. Such a joint is ADC where the same 10 kN force da, which acts on joint ADB, also acts in the opposite direction on joint ADC. We now call it force ad, and the triangle of forces is as shown in Figure 1.15.

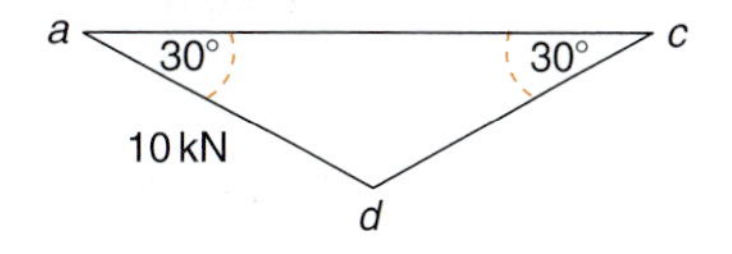

Figure 1.15

Accurate measurement gives,

Force in DC = dc = 10.0 kN

Force in CA = ca = 17.3 kN

All of the internal forces have now been found and also the reaction R_X, which is equal and opposite to *ca*, and can thus be written as *ac*. A final triangle of forces *ABC*, can now be drawn, representing the three external forces (see Figure 1.16). Two of these, the 5 kN load and R_X, are now known in both magnitude and direction together with the angle θ which R_Y makes with the horizontal.

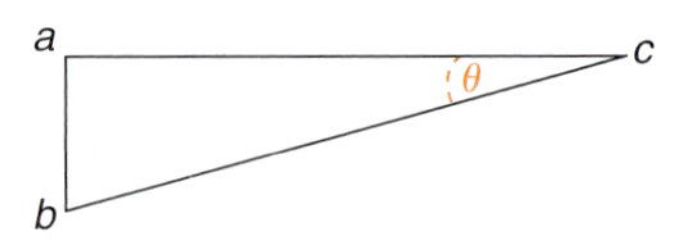

Figure 1.16

Accurate measurement gives,

Reaction $R_Y = bc = 18.0\,\text{kN}$

Angle $\theta = 16.1°$

You will note that the above three vector diagram triangles have sides in common and to save time it is usual to draw the second and third diagrams as additions to the first. The combined vector diagram appears as in Figure 1.17.

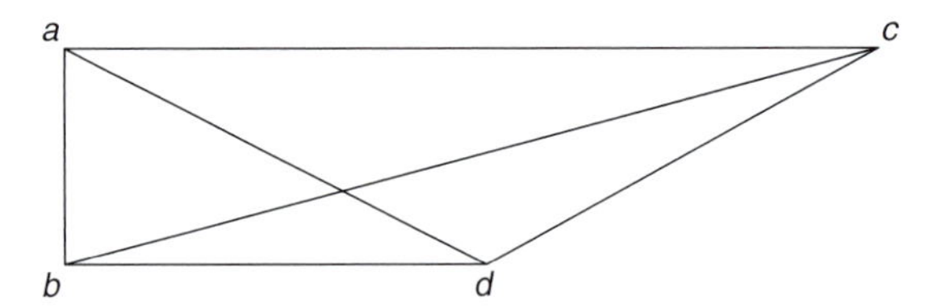

Figure 1.17

The directions of the forces, acting towards or away from the joints on which they act, can be drawn on the space diagram. This will immediately tell you which members are struts and which are ties. The measured values of the support reactions and the forces in the members can then be tabulated (Figure 1.18).

Reaction/member	Force (kN)	Nature
R_X	17.3	
R_Y	18.0	
AC	17.3	Tie
DA	10.0	Tie
BD	8.66	Strut
DC	10.0	Strut

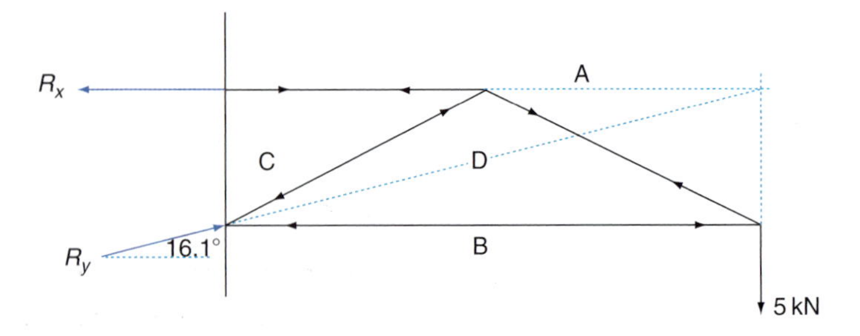

Figure 1.18

Test your knowledge 1.2

1. What are the three different types of member found in framed structures?
2. What are the primary forces which act in structural members?
3. How do you indicate using arrows, which members of a structure are ties and which are struts?
4. How do you letter the space diagram of a coplanar force system using Bow's notation?

Activity 1.1

The jib-crane shown in Figure 1.19 carries a load of 10 kN. Making use of Bow's notation find the reactions of the supports at *X* and *Y* and the magnitude and nature of the force in each member graphically by means of a force vector diagram. Apply the conditions for static equilibrium to the structure and check your results by calculation.

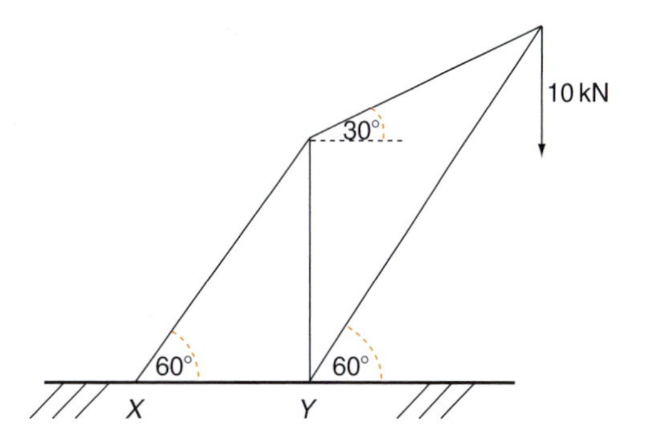

Figure 1.19

To check your understanding of the preceding section, you can solve Review question 4 at the end of this chapter.

Simply supported beams

A simply supported beam is supported at two points in such a way that it is allowed to expand and bend freely. In practice the supports are often rollers. The loads on the beam may be concentrated at different points or uniformly distributed along the beam. Figure 1.20 shows concentrated loads only.

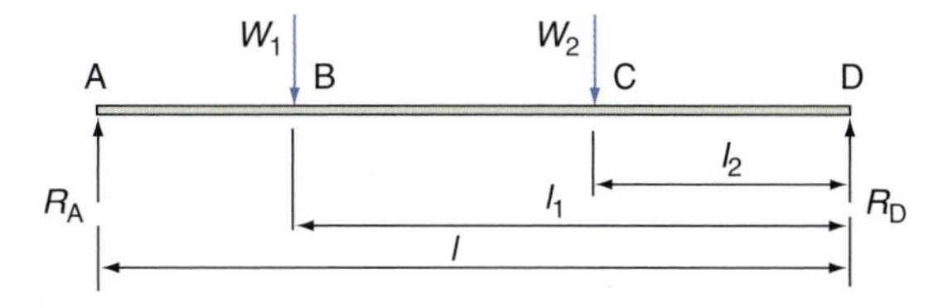

Figure 1.20 Simply supported beam with concentrated loads

The downward forces on a beam are said to be active loads, due to the force of gravity, whilst the loads carried by the supports are said to be reactive. When investigating the effects of loading, we often have to begin by calculating the support reactions.

The beam is in static equilibrium under the action of these external forces, and so we proceed as follows.

1. Equate the sum of the turning moments, taken about the right-hand support D, to zero.

$$\text{i.e. } \Sigma M_D = 0$$
$$\mathbf{R_A l - W_1 l_1 - W_2 l_2 = 0}$$

You can find R_A from this condition.

2. Equate vector sum of the vertical forces to zero.

$$\text{i.e. } \Sigma F_V = 0$$
$$\mathbf{R_A + R_B - W_1 - W_2 = 0}$$

You can find R_D from this condition.

Example 1.5

Calculate the support reactions of the simply supported beam shown in Figure 1.21.

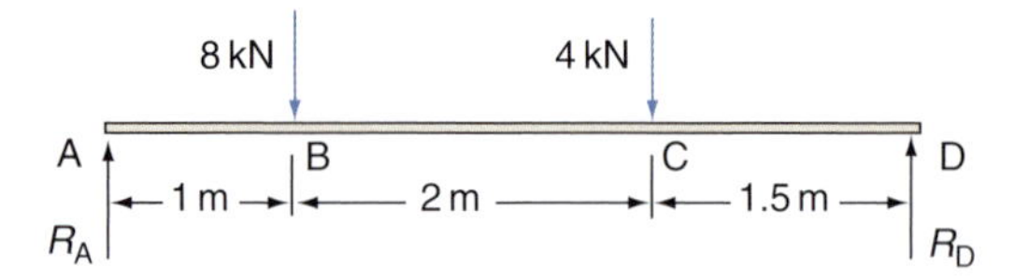

Figure 1.21

Take moments about the point D, remembering to use the sign convention that clockwise moments are positive and anticlockwise moments are negative. For equilibrium, $\Sigma M_D = 0$.

$$(R_A \times 4.5) - (8 \times 3.5) - (4 \times 1.5) = 0$$
$$4.5R_A - 28 - 6 = 0$$
$$\mathbf{R_A = \frac{28 + 6}{4.5} = 7.56\ kN}$$

Equate the vector sum of the vertical forces to zero, remembering the sign convention that upward forces are positive and downward forces are negative. For equilibrium, $\Sigma F_V = 0$.

$$8 + 4 - 7.56 - R_D = 0$$
$$\mathbf{R_D = 8 + 4 - 7.56 = 4.44\ kN}$$

Uniformly distributed loads

Uniformly distributed loads, or UDLs, are evenly spread out along a beam. They might be due to the beam's own weight, paving slabs or an asphalt surface. UDLs are generally expressed in kN per metre length, i.e. $kN\,m^{-1}$. This is also known as the 'loading rate'. UDLs are shown diagrammatically as in Figure 1.22.

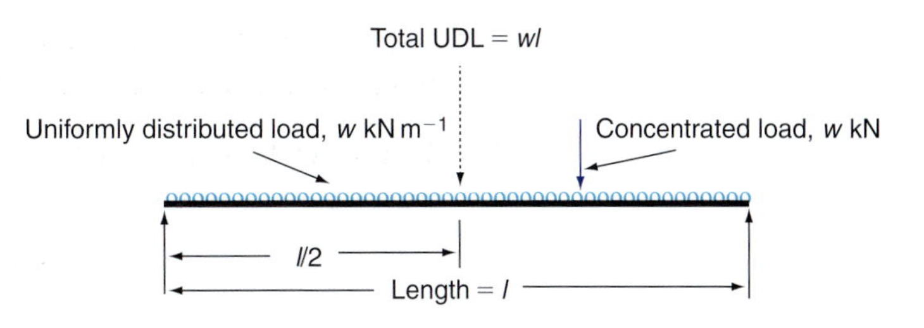

Figure 1.22 Simply supported beam with concentrated and distributed load

The total UDL over a particular length l of a beam is given by the product of the loading rate and the length.

$$\text{i.e.} \quad \text{Total UDL} = wl$$

KEY POINT

A uniformly distributed load may be considered to act at its centroid.

When you are equating moments to find the beam reactions, the total UDL is assumed to act at its centroid, i.e. at the centre of the length l. You can then treat it as just another concentrated load and calculate the support reactions in the same way as before.

Example 1.6

Calculate the support reactions of the simply supported beam shown in Figure 1.23.

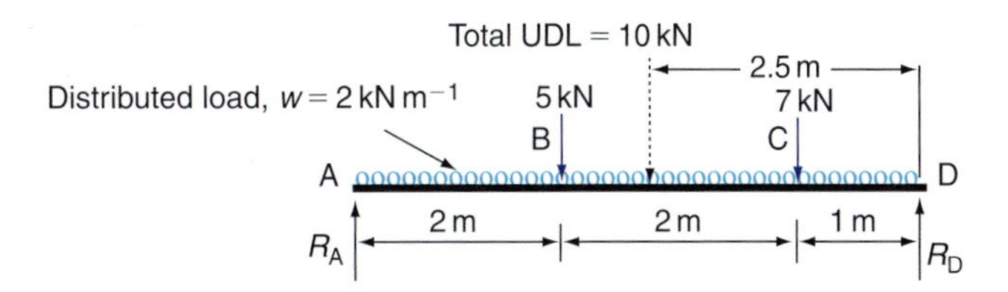

Figure 1.23

Begin by calculating the total UDL. Then, replace it by an equal concentrated load acting at its centroid, which is the midpoint of the span. This is shown with dotted line in the figure.

$$\textbf{Total UDL} = wl = 5 \times 2 = 10 \textbf{ kN}$$

You can now apply the conditions for static equilibrium. Begin by taking moments about the point D. For equilibrium, $\Sigma M_D = 0$.

$$(R_A \times 5) - (5 \times 3) - (10 \times 2.5) - (7 \times 1) = 0$$

$$5R_A - 15 - 25 - 7 = 0$$

$$R_A = \frac{15 + 25 + 7}{5} = 9.4 \text{ kN}$$

Now equate the vector sum of the vertical forces to zero. For equilibrium, $\Sigma F_V = 0$.

$$9.4 + R_D - 5 - 10 - 7 = 0$$

$$R_D = 5 + 10 + 7 - 9.4 = 12.6 \text{ kN}$$

Bending of beams

When structural engineers are designing beams to carry given loads, they have to make sure that the maximum allowable stresses will not be exceeded. On any transverse section of a loaded beam or cantilever, shear stress, tensile stress and compressive stress are all usually present.

As you can be seen in the cantilever in Figure 1.24(a), the load F has a shearing effect and thus sets up shear stress at section Y–Y. The load also has a bending effect and at any section Y–Y, this produces tensile stress in the upper layers of the beam and compressive stress in the lower layers.

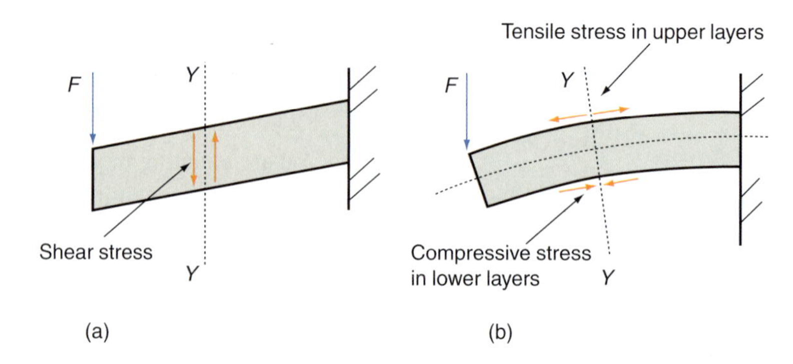

Figure 1.24 Stress in beams: (a) shearing effect of load, (b) bending effect of load

Generally, the stresses and deflection caused by bending greatly exceed those caused by shear and the shearing effects of the loading are usually neglected for all but short and stubby cantilevers. Nevertheless, the distribution of shear loading is of importance because it enables the likely positions of maximum bending stress to be pin-pointed, as will be shown.

Somewhere inside a beam there is a layer called the *neutral layer or neutral axis*. Although this becomes bent, like all the other layers, it is in neither tension nor compression and there is no tensile or compressive stress present. For elastic materials which have the same modulus of elasticity in tension and compression, it is found that the neutral axis is located at the centroid of the cross-section.

Shear force distribution

At any transverse section of a loaded horizontal beam, the *shear force* is defined as being *the algebraic sum of the forces acting to one side of the section*. You might think of it as the upward or downward breaking force at that section. Usually it is forces to the left of the section which are considered and the following sign convention is used:

1. Upward forces to the left of a section are positive and downward forces are negative.
2. Downward forces to the right of a section are positive and upward forces are negative.

This gives rise to the idea of positive and negative shearing as shown in Figure 1.25, and the variation of the shear force along a loaded beam can be plotted on a shear force diagram.

KEY POINT

Upward shearing forces to the left of a section through a loaded beam are positive and downward shearing forces are negative.

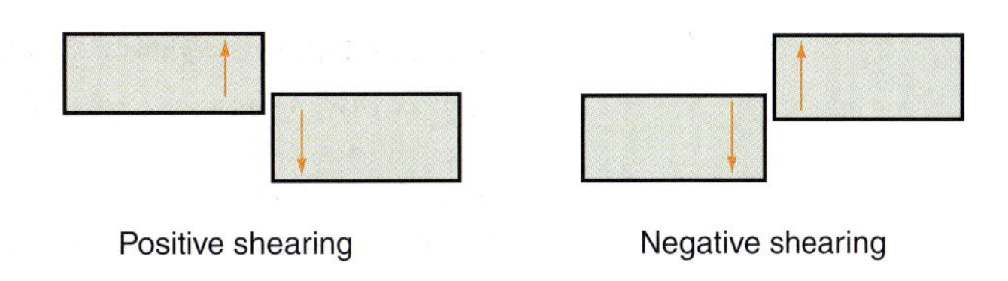

Figure 1.25 Positive and negative shearing

Bending moment distribution

At any transverse section of a loaded beam, the *bending moment* is defined as being *the algebraic sum of the moments of the forces acting to one side of the section*. Usually it is moments to the left of a section which are considered and the following sign convention is used.

1. Clockwise moments to the left of a section are positive and anticlockwise moments are negative.
2. Anticlockwise moments to the right of a section are positive and moments are negative.

> **KEY POINT**
>
> Clockwise bending moments to the left of a section through a loaded beam are positive and anticlockwise moments are negative.

This gives rise to the idea of positive and negative bending as shown in Figure 1.26, and the variation of bending moment along a loaded beam can be plotted on a bending moment diagram.

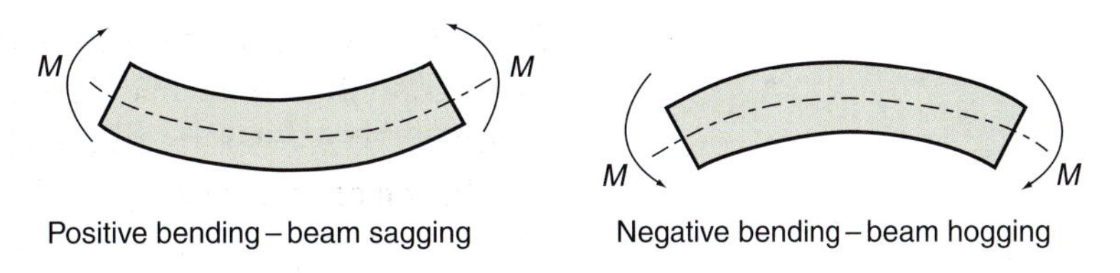

Figure 1.26 Positive and negative bending

When you are plotting shear force and bending moment diagrams you will see that:

1. The maximum bending moment always occurs where the shear force diagram changes sign.
2. The area of the shear force diagram up to a particular section gives the bending moment at that section.
3. Under certain circumstances, the bending moment changes sign and this is said to occur at a point of contraflexure.

> **KEY POINT**
>
> If a beam is to be made up of joined sections, it is good practice when practicable, to position the joints at points of contraflexure where there are no bending stresses.

At a point of contraflexure, where the bending moment is zero, the deflected shape of a beam changes from sagging to hogging or vice versa. The location of these points is of importance to structural engineers since it is here that welded, bolted or riveted joints can be made which will be free of bending stress.

Example 1.7

Plot the shear force and bending moment distribution diagrams for the simple cantilever beam shown in Figure 1.27 and state the magnitude, nature and position of the maximum values of shear force and bending moment.

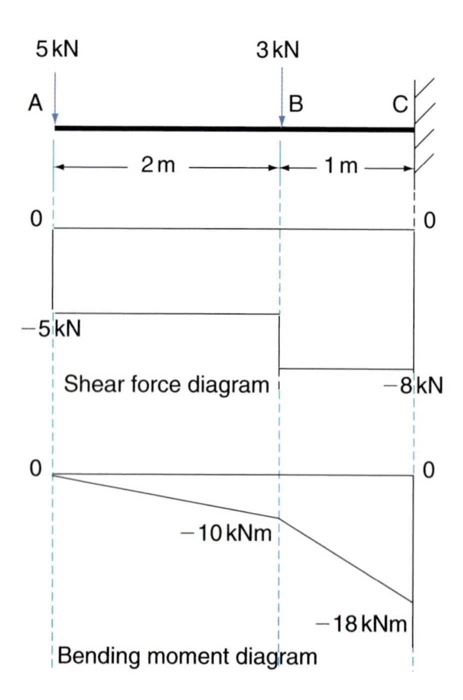

Figure 1.27

Shear force from A to B = −5kN
Shear force from B to C = −8kN
Maximum shear force = −8 kN between B and C

Bending moment at A = 0
Bending moment at B = −(5 × 2) = −10 kNm
Bending moment at C = −(5 × 3) − (3 × 1) = −18 kNm
Maximum bending moment = −18 kNm at C

As you can see, the shear force is negative over the whole length of the cantilever because there is a downward breaking force to the left of any section, i.e. negative shear.

The bending moment is always zero at the free end of a simple cantilever and negative over the remainder of its length. You might think of a cantilever as being half of a hogging beam in which the bending moment at any section is anticlockwise, i.e. negative bending.

Example 1.8

Plot the shear force and bending moment distribution diagrams for the simply supported beam shown in Figure 1.28. State the magnitude, nature and position of the maximum values of shear force and bending moment.

Begin by finding the support reactions.
For equilibrium, $\Sigma M_E = 0$.

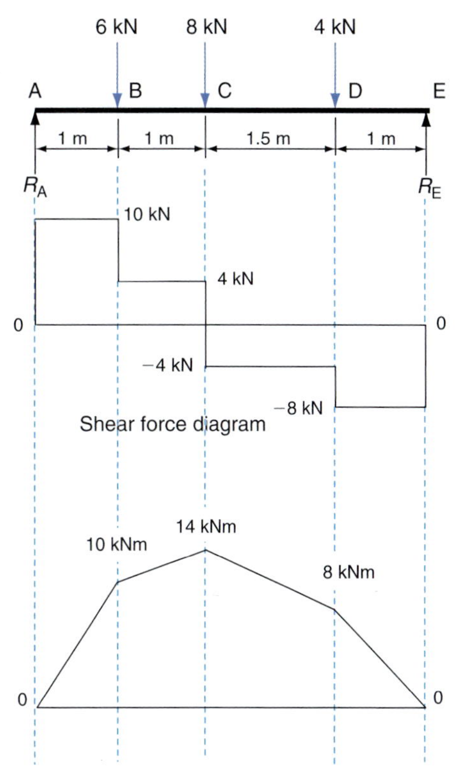

Figure 1.28

$$(R_A \times 4.5) - (6 \times 3.5) - (8 \times 2.5) - (4 \times 1) = 0$$

$$4.5R_A - 21 - 20 - 4 = 0$$

$$\mathbf{R_A = \frac{21 + 20 + 4}{4.5} = 10\ kN}$$

Also, for equilibrium, $\Sigma F_V = 0$.

$$10 + R_E - 6 - 8 - 4 = 0$$

$$\mathbf{R_E = 6 + 8 + 4 - 10 = 8\ kN}$$

Next find the shear force values.

SF from A to B $= +10\,kN$

SF from B to C $= +4\,kN$

SF from C to D $= -4\,kN$

SF from D to E $= -8\,kN$

Maximum shear force = +10 kN between A and B

Now find the bending moment values.

BM at A $= 0$

BM at B $= +(10 \times 1) = +10\,kNm$

BM at C $= +(10 \times 2) - (6 \times 1) = +14\,kNm$

$= 8\,kNm$

BM at D $= +(10 \times 3.5) - (6 \times 2.5) - (8 \times 1.5)$

$= 8\ kNm$

BM at E $= 0$

Maximum bending moment = +14 kNm at C

Example 1.9

Plot the shear force and bending moment distribution diagrams for the simply supported beam shown in Figure 1.29. State the magnitude, nature and position of the maximum values of shear force and bending moment and the position of a point of contraflexure.

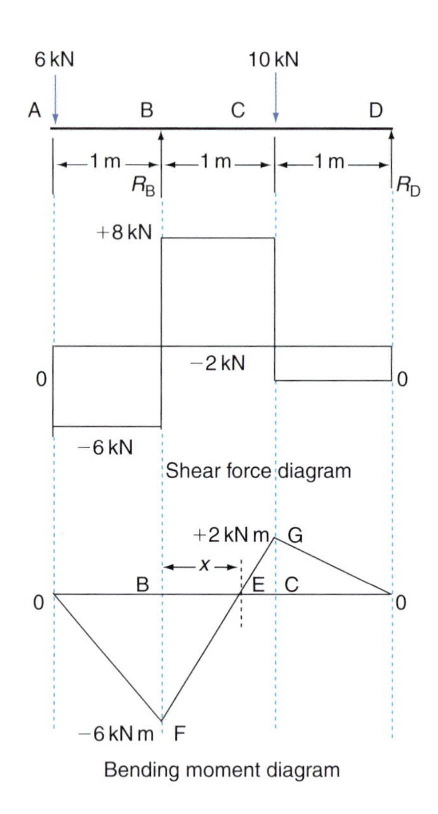

Figure 1.29

Begin by finding the support reactions.
For equilibrium, $\Sigma M_D = 0$.

$$(R_B \times 2) - (6 \times 3) - (10 \times 1) = 0$$

$$2R_B - 18 - 10 = 0$$

$$\mathbf{R_B = \frac{18 + 10}{2} = 14\ kN}$$

Also, for equilibrium, $\Sigma F_V = 0$.

$$14 + R_D - 6 - 10 = 0$$

$$\mathbf{R_D = 6 + 10 - 14 = 2\ kN}$$

Next find the shear force values.

SF from A to B $= -6$ kN
SF from B to C $= +8$ kN
SF from C to D $= -2$ kN
Maximum shear force = +8 kN between B and C

Now find the bending moment values.

BM at A = 0

BM at B = −(6 × 1) = −6 kNm

BM at C = −(6 × 2) + (14 × 1) = +2 kNm

BM at D = 0

Maximum bending moment = −6 kNm at B

There is a point of contraflexure at E, where the bending moment is zero. To find its distance *x*, from *B*, consider the similar triangles BEF and ECG (Figure 1.30).

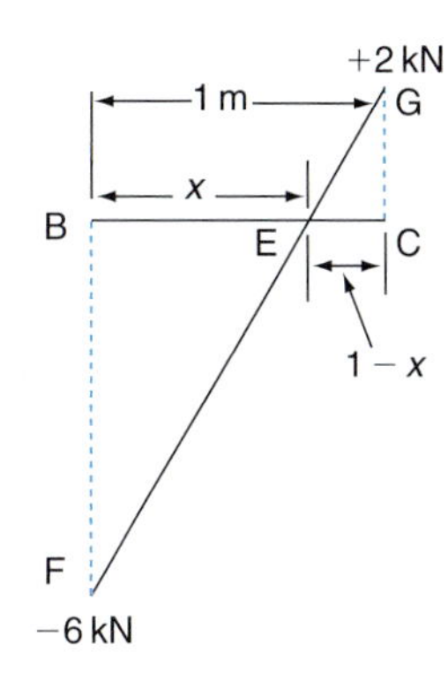

Figure 1.30

$$\frac{x}{6} = \frac{1-x}{2}$$

$$2x = 6(1-x)$$

$$2x = 6 - 6x$$

$$8x = 6$$

$$\mathbf{x = \frac{6}{8} = 0.75\ m}$$

If you examine the shear force and bending moment diagrams in the above examples you will find that there is a relationship between shear force and bending moment. Calculate the area under the shear force diagram from the left-hand side, up to any point along the beam you will find that this is equal to the bending moment at that point. Try it, but remember to use the sign convention that areas above the zero line are positive and those below are negative.

Example 1.10

Plot the shear force and bending moment distribution diagrams for the cantilever shown in Figure 1.31 and state the magnitude and position of the maximum values of shear force and bending moment.

The presence of the UDL produces a gradually increasing shear force between the concentrated loads.

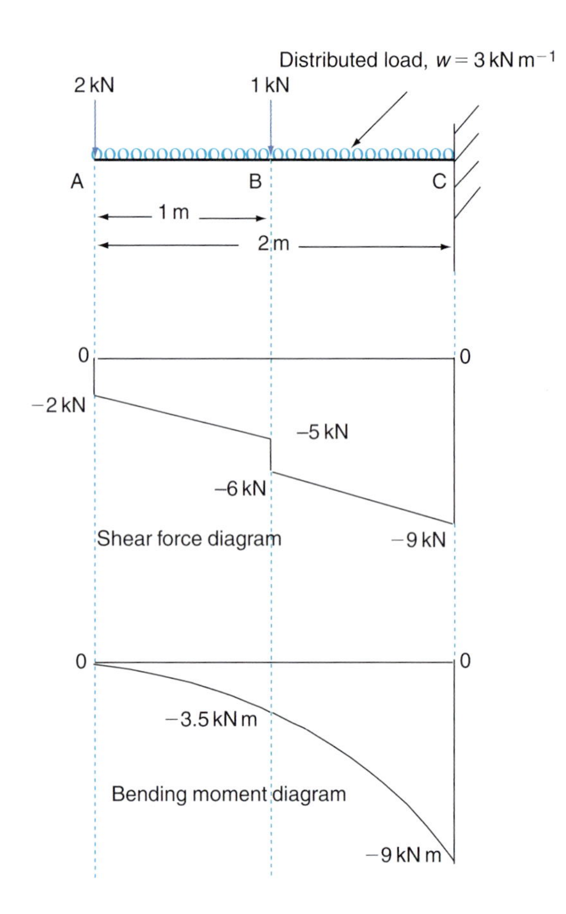

Figure 1.31

Finding shear force values.

SF immediately to right of $A = -2\text{kN}$

SF immediately to left of $B = -2 - (3 \times 1) = 5\text{kN}$

SF immediately to right of $B = -2 - 1 - (3 \times 1) = -6\text{kN}$

SF immediately to left of $C = -2 - 1 - (3 \times 2) = -9\text{kN}$

Maximum SF = −9 kN immediately to left of C

The presence of the UDL produces a bending moment diagram with parabolic curves between the concentrated load positions.

Finding bending moment values.

BM at $A = 0$

BM at $B = -(2 \times 1) - (3 \times 1 \times 0.5) = -3.5\text{kNm}$

BM at $C = -(2 \times 1) - (1 \times 1) - (3 \times 2 \times 1) = -9\text{kNm}$

Maximum bending moment = −9 kNm at C

Example 1.11

Plot the shear force and bending moment distribution diagrams for the simply supported beam shown in Figure 1.32. State the magnitude, nature and position of the maximum values of shear force and bending moment.

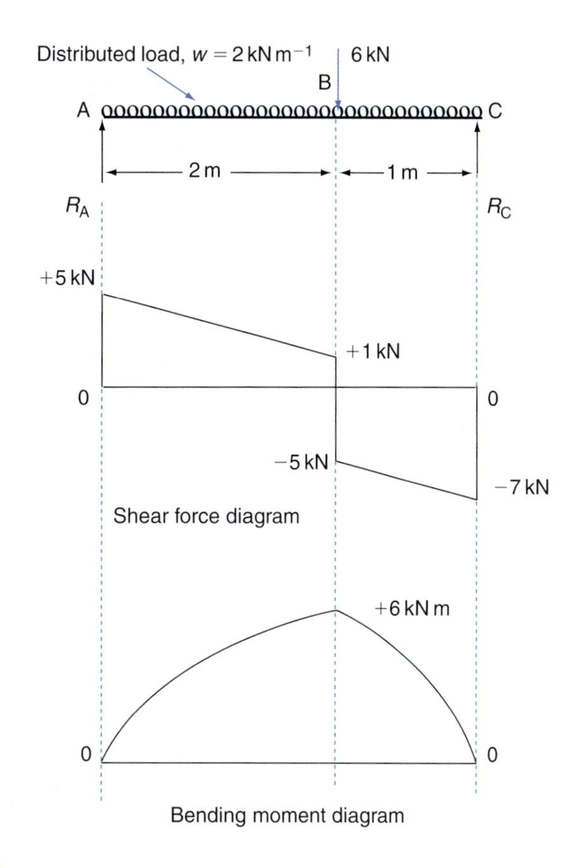

Figure 1.32

Finding support reactions.
For equilibrium, $\Sigma M_C = 0$.

$$(R_A \times 3) - (6 \times 1) - (2 \times 3 \times 1.5) = 0$$
$$3R_A - 6 - 9 = 0$$
$$\mathbf{R_A = \frac{6+9}{3} = 5\ kN}$$

Also for equilibrium, $\Sigma F_V = 0$.

$$5 + R_C - 6 - (2 \times 3) = 0$$
$$5 + R_C - 6 - 6 = 0$$
$$\mathbf{R_C = 6 + 6 - 5 = 7\ kN}$$

Finding the shear force values.

SF immediately to right of A $= +5\text{kN}$
SF immediately to left of B $= 5 - (2 \times 2) = +1\text{kN}$
SF immediately to right of B $= 5 - (2 \times 2) - 6 = -5\text{kN}$
SF immediately to left of C $= 5 - (2 \times 3) - 6 = -7\text{kN}$
Maximum SF = −7 kN immediately to left of C

Finding bending moment values.

BM at A $= 0$
BM at B $= (5 \times 2) - (2 \times 2 \times 1) = +6\text{kNm}$
BM at C $= 0$
Maximum BM = +6 kNm at B

As you can see from the diagrams in Figure 1.32, the effect of the UDL is to produce a shear force diagram that slopes between the supports and the concentrated load. Its slope $2\,kN\,m^{-1}$, which is the UDL rate. The effect on the bending moment diagram is to produce parabolic curves between the supports and the concentrated load.

Test your knowledge 1.3

1. What is the difference between an active and a reactive load?
2. How do you define the shear force at any point on a loaded beam?
3. How do you define the bending moment at any point on a loaded beam?
4. What is a point of contraflexure?

Activity 1.2

The simply supported beam shown in Figure 1.33 is to made in two sections which will be joined together at some suitable point between the supports. Draw the shear force and bending moment diagrams for the beam. State the maximum values of shear force and bending moment and the positions where they occur. Where would be the most suitable point to join the two sections of the beam together?

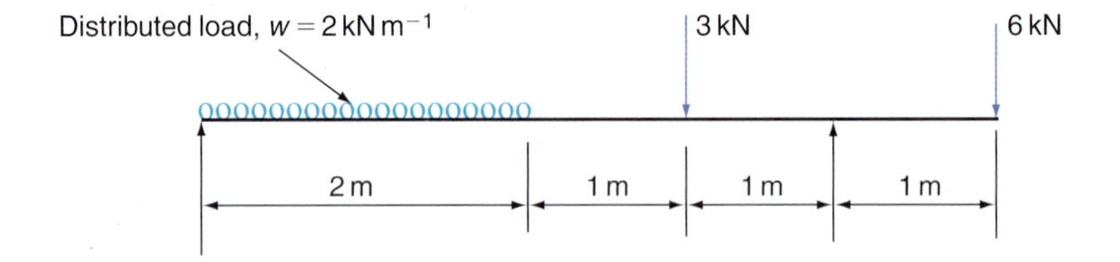

Figure 1.33

To check your understanding of the preceding section, you can solve Review question 5 at the end of this chapter.

Engineering Components

Structural members

The primary forces acting in structural components such as ties and struts are direct tensile and compressive forces. Direct or uni-axial loading occurs when equal and opposite tensile or compressive forces act along the same line of action. The intensity of the loading in the component material is quantified as *direct stress* and the deformation which it produces is quantified as *direct strain*.

Direct stress

Consider a component of original length l, and cross-sectional area A, which is subjected to a direct tensile load F as shown in Figure 1.34. Let the change of length be x.

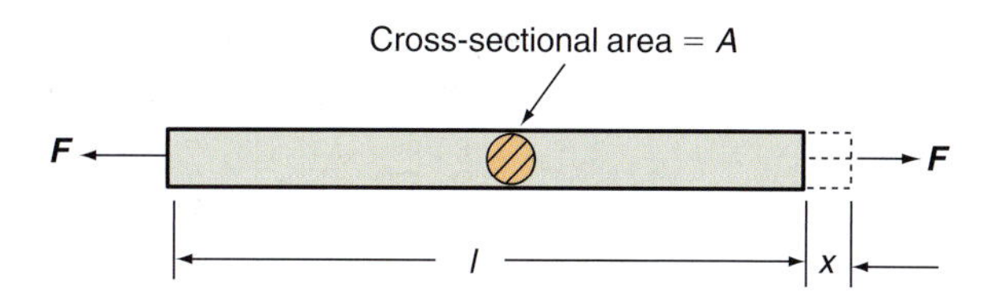

Figure 1.34 Direct loading

It is assumed that the load in the material is distributed evenly over the cross-sectional area A of the component. The direct stress σ in the material is the load carried by each square millimetre or square metre of cross-sectional area.

$$\text{Direct stress} = \frac{\text{Direct load}}{\text{Cross-sectional area}}$$

$$\sigma = \frac{F}{A} \text{ (Pa or N m}^{-2}\text{)} \tag{1.3}$$

In Figure 1.34 the load and the stress are tensile and these are sometimes given a positive sign. Compressive loads produce compressive stress and these are sometimes given a negative sign. You will recall from the work you did in the core unit Mechanical Principles and Applications that the approved SI unit of stress is the Pascal although you will often find its value quoted in N m^{-2} and N mm^{-2}. These are in fact more convenient and because many engineers prefer them, you will still see them used in trade catalogues, British Standards and engineering publications.

Direct strain

KEY POINT

A sign convention which is often used is that tensile stress and strain are positive and compressive stress and strain are negative.

Direct strain ε is a measure of the deformation which the load produces. It is the change in length given as a fraction of the original length.

$$\text{Direct strain} = \frac{\text{Change in length}}{\text{Original length}}$$

$$\varepsilon = \frac{x}{l} \text{ (No units)} \tag{1.4}$$

Modulus of elasticity (Young's modulus)

An elastic material is one in which the change in length is proportional to the load applied and in which the strain is proportional to the stress. Furthermore a perfectly elastic material will immediately return to its original length when the load is removed.

A graph of stress σ against strain ε is a straight line whose gradient is always found to be the same for a given material. Figure 1.35 shows typical graphs for steel, copper and aluminium. The value of the gradient is a measure of the elasticity or 'stiffness' of the material and is known as its modulus of elasticity, E.

$$\text{Modulus of elasticity} = \frac{\text{Direct stress}}{\text{Direct strain}}$$

$$E = \frac{\sigma}{\varepsilon} \text{ (Pa or N m}^{-2}\text{)} \tag{1.5}$$

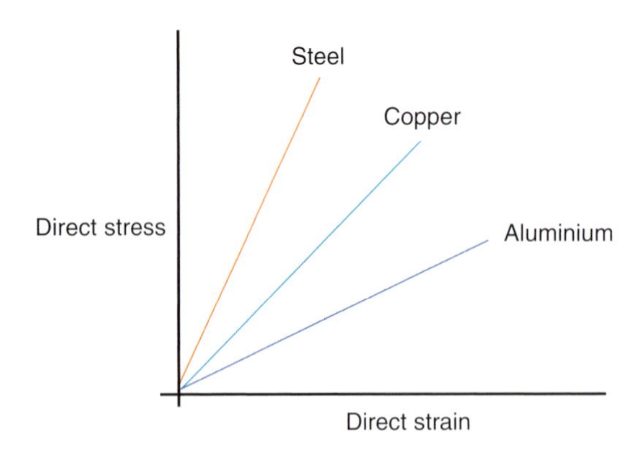

Figure 1.35 Graph of stress vs. strain

Substituting the expressions for stress and strain from equations (1.3) and (1.4) gives an alternative formula:

$$E = \frac{F/A}{x/l}$$

$$\boldsymbol{E = \frac{F}{A} \times \frac{l}{x}} \tag{1.6}$$

KEY POINT

The modulus of elasticity of a material is a measure of it stiffness, i.e. its resistance to being stretched or compressed.

Factor of safety

Engineering components should be designed so that the working stress which they are likely to encounter is well below the ultimate stress at which failure occurs.

$$\textbf{Factor of safety} = \frac{\textbf{Ultimate stress}}{\textbf{Working stress}} \tag{1.7}$$

A factor of safety of at least 2 is generally applied for static structures. This ensures that the working stress will be no more than half of that at which failure occurs. Much lower factors of safety are applied in aircraft design where weight is at a premium and with some of the major components it is likely that failure will eventually occur due to metal fatigue. These are rigorously tested at the prototype stage to predict their working life and replaced periodically in service well before this period has elapsed.

Example 1.12

A strut of diameter 25 mm and length 2 m carries an axial load of 20 kN. The ultimate compressive stress of the material is 350 MPa and its modulus of elasticity is 150 GPa. Find (a) the compressive stress in the material, (b) the factor of safety in operation, (c) the change in length of the strut.

(a) Finding cross-sectional area of strut.

$$A = \frac{\pi d^2}{4} = \frac{\pi \times 0.025^2}{4}$$

$$\boldsymbol{A = 491 \times 10^{-6}\ \mathrm{m^2}}$$

Finding compressive stress.

$$\sigma = \frac{F}{A} = \frac{20 \times 10^3}{491 \times 10^{-6}}$$

$$\boldsymbol{\sigma = 40.7 \times 10^6\ \text{Pa} \quad \text{or} \quad 40.7\ \text{MPa}}$$

(b) Finding factor of safety in operation.

$$\text{Factor of safety} = \frac{\text{Ultimate stress}}{\text{Working stress}} = \frac{350 \times 10^6}{40.7 \times 10^6}$$

Factor of safety = 8.6

(c) Finding compressive strain.

$$E = \frac{\sigma}{\varepsilon}$$

$$\varepsilon = \frac{\sigma}{E} = \frac{40.7 \times 10^6}{150 \times 10^9}$$

$$\boldsymbol{\varepsilon = 271 \times 10^{-6}}$$

Finding change in length.

$$\varepsilon = \frac{x}{l}$$

$$x = \varepsilon l = (271 \times 10^{-6}) \times 2$$

$$\boldsymbol{x = 0.543 \times 10^{-3}\ \text{m} \quad \text{or} \quad 0.543\ \text{mm}}$$

Test your knowledge 1.4

1. What is meant by uni-axial loading?
2. What is the definition of an elastic material?
3. A tie bar of cross-sectional area 50 mm^2 carries a load of 10 kN. What is the tensile stress in the material measured in Pascals?
4. If the ultimate tensile stress at which failure occurs in a material is 550 MPa and a factor of safety of 4 is to apply, what will be the allowable working stress?

Activity 1.3

A tube of length 1.5 m has an inner diameter 50 mm and wall thickness of 6 mm. When subjected to a direct tensile load of 75 kN the length is seen to increase by 0.55 mm. Determine (a) the tensile stress in the material, (b) the factor of safety in operation if the ultimate tensile strength of the material is 350 MPa, (c) the modulus of elasticity of the material.

Thermal loading

Most materials expand when their temperature rises. This is certainly true of the more commonly used metals in engineering and the effect is measured as the *linear expansivity* of a material. It is defined as *the change in length per unit of length per degree of temperature change* and its symbol is α. It is also known as the *coefficient of linear expansion* and its units are °C^{-1}.

To find the change in length x of a component of original length l and linear expansivity α when its temperature changes by t°C, we use the formula,

$$x = l\alpha t \tag{1.8}$$

Some typical values of linear expansivity α are given in Table 1.1.

Table 1.1 Linear expansivity values

Material	Linear expansivity (°C^{-1})
Aluminium	24×10^{-6}
Brass and bronze	19×10^{-6}
Copper	17×10^{-6}
Nickel	13×10^{-6}
Carbon steel	12×10^{-6}
Cast iron	10×10^{-6}
Platinum	9×10^{-6}
Invar	1.6×10^{-6}

Invar is an alloy steel containing around 36% nickel. The combination results in a material with a very low expansivity. It is used in applications such as instrumentation systems where expansion of the components could result in output errors.

Example 1.13

A steel bar of length 2.5 m and linear expansivity 12×10^{-6} °C^{-1} undergoes a rise in temperature from 15°C to 250°C. What will be its change in length?

$$x = l\alpha t$$
$$x = 2.5 \times 12 \times 10^{-6} \times (250 - 15)$$
$$\mathbf{x = 7.05 \times 10^{-3}\ m \quad or \quad 7.05\,mm}$$

Equation (1.8) can be rearranged to give the thermal strain ε which has resulted from the temperature change. It is measured in just the same way as the strain due to direct loading.

$$\varepsilon = \frac{x}{l} = \alpha t \tag{1.9}$$

If a material is allowed to expand freely, there will be no stress produced by the temperature change. If however the material is securely held and the change in length is prevented, thermal stress σ will be induced.

The above equation gives the virtual mechanical strain to which it is proportional and if the modulus of elasticity E for the material is known, the value of the stress can be calculated as follows:

$$E = \frac{\sigma}{\varepsilon} = \frac{\sigma}{\alpha t}$$

$$\boldsymbol{\sigma = E\alpha t} \qquad (1.10)$$

KEY POINT

Thermal stress is independent of the dimensions of a restrained component. It is dependent only on its modulus of elasticity, its linear expansivity and the temperature change which takes place.

You will see from the above equation that thermal stress depends only on the material properties and the temperature change.

Example 1.14

A steam pipe made from steel is fitted at a temperature of 20°C. If expansion is prevented, what will be the compressive stress in the material when steam at a temperature of 150°C flows through it? Take $\alpha = 12 \times 10^{-6}\,°\mathrm{C}^{-1}$ and $E = 200\,\mathrm{GN\,m^{-2}}$.

$$\sigma = E\alpha t$$

$$\sigma = 200 \times 10^{9} \times 12 \times 10^{-6} \times (150 - 20)$$

$$\boldsymbol{\sigma = 312 \times 10^{6}\,\mathrm{Pa} \quad \text{or} \quad 312\,\mathrm{MPa}}$$

You should note that this is quite a high level of stress which could very easily cause the pipe to buckle. This is why expansion loops and expansion joints are included in steam pipe systems, to reduce the stress to an acceptably low level.

Combined direct and thermal loading

It is of course quite possible to have a loaded component which is rigidly held and which also undergoes temperature change. This often happens with aircraft components and with components in process plant. Depending on whether the temperature rises or falls, the stress in the component may increase or decrease.

When investigating the effects of combined direct and thermal stress it is useful to adopt the sign convention that tensile stress and strain are positive and compressive stress and strain are negative. The resultant strain and stress is the algebraic sum of the direct and thermal values.

$$\text{Resultant stress} = \text{Direct stress} + \text{Thermal stress}$$

$$\sigma = \sigma_{\mathrm{D}} + \sigma_{\mathrm{T}}$$

$$\boldsymbol{\sigma = \frac{F}{A} + E\alpha t} \qquad (1.11)$$

KEY POINT

Residual thermal stresses are sometimes present in cast and forged components which have cooled unevenly. They can cause the component to become distorted when material is removed during machining. Residual thermal stresses can be removed by heat treatment and this will be described in Chapter 4.

$$\text{Resultant strain} = \text{Direct strain} + \text{Thermal strain}$$

$$\varepsilon = \varepsilon_{\mathrm{D}} + \varepsilon_{\mathrm{T}}$$

$$\boldsymbol{\varepsilon = \frac{x}{l} + \alpha t} \qquad (1.12)$$

Having calculated either the resultant stress or the resultant strain, the other can be found from the modulus of elasticity of the material.

$$\text{Modulus of elasticity} = \frac{\text{Resultant stress}}{\text{Resultant strain}}$$

$$E = \frac{\sigma}{\varepsilon} \qquad (1.13)$$

Example 1.15

A rigidly held tie bar in a heating chamber has a diameter of 60 mm and is tensioned to a load of 150 kN at a temperature of 15°C. What is the initial stress in the bar and what will be the resultant stress when the temperature in the chamber has risen to 50°C?

Take $E = 200\,\text{GN}\,\text{m}^{-2}$ and $\alpha = 12 \times 10^{-6}\,°\text{C}^{-1}$.

Note: The initial tensile stress will be positive but the thermal stress will be compressive and negative. It will thus have a cancelling effect.

Finding cross-sectional area.

$$A = \frac{\pi d^2}{4} = \frac{\pi \times 0.06^2}{4}$$

$$\mathbf{A = 2.83 \times 10^{-3}\,m^2}$$

Finding initial direct tensile stress at 15°C.

$$\sigma_D = \frac{F}{A} = +\frac{150 \times 10^3}{2.83 \times 10^{-3}}$$

$$\boldsymbol{\sigma_D = +53.0 \times 10^6\,\text{Pa} \quad \text{or} \quad +53.0\,\text{MPa}}$$

Finding thermal compressive stress at 180°C.

$$\sigma_T = E\alpha t = -200 \times 10^9 \times 12 \times 10^{-6} \times (50 - 15)$$

$$\boldsymbol{\sigma_T = -84.0 \times 10^{-6}\,\text{Pa} \quad \text{or} \quad -84.0\,\text{MPa}}$$

Finding resultant stress at 180°C.

$$\sigma = \sigma_D + \sigma_T = +53.0 + (-84.0)$$

$$\boldsymbol{\sigma = -31.0\,\text{MPa}}$$

i.e. the initial tensile stress has been cancelled out during the temperature rise and the final resultant stress is compressive.

Test your knowledge 1.5

1. How do you define the linear expansivity of a material?
2. If identical lengths of steel and aluminium bar undergo the same temperature rise, which one will expand the most?
3. What effect do the dimensions of a rigidly clamped component have on the thermal stress caused by temperature change?

Activity 1.4

A rigidly fixed strut in a refrigeration system carries a compressive load of 50 kN when assembled at a temperature of 20°C. The initial length of the strut is 0.5 m and its diameter is 30 mm. Determine the initial stress in the strut

and the amount of compression under load. At what operating temperature will the stress in the material have fallen to zero? Take $E = 150\,\text{GN m}^{-2}$ and $\alpha = 16 \times 10^{-6}$.

To check your understanding of the preceding section, you can solve Review questions 6–10 at the end of this chapter.

Compound members

Engineering components are sometimes fabricated from two different materials. Materials joined end to end form series compound members. Materials which are sandwiched together or contained one within another form parallel compound members.

With a series compound member as shown in Figure 1.36, the load F is transmitted through both materials.

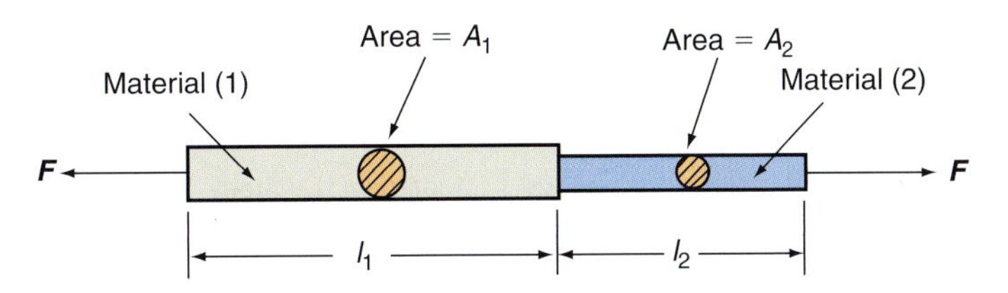

Figure 1.36 Series compound member

Let the moduli of elasticity or the materials be E_1 and E_2.
Let the cross-sectional areas be A_1 and A_2.
Let the stresses in the materials be σ_1 and σ_2.
Let the strains in the materials be ε_1 and ε_2.
Let the changes in length be x_1 and x_2.

The stress in each material is given by:

$$\sigma_1 = \frac{F}{A_1} \tag{1.14}$$

and

$$\sigma_2 = \frac{F}{A_2} \tag{1.15}$$

If the modulus of elasticity of the material (1) is E_1, the strain in the material will be:

$$\varepsilon_1 = \frac{\sigma_1}{E_1} \tag{1.16}$$

and the change in length will be:

$$x_1 = \varepsilon_1 l_1$$

$$x_1 = \frac{\sigma_1}{E_1} l_1 \tag{1.17}$$

If the modulus of elasticity of the material (2) is E_2, the strain in the material will be:

$$\varepsilon_2 = \frac{\sigma_2}{E_2} \tag{1.18}$$

CHAPTER 1

and the change in length will be:

$$x_2 = \varepsilon_2 l_2$$

$$x_2 = \frac{\sigma_2}{E_2} l_2 \qquad (1.19)$$

KEY POINT

The load is transmitted through both materials in series compound members, i.e. both materials carry the same load.

The total change in length will thus be:

$$x = x_1 + x_2$$

$$x = \frac{\sigma_1}{E_1} l_1 + \frac{\sigma_2}{E_2} l_2 \qquad (1.20)$$

Example 1.16

The compound member, shown in Figure 1.37 consists of a steel bar of length 2 m and diameter 30 mm, to which is brazed a copper tube of length 1.5 m, outer diameter 30 mm and inner diameter 15 mm. If the member carries a tensile load of 15 kN, determine the stress in each material and the overall change in length. For steel, $E = 200\,\text{GN m}^{-2}$, for copper $E = 120\,\text{GN m}^{-2}$.

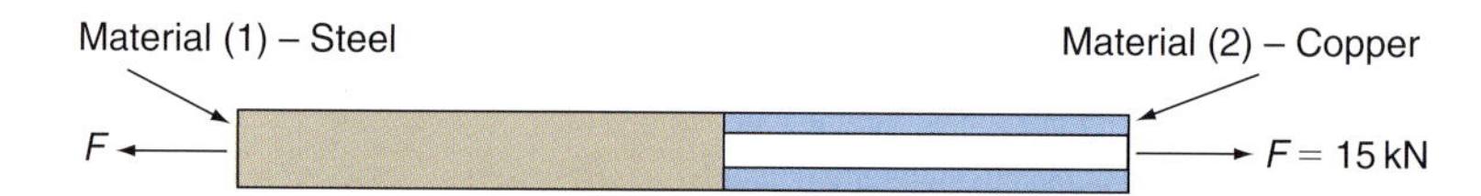

Figure 1.37

Finding cross-sectional area of steel.

$$A_1 = \frac{\pi D_I^2}{4} = \frac{\pi \times 0.03^2}{4}$$

$$\mathbf{A_1 = 707 \times 10^{-6}\,m^2}$$

Finding cross-sectional area of copper.

$$A_2 = \frac{\pi(D_1^2 - d_2^2)}{4} = \frac{\pi \times (0.03^2 - 0.015^2)}{4}$$

$$\mathbf{A_1 = 530 \times 10^{-6}\,m^2}$$

Finding stress in steel.

$$\sigma_1 = \frac{F}{A_1} = \frac{15 \times 10^3}{707 \times 10^{-6}}$$

$$\boldsymbol{\sigma_1 = 21.2 \times 10^6\,\text{Pa} \quad \text{or} \quad 21.2\,\text{MPa}}$$

Finding stress in copper.

$$\sigma_2 = \frac{F}{A_2} = \frac{15 \times 10^3}{530 \times 10^{-6}}$$

$$\boldsymbol{\sigma_1 = 28.3 \times 10^6\,\text{Pa} \quad \text{or} \quad 28.3\,\text{MPa}}$$

Finding overall change in length using equation (1.19).

$$x = \frac{\sigma_1}{E_1} l_1 + \frac{\sigma_2}{E_2} l_2 = \frac{(21.2 \times 10^6 \times 2)}{200 \times 10^9} + \frac{(28.3 \times 10^6 \times 1.5)}{120 \times 10^9}$$

$$x = (212 \times 10^{-6}) + (354 \times 10^{-6})$$

$$\mathbf{x = 0.566 \times 10^{-3}\,m \quad or \quad 0.566\,mm}$$

With parallel compound members, the loads in each material will most likely be different but when added together they will equal the external load. The change in length of each material will be the same. When you are required to find the stress in each material, this information enables you to write down two equations, each of which contains the unknown stresses. You can then solve these by substitution and use one of the stress values to find the common change in length.

Consider a member made up of two plates of the same material between which is bonded a plate of another material as shown in Figure 1.38. It is a good idea to let the material with the larger modulus of elasticity be material (1) and the that with the lower value be material (2).

Let the moduli of elasticity of the materials be E_1 and E_2.
Let the loads carried be F_1 and F_2.
Let the cross-sectional areas be A_1 and A_2.
Let the stresses in the materials be σ_1 and σ_2.

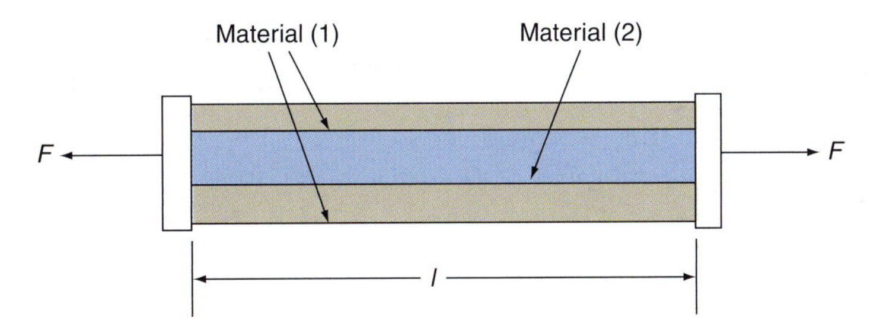

Figure 1.38 Parallel compound member

The external load is equal to the sum of the loads carried by the two materials.

$$F = F_1 + F_2$$

Now $F_1 = \sigma_1 A_1$ and $F_2 = \sigma_2 A_2$:

$$\mathbf{F = \sigma_1 A_1 + \sigma_2 A_2} \quad (1.21)$$

The strain in each material is the same, and so

$$\varepsilon_1 = \varepsilon_2$$

Now $\varepsilon_1 = \dfrac{\sigma_1}{E_1}$ and $\varepsilon_2 = \dfrac{\sigma_2}{E_2}$

$$\mathbf{\frac{\sigma_1}{E_1} = \frac{\sigma_2}{E_2}} \quad (1.22)$$

KEY POINT

Both materials undergo the same strain in parallel compound members and the sum of the loads carried by each material is equal to the external load.

Equations (1.21) and (1.22) can be solved simultaneously to find σ_1 and σ_2. The common strain can then be found from either of the above two expressions for strain, and this can then be used to find the common change in length.

Example 1.17

A concrete column 200 mm square is reinforced by nine steel rods of diameter 20 mm. The length of the column is 3 m and it is required to support an axial compressive load of 500 kN. Find the stress in each material and the change in length of the column under load. For steel $E = 200\,\text{GN m}^{-2}$ and for concrete, $E = 20\,\text{GN m}^{-2}$.

Because the steel has the higher modulus of elasticity, let it be material (1) and the concrete be material (2).

Begin by finding the cross-sectional areas of the two materials.

$$A_1 = 9 \times \frac{\pi d^2}{4} = 9 \times \frac{\pi \times 0.02^2}{4}$$

$$\mathbf{A_1 = 2.83 \times 10^{-3}\,m^2}$$

$$A_2 = (0.2 \times 0.2) - A_1 = 0.04 - (2.83 \times 10^{-3})$$

$$\mathbf{A_2 = 37.2 \times 10^{-3}\,m^2}$$

Now equate the external force to the sum of the forces in the two materials using equation (1.21).

$$F = \sigma_1 A_1 + \sigma_2 A_2$$

$$500 \times 10^3 = (2.83 \times 10^{-3})\sigma_1 + (37.2 \times 10^{-3})\sigma_2$$

This can be simplified by dividing both sides by 10^{-3}:

$$\mathbf{500 \times 10^6 = 2.83\sigma_1 + 37.2\sigma_2} \qquad (1)$$

Now equate the strains in each material using equation (1.21):

$$\frac{\sigma_1}{E_1} = \frac{\sigma_2}{E_2}$$

$$\sigma_1 = \frac{E_1}{E_2}\sigma_2$$

$$\sigma_1 = \frac{200 \times 10^9}{20 \times 10^9} \times \sigma_2$$

$$\boldsymbol{\sigma_1 = 10\sigma_2} \qquad (2)$$

Substitute in (1) for σ_1:

$$500 \times 10^6 = 2.83(10\sigma_2) + 37.2\sigma_2$$

$$500 \times 10^6 = 28.3\sigma_2 + 37.2\sigma_2 = 65.5\sigma_2$$

$$\sigma_2 = \frac{500 \times 10^6}{65.5}$$

$$\boldsymbol{\sigma_2 = 7.63 \times 10^6\,\text{Pa} \quad \text{or} \quad 7.63\,\text{MPa}}$$

Finding σ_1 from (2):

$$\boldsymbol{\sigma_1 = 10 \times 7.63 = 7.63\,\text{MPa}}$$

Now find the common strain using the stress and modulus of elasticity for material (2):

$$\varepsilon = \frac{\sigma_2}{E_2} = \frac{7.63 \times 10^6}{20 \times 10^9}$$

$$\boldsymbol{\varepsilon = 382 \times 10^{-6}}$$

Finally find the common change in length:

$$\varepsilon = \frac{x}{l}$$

$$x = \varepsilon l = 382 \times 10^{-6} \times 3$$

$$\mathbf{x = 1.15 \times 10^{-3}\,m \quad or \quad 1.15\,mm}$$

Test your knowledge 1.6

1. What is the parameter that the two materials in a series compound member have in common when it is under load?
2. What is the parameter that the two materials in a parallel compound member have in common when it is under load?
3. On which material property does the ratio of the stresses in the materials of a parallel compound member depend?

Activity 1.5

The tie bar shown in Figure 1.39 consists of a steel core 25 mm in diameter on to which is cast an aluminium outer case of outer diameter 40 mm. The compound section is 1.5 m long carries a tensile load of 30 kN. The load is applied to the protruding steel core as shown below.

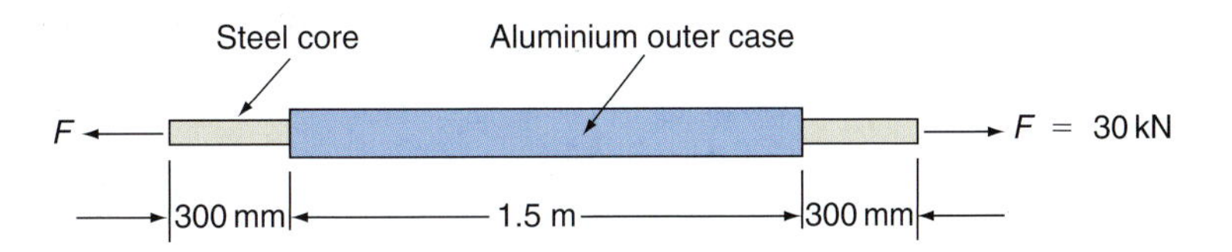

Figure 1.39

Find (a) the stress in the protruding steel which carries the full load, (b) the stress in the two materials in the compound section, (c) the overall change in length.

For steel, $E = 200\,\text{GN}\,\text{m}^{-2}$ and for aluminium, $E = 90\,\text{GN}\,\text{m}^{-2}$.

To check your understanding of the preceding section, you can solve Review questions 11–15 at the end of this chapter.

Fastenings

Engineering fastenings such as rivets, bolts, the different kinds of machine screws and setscrews, self-tapping screws and hinge pins are frequently subjected to shear loading. Shear loading occurs when equal and opposite parallel forces act on a component. Direct loading tends to cause failure perpendicular to the direction of loading whereas shear loading tends to cause failure parallel to the direction of loading. Direct loading and its effects have already been dealt with and we will now examine the effects of shear loading (Figure 1.40).

KEY POINT

Direct loading tends to cause failure in a plane which is at right angles to the direction of loading whereas shear loading tends to cause failure in a plane which is parallel to the direction of loading.

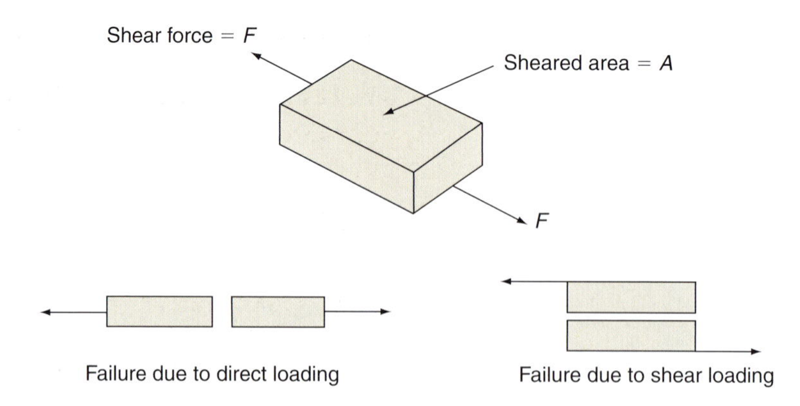

Figure 1.40 Direct and shear loading

Shear stress

Shear stress τ is a measure of the intensity of loading over the sheared area A.

$$\text{Shear stress} = \frac{\text{Shear force}}{\text{Sheared area}}$$

$$\boldsymbol{\tau = \frac{F}{A}\,(\text{Pa or N m}^{-2})} \qquad (1.23)$$

Shear strain

Shearing forces tend to distort the shape of a component as shown in Figure 1.41.

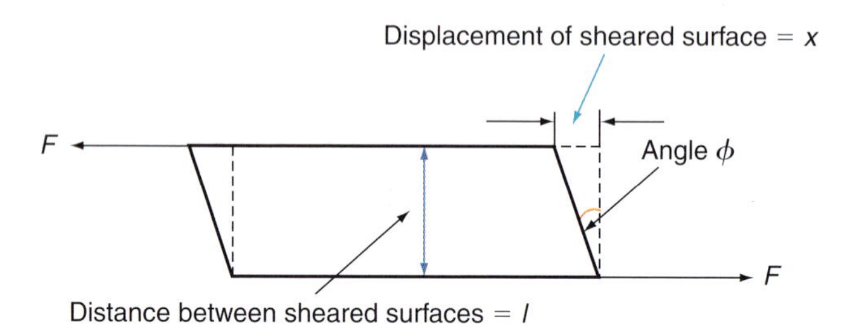

Figure 1.41 Deformation due to shearing

Shear strain γ is a measure of the deformation which the shearing force produces. It is the ratio of the displacement of the sheared surfaces to the distance between them.

$$\text{Shear strain} = \frac{\text{Displacement of sheared surfaces}}{\text{Distance between sheared surfaces}}$$

$$\boldsymbol{\gamma = \frac{x}{l}\,(\text{No units})} \qquad (1.24)$$

The angle ϕ is called the angle of shear. Its tangent is equal to the shear strain.

$$\tan\phi = \frac{x}{l}$$

$$\boldsymbol{\tan\phi = \gamma} \qquad (1.25)$$

Shear modulus (or modulus of rigidity)

When an elastic material is subjected to shear loading, the displacement x of the sheared surfaces is proportional to the load F which is applied. Also the shear stress τ is proportional to the shear strain γ.

A graph of shear stress against shear strain is a straight line, as shown in Figure 1.42, whose gradient for a given material is always found to be the same. It gives a measure of the elasticity or 'stiffness' of the material in shear and is known as its *shear modulus* G. In older text books you might find that it is called the *modulus of rigidity*.

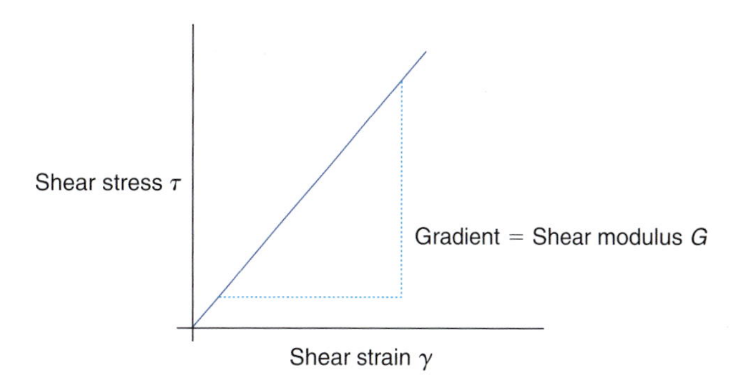

Figure 1.42 Graph of shear stress vs. shear strain

$$G = \frac{\tau}{\gamma} \text{ (Pa or N m}^{-2}\text{)} \tag{1.26}$$

Substituting the expressions for shear stress and shear strain from equations (1.23) and (1.24) gives an alternative formula:

$$G = \frac{F/A}{x/l}$$

$$G = \frac{F}{A} \times \frac{l}{x} \tag{1.27}$$

It will be noted that several of the above formulae are similar to those derived for direct stress and strain and the modulus of elasticity but they should not be confused. The symbols F, A, l and x have different meanings when used to calculate shear stress, shear strain and shear modulus. Furthermore, the values of modulus of elasticity E and shear modulus G are not the same for any given material. With mild steel for example, $E = 210\,\text{GN m}^{-2}$ whilst $G = 85\,\text{GN m}^{-2}$.

KEY POINT

Shear modulus is a measure of the shear stiffness of a material, i.e. its resistance to being deformed by shearing forces.

Example 1.18

A block of an elastic material is subjected to a shearing force of 50 kN which deforms its shape as shown in Figure 1.43. Find (a) the shear stress, (b) the shear strain, (c) the shear modulus for the material.

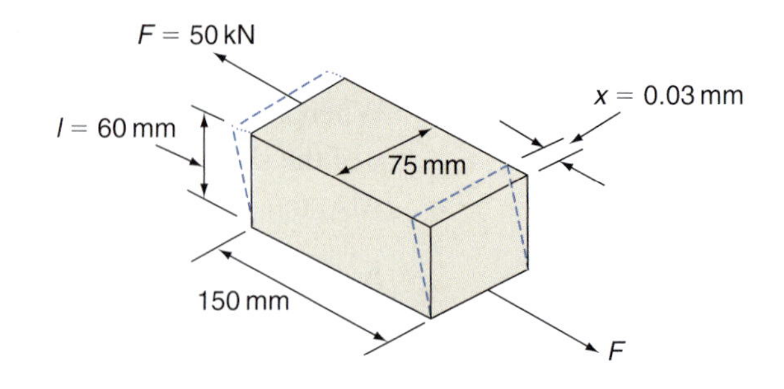

Figure 1.43

(a) Finding shear stress.

$$\tau = \frac{F}{A} = \frac{50 \times 10^3}{0.15 \times 0.075}$$

$$\boldsymbol{\tau = 4.44 \times 10^6 \text{ Pa} \quad \text{or} \quad 4.44 \text{ MPa}}$$

(b) Finding shear strain.

$$\gamma = \frac{x}{l} = \frac{0.03}{60}$$

$$\boldsymbol{\gamma = 500 \times 10^{-6}}$$

(c) Finding shear modulus.

$$G = \frac{\tau}{\gamma} = \frac{4.44 \times 10^6}{500 \times 10^{-6}}$$

$$\boldsymbol{G = 8.88 \times 10^9 \text{ Pa} \quad \text{or} \quad 8.88 \text{ GPa}}$$

Fastenings in single shear

Riveted lap joints and joints employing screwed fastenings are often subjected to shearing forces. Tensile forces may also be present and these are very necessary to hold the joint surfaces tightly together. It is very likely however that the external loads will have a shearing effect and it is assumed that this will be carried entirely by the fastenings. The tendency is to shear the fastenings at the joint interface and this is known as *single shear* as shown in Figure 1.44.

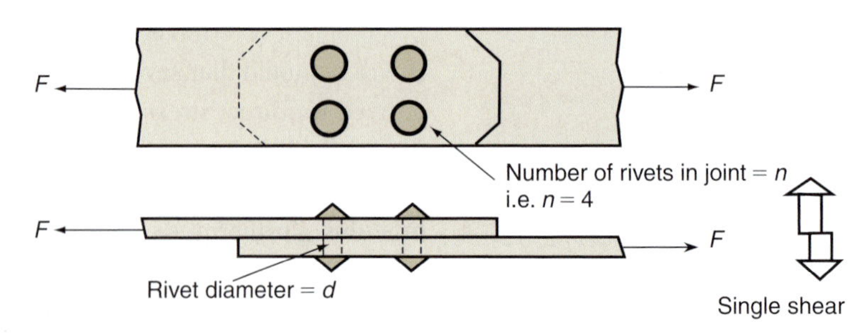

Figure 1.44 Lap joint with rivets in single shear

The total sheared cross-sectional area of the fastenings is:

$$A = \frac{n\,\pi d^2}{4}$$

The shear stress in the fastenings will be:

$$\tau = \frac{F}{A} = \frac{F}{n\pi d^2/4}$$

$$\tau = \frac{4F}{n\pi d^2} \tag{1.28}$$

In design problems you will probably know the safe working stress and the recommended rivet or bolt diameter for the thickness of the materials being joined. The task will then be to calculate the number of fastenings required and to decide on their spacing. Transposing the above formula gives:

$$n = \frac{4F}{\tau\pi d^2} \tag{1.29}$$

KEY POINT

The allowable working stress is the stress at which a component is considered to have failed divided by the factor of safety which is required. It is calculated for shear loading in exactly the same way as for direct loading.

The fastenings should not be too close together or too near to the edge of the material being joined. You can find the rules for particular applications in British and international standard specifications, and also in design code handbooks which are based on them.

Example 1.19

A lap joint is required to join plates using rivets of diameter 6 mm. The shearing force to be carried by the joint is 12 kN and the shear strength of the rivet material is 300 MPa. If a factor of safety of 8 is to apply, find the number of rivets required for the joint.

Finding allowable shear stress in rivets.

$$\tau = \frac{\text{Shear strength}}{\text{Factor of safety}} = \frac{300}{8}$$

$$\tau = \mathbf{37.5\ MPa}$$

Finding number of rivets required.

$$n = \frac{4F}{\tau\pi d^2}$$

$$n = \frac{4 \times 12 \times 10^3}{37.5 \times 10^6 \times \pi \times (6 \times 10^{-3})^2}$$

$\mathbf{n = 11.3}$, i.e. use **12 rivets**

Fastenings in double shear

In joints where the plates must be butted together, connecting plates are used above and below the joint. There is then a tendency to shear the rivets in two places and they are said to be in *double shear*.

There are in fact two joints in Figure 1.45, where each of the butted plates is riveted to the connecting plates. As a result the number of rivets per joint is half the total number shown. i.e. $n = 3$ rivets in this particular example.

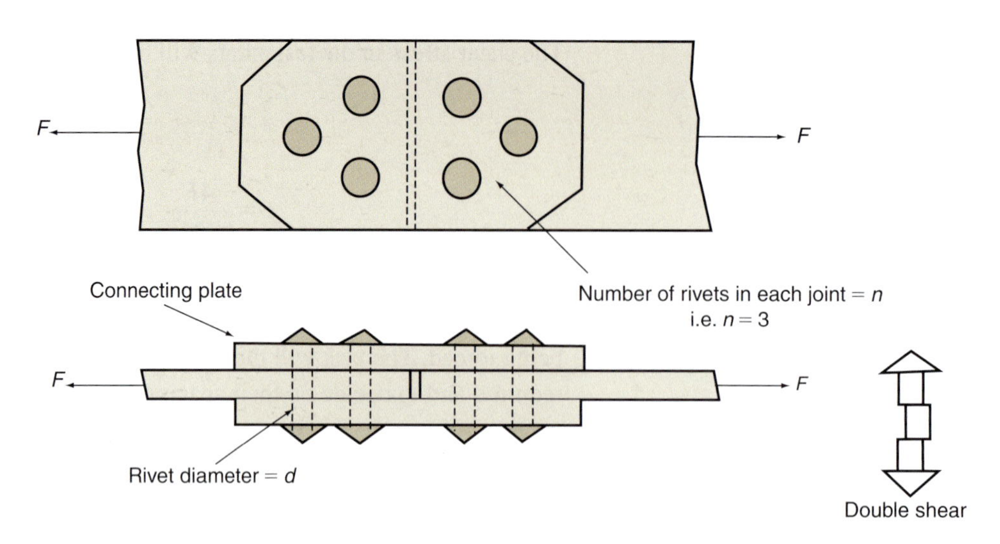

Figure 1.45 Butt joint with rivets in double shear

The total sheared cross-sectional area of the fastenings is double that for single shear.

$$\text{i.e. } A = \frac{2n\,\pi d^2}{4}$$

The shear stress in the fastenings will be:

$$\tau = \frac{F}{A} = \frac{F}{2n\pi d^2/4}$$

$$\tau = \frac{2F}{n\,\pi d^2} \qquad (1.30)$$

If you compare this with equation (1.28) you will note that for the same load, diameter and number of rivets, the shear stress for double shear is half the value for single shear. A joint in double shear can therefore carry twice as much load as the equivalent joint in single shear. If it is required to find the number of rivets required, the above equation can be transposed to give:

KEY POINT

A rivet in double shear can carry twice as much load as in single shear before failure occurs.

$$n = \frac{2F}{\tau\pi d^2} \qquad (1.31)$$

Example 1.20

Two aluminium plates are to be joined by means of a double strap riveted butt joint. The joint is required to support a load of 15 kN and the total number of rivets is 12. The rivet material has a shear strength of 200 MPa and a factor of safety of 6 is to apply. Determine the required diameter of the rivets.

Finding allowable shear stress in rivets.

$$\tau = \frac{\text{Shear strength}}{\text{Factor of safety}} = \frac{200}{6}$$

$$\boldsymbol{\tau = 33.3\,\text{MPa}}$$

Finding required rivet diameter.

$$\tau = \frac{2F}{n\pi d^2}$$

$$d^2 = \frac{2F}{n\pi\tau}$$

$$d = \sqrt{\frac{2F}{n\pi\tau}} \qquad \text{where } n = \frac{12}{2} = 6 \text{ rivets}$$

$$d = \sqrt{\frac{2\times10\times10^3}{6\times\pi\times33.3\times10^6}}$$

$$\mathbf{d = 5.64\times10^{-3}\,m \quad or \quad 5.64\,mm,}$$

i.e. use 6.0 mm rivets

Test your knowledge 1.7

1. What is the difference between direct loading and shear loading?
2. Define shear stress and shear strain.
3. What is the angle of shear?
4. Define shear modulus.
5. Compare the load carrying capacity for rivets of a given diameter when subjected to single shear and double shear.

Activity 1.6

Two steel plates are riveted together by means of a lap joint. Ten rivets of diameter 9 mm are used and the load carried is 20 kN. What is the shear stress in the rivet material and the factor of safety in operation if the shear strength is 300 MPa? If the lap joint were replaced by a double strap butt joint with 10 rivets per joint and each carrying the same shear stress as above, what would be the required rivet diameter?

To check your understanding of the preceding section, you can solve Review questions 16–19 at the end of this chapter.

Rotating Systems with Uniform Angular Acceleration

By uniform angular motion we mean rotation at a steady speed or rotation with uniform angular acceleration or retardation. If you have completed the core unit Mechanical Principles and Applications you will be familiar with uniform linear motion and the equations that connect linear displacement, time, velocity and acceleration. You may also recall the expressions for work power, momentum and energy. We shall now develop a similar set of equations for angular motion.

The symbols that we use for linear and angular parameters are shown in Table 1.2.

The unit of angular displacement or angle turned is the radian. There are other ways of measuring angles you will be familiar with, such as degrees of arc and number of revolutions turned. However, it is the

Table 1.2 Linear and angular motion parameters

Parameter	Linear symbols	Angular symbols
Displacement	S (m)	θ (rad)
Time	T (s)	t (s)
Linear/angular velocity	V (m s^{-1})	ω (rad s^{-1})
Linear/angular acceleration	A (m s^{-2})	α (rad s^{-2})

radian that we must use in our formulae because it is a natural way of measuring angles and specified as the *supplementary* SI unit for angular measurement. A radian is the angle subtended at the centre of a circle by an arc equal in length to the radius (Figure 1.46).

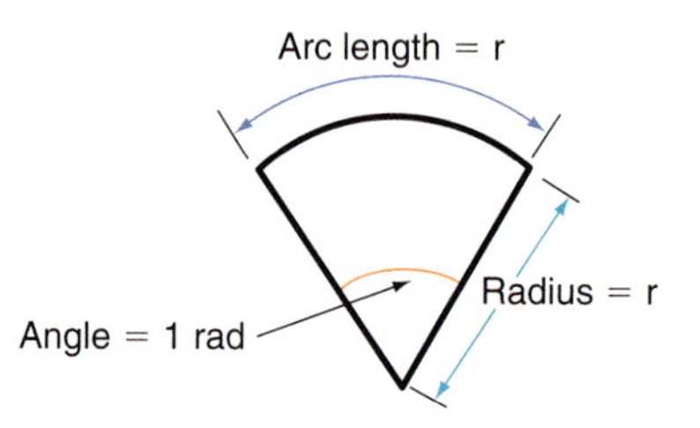

Figure 1.46 The radian

KEY POINT

In rotational dynamics calculations, angular displacement must be measured in radians, angular velocity in radians per second (rad s^{-1}) and angular acceleration in radians per second per second (rad s^{-2}).

There are 2π radians in a full circle and if you should need to convert from revolutions to radians or from revolutions per minute (rpm) to radians per second (rad s^{-1}), or from degrees to radians and vice versa, the formulae are as follows.

To convert from degrees to radians and vice versa use:

$$\theta \text{ rad} = \theta^\circ \times \frac{2\pi}{360} \quad \text{or} \quad \theta^\circ = \theta \text{ rad} \times \frac{360}{2\pi}$$

To convert from number of revolutions (n) to radians (θ) and vice versa use:

$$\theta = n \times 2\pi \quad \text{or} \quad n = \frac{\theta}{2\pi}$$

To convert from number of revolutions per minute (N) to radians per second (θ) use:

$$\omega = N \times \frac{2\pi}{60} \quad \text{or} \quad N = \omega \times \frac{60}{2\pi}$$

Angular velocity is the change of angular displacement per second and angular acceleration is the change of angular velocity per second. They are often written in calculus notation as:

$$\omega = \frac{d\theta}{dt} \quad \text{and} \quad \alpha = \frac{d\omega}{dt} = \frac{d^2\theta}{dt^2}$$

Consider now the graph of angular velocity against time for a body that accelerates uniformly from an initial angular velocity of ω_1 to a final angular velocity ω_2 in a time of t seconds whilst rotating through an angle of θ radians (Figure 1.47).

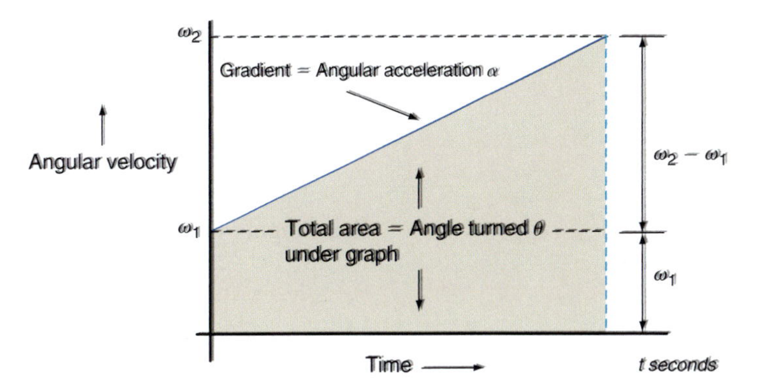

Figure 1.47 Angular velocity vs. time graph

The gradient of the graph is the angular acceleration which is given by:

$$\alpha = \frac{\omega_2 - \omega_1}{t} \tag{1.32}$$

This can be transposed to give the final angular velocity:

$$\omega_2 = \omega_1 + \alpha t \tag{1.33}$$

The angular displacement, or angle turned whilst accelerating, is the area under the graph. You can find this by adding together the area of the triangle and rectangle under the graph. You can also find it using calculus by integrating the equation of the graph, which is equation (1.33), between the limits of 0 and t as follows:

$$\omega_2 = \omega_1 + \alpha t$$

$$\theta = \int_0^t (\omega_1 + \alpha t)\,dt$$

$$\theta = \left(\omega_1 t + \frac{\alpha t^2}{2} + c\right)_0^t$$

$$\theta = \left(\omega_1 t + \frac{\alpha t^2}{2} + c\right) - (0 + 0 + c)$$

$$\theta = \omega_1 t + \frac{1}{2}\alpha t^2 \tag{1.34}$$

An alternative formula for finding the angle turned is as follows:

$$\theta = \text{Average angualr velocity} \times \text{Time taken}$$

$$\theta = \frac{1}{2}(\omega_2 + \omega_1)t \tag{1.35}$$

In all of the above formulae you need to know the time taken during the period of acceleration but this can be eliminated by obtaining an expression for t from equation (1.31) and substituting it for t in equation (1.34).

$$\alpha = \frac{\omega_1 - \omega_2}{t}$$
$$t = \frac{\omega_1 - \omega_2}{\alpha}$$

Substituting for t in equation (1.34) gives:

$$\theta = \frac{1}{2}(\omega_2 + \omega_1)\left(\frac{\omega_1 - \omega_2}{\alpha}\right)$$
$$2\alpha\theta = (\omega_2 + \omega_1)(\omega_1 - \omega_2)$$

Rearranging gives:

$$2\alpha\theta = \omega_2^2 + \omega_2\omega_1 - \omega_2\omega_1 - \omega_1^2$$
$$2\alpha\theta = \omega_2^2 - \omega_1^2$$
$$\boldsymbol{\omega_2^2 = \omega_1^2 + 2\alpha\theta} \qquad (1.36)$$

As you will see in Table 1.3, the equations for linear and angular motion are very similar.

Table 1.3 Equations for uniform linear and angular motion

Linear motion	Angular motion
$v = u + at$	$\omega_2 = \omega_1 + \alpha t$
$s = ut + ½at^2$	$\theta = \omega_1 t + ½\alpha t^2$
$s = ½(u + v)t$	$\theta = ½(\omega_2 + \omega_1)t$
$v^2 = u^2 + 2as$	$\omega_2^2 = \omega_1^2 + 2\alpha\theta$

KEY POINT

The gradient of a graph of angular velocity against time is the angular acceleration and the area under the graph is the angular displacement.

Here are a few more useful formulae for converting arc length s to angle turned θ, tangential linear velocity v to angular velocity ω and tangential linear acceleration α to angular acceleration α and vice versa.

$$\theta = \frac{s}{r} \qquad \omega = \frac{v}{r} \qquad \alpha = \frac{a}{r}$$

$$\text{or} \qquad s = r\theta \qquad v = \omega r \qquad a = \alpha r$$

Example 1.21

The winding drum of a hoist is 250 mm in diameter and is accelerated uniformly from rest to a speed of 120 rpm in a time of 45 s. Determine (a) the final angular velocity of the drum, (b) the angular acceleration, (c) the number of revolutions turned during the period of acceleration, (d) the height that the load on the hoist is raised, (e) the final velocity of the load.

(a) Finding final angular velocity of drum in rad s^{-1}.

$$\omega_2 = N_2 \times \frac{2\pi}{60} = 120 \times \frac{2\pi}{60}$$
$$\boldsymbol{\omega_2 = 12.6 \text{ rad s}^{-1}}$$

(b) Finding angular acceleration of drum.

$$\alpha = \frac{\omega_2 - \omega_1}{t} = \frac{12.6 - 0}{45}$$

$$\boldsymbol{\alpha = 0.28\ \mathrm{rad\,s^{-2}}}$$

(c) Finding angle turned by drum.

$$\theta = \tfrac{1}{2}(\omega_2 + \omega_1)t = \tfrac{1}{2}(12.6 + 0) \times 45$$

$$\boldsymbol{\theta = 283.5\ \mathrm{rad}}$$

or alternatively $\theta = \omega_1 t + \tfrac{1}{2}\alpha t^2 = 0 + (\tfrac{1}{2} \times 0.28 \times 45^2)$

$$\boldsymbol{\theta = 283.5\ \mathrm{rad}}$$

Finding number of revolutions turned by drum.

$$n = \frac{\theta}{2\pi} = \frac{283.5}{2\pi}$$

$$\boldsymbol{n = 45.1\ \mathrm{revolutions}}$$

(d) Finding height raised by load.

$$s = r\theta = 0.125 \times 283.5$$

$$\boldsymbol{s = 35.4\ \mathrm{m}}$$

(e) Finding final velocity of load.

$$v = \omega_2 r = 12.6 \times 0.125$$

$$\boldsymbol{v = 1.58\ \mathrm{m\,s^{-1}}}$$

Work done and power developed for angular motion

When a tangential force is applied to a body, causing it to rotate, work is done. Work is the product of the force and the distance around the circular path through which it moves (Figure 1.48).

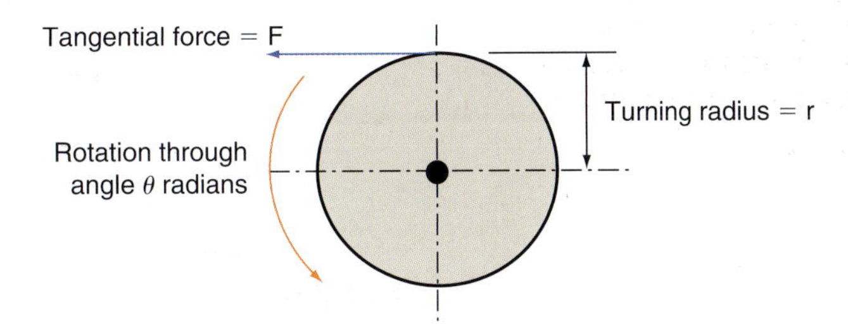

Figure 1.48 Rotating body

The applied torque will be,

$$\text{Torque} = \text{Tangential force} \times \text{Turning radius}$$

$$\boldsymbol{T = Fr\ (\mathrm{Nm})}$$

The work done is given by:

$$\text{Work done} = \text{Tangential force} \times \text{Distance moved}$$

$$W = Fs$$

Now from the conversion formula that we have listed, the distance moved is given by $s = r\theta$, giving:

$$W = Fr\theta$$

But $Fr = T$, the applied torque, and so:

$$\mathbf{W = T\theta} \tag{1.37}$$

You will recall that power is the rate of doing work, or the work done per second.

$$\text{Power} = \frac{\text{Work done}}{\text{Time taken}}$$

$$\textbf{Power} = \frac{T\theta}{t}\ \textbf{(W)} \tag{1.38}$$

Equation (1.38) is the one you should use when asked to calculate the average power developed over a period of time. But θ/t in the equation can be written as ω, the angular velocity, giving:

$$\textbf{Power} = T\omega\ \textbf{(W)} \tag{1.39}$$

KEY POINT

The power being developed at any instant during a period of angular acceleration is the product of the driving torque and the angular velocity at that point in time.

Equation (1.39) is the one you should use when asked to calculate the power that is being developed at the instant in time when the angular velocity is $\omega\,\text{rad}\,\text{s}^{-1}$.

Example 1.22

The drum of a mixer is accelerated uniformly from rest by a driving torque of 12 Nm and rotates through 50 revolutions in the first 30 s. The acceleration continues for a further 20 s to reach the steady operating speed of the mixer. Determine (a) the angular acceleration, (b) the final speed attained, (c) the work done, (d) the average power developed by the driving motor.

It may be useful to display the information on an angular velocity–time graph as shown in Figure 1.49.

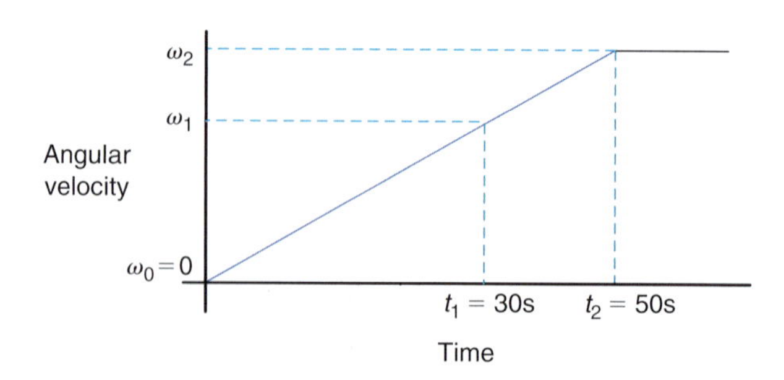

Figure 1.49

(a) Finding angle in radians, turned in first 30 s.

$$\theta_1 = n \times 2\pi = 50 \times 2\pi$$

$$\boldsymbol{\theta_1 = 314\ \text{rad}}$$

Finding angular velocity after first 30 s.

$$\theta_1 = \frac{1}{2}(\omega_1 + \omega_0)\, t_1$$

$$314 = \frac{1}{2} \times (\omega_1 + 0) \times 30 = 15\omega_1$$

$$\omega_1 = \frac{\mathbf{314}}{\mathbf{15}} = \mathbf{20.9\,rad\,s^{-1}}$$

Finding angular acceleration.

$$\alpha = \frac{\omega_1 - \omega_0}{t_1} = \frac{20.9 - 0}{30}$$

$$\boldsymbol{\alpha} = \mathbf{0.697\,rad\,s^{-2}}$$

(b) Finding final angular velocity.

$$\omega_2 = \omega_0 + \alpha t = 0 + (0.697 \times 50)$$

$$\boldsymbol{\omega_2} = \mathbf{34.9\,rad\,s^{-1}}$$

Finding final speed in rpm.

$$N = \omega_2 \times \frac{60}{2\pi} = 34.9 \times \frac{60}{2\pi}$$

$$\boldsymbol{N} = \mathbf{333\ rpm}$$

(c) Finding total angle turned, θ_2.

$$\theta_2 = \omega_0 + \frac{1}{2}\alpha t_2^2 = 0 + (0.5 \times 0.697 \times 50^2)$$

$$\boldsymbol{\theta_2} = \mathbf{871\,rad}$$

Finding total work done.

$$W = T\theta_2 = 12 \times 871$$

$$\boldsymbol{W} = \mathbf{10.5 \times 10^3\,J} \quad \textbf{or} \quad \mathbf{10.5\,kJ}$$

(d) Finding average power developed by driving motor.

$$\text{Average power} = \frac{W}{t_2} = \frac{10.5 \times 10^3}{50}$$

$$\textbf{Average power} = \mathbf{210\,W}$$

Test your knowledge 1.8

1. What is the definition of a radian?
2. How do you convert revolutions to radians?
3. How do you convert rotational speed in revolutions per minute (rpm) to angular velocity in radians per second ($rad\,s^{-1}$)?
4. How do you calculate the average power developed during a period of angular acceleration?
5. How do you calculate the power being developed at a given instant during a period of angular acceleration?

Activity 1.7

A driving torque of 15 Nm is applied to the winding drum of a hoist whose diameter is 500 mm. This causes the drum to accelerate uniformly from rest and raise the load on the hoist cable through a distance of 8 m in a time of 20 s. Determine (a) the angle turned by the drum, (b) the final angular velocity of the drum, (c) the angular acceleration of the drum, (d) the maximum power developed by the driving motor.

To check your understanding of the preceding section, you can solve Review questions 20–25 at the end of this chapter.

Moment of Inertia

Newton's second law of motion gives us a means of calculating the inertial resistance to linear acceleration using the well-known formula:

$$\textbf{Inertial resistance} = \textbf{Mass} \times \textbf{Acceleration}$$

$$F = ma$$

There is also inertial resistance to angular acceleration which depends in part on the mass of the rotating body. Here it is more convenient to measure it as a resisting torque rather than a resisting force and it is not only the mass but also its distribution about the axis of rotation that determines its value.

Consider a rigid body of mass m kg which is rotating about an axis XX with angular acceleration α rad s^{-2}. Let the inertial resisting torque be T Nm (Figure 1.50).

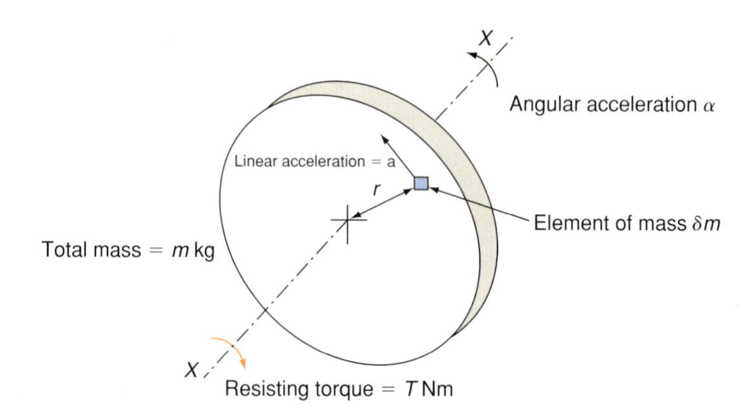

Figure 1.50 Inertia of a rotating body

At the instant shown, the element δm at radius r from the axis of rotation has a linear acceleration a in the direction shown. The inertial resistance δF of the element is given by the above formula as:

$$\delta F = \delta ma$$

The resisting torque δT will be:

$$\delta T = \delta Fr$$

or $$\delta T = \delta mar$$

Now all such elements have the same angular acceleration α rad s^{-2} and the linear acceleration can be written as $\alpha = ar$. The resisting torque can thus be written as:

$$\delta T = \delta m \alpha r r$$
$$\delta T = \delta m \alpha r^2$$

The total inertial resisting torque T is the sum of all such elements:

$$T = \sum (\delta m \alpha r^2)$$
$$\boldsymbol{T = \alpha \sum (\delta m r^2)} \quad (1.40)$$

Now the sum $\Sigma(\delta mr^2)$ is known as the *second moment of mass* or *mass moment of inertia* of the body. Often it is just called *moment of inertia* and is given the symbol I. Its units are kgm^2. The torque that resists the angular acceleration can thus be written as:

$$\boldsymbol{T = Ia} \quad (1.41)$$

Although the elements of mass might lie at different radii, it is convenient to imagine an intermediate value of radius around which the whole mass m of a rotating body is concentrated. This is called the *radius of gyration* and is usually given the symbol k. If the value of k is known, the moment of inertia can be calculated using the formula:

$$\boldsymbol{I = mk^2} \quad (1.42)$$

For complex rotating objects such as motor vehicle wheels, motor armatures and turbine rotors the value of radius of gyration is generally found from experimental test data. For more simple systems such as a concentrated mass, a rotating ring, a disc and a rotating link, the values of k can be found from simple reasoning or by applying integral calculus as follows.

(i) Concentrated mass (Figure 1.51)

Radius of gyration $k = r$

Moment of inertia $I = mr^2$

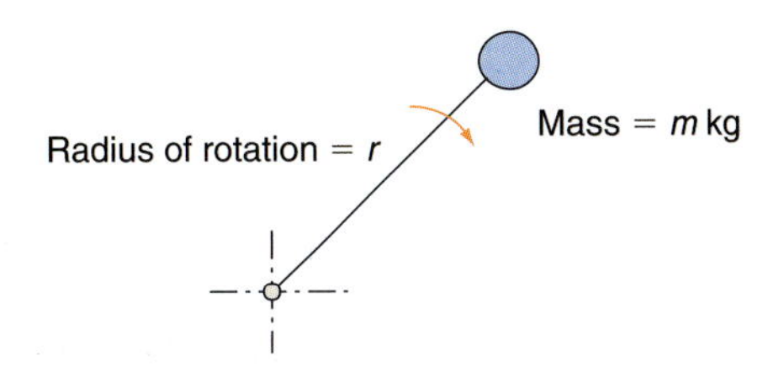

Figure 1.51

(ii) Thin rotating ring (Figure 1.52)

Radius of gyration $k = r$

Moment of inertia $I = mr^2$

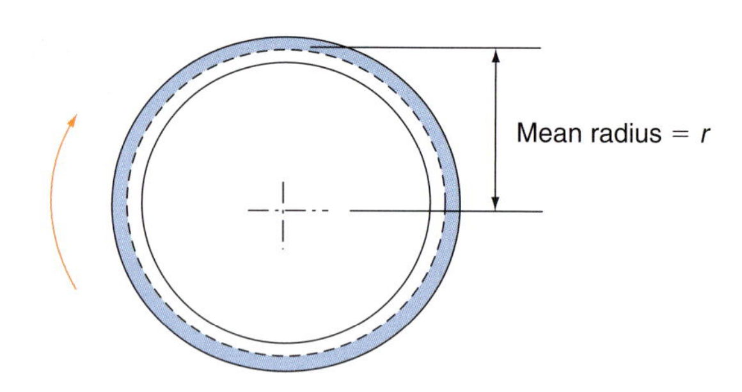

Figure 1.52

(iii) Rotating disc (Figure 1.53)

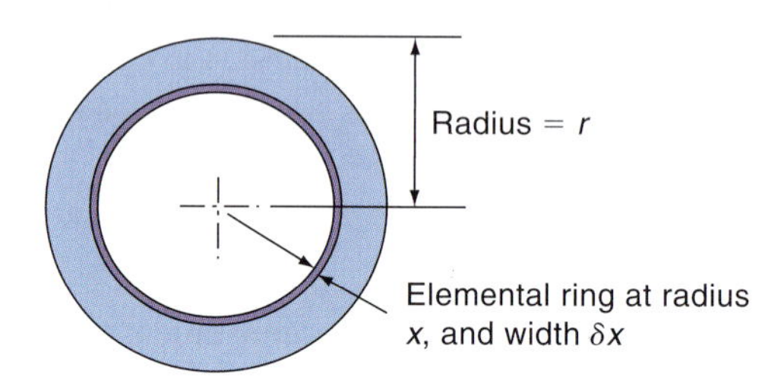

Figure 1.53

Let the thickness of the disc be t and the density of its material be ρ $\mathrm{kg\,m^{-3}}$. The mass δm of the elemental ring at radius x and width δx will be:

$$\delta m = 2\pi xt\ \delta x\ \rho$$

The moment of inertia δI of the elemental ring will be:

$$\delta I = \delta m\ x^2 = 2\pi xt\ \delta x\ \rho\ x^2$$
$$\delta I = 2\pi\rho t x^3\ \delta x$$

The moment of inertia I of the body will be the sum of all such elements:

$$I = \sum 2\pi\rho t x^3\ \delta x$$

In the limit as $\delta r \to 0$

$$I = \int_0^r 2\pi\rho t x^3\ \mathrm{d}x = 2\pi\rho t \int_0^r x^3\ \mathrm{d}x$$

$$I = 2\pi\rho t \left(\frac{x^4}{4}\right)_0^r = 2\pi\rho t\,\frac{r^4}{4}$$

$$I = \pi\rho t\,\frac{r^4}{2}$$

Now the total mass of the disc is given by $m = \pi r^2\rho$ and so the moment of inertia can be written as:

$$\boldsymbol{I = m\,\frac{r^2}{2}} \qquad (1.43)$$

Because $I = mk^2$, the radius of gyration k of a disc is given by:

$$k^2 = \frac{r^2}{2}$$

$$\mathbf{k = \frac{r}{\sqrt{2}} = 0.7071r} \qquad (1.44)$$

You may also have encountered this figure of 0.7071 in your electrical principles studies where it gives the root-mean square value of an alternating current when multiplied by the peak value. This is very much the same idea, since the radius of gyration is the square root of the sum of all the squared values of the elemental radii.

(iv) Rod rotating about a perpendicular axis through its centre (Figure 1.54)

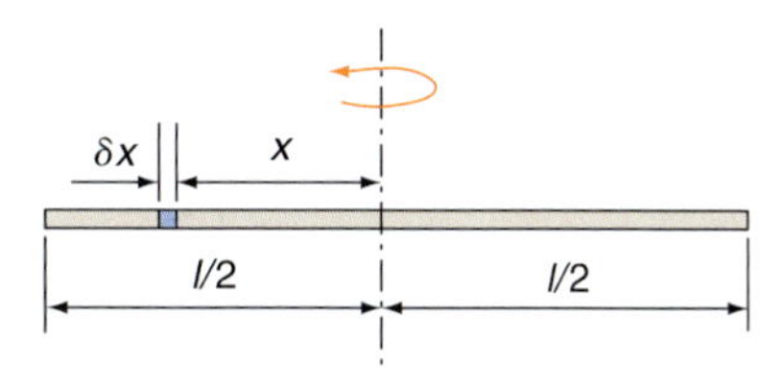

Figure 1.54

Let a be the mass per unit length of the rod. The mass δm of the element of length δx distance x from the central axis will be:

$$\delta m = a\delta x$$

The moment of inertia δI of the element will be:

$$\delta I = \delta m x^2 = a\delta x x^2$$
$$\delta I = a x^2 \delta x$$

The total moment of inertia is the sum of all such elements:

$$I = \sum a x^2 \delta x$$
$$I = a \sum x^2 \delta x$$

In the limit as $\delta x \to 0$

$$I = a \int_{-l/2}^{+l/2} x^2 \, \mathrm{d}x$$

$$I = a \left(\frac{x^3}{3} \right)_{-l/2}^{+l/2} = a \left(\frac{l^3}{24} + \frac{l^3}{24} \right)$$

$$I = a \frac{l^3}{12}$$

Now the mass m of the rod is given by $m = al$ and so the moment of inertia of the rod can be written as:

$$\mathbf{I = m \frac{l^2}{12}} \qquad (1.45)$$

Because $I = mk^2$, the radius of gyration k of the rod when rotating about a perpendicular axis through its centre is given by:

$$k^2 = \frac{l^2}{12}$$

$$\mathbf{k = \frac{l}{\sqrt{12}} = 0.289l} \qquad (1.46)$$

(v) Rod rotating about a perpendicular axis through one end (Figure 1.55)

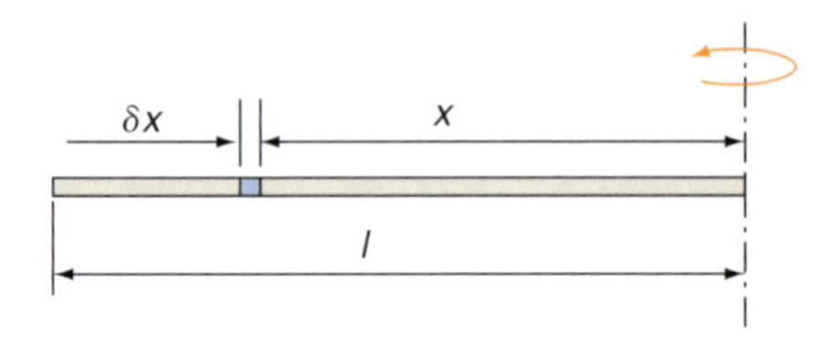

Figure 1.55

The procedure for this configuration is the same as that for the rod rotating about an axis through its centre except for the limits of the integral. This now becomes:

$$I = a\int_{0}^{1} x^2\,\mathrm{d}x$$

$$I = a\left(\frac{x^3}{3}\right)_0^1 = a\left(\frac{l^3}{3} + \frac{l^3}{3}\right)$$

$$I = a\frac{l^3}{3}$$

Now the mass m of the rod is given by $m = al$, and so the moment of inertia of the rod can be written as:

$$\mathbf{I = m\frac{l^2}{3}} \qquad (1.47)$$

Because $I = mk^2$, the radius of gyration k of the rod when rotating about a perpendicular axis through one end is given by:

$$k^2 = \frac{l^2}{3}$$

$$\mathbf{k = \frac{l}{\sqrt{3}} = 0.577l} \qquad (1.48)$$

KEY POINT

The radius of gyration of a rotating body is the radius at which a concentrated mass, the same as that of the body, would rotate with the same of the same dynamic characteristics as the body. It is also the root-mean square radius of the elemental particles that go to make up the body, measured from the axis of rotation.

Example 1.23

A flywheel of mass 125 kg and radius of gyration 150 mm is accelerated from rest to a speed of 1250 rpm whilst making 500 revolutions. The motion is resisted by a friction torque of 2.5 Nm. Determine (a) the angular acceleration, (b) the applied torque, (c) the work done, (d) the maximum power developed by the driving motor.

(a) Finding angle turned in radians.

$$\theta = n \times 2\pi = 500 \times 2\pi$$

$$\boldsymbol{\theta = 3142\ \text{rad}}$$

Finding final angular velocity.

$$\omega_2 = N \times \frac{2\pi}{60} = 1250 \times \frac{2\pi}{60}$$

$$\boldsymbol{\omega_2 = 131\ \text{rad s}^{-1}}$$

Finding angular acceleration.

$$\omega_2^2 = \omega_1^2 + 2\alpha\theta$$

$$\alpha = \frac{\omega_2^2 - \omega_1^2}{2\theta} = \frac{131^2 - 0}{2 \times 3142}$$

$$\boldsymbol{\alpha = 2.73\ \text{rad s}^{-2}}$$

(b) Finding moment of inertia of flywheel.

$$I = mk^2 = 125 \times 0.15^2$$

$$\boldsymbol{I = 2.81\ \text{kgm}^2}$$

Finding applied torque (Figure 1.56).}

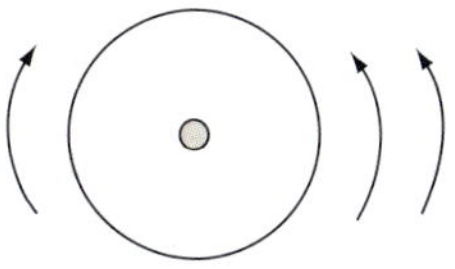

Figure 1.56

As with linear motion, D'Alembert's principle can be applied to a system with uniform angular motion, i.e.:

$$\text{Applied torque} = \text{Inertia torque} + \text{Friction torque}$$

$$T = I\alpha + T_f = 0.313$$

$$T = (2.81 \times 2.73) + 2.5$$

$$\boldsymbol{T = 10.2\ \text{Nm}}$$

KEY POINT

D'Alembert's principle, by which a free body diagram can be drawn to display the forces acting on a body moving with uniform linear motion, may be applied to a body rotating with uniform angular motion to display the driving and resisting torques acting on it.

(c) Finding work done.

$$W = T\theta = 10.2 \times 3142$$

$$\boldsymbol{W = 32 \times 10^3\ \text{J} \quad \text{or} \quad 32\ \text{kJ}}$$

(d) Finding maximum power developed by driving motor.

$$\text{Maximum power} = T\omega_2 = 10.2 \times 131$$

$$\textbf{Maximum power} = \boldsymbol{1.34 \times 10^3\ \text{W} \quad \text{or} \quad 1.34\ \text{kW}}$$

Test your knowledge 1.9

1. What is the definition of the radius of gyration of a rotating body?
2. What is the general formula used to calculate moment of inertia?
3. What is the formula for calculating the radius of gyration of a disc rotating about a polar axis through its centre?
4. How do you calculate the torque which must be applied to overcome the inertia of a body during a period of uniform angular acceleration?

Activity 1.8

A cast iron flywheel of diameter 750 mm and thickness 100 mm may be regarded as a disc. The density of the material is 7200 kg m^{-3}. The flywheel is accelerated uniformly from a speed of 100 rpm to 1000 rpm in a time of 1 min against a bearing friction torque of 4.5 Nm. Determine (a) the moment of inertia of the flywheel, (b) the angle turned during the period of acceleration, (c) the angular acceleration, (d) the driving torque, (e) the average power developed by the driving motor.

To check your understanding of the preceding section, you can solve Review questions 26–31 at the end of this chapter.

Linear and rotational kinetic energy

The kinetic energy of a body when travelling at a particular speed is the work that has been done during the period of uniform acceleration to overcome its inertia. If friction is neglected, the force F required to accelerate a body with uniform linear motion is given by Newton's second law, i.e.:

$$F = ma$$

If the acceleration takes place whilst travelling though a distance s, the stored kinetic energy or work done is given by:

$$KE = Fs$$

$$\text{or} \quad KE = mas \qquad \text{(i)}$$

Now you will recall that one of the equations for uniform linear motion is:

$$v^2 = u^2 + 2as$$

If the acceleration is from rest, then $u = 0$ giving:

$$v^2 = 2as$$

$$\text{or} \quad as = \frac{1}{2}v^2 \qquad \text{(ii)}$$

Substituting for as in equation (i) gives:

$$\mathbf{KE = \frac{1}{2}mv^2 \ (J)} \qquad (1.49)$$

The same applies for a body that is accelerated from rest with uniform angular acceleration. Here the torque required to overcome inertia is given by:

$$T = I\alpha$$

If the angular acceleration takes place whilst rotating through an angle θ rad, the stored kinetic energy or work done is given by:

$$\text{KE} = I\alpha\theta \qquad \text{(iii)}$$

One of the equations for uniform angular acceleration is:

$$\omega_2^2 = \omega_1^2 + 2\alpha\theta$$

If the acceleration is from rest, then $\omega_1 = 0$ giving:

$$\omega_2^2 = 2\alpha\theta$$

$$\text{or} \quad \alpha\theta = \frac{1}{2}\omega_2^2 \qquad \text{(iv)}$$

Substituting for $\alpha\,\theta$ in equation (iii) gives:

$$\text{KE} = \frac{1}{2} I\omega_2^2$$

For a body rotating with angular velocity ω rad s^{-1} this can just be written as:

$$\mathbf{KE} = \frac{1}{2} I\omega^2 \text{ (J)} \qquad (1.50)$$

A body can of course also have gravitational potential energy. This is the work that has been done in raising the body to a height h above some given datum level.

i.e. Gravitational potential energy = Weight × Height raised

$$\mathbf{PE} = mgh \qquad (1.51)$$

The principle of conservation of energy is particularly useful when solving problems involving a combination of linear and angular motion. A typical example is a hoist, where a load is raised by a cable on a winding drum. The driving motor must supply a torque that is sufficient to overcome the inertia of both the winding drum and the load, the weight of the load and also frictional resistance. Such a mechanical system can be represented diagrammatically in Figure 1.57.

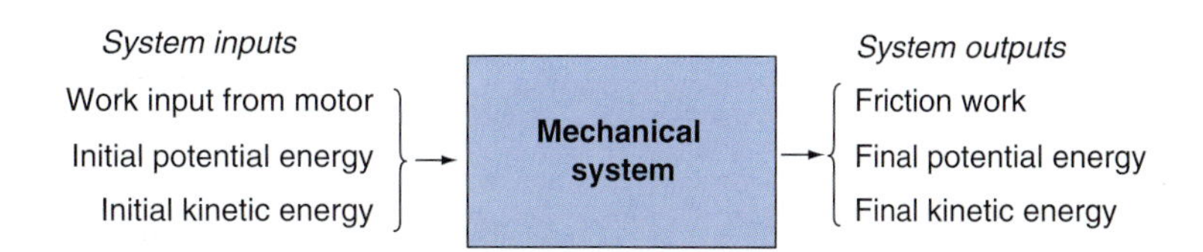

Figure 1.57 Mechanical system

Total work and energy input = Total work and energy output

Work input + Initial PE + Initial KE = Final PE + Final KE + Friction work

$$W_i + mgh_1 + \tfrac{1}{2}mv_1^2 + \tfrac{1}{2}I\omega_1^2 = mgh_2 + \tfrac{1}{2}mv_2^2 + \tfrac{1}{2}I\omega_2^2 + W_f$$

This can be transposed to give an expression for work input.

$$W_i = (mgh_2 - mgh_1) + (\tfrac{1}{2}mv_2^2 - \tfrac{1}{2}mv_1^2) + (\tfrac{1}{2}I\omega_2^2 - \tfrac{1}{2}I\omega_1^2) + W_f$$

$$\boldsymbol{W_i = mg(h_2 - h_1) + \tfrac{1}{2}m(v_2^2 - v_1^2) + \tfrac{1}{2}I(\omega_2^2 - \omega_1^2) + W_f} \qquad (1.52)$$

If the frictional resistance force F_f is known, the work done in overcoming friction can be calculated using the formula, $W_f = F_f s$, where s is the linear distance travelled.

KEY POINT

The principle of conservation of energy states that the energy cannot be created or destroyed but changes may take place from one form to another.

If the friction torque T_f in the bearings of a system is known, the work done in overcoming friction can be calculated using the formula $W_f = T_f\theta$, where θ is the angle turned.

Example 1.24

A hoist raises a load of 75 kg from rest to a speed of 2.5 m s^{-1} in a distance of 6 m. The load is attached to a light cable wrapped around a winding drum of diameter 500 mm. The mass of the drum is 60 kg and its radius of gyration is 180 mm. The bearing friction torque is 5 Nm. Determine (a) the angle turned by the drum, (b) the final angular velocity of the drum, (c) the work input from the driving motor, (d) input torque to the winding drum, (e) the maximum power developed by the motor.

(a) Finding angle turned by drum.

$$\theta = \frac{h_2 - h_1}{r} = \frac{6}{0.25}$$

$$\boldsymbol{\theta = 24\ \text{rad}}$$

(b) Finding final angular velocity of drum.

$$\omega_2 = \frac{v_2}{r} = \frac{2.5}{0.25}$$

$$\boldsymbol{\omega_2 = 10\ \text{rad s}^{-1}}$$

(c) Finding moment of inertia of winding drum.

$$I = Mk^2 = 60 \times 0.18^2$$

$$\boldsymbol{I = 1.94\ \text{kgm}^2}$$

Finding work input from driving motor.

$$W_i = mg(h_2 - h_1) + \tfrac{1}{2}m(v_2^2 - v_1^2) + \tfrac{1}{2}(\omega_2^2 - \omega_1^2) + T_f\theta$$

$$W_i = [75 \times 9.81 \times 6] + [\tfrac{1}{2} \times 75 \times (2.5^2 - 0)] + [\tfrac{1}{2} \times (10^2 - 0)] + [5 \times 24]$$

$$W_i = 4415 + 234 + 50 + 120$$

$$\boldsymbol{W_i = 4819\ \text{J} \quad \text{or} \quad 4.82\ \text{kJ}}$$

(d) Finding input torque from driving motor.

$$W_i = T_i\theta$$

$$T_i = \frac{W_i}{\theta} = \frac{4819}{24}$$

$$\boldsymbol{T_i = 201\text{Nm}}$$

(e) Finding maximum power developed by motor.

$$\text{Maximum power} = T_i\omega_2 = 201 \times 10$$

$$\textbf{Maximum power} = \boldsymbol{2010\ \text{W} \quad \text{or} \quad 2.01\ \text{kW}}$$

Example 1.25

The combined mass of a motor cycle and rider is 300 kg. The wheels are each of mass of 25 kg, diameter 500 mm and radius of gyration

200 mm. The rider accelerates from a speed of 5 m s^{-1} to 20 m s^{-1} over a distance of 100 m whilst ascending an incline of slope 1 in 20 (sine). The average resistance to motion including air resistance is 110 N. Determine (a) the total work done, (b) the average power developed (Figure 1.58).

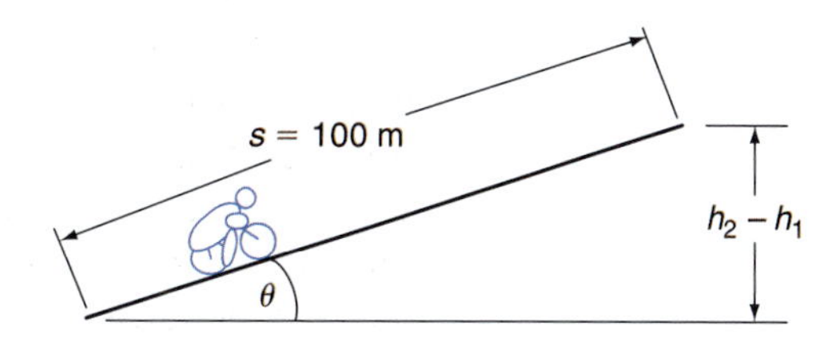

Figure 1.58

(a) Finding vertical height gained.

$$h_2 - h_1 = s \sin\theta = 100 \times \frac{1}{20}$$

$$\mathbf{h_2 - h_1 = 5\ m}$$

Finding initial and final angular velocities of wheels.

$$\omega_1 = \frac{u}{r} = \frac{5}{0.25} = \mathbf{20\ rad\,s^{-1}}$$

$$\omega_2 = \frac{v}{r} = \frac{20}{0.25} = \mathbf{80\ rad\,s^{-1}}$$

Finding combined moment of inertia of wheels.

$$I = 2 \times m_w k^2 = 2 \times 25 \times 0.2^2$$

$$\mathbf{I = 2\ kgm^2}$$

Finding total work done.

$$W = mg(h_2 - h_1) + \tfrac{1}{2}m(v_2^2 - v_1^2) + \tfrac{1}{2}(\omega_2^2 - \omega_1^2) + F_f s$$

$$W = [300 \times 9.81 \times 5] + [\tfrac{1}{2} \times 300 \times (20^2 - 5^2)] + [\tfrac{1}{2} \times 2 \times (80^2 - 20^2)] + [110 \times 100]$$

$$W = 14715 + 56250 + 6000 + 11000$$

$$\mathbf{W = 88\,000\ J \quad or \quad 88\ kJ}$$

(b) Finding time taken.

$$s = \tfrac{1}{2}(v + u)t$$

$$t = \frac{2s}{(v+u)} = \frac{2 \times 100}{(20+5)}$$

$$\mathbf{t = 8\ s}$$

Finding average power developed.

$$\text{Average power} = \frac{W}{t} = \frac{88000}{8}$$

$$\textbf{Average power} = \mathbf{11000\ W \quad or \quad 11\ kW}$$

Test your knowledge 1.10

1. What is gravitational potential energy?
2. What is kinetic energy?
3. What is the principle of conservation of energy?
4. How can you calculate the work done in overcoming a friction torque?
5. Write down the energy equation for a mechanical system?

Activity 1.9

A mine cage of mass 3 tonnes is accelerated uniformly upwards from a speed of 1.0 m s^{-1} to 4.5 m s^{-1} in a time of 20 s. The winding drum which is 1.5 m in diameter has a mass of 1 tonne and has the dynamic characteristics of a disc. The average frictional resistance in the cage guides is 500 N and there is a bearing friction torque of 45 Nm. The mass of the lift cable can be neglected. Determine (a) the angle turned by the winding drum, (b) the moment of inertia of the winding drum, (c) the total work done, (d) the input torque to the winding drum, (e) the maximum power developed by the driving motor.

To check your understanding of the preceding section, you can solve Review questions 32–35 at the end of this chapter.

Experiments to determine moment of inertia and radius of gyration

There are a number of experimental methods which may be used to determine the moment of inertia and radius of gyration of a rotating body. Two methods, which are within the scope of this chapter, will be described. The first method may be applied to bodies which rotate on an axle which is supported in bearings. The second method makes use of an inclined plane or track.

Method 1

Apparatus

Rotating body under investigation which is mounted on an axle. The axle may be supported in bearings or between free-running centres at some distance above ground level as shown in Figure 1.59. Length of cord, hanger and slotted masses, metre rule, stopwatch.

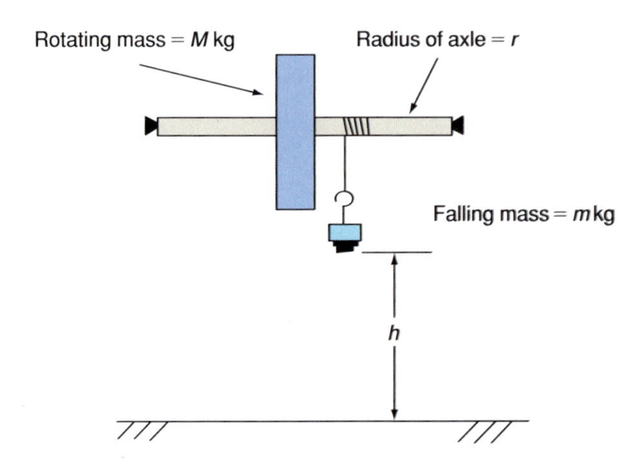

Figure 1.59 Arrangement of apparatus

Procedure

1. Measure the diameter of the axle and record the value of its radius r.
2. Wind the cord around the axle and attach the hanger to its free end.
3. Place small slotted masses on the hanger until bearing friction is just overcome. The hanger and masses should then fall with a slow uniform velocity.
4. Rewind the cord and place an additional mass of m kg on the hanger.
5. Measure and record the distance h from the base of the hanger to the ground.
6. Release the hanger and record the time for it to fall to the ground. Repeat this operation a number of times and obtain the average time of fall.
7. Calculate the moment of inertia and radius of gyration of the rotating body using the following derived formulae.

Theory

Finding final velocity of falling mass.

$$h = \frac{1}{2}(v_2 + v_1)t$$

But $v_1 = 0$,

$$h = \frac{1}{2}v_2 t$$

$$\mathbf{v_2 = \frac{2h}{t}} \qquad \text{(i)}$$

Finding final angular velocity of rotating body.

$$\omega_2 = \frac{v_2}{r}$$

$$\boldsymbol{\omega_2 = \frac{2h}{rt}} \qquad \text{(ii)}$$

The energy equation for the system is:

$$W = mg(h_2 - h_1) + \frac{1}{2}m(v_2^2 - v_1^2) + \frac{1}{2}I(\omega_2^2 - \omega_1^2) + F_f s + W_f$$

There is no work input and so W is zero. Also v_1 and ω_1 are zero and the work done in overcoming friction W_f has been compensated for by the small slotted mass initially placed on the hanger. This leaves

$$0 = mg(h_2 - h_1) + \frac{1}{2}mv_2^2 + \frac{1}{2}I\omega_2^2$$

Now $h_2 - h_1 = -h$

$$0 = -mgh + \frac{1}{2}mv_2^2 + \frac{1}{2}I\omega_2^2$$

$$\boldsymbol{\frac{1}{2}I\omega_2^2 = mgh - \frac{1}{2}mv_2^2} \qquad \text{(iv)}$$

Substituting for v_2 and ω_2 gives:

$$\frac{1}{2}I\left(\frac{2h}{rt}\right)^2 = mgh - \frac{1}{2}m\left(\frac{2h}{t}\right)^2$$

$$\frac{1}{2}I\left(\frac{4h^2}{r^2t^2}\right) = mgh - \frac{1}{2}m\left(\frac{4h^2}{t^2}\right)$$

$$I\left(\frac{2h^2}{r^2t^2}\right) = mh\left(g - \frac{2h}{t^2}\right)$$

$$\boldsymbol{I = \frac{mr^2t^2}{2h}\left(g - \frac{2h}{t^2}\right)} \qquad \text{(v)}$$

Finding radius of gyration.

$$I = mk^2$$

$$\boldsymbol{k = \sqrt{\frac{I}{m}}} \qquad \text{(vi)}$$

Method 2

Apparatus

Rotating body under investigation, inclined plane or track, metre rule, stopwatch, clinometer (if available).

If the body under investigation is mounted on an axle or shaft it may be positioned to roll down inclined parallel rails, as shown in Figure 1.60(a). If the body is cylindrical, it may be positioned to roll down an inclined plane as shown in Figure 1.60(b).

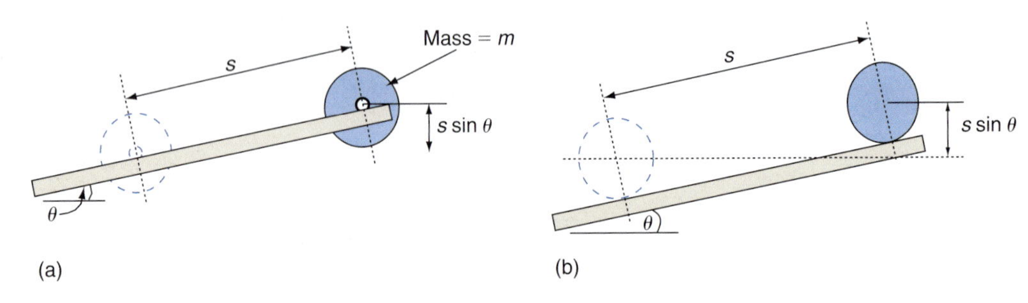

Figure 1.60 Arrangement of apparatus: (a) body rotating on axle, (b) body rolling on its cylindrical surface.

Procedure

1. Record the mass of the body under investigation. If it is to roll on an axle, record the axle diameter. If it is to roll on its cylindrical surface, record its outer diameter.
2. Set the track or inclined plane to a small angle, i.e. less than 10°. It is best to use a clinometer for this if one is available, even if the track is fitted with a protractor.
3. Mark the start and finish positions on the track and record s, the distance between them.
4. Place the body in the start position, release it and at the same time start the stopwatch.

5. Record the time taken for the body to roll through the distance s. Repeat the operation several times and obtain an average value of the time t.
6. Use the collected data to calculate the values of moment of inertia and radius of gyration of the body.

Theory
Finding vertical distance travelled by body. (This will be negative since the body loses height.)

$$\boldsymbol{h_1 - h_2 = -s \sin \theta} \qquad \text{(i)}$$

Finding the final linear velocity of the body.

$$v_2 = 2 \times \text{Average velocity}$$

$$\boldsymbol{v_2 = \frac{2s}{t}} \qquad \text{(ii)}$$

The final angular velocity is given by:

$$\omega_2 = \frac{v_2}{r}$$

$$\boldsymbol{\omega_2 = \frac{2s}{rt}} \qquad \text{(iii)}$$

The energy equation for the system is:

$$W = mg(h_2 - h_1) + \tfrac{1}{2}m(v_2^2 - v_1^2) + \tfrac{1}{2}I(\omega_2^2 - \omega_1^2) + F_f s + T_f \theta$$

There is no work input and so W is zero. Also v_1 and ω_1 are zero. The effects of rolling friction will be small and so the friction work terms may also be neglected. This leaves

$$0 = mg(h_2 - h_1) + \frac{1}{2}mv_2^2 + \frac{1}{2}I\omega_2^2$$

$$\boldsymbol{\frac{1}{2}I\omega_2^2 = -mg(h_2 - h_1) - \frac{1}{2}mv_2^2} \qquad \text{(iv)}$$

Substituting for $h_1 - h_2$ and v_2 and ω_2 gives:

$$\frac{1}{2}I\left(\frac{2s}{rt}\right)^2 = -mg(-s\sin\theta) - \frac{1}{2}m\left(\frac{2s}{t}\right)^2$$

$$\frac{1}{2}I\left(\frac{4s^2}{r^2t^2}\right) = mgs\sin\theta - \frac{1}{2}m\left(\frac{4s^2}{t^2}\right)$$

$$I\left(\frac{2s^2}{r^2t^2}\right) = ms\left(g\sin\theta - \frac{2s}{t^2}\right)$$

$$I = \frac{msr^2t^2}{2s^2}\left(g\sin\theta - \frac{2s}{t^2}\right)$$

$$\boldsymbol{I = \frac{mr^2t^2}{2s}\left(g\sin\theta - \frac{2s}{t^2}\right)} \qquad \text{(v)}$$

Finding radius of gyration.

$$I = mk^2$$

$$k = \sqrt{\frac{I}{m}} \quad \text{(vi)}$$

Note: The expressions for moment of inertia may also be obtained by applying D'Alembert's principle and Newton's second law of motion to the systems.

Centripetal Acceleration and Centripetal Force

Newton's first law of motion states that *a moving body will continue in a straight line unless acted upon by some external force*. It follows that when a body is travelling in a circular path, there must be some sideways force pushing, or pulling it, towards the centre of rotation. This is known as *centripetal* force, which means 'centre seeking'. With a car travelling round a curve, the centripetal force is provided by friction between the tyres and the road surface and with an orbiting satellite, it is provided by the earth's gravitational pull.

Vector change of velocity

You will recall that velocity is a vector quantity. This means that it has both magnitude and direction. For a body travelling in a straight path we very often only state the magnitude of its velocity, i.e. its speed. We take its direction for granted but strictly speaking, we should state both magnitude and direction when we write down the velocity of a body.

KEY POINT

Velocity is a vector quantity and if either the speed or the direction of a body changes, its velocity will change.

Because velocity is a vector quantity, it can be represented in magnitude, direction and sense by a line v_1 drawn to a suitable scale on a vector diagram. If either the speed or direction of a body should change, then its velocity will change. The new vector v_2 will be drawn from the same starting point but to a different length or in a different direction. There is then said to have been a vector change of velocity which can be measured as the distance between the end of the initial vector to the end of the final vector.

You will also recall that acceleration is the rate of change of velocity, i.e. the change of velocity per second. This not only applies for changes of speed, as in Figure 1.61(a), but also for vector changes of velocity as in Figure 1.61(b). If the time taken for the vector change is known, the acceleration which has taken place can be calculated. This is also a vector quantity whose direction is the same as the vector change of velocity.

$$\textbf{Acceleration} = \frac{\textbf{Vector change of velocity}}{\textbf{Time taken for change}}$$

$$a = \frac{\Delta v}{t} \quad (1.53)$$

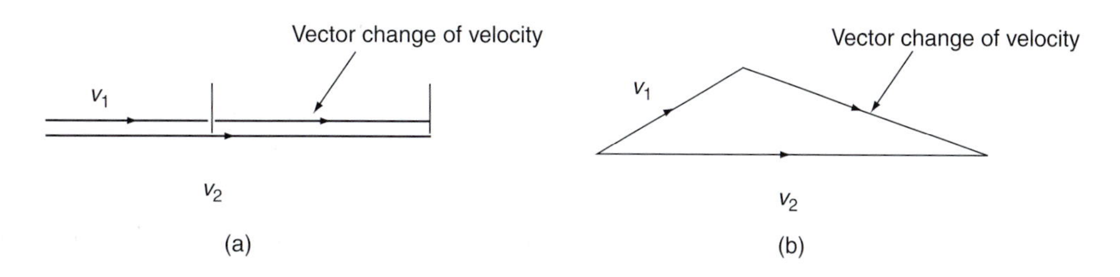

Figure 1.61 Vector change of velocity: (a) change of speed, (b) change of speed and direction

KEY POINT

Acceleration is the rate of change of velocity which can result from a change of speed or a change of direction.

When a body picks up speed or changes direction there must be some external force acting to cause the change. This force is given by Newton's second law of motion which states that *the rate of change of momentum of a body is equal to the force which is causing the change*. It is from this law that we obtain the formula:

$$\text{Force} = \text{Mass} \times \text{Acceleration}$$

$$F = ma \tag{1.54}$$

Having found the vector change of velocity of a body from a vector diagram, you can then calculate the acceleration which has occurred using expression (1.53). If you know the mass m of the body you can then calculate the force which has produced the change using expression (1.54). The direction in which the force acts on the body is the same as that of the vector change of velocity and the acceleration which it produces.

Example 1.26

A vehicle of mass 750 kg travelling east at a speed of 50 km h^{-1} accelerates around a bend in the road and emerges 10 s later at a speed of 90 km h^{-1} travelling in a north easterly direction (Figure 1.62). What is its vector change of velocity, the acceleration and the force which produces the change?

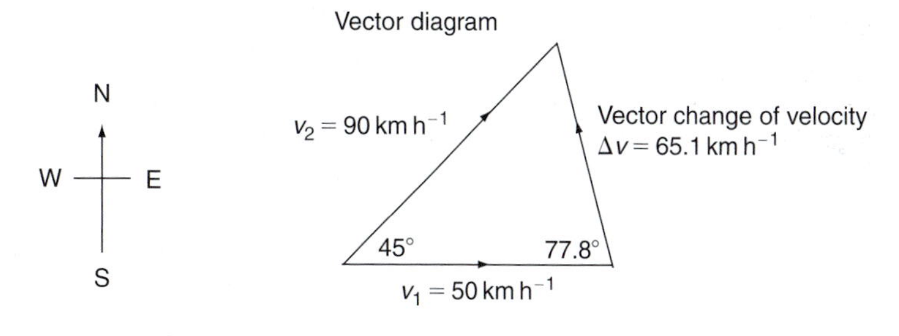

Figure 1.62

The vector change of velocity Δv is **65.1 km h^{-1}** in a direction which is **77.8° north of west**.

Finding velocity change in m s^{-1}.

$$\Delta v = \frac{65.1 \times 1000}{60 \times 60}$$

$$\boldsymbol{\Delta v = 18.1\,\text{m s}^{-1}}$$

Finding acceleration produced.

$$a = \frac{\Delta v}{t} = \frac{18.1}{10}$$

$$\mathbf{a = 1.81\,m\,s^{-2}}$$

Finding force which has produced the change of speed and direction.

$$F = ma = 750 \times 1.81$$

$$\mathbf{F = 13.6 \times 10^3\,N \quad or \quad 13.6\,kN}$$

This force is the resultant of the force between the driving wheels and the road, which is producing the increase in speed, and the side thrust on the vehicle, which results from turning the steering wheel into the bend. Its direction is the same as the vector change which it produces.

If the vehicle in Example 1.26 had been travelling around the bend without increasing its speed there would still have been a vector change in velocity and a resulting acceleration. It is thus possible to have acceleration at constant speed. This may sound a bit odd but you must remember that velocity is a vector quantity and if either the speed or the direction of a body changes, there will be a change of velocity and acceleration.

Centripetal acceleration and force

Consider now what happens when a body is travelling at a constant speed v m s^{-1} and angular velocity ω rad s^{-1} around a circular path of radius r. Its direction, and also its velocity, is continually changing.

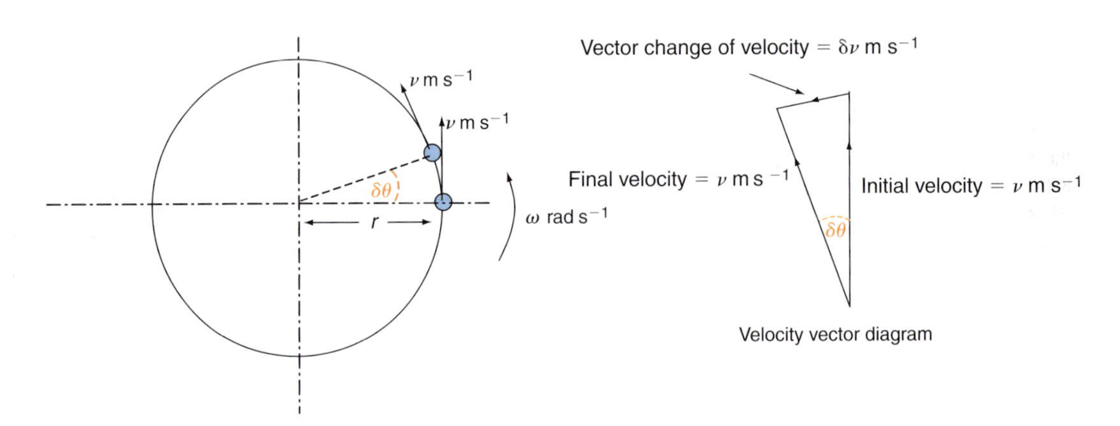

Figure 1.63 Uniform circular motion

Let the body move through a small angle $\delta\theta$ radians in time δt seconds, as shown in Figure 1.63. There will be a vector change in velocity during this period, as shown in the velocity vector diagram. If the angle $\delta\theta$ is very small, the velocity vector diagram will be a long, thin isosceles triangle and the vector change in velocity will be given by:

$$\delta v = v\,\delta\theta$$

The acceleration which occurs will be:

$$a = \frac{\delta v}{\delta t} = v\frac{\delta \theta}{\delta t}$$

But $\frac{\delta \theta}{\delta t} = \omega$, the angular velocity of the body.

$$\boldsymbol{a = v\omega} \qquad (1.55)$$

This is known as the *centripetal acceleration* of the body and whatever its position, its centripetal acceleration is always towards the centre of the circular path.

Now $v = \omega r$ and $\omega = \frac{v}{r}$. Substituting for v and ω in equation (1.55) gives:

$$\boldsymbol{a = \omega^2 r} \qquad (1.56)$$

$$\textbf{and} \quad \boldsymbol{a = \frac{v^2}{r}} \qquad (1.57)$$

It is more usual to use these expressions when calculating centripetal acceleration than that given by equation (1.55).

KEY POINT

When a body is travelling in a circular path its direction is continually changing. The change of velocity gives rise to centripetal acceleration which directed towards the centre of rotation.

Wherever acceleration occurs, there must be a force present. For a body travelling in a circular path, this is the centripetal force pulling or pushing it towards the centre of the path. If the body shown in Figure 1.63 were attached to the centre of its path by a cord, this would all the time be pulling it inwards and the tension in the cord would be the centripetal force. For a body of mass m kg the centripetal force is obtained using expression (1.56) or (1.57) and applying the formula $F = ma$:

$$\boldsymbol{F = m\omega^2 r} \qquad (1.58)$$

$$\textbf{or} \quad \boldsymbol{F = \frac{mv^2}{r}} \qquad (1.59)$$

KEY POINT

Centripetal force is an active force which pulls or pushes a body towards the centre of rotation. Centrifugal force is the equal and opposite reaction whose direction is away from the centre of rotation.

Example 1.27

A body of mass 5 kg travels in a horizontal circular path of radius 2 m at a speed of 100 rpm and is attached to the centre of rotation by a steel rod of diameter 3 mm. Determine (a) the centripetal acceleration of the mass, (b) the centripetal force, (c) the tensile stress in the rod.

(a) Finding angular velocity of rotation.

$$\omega = N\,(\text{rpm}) \times \frac{2\pi}{60} = 100 \times \frac{2\pi}{60}$$

$$\boldsymbol{\omega = 10.5\ \text{rad s}^{-1}}$$

Finding centripetal acceleration of mass.

$$a = \omega^2 r = 10.5^2 \times 2$$

$$\boldsymbol{a = 219\ \text{m s}^{-2}}$$

(b) Finding centripetal force.

$$F = ma = 5 \times 219$$

$$\boldsymbol{F = 1.10 \times 10^3\ \text{N} \quad \text{or} \quad 1.10\ \text{kN}}$$

(c) Finding cross-sectional area of rod.

$$A = \frac{\pi d^2}{4} = \frac{\pi \times 0.003^2}{4}$$

$$\mathbf{A = 7.07 \times 10^{-6}\,m^2}$$

Finding tensile stress in rod.

$$\sigma = \frac{F}{A} = \frac{1.10 \times 10^3}{7.07 \times 10^{-6}}$$

$$\boldsymbol{\sigma = 155 \times 10^6\,\mathrm{Pa} \quad \text{or} \quad 155\,\mathrm{kPa}}$$

Test your knowledge 1.11

1. What is a vector quantity?
2. How is it possible for a body to have acceleration whilst travelling at constant speed?
3. In what direction does centripetal force act?
4. What is centrifugal reaction?

Activity 1.10

A body of mass 0.6 kg rotates on the end of a helical spring of stiffness 1.2 kN m^{-1} in a horizontal plane. The initial distance from the centre of rotation to the centre of the mass is 200 mm. What will be the radius of rotation of the mass and the stretch in the spring when the rotational speed is 300 rpm.

Centrifugal clutches

Centrifugal clutches are to be found on motor-driven equipment such as lawn mowers, mixers and pumps. They engage automatically when the motor speed reaches a predetermined level and do not depend on the skill of the operator.

In its simplest form a centrifugal clutch consists of a drive shaft to which are attached two or more spring loaded masses. These are lined with a friction material on their outer surfaces and rotate inside a drum which is fixed to the output shaft (see Figure 1.64).

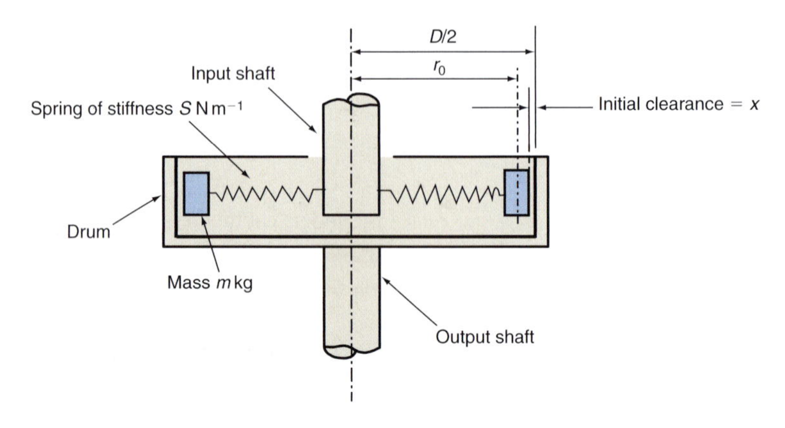

Figure 1.64 Centrifugal clutch

The masses, or 'bobs', are free to slide outwards in guides which are rigidly fixed to the input shaft. These are not shown in the above diagram. As the angular velocity of the input shaft increases, the masses slide outwards and eventually make contact with the drum. A further increase in speed causes the drive to be transmitted to the output shaft through friction between the masses and the drum.

Let initial clearance between the masses and the drum be x.
Let the initial radius of rotation from the centre of the shafts to the centre of gravity of the masses be r_0.
Let the diameter of the drum be D.
Let the coefficient of friction between the drum and the masses be μ.
Let the angular velocity at which the masses just make contact with the drum be ω_0.

When the masses just make contact with the drum, the tension F_0 in the springs is given by:

$$\text{Spring tension} = \text{Spring stiffness} \times \text{Extension}$$

$$\boldsymbol{F_0 = Sx} \qquad (1.60)$$

This is also the centripetal force acting on the masses whose radius of rotation is now $r_0 + x$. Equating spring tension and centripetal force enables the angular velocity ω_0, at which the masses engage with the drum, to be found.

$$F_0 = m\omega_0^2(r_0 + x)$$

$$\boldsymbol{\omega_0 = \sqrt{\frac{F_0}{m(r_0 + x)}}} \qquad (1.61)$$

Consider now the normal force and the tangential force between the masses and the drum at some higher speed ω when the drive is being transmitted to the output shaft. These are shown in Figure 1.65.

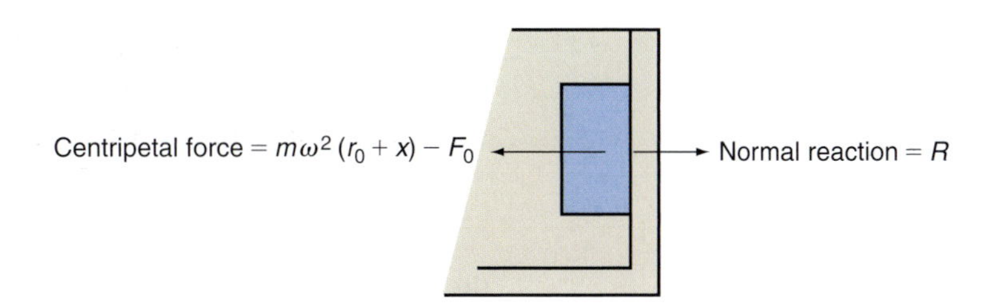

Figure 1.65 Forces acting on clutch drum

The normal force between each mass and the drum is equal to the centripetal force, part of which is supplied by the spring tension F_0.

$$\boldsymbol{R = m\omega^2(r_0 + x) - F_0} \qquad (1.62)$$

The tangential friction force between each mass and the drum will be:

$$F = \mu R$$

$$\boldsymbol{F = \mu[m\omega^2(r_0 + x) - F_0]} \qquad (1.63)$$

KEY POINT

The limiting coefficient of friction, k, between two surfaces which can slide over each other is the ratio of the force required to overcome frictional resistance and the normal force between the surfaces.

If the number of masses is n, the torque transmitted will be:

$$T = nF\frac{D}{2}$$

$$T = n\mu[m\omega^2(r_0 + x) - F_0]\frac{D}{2} \qquad (1.64)$$

Having calculated the torque transmitted, the power transmitted can be calculated.

$$\text{Power} = T\omega \qquad (1.65)$$

Example 1.28

A centrifugal clutch has two rotating contact masses of 0.25 kg. These are able to move in guides attached to the input shaft and are held by springs of stiffness 8 kN m^{-1}. The initial clearance between the masses and the internal surface of a drum on the output shaft is 5 mm. The inner diameter of the drum is 300 mm and the distance from the centre of rotation to the centre of gravity of the masses is 135 mm. The coefficient of friction between the masses and the drum is 0.35. Determine (a) the rotational speed in revolutions per minute at which the clutch begins to transmit power, (b) the torque and power transmitted at a speed of 1000 rpm.

(a) Finding tension in springs as masses make contact with drum.

$$F_0 = Sx = 8 \times 10^3 \times 5 \times 10^{-3}$$

$$\mathbf{F_0 = 40\ N}$$

Finding angular velocity at which masses just make contact with the drum.

$$\omega_0 = \sqrt{\frac{F_0}{m(r_0 + x)}} = \sqrt{\frac{40}{0.25(0.135 + 0.005)}}$$

$$\omega_0 = 33.8 \text{ rad s}^{-1}$$

Change to revolutions per minute.

$$N_0 = \omega_0 \times \frac{60}{2\pi} = 33.8 \times \frac{60}{2\pi}$$

$$\mathbf{N_0 = 323\ rpm}$$

(b) Finding angular velocity at rotational speed of 1000 rpm.

$$\omega = N \times \frac{2\pi}{60} = 1000 \times \frac{2\pi}{60}$$

$$\boldsymbol{\omega = 105\ \mathbf{rad\,s^{-1}}}$$

Finding torque transmitted at this speed.

$$T = n\mu[m\omega^2(r_0 + x) - F_0]\frac{D}{2}$$

$$T = 2 \times 0.35 \times \big[[0.25 \times 105^2 \times (0.135 + 0.005)] - 40\big] \times 0.15$$

$$\mathbf{T = 36.3\ Nm}$$

Finding power transmitted at this speed.

$$\text{Power} = T\omega = 36.3 \times 105$$

$$\textbf{Power} = \mathbf{3.81 \times 10^3\ W} \quad \textbf{or} \quad \mathbf{3.81\ kW}$$

Test your knowledge 1.12

1. What are the units in which spring stiffness is measured?
2. If the normal force between two surfaces is 100 N and their coefficient of friction is 0.3, what force will be required to make them slide over each other?
3. What determines the rotational speed at which a centrifugal clutch starts to engage?
4. How do you convert rotational speed in revolutions per minute to angular velocity measured in radians per second?

Activity 1.11

A centrifugal clutch has four bobs of mass 0.25 kg which can slide outwards in guides and are attached to the input shaft by restraining springs. The drum attached to the output shaft has an inner diameter of 350 mm. The static clearance between the bobs and the drum is 6 mm and the distance from the centre of rotation to the centre of gravity of the bobs is initially 155 mm. The coefficient of friction between the bobs and the drum is 0.3. If the clutch start engage at a speed of 500 rpm what is the stiffness of the restraining springs? What will be the power transmitted by the clutch at a speed of 1200 rpm?

To check your understanding of the preceding section, you can solve Review questions 36–39 at the end of this chapter.

Stability of vehicles

When you are travelling around a bend on a motor cycle or in a car, the centripetal force is supplied by friction between the wheels and the road surface as you turn into the curve. If you are travelling too fast for the road conditions, or the condition of your vehicle, one of two things may happen:

- The friction force between the wheels and the road surface will be insufficient and you will skid. The direction of your skid will be at a tangent to the curve.
- If the centre of gravity of your vehicle is high it may overturn before starting to skid.

With a motor cycle only the first option is possible. You may of course skid into the kerb and then overturn but it is skidding which will initiate the problem. Cars, buses and trucks are designed so that even when fully loaded, they should skid before reaching the speed at which overturning would occur. Large heavy items stacked on a roof rack may raise the centre of gravity of a car to such a height that overturning is a possibility and of course this would also increase the possibility of overturning should the vehicle skid into the kerb.

Consider a four-wheeled vehicle of mass m, travelling at a speed v, round a level unbanked curve of radius r as shown in Figure 1.66.

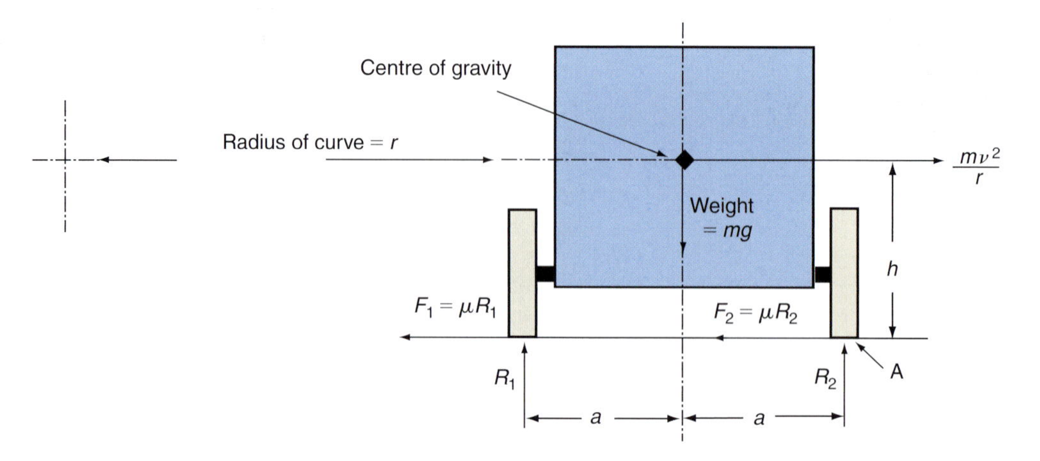

Figure 1.66 Vehicle on an unbanked horizontal curve

Although the vehicle is moving, it can be considered to be in a state of dynamic equilibrium. In the vertical direction, the active force is the weight W acting downwards which is balanced by the upward reactions R_1 and R_2 of the wheels. The active forces in the horizontal direction are the friction forces F_1 and F_2 which provide the centripetal force. The equal and opposite reaction to them is the centrifugal force given by mv^2/r. As has been explained, this method of displaying the forces on a free body for a moving object is known as D'Alembert's principle.

Let the height of the centre of gravity be h.
Let the track width be $2a$.
Let the limiting coefficient of friction between the wheels and the road be μ.

To find the speed that skidding is likely to occur, equate horizontal forces as the vehicle is about to skid.

$$\text{Centripetal force} = \text{Centrifugal reaction}$$

$$F_1 + F_2 = \frac{mv^2}{r}$$

In the limit as the vehicle is about to skid, $F_1 = \mu R_1$ and $F_2 = \mu R_2$. Substituting for these gives:

$$\mu R_1 + \mu R_2 = \frac{mv^2}{r}$$

$$\mu(R_1 + R_2) = \frac{mv^2}{r}$$

But $R_1 + R_2 = mg$, the weight of the vehicle.

$$\mu \cancel{m} g = \frac{\cancel{m} v^2}{r}$$

$$\mu r g = v^2$$

$$\mathbf{v = \sqrt{\mu r g}} \qquad (1.66)$$

KEY POINT

When a motor vehicle is travelling round a curve, centripetal force is provided by friction between its tyres and the road surface. When skidding occurs, the vehicle continues in a straight line at a tangent to the curve.

To find the speed at which overturning is likely to occur equate moments about the point A in the limit as the vehicle is about to overturn. In this condition R_1 is zero as the nearside wheels are about to lift off the road and all of the weight will be carried on the offside wheels.

$$\text{Clockwise overturning moment} = \text{Anticlockwise righting moment}$$

$$\frac{mv^2}{r}h = mga$$

$$v^2 = \frac{gar}{h}$$

$$\mathbf{v = \sqrt{\frac{a}{h}rg}} \tag{1.67}$$

Whichever of the equations (1.66) and (1.67) gives the lower value velocity, that will be the limiting value. You should note that both values of limiting velocity are independent of the mass of the vehicle. Comparing the two equations shows that:

If $\mu < \frac{a}{h}$, the vehicle will skid before the overturning speed is reached.

If $\mu > \frac{a}{h}$, the vehicle will overturn before the skidding speed is reached.

The reaction of the outer wheels will always be greater than that of the inner wheels. To find these reactions at any speed below that at which overturning is likely take moments about the point A, again but this time include the moment of the reaction R_1.

$$\text{Clockwise overturning moments} = \text{Anticlockwise righting moment}$$

$$\frac{mv^2}{r}h + R_1 2a = mga$$

$$R_1 2a = mga - \frac{mv^2}{r}h = m\left(ga - \frac{v^2 h}{r}\right)$$

$$\mathbf{R_1 = \frac{m}{2a}\left(ga - \frac{v^2}{r}h\right)} \tag{1.68}$$

KEY POINT

Properly loaded motor vehicles will skid rather than overturn if their speed around a curve is excessive. Overloading a roof rack can raise the centre of gravity to such a degree that overturning becomes a danger.

Now equate vertical forces to find the reaction of the outer wheels.

$$R_1 + R_2 = mg$$

$$\mathbf{R_2 = mg - R_1} \tag{1.69}$$

Example 1.29

A car of mass 900 kg travels round an unbanked horizontal curve of radius 30 m. The track width of the car is 1.5 m and its centre of gravity is central and at a height of 0.9 m above the road surface. The limiting coefficient of friction between the tyres and the road is 0.7. Show that the car is likely to skid rather than overturn if the speed is excessive and calculate the maximum speed in km h^{-1} at which it can travel round the curve. Determine also the reactions of the inner and outer wheels at this speed.

Finding the ratio of half the track width to height of centre of gravity:

$$\frac{a}{h} = \frac{0.75}{0.9} = 0.833$$

Comparing this with the coefficient of friction whose value is $\mu = 0.7$ shows that:

$$\mu < \frac{a}{h}$$

This indicates that the car will skid before the overturning speed is reached.

Finding speed at which car can travel round the curve, i.e. when skidding is likely to occur.

$$v = \sqrt{\mu r g} = \sqrt{0.7 \times 30 \times 9.81}$$

$$\mathbf{v = 14.35\ m\,s^{-1}}$$

Change to $\text{km}\,\text{h}^{-1}$

$$v = \frac{14.4 \times 60 \times 60}{1000}$$

$$\mathbf{v = 51.7\ km\,h^{-1}}$$

Finding reaction R_1 of inner wheels by taking moments about point of contact of outer wheels and road.

$$R_1 = \frac{m}{2a}\left(ga - \frac{v^2}{r}h\right)$$

$$R_1 = \frac{900}{1.5}\left[(9.81 \times 0.75) - \left(\frac{14.35^2 \times 0.9}{30}\right)\right]$$

$$\mathbf{R_1 = 708\ N}$$

Finding R_2 by equating vertical forces.

$$R_2 = mg - R_1 = (900 \times 9.81) - 708$$

$$\mathbf{R_2 = 8121\ N \quad or \quad 8.12\ kN}$$

In practice the bends on major roads are banked to reduce the tendency for side-slip. When driving at the speed where there is no side-slip, the forces acting on a vehicle are as shown in Figure 1.67.

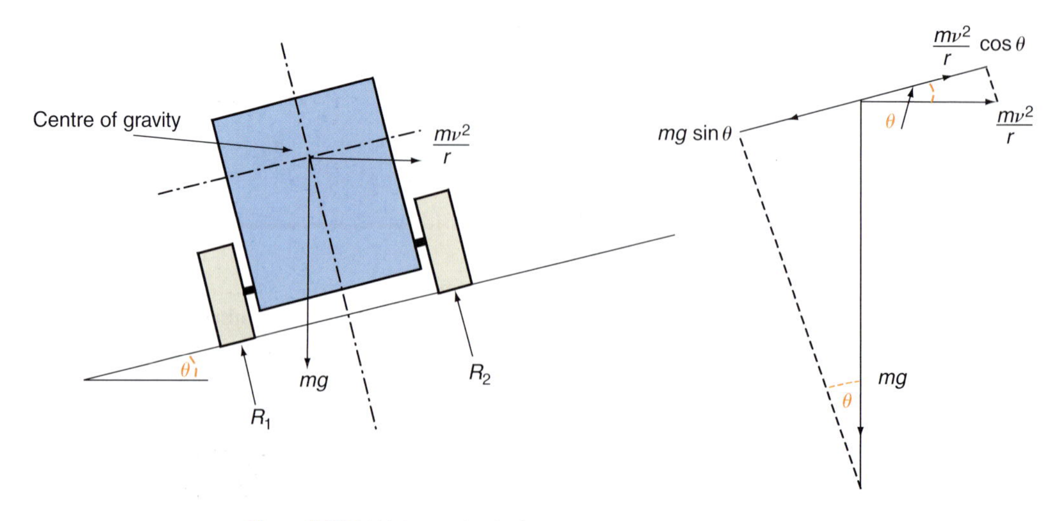

Figure 1.67 Vehicle on a banked curve

KEY POINT

In the condition where there is no tendency for side-slip, the reactions of the inner and outer wheels are equal.

The speed at which there is no side-slip can be found by resolving forces parallel to the road surface. The two wheel reactions are perpendicular to the road surface and are not involved. This just leaves the component of the weight down the slope, which is the active centripetal force, and the component of the centrifugal reaction up the slope. For no side-slip, these must be equal and opposite.

$$\frac{\cancel{m}v^2}{r}\cos\theta = \cancel{m}g\sin\theta$$

$$v^2 = rg\frac{\sin\theta}{\cos\theta} = rg\tan\theta$$

$$\mathbf{v = \sqrt{rg\tan\theta}} \qquad (1.70)$$

Example 1.30

A vehicle has a track width of 1.4 m and the height of its centre of gravity is 1.2 m above the road surface. The limiting coefficient of friction between the wheels and the road surface is 0.7.

(a) What is the maximum speed at which the vehicle can travel round a curve of radius 75 m?

(b) To what angle would the curve have to be banked for the vehicle to travel round it at a speed of 50 km h^{-1} without any tendency to side-slip?

(a) Finding the ratio of half the track width to height of centre of gravity.

$$\frac{a}{h} = \frac{0.7}{1.2} = 0.583$$

Comparing this with the coefficient of friction whose value is k = 0.7 shows that:

$$\mu > \frac{a}{h}$$

This indicates that the car will overturn before the skidding speed is reached.

Finding speed at which car can travel round the curve, i.e. when overturning is likely to occur.

$$v = \sqrt{\frac{a}{h}rg} = \sqrt{\frac{0.7}{1.2}\times 75\times 9.81}$$

$$\mathbf{v = 20.7\ m\,s^{-1}}$$

Change to km h^{-1}

$$v = \frac{20.7\times 60\times 60}{1000}$$

$$\mathbf{v = 74.6\ km\,h^{-1}}$$

Changing 50 km h^{-1} to m s^{-2}

$$v = \frac{50\times 1000}{60\times 60}$$

$$\mathbf{v = 13.9\ m\,s^{-1}}$$

(b) Finding angle to which curve would need to be banked for no side-slip at this speed.

$$v = \sqrt{rg\tan\theta}$$
$$v^2 = rg\tan\theta$$
$$\tan\theta = \frac{v^2}{rg} = \frac{13.9^2}{75 \times 9.81}$$
$$\tan\theta = 0.262$$
$$\boldsymbol{\theta = 14.7^\circ}$$

Test your knowledge 1.13

1. What effect does the mass of a vehicle have on the maximum speed at which it can travel round a curve?
2. If $\mu < \frac{a}{h}$, is a vehicle likely to skid or overturn if its speed is excessive?
3. How is the centripetal force provided when a car travels round an unbanked curve?
4. How is the centripetal force provided when a car travels round a banked curve?

Activity 1.12

A vehicle of mass 1 tonne travels round a horizontal unbanked curve of radius 35 m. The track width of the vehicle is 1.8 m and the height of its centre of gravity above the road surface is 0.95 m. The coefficient of friction between the tyres and the road is 0.65.

(a) Show that if the speed of the vehicle is excessive, it is likely to skid rather than overturn.
(b) Calculate the speed at which skidding is likely to occur.
(c) Calculate the reactions of the inner and outer wheels at this speed.
(d) If the curve were to be banked at an angle of 3.5° what would be the speed at which the vehicle could travel around it without any tendency for side-slip?

To check your understanding of the preceding section, you can solve Review questions 40–44 at the end of this chapter.

Simple Machines

A simple machine is an arrangement of moving parts whose purpose is to transmit motion and force. The ones which we will consider are those in which a relatively small input force is used to raise a heavy load. They include lever systems, inclined planes, screw jacks, wheel and axle arrangements and gear trains (Figure 1.68).

For all simple machines the *mechanical advantage* or *force ratio* is the ratio of the load W raised to the input effort E.

$$\text{Mechanical advantage} = \frac{\text{Load}}{\text{Effort}}$$

$$\mathbf{MA} = \frac{\boldsymbol{W}}{\boldsymbol{E}} \qquad (1.71)$$

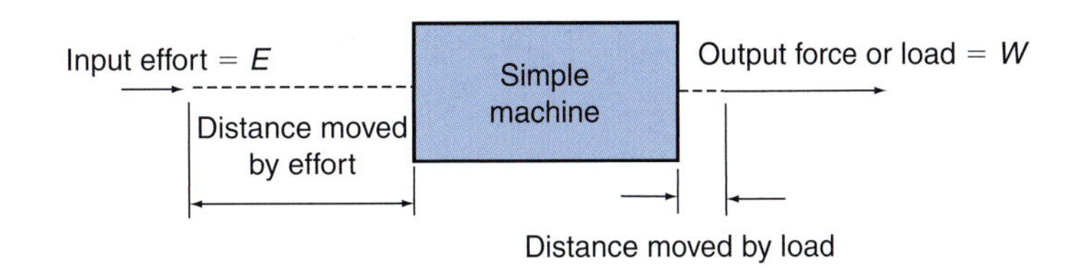

Figure 1.68 Block diagram of a simple machine

A characteristic of a simple machine is that the distance moved by the load is much smaller than the distance moved by the input effort. The *velocity ratio* or *movement ratio* is used to measure this effect.

$$\textbf{Velocity ratio} = \frac{\textbf{Distance moved by effort}}{\textbf{Distance moved by load}} \qquad (1.72)$$

$$\textbf{or} \quad \textbf{Velocity ratio} = \frac{\textbf{Velocity at which effort moves}}{\textbf{Velocity at which load moves}} \qquad (1.73)$$

The efficiency η of a simple machine is the ratio of the work output to the work input. It is usually given as a percentage.

$$\text{Efficiency} = \frac{\text{Work output}}{\text{Work input}}$$

$$\eta = \frac{\text{Load} \times \text{Distance moved by load}}{\text{Effort} \times \text{Distance moved by effort}} \times 100\%$$

$$\eta = \text{MA} \times \frac{1}{\text{VR}} \times 100\%$$

$$\boldsymbol{\eta} = \frac{\textbf{MA}}{\textbf{VR}} \times \textbf{100\%} \qquad (1.74)$$

KEY POINT

The mechanical advantage is always less than the velocity ratio.

There is always some friction between the moving parts of a machine. Some of the work input must be used to overcome friction and so the work output is always less than the input. As a result, the efficiency can never be 100% and it is very often a great deal less. If there were no friction present, the mechanical advantage would be equal to the velocity ratio. In practice it is always a lower figure.

Example 1.31

A jack requires an input effort of 25 N to raise a load of 150 kg. If the distance moved by the effort is 500 mm and the load is raised through a height of 6 mm find (a) the mechanical advantage, (b) the velocity ratio, (c) the efficiency, (d) the work done in overcoming friction.

(a) Finding mechanical advantage.

$$\text{MA} = \frac{\text{Load}}{\text{Effort}} = \frac{150 \times 9.81}{25}$$

$$\textbf{MA = 58.9}$$

(b) Finding velocity ratio.

$$\text{VR} = \frac{\text{Distance moved by effort}}{\text{Distance moved by load}} = \frac{500}{6}$$

$$\textbf{VR = 83.3}$$

(c) Finding efficiency.

$$\eta = \frac{\text{MA}}{\text{VR}} = \frac{58.9}{83.3} \times 100\%$$

$$\boldsymbol{\eta = 70.7\%}$$

(d) Finding work done in overcoming friction.

The work output is 70.7% of the work input. The remaining 29.3% of the work input is the work done in overcoming friction.

$$\text{Work done in overcoming friction} = \text{Work input} \times \frac{29.3}{100}$$

$$\text{Work done in overcoming friction} = \text{Effort} \times \text{Distance effort moves} \times \frac{29.3}{100}$$

$$\text{Work done in overcoming friction} = 25 \times 0.5 \times \frac{29.3}{100}$$

Work done in overcoming friction = 3.66 J

Velocity ratio formulae

The expressions (1.72) and (1.73) are general formulae for calculating velocity ratio and these can now be applied to a range of devices. Each will then have its own particular formula for velocity ratio which is dependent upon its dimensions and its mode of operation. For each device let the distance moved by the effort be a and the corresponding distance moved by the load be b.

$$\text{i.e. Velocity ratio} = \frac{\text{Distance moved by effort}}{\text{Distance moved by load}}$$

$$\textbf{VR} = \frac{a}{b} \qquad (1.75)$$

1. Simple lever systems

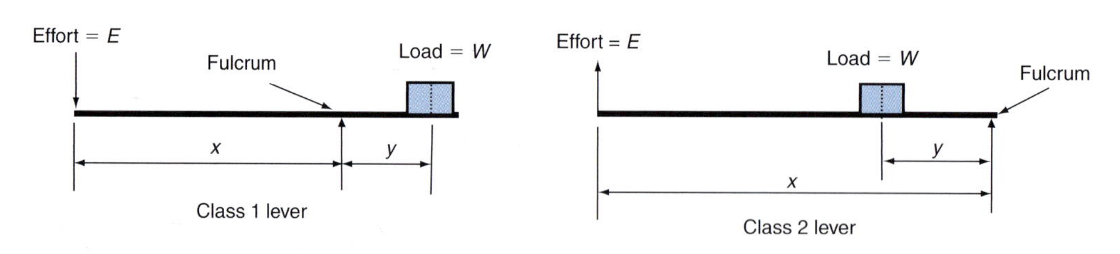

Figure 1.69 Lever systems

For both types of lever shown in Figure 1.69 the distances moved by the effort and load are proportional to the distances x and y from the fulcrum of the lever.

$$\textbf{VR} = \frac{a}{b} = \frac{x}{y} \qquad (1.76)$$

2. Inclined plane

In pulling the load up the incline the effort moves through distance a whilst lifting the load through a vertical height b (Figure 1.70).

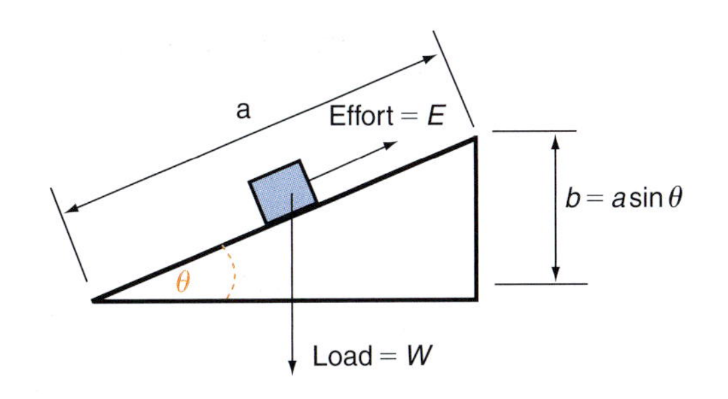

Figure 1.70 Inclined plane

$$\text{VR} = \frac{a}{b} = \frac{\cancel{a}}{\cancel{a}\sin\theta}$$

$$\mathbf{VR = \frac{1}{\sin\theta}} \quad (1.77)$$

3. Screw jack

The screw jack is a practical form of the inclined plane. The plane is now in the form of a spiral and the effort is applied horizontally to the end of the operating handle. One complete turn of the handle causes the load to rise through a distance equal to the pitch of the screw thread (Figure 1.71).

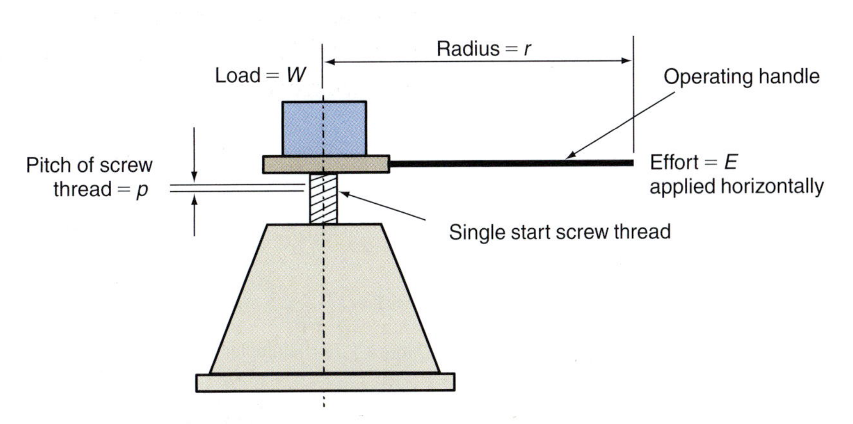

Figure 1.71 Screw jack

$$\text{VR} = \frac{\text{Distance moved by effort in one revolution of screw thread}}{\text{Pitch of screw thread}}$$

$$\mathbf{VR = \frac{2\pi r}{p}} \quad (1.78)$$

4. Pulley blocks

Here the velocity ratio is equal to the number of pulleys in operation or, alternatively, the number of ropes lengths connecting the pulleys, excluding the effort rope (Figure 1.72).

$$\mathbf{VR = Number\ of\ pulleys} \quad (1.79)$$

or

$$\mathbf{VR = Number\ of\ connecting\ rope\ lengths} \quad (1.80)$$

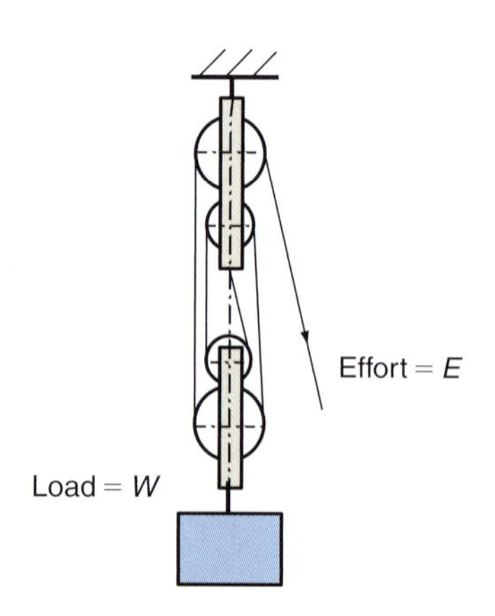

Figure 1.72 Pulley blocks

5. Weston differential pulley block

This comprises a compound pulley with diameters D_1 and D_2 and a snatch block which carries the load. They are connected by an endless chain to which the effort is applied as shown in Figure 1.73.

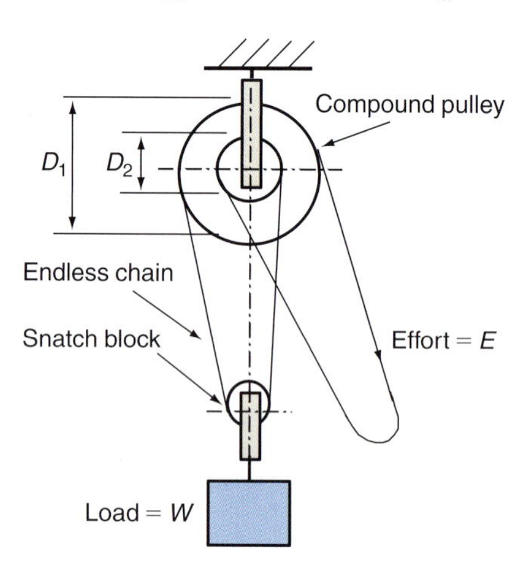

Figure 1.73 Differential pulley block

For one rotation of the compound pulley the chain is wound onto the larger diameter and off the smaller diameter. That part of the chain passing round the snatch block shortens by a length equal to the difference between the circumferences of the compound pulley. The load is raised by half of this distance.

$$\text{VR} = \frac{\text{Circumference of larger compound pulley}}{\text{Half the difference of the compound pulley circumferences}}$$

$$\text{VR} = \frac{\pi D_1}{\dfrac{\pi D_1 - \pi D_2}{2}} = \frac{\pi D_1}{\dfrac{\pi(D_1 - D_2)}{2}}$$

$$\mathbf{VR} = \frac{\mathbf{2D_1}}{\mathbf{(D_1 - D_2)}} \qquad (1.81)$$

The links of the chain engage on teeth or 'flats' which are cut on the compound pulley. If the numbers of flats n_1 and n_2 are known, they too can be used to calculate the velocity ratio.

$$\mathbf{VR} = \frac{2n_1}{(n_1 - n_2)} \tag{1.82}$$

You will note that it is the difference between the diameters on the compound pulley which determines the velocity ratio. The closer the two diameters, the higher will be the velocity ratio.

6. Simple wheel and axle

For one revolution of the wheel and axle, the effort moves a distance equal to the wheel circumference and the load is raised through a distance equal to the axle circumference (Figure 1.74).

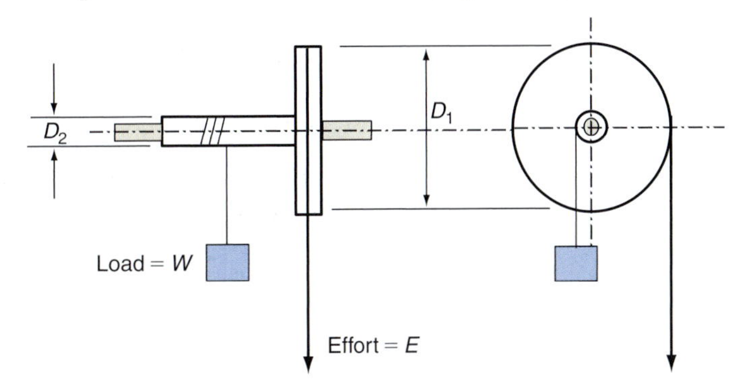

Figure 1.74 Simple wheel and axle

$$\text{VR} = \frac{\text{Circumference of wheel}}{\text{Circumference of axle}} = \frac{\pi D_1}{\pi D_2}$$

$$\mathbf{VR} = \frac{D_1}{D_2} \tag{1.83}$$

7. Differential wheel and axle

Here the wheel, whose diameter is D_1, has a compound axle with two diameters D_2 and D_3. The two axle diameters are wound in opposite directions with the same cord which also passes round the snatch block. For one revolution of the wheel, the effort moves a distance equal to the wheel's circumference. At the same time the length of cord around the snatch block shortens by a length equal to the difference between the two axle circumferences, and the load is raised by a distance which is half of this difference (Figure 1.75).

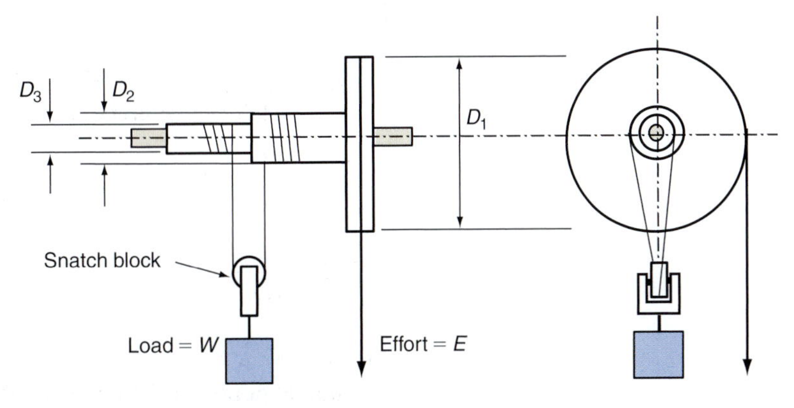

Figure 1.75 Differential wheel and axle

$$\text{VR} = \frac{\text{Circumference of wheel}}{\text{Half the difference of the axle circumferences}}$$

$$\text{VR} = \frac{\pi D_1}{\pi D_2 - \pi D_3/2} = \frac{\pi D_1}{\pi(D_2 - D_3)/2}$$

$$\boldsymbol{VR = \frac{2D_1}{(D_2 - D_3)}} \tag{1.84}$$

You will note that it is not only the wheel diameter, but also the difference between the two axle diameters which determines the velocity ratio. The closer the two axle diameters, the higher will be the velocity ratio for a given wheel size.

8. Simple gear winch

The number of teeth on the gears A and B in the speed reduction gear train are t_A and t_B, respectively. The gear ratio or velocity ratio of the simple gear train alone will be:

$$\text{Gear ratio} = \frac{t_A}{t_B}$$

For one complete revolution of the operating handle, the effort will move through a distance equal to the circumference of its turning circle (Figure 1.76). At the same time the load will rise through a distance equal to the circumference of the winding drum multiplied by the gear ratio.

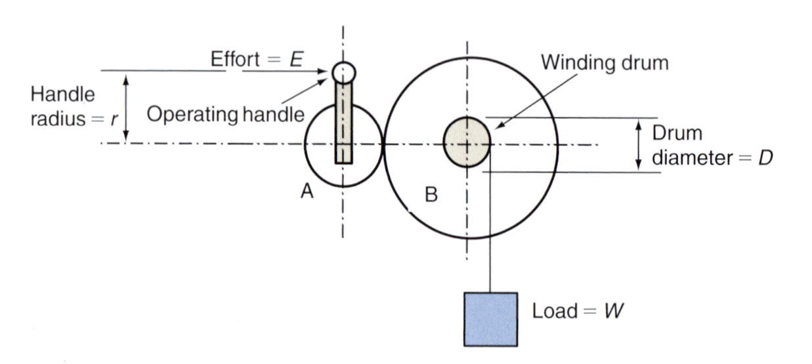

Figure 1.76 Simple gear winch

$$\text{VR} = \frac{\text{Circumference of handle turning circle}}{\text{Circumference of winding drum} \times \text{Gear ratio}}$$

$$\text{VR} = \frac{2\pi r}{\pi D \times (t_A/t_B)}$$

$$\mathbf{VR} = \frac{\boldsymbol{2rt_B}}{\boldsymbol{Dt_A}} \tag{1.85}$$

9. Compound gear winch or crab winch

The numbers of teeth on gears A, B, C and D are t_A, t_B, t_C and t_D, respectively. Gears B and C are keyed on the same shaft and rotate together to form a compound gear. The gear ratio or velocity ratio of the compound reduction gear train alone will be:

$$\text{Gear ratio} = \frac{t_A \times t_C}{t_B \times t_D}$$

As with the simple gear winch, one complete revolution of the operating handle will cause the effort to move through a distance equal to the circumference of its turning circle (Figure 1.77). At the same time the load will rise through a distance equal to the circumference of the winding drum multiplied by the gear ratio.

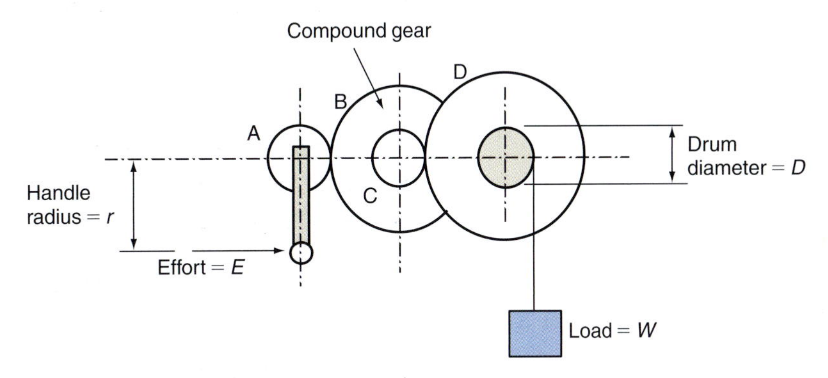

Figure 1.77 Crab winch

$$VR = \frac{\text{Circumference of handle turning circle}}{\text{Circumference of winding drum} \times \text{Gear ratio}}$$

$$VR = \frac{2\pi r}{\pi D \times \left(\dfrac{t_A}{t_B} \times \dfrac{t_C}{t_D}\right)}$$

$$\boldsymbol{VR = \frac{2r}{D}\left(\frac{t_B}{t_A} \times \frac{t_D}{t_C}\right)} \qquad (1.86)$$

Law of a machine

When a range of load and the corresponding effort values are tabulated for any of the above machines, a graph of effort against load is found to have the straight line form shown in Figure 1.78(a).

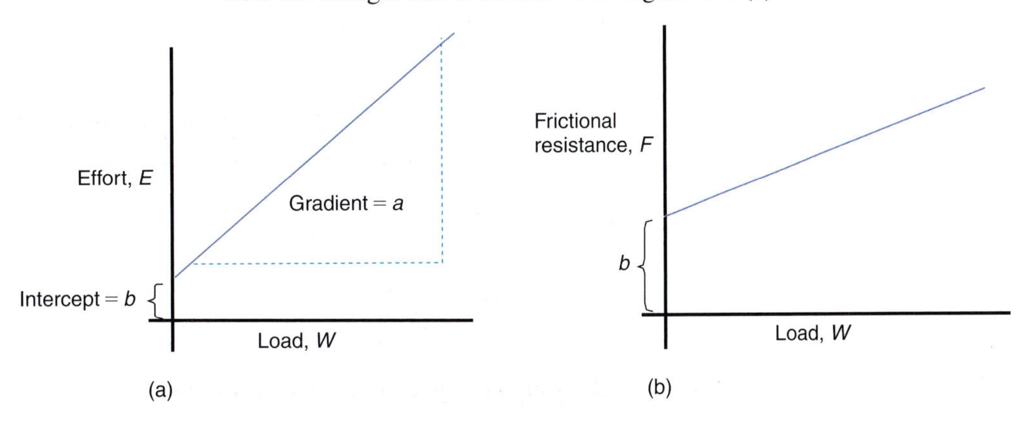

Figure 1.78 Graphs of effort and frictional resistance vs. load: (a) graph of effort vs. load, (b) graph of frictional resistance vs. load

KEY POINT

Frictional resistance increases with load and some of the effort is required to overcome it. As a result the work output is always less than the work input and the efficiency can never be 100%.

The equation of graph in Figure 1.78(a) is:

$$E = aW + b \tag{1.87}$$

This is known as the *law of the machine*. The constant a is the gradient of the straight line graph. The constant b is the intercept on the effort axis. This is the effort initially required to overcome friction before any load can be lifted. It is found that as the load is increased, the frictional resistance in the mechanism increases from the initial value b, in a linear fashion as shown in Figure 1.78(b).

Example 1.32

A differential wheel and axle has a wheel diameter of 275 mm and axle diameters of 50 and 100 mm. The law of the machine is $E = 0.11W + 4.5$. Determine (a) the velocity ratio of the device, (b) mechanical advantage and efficiency when raising a load of 25 kg, (c) the work input which is required to raise the load through a height of 1.5 m.

(a) Finding velocity ratio.

$$\text{VR} = \frac{2D_1}{(D_2 - D_3)} = \frac{2 \times 275}{(100 - 50)}$$

$$\mathbf{VR = 11}$$

(b) Finding effort required to raise a mass of 25 kg.

$$E = 0.15W + 4.5 = (0.11 \times 25 \times 9.81) + 4.5$$

$$\mathbf{E = 31.5\,N}$$

Finding mechanical advantage when raising this load.

$$\text{MA} = \frac{W}{E} = \frac{25 \times 9.81}{31.5}$$

$$\mathbf{MA = 7.79}$$

Finding efficiency when raising this load.

$$\eta = \frac{\text{MA}}{\text{VR}} = \frac{7.79}{11}$$

$$\mathbf{\eta = 0.708 \quad or \quad 70.8\%}$$

(c) Finding work input.

$$\eta = \frac{\text{Work output}}{\text{Work input}}$$

$$\text{Work input} = \frac{\text{Work output}}{\eta} = \frac{\text{Load} \times \text{Distance moved by load}}{\eta}$$

$$\text{Work input} = \frac{25 \times 9.81 \times 1.5}{0.708}$$

Work input = 520 J

Limiting efficiency and mechanical advantage

It is found that the efficiency of a simple machine increases with load but not in a linear fashion.

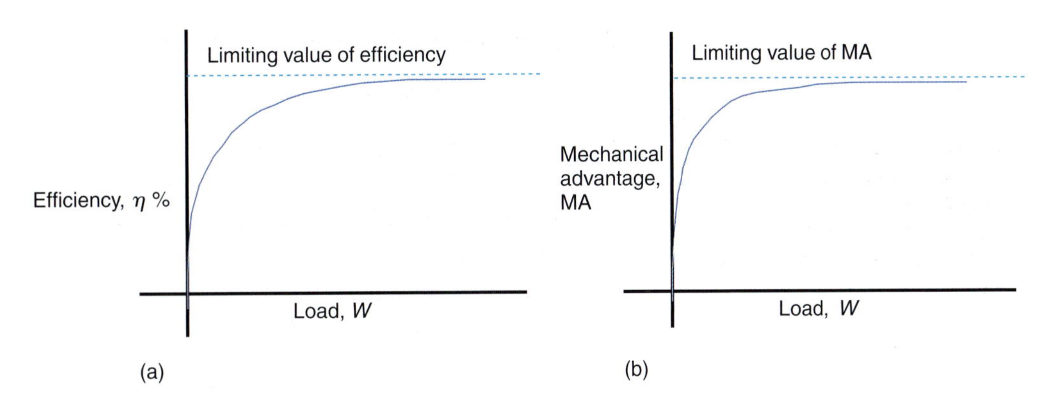

Figure 1.79 Variation of efficiency and mechanical advantage with load: (a) graph of efficiency vs. load, (b) graph of mechanical advantage vs. load

As can be seen in Figure 1.79, the efficiency eventually levels off at a limiting value. This is found to depend on the constant a, in the law of the machine and its velocity ratio.

$$\text{Efficiency} = \frac{\text{Mechanical advantage}}{\text{Velocity ratio}}$$

$$\eta = \frac{\text{MA}}{\text{VR}} = \frac{W}{E \times \text{VR}}$$

Now from the law of the machine, $E = aW + b$

$$\eta = \frac{W}{(aW + b) \times \text{VR}}$$

Dividing numerator and denominator by W gives:

$$\eta = \frac{1}{\left(a + \dfrac{b}{W}\right) \times \text{VR}}$$

$$\boldsymbol{\eta = \frac{1}{a\ \text{VR} + \dfrac{b}{W}\ \text{VR}}} \qquad (1.88)$$

Examination of this expression shows that as the load W increases, the second term in the denominator becomes smaller and smaller. When the load becomes very large, this term tends to zero and the limiting value of efficiency is:

$$\boldsymbol{\eta = \frac{1}{a\ \text{VR}}} \qquad (1.89)$$

In this limiting condition the mechanical advantage levels off to a value given by:

$$\eta = \frac{\text{MA}}{\text{VR}}$$

$$\text{MA} = \eta \times \text{VR}$$

$$\boldsymbol{\text{MA} = \frac{1}{a}} \qquad (1.90)$$

KEY POINT

The mechanical advantage and the efficiency of a simple machine increase with load but eventually level off at a limiting value.

The limiting value of mechanical advantage is thus the reciprocal of the gradient of the effort vs. load graph.

Overhauling

A simple machine is said to *overhaul* if when the effort is removed, the load falls under the effects of gravity. If a machine does not overhaul, then friction alone must be sufficient to support the load. If you have changed a wheel on a car you probably used a screw jack to raise the wheel off the road surface. Friction in the screw thread is then sufficient to support the car. This means, of course, that there must be quite a lot of friction present which results in a low value of efficiency. This however is the price that often has to be paid for safety in lifting devices such as car jacks and engine hoists.

Frictional resistance can be considered as an additional load which the effort must overcome.

$$\begin{aligned}\text{Total effort} &= \text{Effort to overcome friction load, } F \\ &\quad + \text{Effort to overcome actual load, } W\end{aligned}$$

Now in an ideal machine where there is no friction, the mechanical advantage is equal to the velocity ratio and,

$$\text{VR} = \text{MA} = \frac{\text{Load}}{\text{Effort}}$$

$$\text{and} \qquad \text{Effort} = \frac{\text{Load}}{\text{VR}} = \frac{W}{\text{VR}}$$

If friction force F is being considered as a separate load, the total effort can be written as:

$$\text{Total effort} = \frac{F}{\text{VR}} + \frac{W}{\text{VR}}$$

$$\boldsymbol{E = \frac{F + W}{\text{VR}}} \tag{1.91}$$

Now the efficiency is given by:

$$\eta = \frac{\text{MA}}{\text{VR}} = \frac{W}{E\ \text{VR}}$$

Substituting for E gives:

$$\eta = \frac{W}{\left(\dfrac{F + W}{\text{VR}}\right)\text{VR}}$$

$$\boldsymbol{\eta = \frac{W}{F + W}} \tag{1.92}$$

When the effort is removed there is only the frictional resistance F to oppose the load W. To stop it overhauling the friction force must be

KEY POINT

A simple machine will overhaul if the load falls under the effects of gravity when the effort is removed. Overhauling is likely to occur if the efficiency of a machine is greater than 50%.

equal or greater than the load. In the limit when $F = W$, the efficiency will be:

$$\eta = \frac{W}{2W} = 0.5 \quad \text{or} \quad 50\% \qquad (1.93)$$

It follows that a machine will overhaul if its efficiency is greater than 50%.

Example 1.33

The law of the gear winch shown in Figure 1.80 is $E = 0.0105\,W + 5.5$ and it is required to raise a load of 150 kg. Determine (a) its velocity ratio, (b) the effort required at the operating handle, (c) the mechanical advantage and efficiency when raising this load, (d) the limiting mechanical advantage and efficiency. (e) State whether the winch is likely to overhaul when raising the 150 kg load.

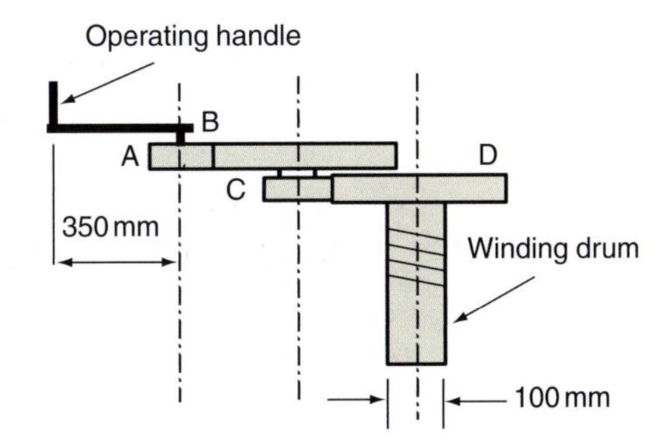

Figure 1.80

Teeth on gear A = 16
Teeth on gear B = 72
Teeth on gear C = 18
Teeth on gear D = 6(

(a) Finding velocity ratio.

$$\text{VR} = \frac{2r}{D}\left(\frac{t_B}{t_A} \times \frac{t_D}{t_C}\right) = \frac{2 \times 350}{100}\left(\frac{72 \times 60}{16 \times 18}\right)$$

$$\mathbf{VR = 105}$$

(b) Finding effort required.

$$E = 0.0105\,W + 5.5 = (0.0105 \times 150 \times 9.81) + 5.5$$

$$\boldsymbol{E = 21.0\,\text{N}}$$

Finding mechanical advantage.

$$\text{MA} = \frac{W}{E} = \frac{150 \times 9.81}{21.0}$$

$$\mathbf{MA = 70.1}$$

Finding efficiency.

$$\eta = \frac{\text{MA}}{\text{VR}} = \frac{70.1}{105}$$

$$\boldsymbol{\eta = 0.668 \quad \text{or} \quad 66.8\%}$$

(c) Finding limiting mechanical advantage.

$$\text{Limiting MA} = \frac{1}{a} \text{ (where } a = 0.0105\text{, from the law of the machine)}$$

$$\textbf{Limiting MA} = \frac{1}{0.0105} = 95.2$$

Finding limiting efficiency.

$$\text{Limiting } \eta = \frac{1}{a\text{ VR}}$$

$$\textbf{Limiting } \eta = \frac{1}{0.0105 \times 105} = 0.907 \quad \text{or} \quad 90.7\%$$

(d) When raising the 150 kg load the efficiency is greater than 50% and so the winch will overhaul under these conditions.

Test your knowledge 1.14

1. How does the mechanical advantage of a simple machine vary with the load raised?
2. How does the frictional resistance in a simple machine vary with the load raised?
3. How is the efficiency of a simple machine defined?
4. What information does the law of a machine contain?
5. What is meant by overhauling and how can it be predicted?

Activity 1.13

A screw jack has a single start thread of pitch 6 mm and an operating handle of radius 450 mm. The following readings of load and effort were taken during a test on the jack (Table 1.4).

Table 1.4

Load (kN)	0	1	2	3	4	5	6	7	8	9	10
Effort (N)	2.2	6.6	11.8	17.0	22.2	27.2	32.2	36.8	41.4	46.9	52.4

(a) Plot a graph of effort against load and from it determine the law of the machine.

(b) Plot a graph of mechanical advantage against load.

(c) Plot a graph of efficiency against load and state whether the machine is likely to overhaul.

(d) Calculate the theoretical limiting values of mechanical advantage and efficiency and state whether your graphs tend towards these values.

(e) Calculate the work input and the work done against friction when raising the 10 kN load through a height of 50 mm.

To check your understanding of the preceding section, you can solve Review questions 45–48 at the end of this chapter.

Review questions

1. Determine the magnitude and direction of the resultant force for the coplanar force system shown in Figure 1.81.

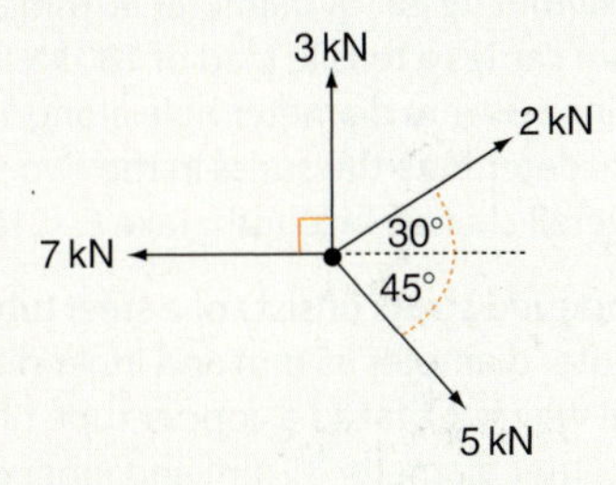

Figure 1.81

2. Determine the magnitude and direction of the equilibrant required for the coplanar force system shown in Figure 1.82.

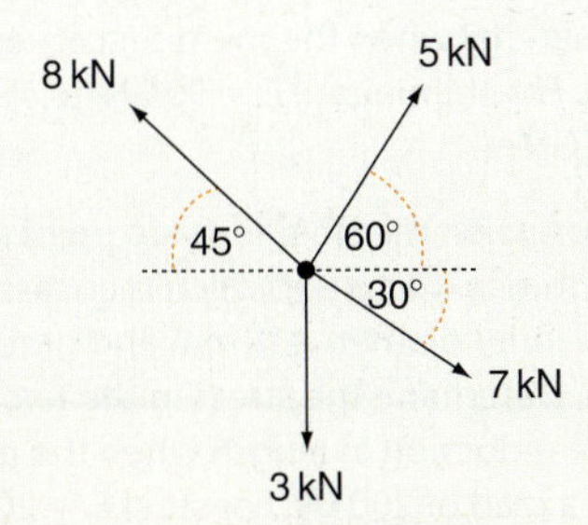

Figure 1.82

3. Determine the magnitude and direction of the equilibrant required for the coplanar force system shown in Figure 1.83.

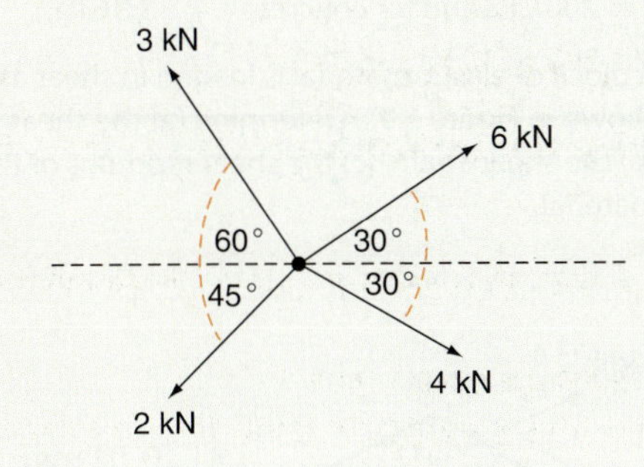

Figure 1.83

4. For each of the following framed structures, determine the support reactions and the magnitude and nature of the force in each member (Figures 1.84–86).

(a)

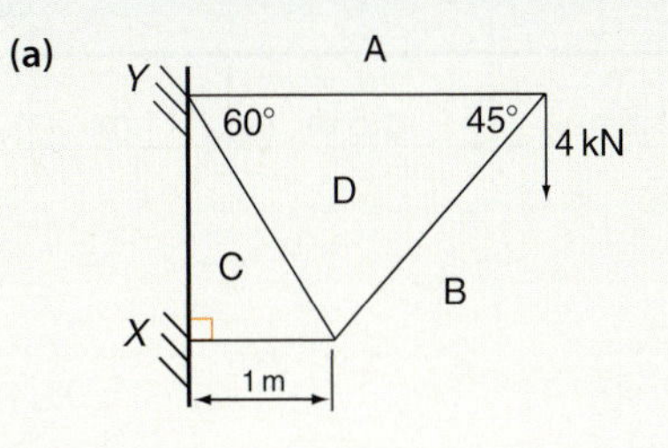

Figure 1.84

(b)

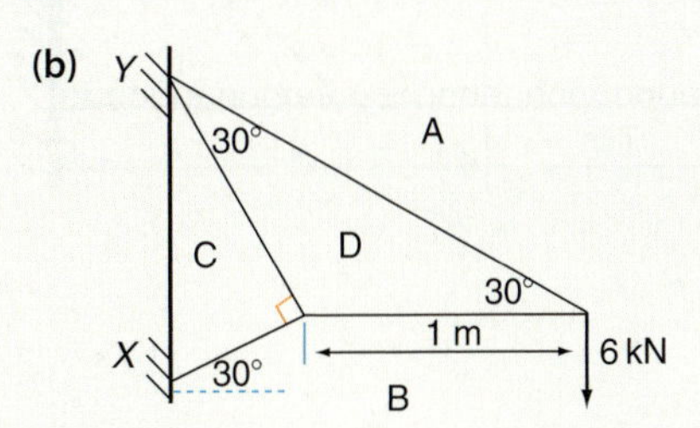

Figure 1.85

(c)

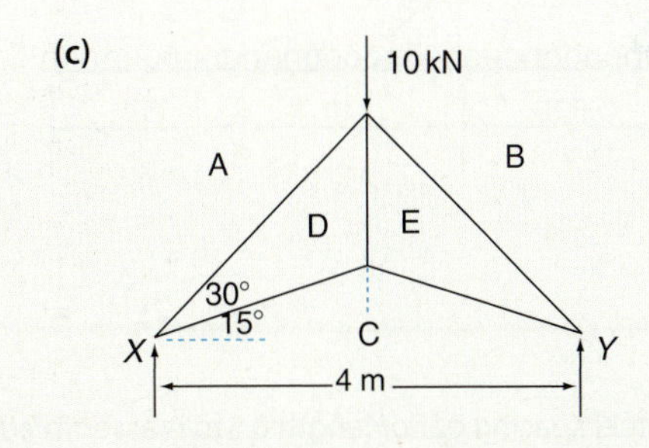

Figure 1.86

5. Sketch the shear force and bending moment diagrams for the simply supported beams and cantilevers shown in Figures 1.87–1.91. Indicate the magnitude and nature of the maximum shear force and bending moment and the positions where they occur. Indicate also the position of any point of contraflexure.

(a)

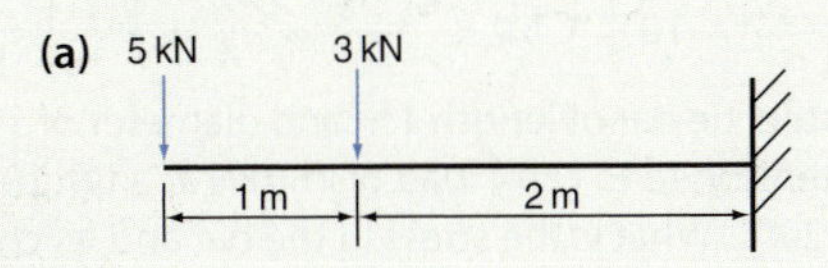

Figure 1.87

(b)

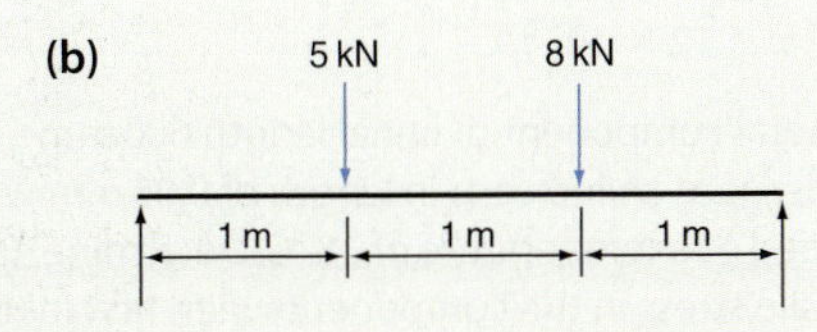

Figure 1.88

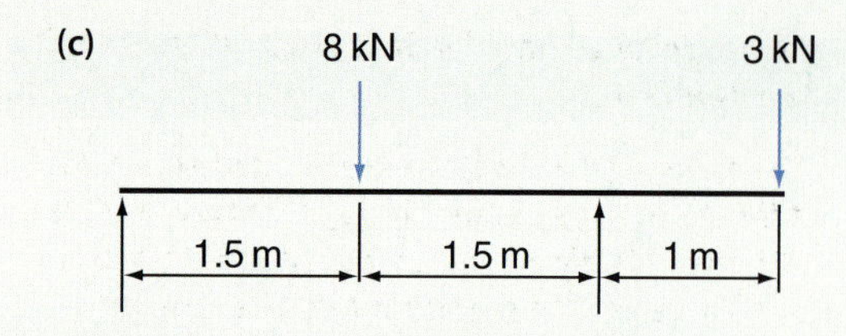

Figure 1.89

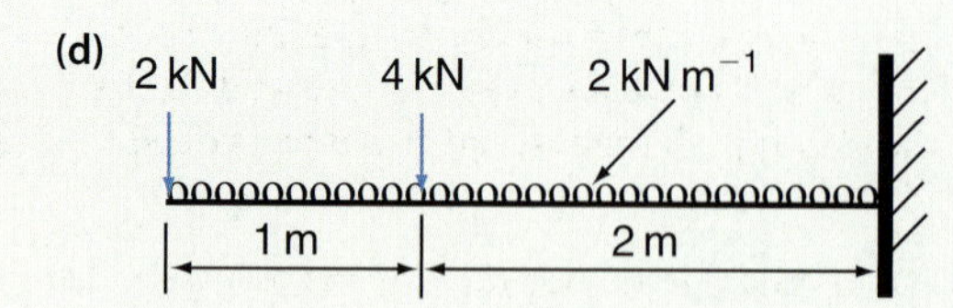

Figure 1.90

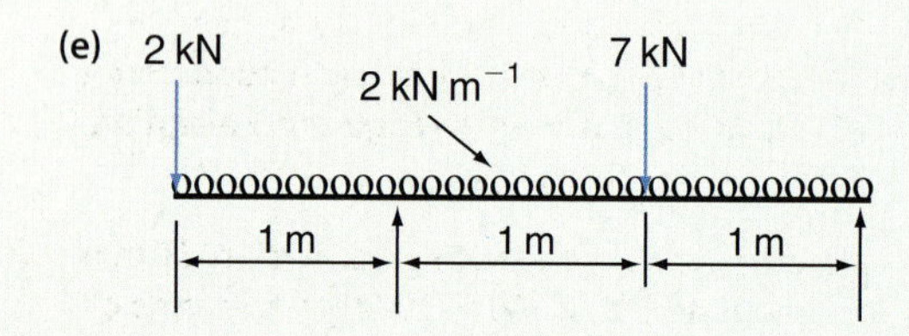

Figure 1.91

6. A steel spacing bar of length 0.5 m is assembled in a structure whilst the temperature is 20°C. What will be its change in length if the temperature rises to 50°C? What will be the stress in the bar if free expansion is prevented? Take $E = 200$ GPa and $\alpha = 12 \times 10^{-1}$ °C^{-1}.

7. A copper bar is rigidly fixed at its ends at a temperature of 15°C. Determine the magnitude and nature of the stress induced in the material (a) if the temperature falls to −25 °C, (b) if the temperature rises to 95°C. Take $E = 96$ GPa and $\alpha = 17 \times 10^{-1}$ °C^{-1}.

8. A steel tie bar of length 1 m and diameter of 30 mm is tensioned to carry load of 180 kN at a temperature of 20 °C. What is the stress in the bar and its change in length when tensioned? What will be the stress in the bar if the temperature rises to 80°C and any change in length is prevented? Take $E = 200$ GN m^{-2} and $\alpha = 12 \times 10^{-6}$ °C^{-1}.

9. A metal component of initial length 600 mm undergoes an increase in length of 0.05 mm when loaded at a temperature of 18 °C. Determine, (a) the tensile stress in the component when first loaded, (b) the temperature at which the stress will be zero if the component is rigidly held at its extended length. Take $E = 120$ GPa and $\alpha = 15 \times 10^{-6}$ °C^{-1} for the component material.

10. A brittle cast iron bar is heated to 140°C and then rigidly clamped so that it cannot contract when cooled. The bar is seen to fracture when the temperature has fallen to 80°C. What is the ultimate tensile strength of the cast iron? Take $E = 120$ GPa and $\alpha = 10 \times 10^{-6}$ °C^{-1}.

11. A duralumin tie bar of diameter 40 mm and length 600 mm carries a tensile load of 180 kN. If the bar contains a 30 mm diameter hole along 100 mm of its length, determine the stress in the two sections and the overall change in length. Take $E = 180$ GPa.

12. A compound strut consists of a steel tube of length 1 m, outer diameter 35 mm and inner diameter 25 mm which is brazed a copper tube of length 1.5 m, outer diameter 35 mm and inner diameter 20 mm. If the member carries a compressive load of 20 kN, determine the stress in each material and the overall change in length. For steel, $E = 200$ GN m^{-2}, for copper $E = 120$ GN m^{-2}.

13. A rectangular timber strut of cross-section 125 mm × 105 mm is reinforced by two aluminium bars of diameter 60 mm. Determine the stress in each material when the member carries a load of 300 kN. For aluminium, $E = 90$ GN m^{-2}, for timber $E = 15$ GN m^{-2}.

14. A compression member is made up of a mild steel bar, 38 mm in diameter, which is encased in a brass tube of inner diameter 38 mm and outer diameter 65 mm. Determine the stress in the two materials and the reduction in length when the member carries a load of 200 kN. For steel $E = 200$ GPa and for brass $E = 96$ GPa.

15. A steel reinforced concrete column of height 3 m has a square cross-section with sides of 375 mm. If the column contains four steel reinforcing rods of diameter 25 mm, determine the stress in each material and the amount of compression when carrying a compressive load of 600 kN. For steel $E = 200$ GPa and for concrete $E = 13.8$ GPa.

16. A block of elastic material is loaded in shear as shown in Figure 1.92. Determine (a) the shear stress, (b) the shear strain, (c) the shear modulus of the material.

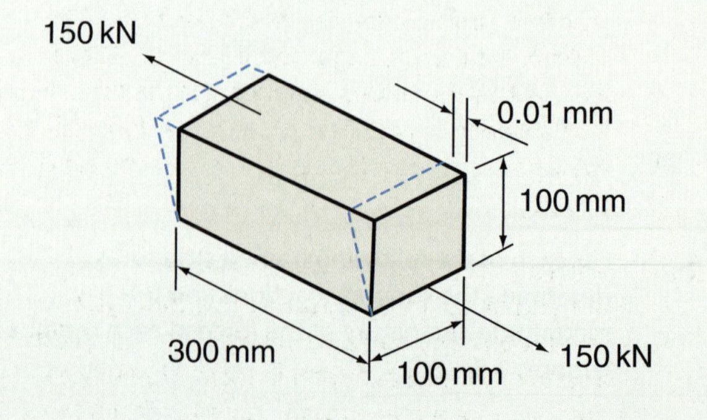

Figure 1.92

17. A mild steel plate of thickness 15 mm has a 50 mm diameter hole punched in it. If the ultimate shear stress of the steel is 275 MPa, determine (a) the punching force required, (b) the compressive stress in the punch.

18. Figure 1.93 shows a riveted lap joint. Determine a suitable diameter for the rivets if the ultimate shear strength of the rivet material is 325 MPa and a factor of safety of 6 is to apply.

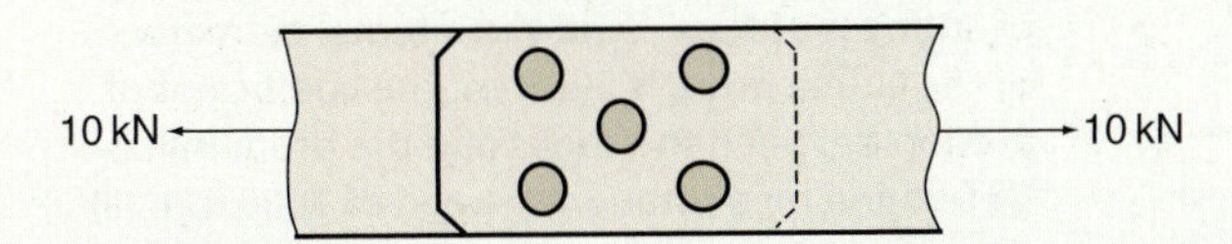

Figure 1.93

19. Determine the load F which the double strap butt joint shown in Figure 1.94 can carry. The rivets are 6 mm in diameter with an ultimate shear strength of 325 MPa. A factor of safety of 8 is to apply.

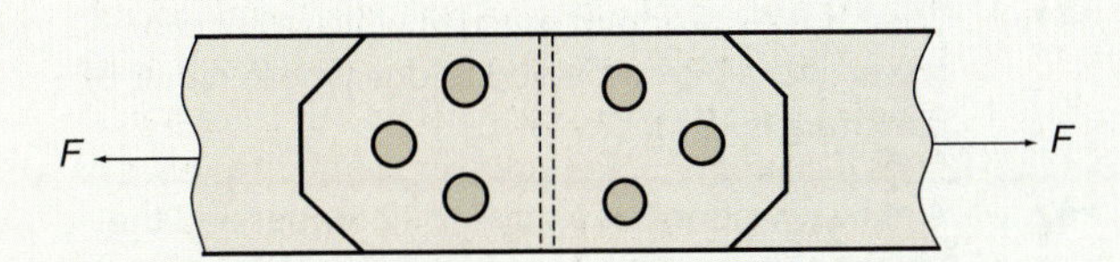

Figure 1.94

20. The speed of an electric motor is increased uniformly from 1200 to 1350 rpm in a time of 1.5 s. Find (a) the angular acceleration, (b) the number of revolutions turned during this time.

21. A winding drum is accelerated from 480 to 840 rpm in a time of 2 s. Acceleration continues at the same rate for a further 4 s after which the drum continues rotating at a steady speed. Calculate the time taken to complete the first 200 revolutions.

22. An applied torque of 6 Nm causes the speed of a machine part to increases uniformly from 300 to 500 rpm whilst rotating through 18 revolutions. Find (a) the angular acceleration, (b) the time taken for the speed to increase, (c) the work done, (d) the average power developed, (e) the maximum power developed.

23. A flywheel is uniformly accelerated from rest to a speed of 120 rpm whilst rotating through 60 revolutions. The applied driving torque is 15 Nm and the flywheel diameter is 1.2 m. Find (a) the angular acceleration, (b) the time taken, (c) the work done, (d) the maximum power developed, (e) the distance travelled by a point on the rim of the flywheel.

24. The drive to a flywheel is disconnected and in 30 s it is seen to have rotated through 120 revolutions; 30 s later it finally comes to rest. Assuming the angular retardation to be uniform, determine (a) its initial angular velocity, (b) the angular retardation.

25. A mixing drum is accelerated from 50 to 160 rpm in a time of 20 s by a driving torque of 25 Nm. Determine (a) the angular acceleration, (b) the number of complete revolutions turned, (c) the work done, (d) the average power developed, (e) the maximum power developed.

26. Determine the applied torque required to accelerate a rotor of mass 100 kg and radius of gyration 500 mm from a speed of 50–300 rpm in a time of 12 s. Neglect bearing friction.

27. A flywheel of mass 50 kg and radius of gyration 175 mm is accelerated from rest to a speed of 500 rpm in 10 s. There is a bearing friction torque of 2 Nm. Determine (a) the angular acceleration, (b) the applied torque.

28. A rotating disc of mass 25 kg and radius 150 mm is accelerated from 500 to 3000 rpm in a time of 25 s. If there is a constant friction torque of 5 Nm, determine (a) the applied torque, (b) the work done, (c) the maximum power developed.

29. When the steam supply is cut off a turbine rotor of mass 850 kg is retarded uniformly under the effects of a 125 Nm bearing friction torque. If its speed falls from 1000 to 500 rpm in a time of 5 min, determine (a) the angular retardation, (b) the radius of gyration of the flywheel.

30. A period of 10 min is required for an applied torque of 1.2 kNm to accelerate a flywheel of mass 2 tonnes and radius of gyration 600 mm from rest to a speed of 6000 rpm. Determine (a) the friction torque in the bearings, (b) the time taken for the flywheel to come to rest when the driving torque is removed.

31. An open-ended mixing drum is a thin cylinder of length 1.3 m, diameter 600 mm and wall thickness 5 mm. The drum is made from steel of density 7800 kg m^3. A driving torque of 55 Nm accelerates the drum uniformly from rest to a speed of 3600 rpm against a friction torque of 5 Nm. Determine (a) the moment of inertia of the drum, (b) the angular acceleration, (c) the time taken, (d) the work done, (e) the average power developed by the driving motor.

32. A lift cage of mass 4000 kg is raised by a hoist drum of mass 750 kg, diameter 1.5 m and radius of gyration 600 mm. The cage is accelerated from rest through a distance of 12 m at the rate of 1.5 m s^{-2} against a friction torque of 3 kNm. Calculate (a) the final angular velocity of the drum, (b) the angular acceleration of the drum, (c) the work done by the driving motor, (d) the input torque to the drum, (e) the maximum power developed by the driving motor.

33. A motor car of total mass 1 tonne is accelerated uniformly from rest to a speed of 60 km h^{-1} up an incline of gradient 10%. The diameter of the wheels is 700 mm and the distance travelled is 100 m. The wheels and transmission have a mass of 120 kg and radius of gyration 250 mm. Frictional resistance to motion is equivalent to a force of 500 N which can be considered to be constant. Determine (a) the time taken, (b) the total work done by the engine, (c) the average power developed.

34. A mass of 100 kg is uniformly accelerated upwards on a light cable by a hoist for a period of 5 s to a speed of 1.75 m s^{-1}. The hoist has a winding drum of mass 220 kg, diameter 1 m and radius of gyration 300 mm. Calculate (a) the vertical height raised, (b) the final angular velocity of the drum, (c) the work done, (d) the input torque to the driving motor, (e) the average power developed.

35. The shaft of the flywheel shown in Figure 1.95 is supported on bearings. A light cord is wound around the shaft from which is hung a 2 kg mass. When allowed to fall freely the mass accelerates uniformly through a distance of 2 m in 3 s. The friction couple in the bearings is estimated to be 0.35 Nm. Calculate the moment of inertia and radius of gyration of the assembly.

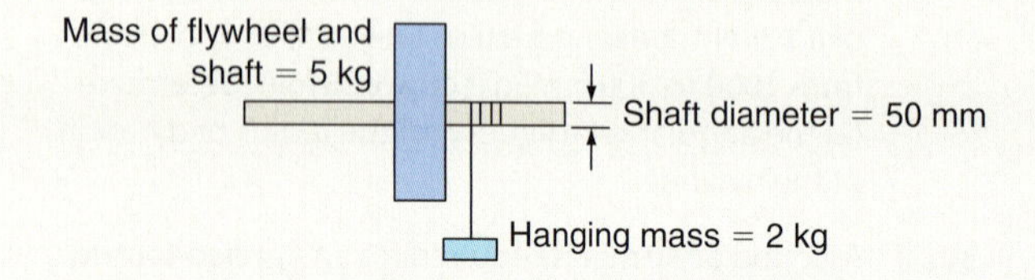

Figure 1.95

36. A body of mass 3 kg is whirled round in a horizontal plane at the end of a steel wire. The distance from the centre of rotation to the centre of the mass is 1.5 m and the diameter of the wire is 2 mm. If the ultimate tensile strength of the steel is 500 MPa, determine the rotational speed at which the wire will break.

37. A body of mass 1 kg is whirled in a horizontal plane at the end of a spring. The spring stiffness is 5 kN m^{-1} and the initial distance from the centre of rotation to the centre of the body is 200 mm. What will be the extension of the spring at a rotational speed of 300 rpm?

38. A centrifugal clutch has four shoes, each of mass 2 kg which are held on retaining springs of stiffness 10 kN m^{-1}. The internal diameter of the clutch drum is 600 mm and the radius of the centre of gravity of the shoes when in contact with the drum is 250 mm. The stationary clearance between the shoes and the drum is 20 mm and the coefficient of friction is 0.25. Determine (a) the speed at which the clutch starts to engage, (b) the power transmitted at a speed of 1500 rpm.

39. A centrifugal clutch has four shoes, each of mass 0.3 kg. When at rest the radial clearance between the shoes and the clutch drum is 5 mm and the radius to the centre of gravity of the shoes is 195 mm. The shoes are held on retaining springs of stiffness 8 kN m^{-1} and the internal diameter of the clutch drum is 500 mm. The coefficient of friction between the shoes and the drum is 0.2. Determine for a rotational speed of 1000 rpm (a) the radial force which each shoe exerts on the drum, (b) the torque transmitted, (c) the power transmitted.

40. A four-wheeled vehicle has a track width of 1.15 m and a mass of 750 kg. Its centre of gravity is 1.2 m above the road surface and equidistant from each of the wheels. What will be the reactions of the inner and outer wheels when travelling at a speed of 60 km h^{-1} around a level curve of radius 80 m.

41. What is the maximum speed at which a car can travel over a hump-backed bridge whose radius of curvature is 18 m?

42. A railway wagon has a mass of 12 tonnes and the height of its centre of gravity is 1.5 m above the rails. The track width is 1.37 m and the wagon travelling at 70 km h^{-1}. What is minimum radius of level curve that the wagon can negotiate at this speed without overturning?

43. A car of mass 850 kg travels round an unbanked horizontal curve of radius 35 m. The limiting coefficient of friction between the tyres and the road is 0.65. The track width of the car is 1.65 m and its centre of gravity is central and at a height of 0.8 m above the road surface. Show that the car is likely to skid rather than overturn if the speed is excessive and calculate the maximum speed in km h^{-1} at which it can travel round the curve. What will be the reactions of the inner and outer wheels at this speed?

44. The height of its centre of gravity of a four-wheeled vehicle is 1.1 m above the road surface and equidistant from each of the wheels. The track width of 1.5 m and the limiting coefficient of friction between the wheels and the road surface is 0.7. Determine (a) the maximum speed at which the vehicle can travel round a curve of radius 90 m, (b) the angle to which the curve would have to be banked for the vehicle to travel round it at a speed of 60 km h^{-1} without any tendency to side-slip?

45. A screw jack has a single start thread of pitch 10 mm and is used to raise a load of 900 kg. The required effort is 75 N, applied at the end of an operating handle 600 mm long. For these operating conditions determine (a) the mechanical advantage, (b) the velocity ratio, (c) the efficiency of the machine.

46. The top pulleys of a Weston differential pulley block have diameters of 210 and 190 mm. Determine the effort required to raise a load of 150 kg if the efficiency of the system is 35%. What is the work done in overcoming friction when the load is raised through a height of 2.5 m?

47. A differential wheel and axle has a wheel diameter of 300 mm and axle diameters of 100 and 75 mm. The effort required to raise a load of 50 kg is 55 N and the effort required to raise a load of 200 kg is 180 N. Determine (a) the velocity ratio, (b) the law of the machine, (c) the efficiency of the machine when raising a load of 100 kg.

48. The crab winch shown in Figure 1.96has a law of the form $E = 0.01W + 7.5$. Determine (a) its velocity ratio, (b) its mechanical advantage and efficiency when raising a load of 50 kg, (c) the limiting values of its mechanical advantage and efficiency.

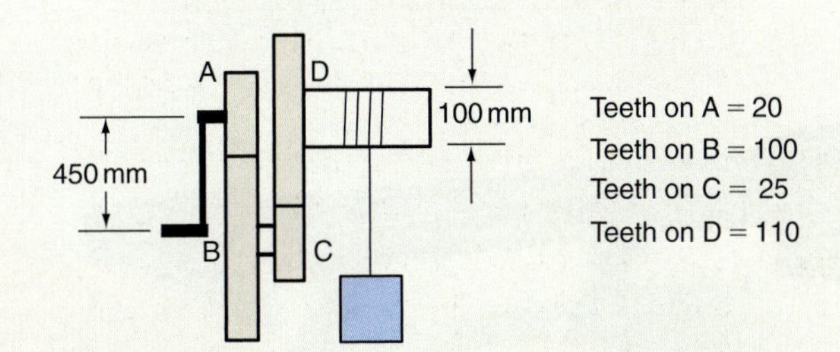

Figure 1.96

Successful Formula 1 and other racing cars incorporate advances in mechanical design that is complimented by skilful driving. Advanced mechanical principles are applied in the design of the steering, suspension and power transmission systems. The drivers must also have an appreciation of mechanics and recognise the limits to which they can safely push their vehicles.

Photo courtesy of iStockphoto, Jan Paul Schrage, Image # 4955692

Chapter 2

Advanced Mechanical Principles and Applications

The aim of this chapter is to build on the work that you have done in the core unit *Mechanical Principles and Applications* and the optional unit *Further Mechanical Principles and Applications*. The first section of this chapter will introduce you to Poisson's ratio and its application in analysing the effects of uni-axial and complex loading on engineering components. The second section builds on the work that you have done on shear force and bending moment distribution in simply supported beams to cover the calculation of stress due to bending. The section on resultant and relative velocity will introduce you to the characteristics of plane mechanisms, in particular the slider-crank and four-bar linkage mechanisms and their inversions.

The final section investigates the occurrence of natural vibrations in mechanical systems. It will introduce you to the concept of simple harmonic motion (SHM) and the characteristics of some common mechanical systems in which a disturbing force produces mechanical oscillations.

Uni-axial and complex loading

Poisson's ratio

When a material is loaded in tension, as shown in Figure 2.1, there is a tendency for its length to increase and its width or thickness to decrease. When loaded in compression, the opposite tends to occur. The deformation is measured as *longitudinal strain* in the direction of loading and *lateral strain* at right angles to the direction of loading.

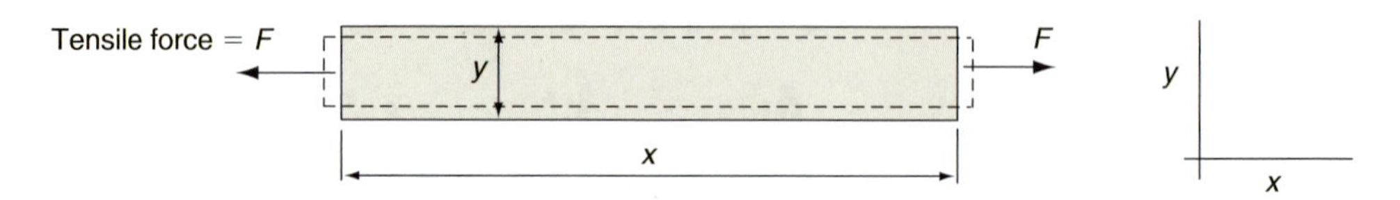

Figure 2.1 Longitudinal and lateral deformation

You will already be familiar with longitudinal strain ε_x which is also sometimes called *axial strain*. It is shown as acting in the x-direction in Figure 2.1 and is calculated using the formula:

$$\text{Longitudinal strain} = \frac{\text{Change in length}}{\text{Original length}}$$

$$\varepsilon_x = \frac{\Delta x}{x} \tag{2.1}$$

The lateral or transverse strain is calculated in a similar way:

$$\text{Lateral strain} = \frac{\text{Change in thickness}}{\text{Original thickness}}$$

$$\varepsilon_y = \frac{\Delta y}{y} \tag{2.2}$$

It is found that the two strains are unequal but proportional to each other. The constant of proportionality is called *Poisson's ratio* whose symbol is the Greek letter v (Nu):

$$\text{Poisson's ratio} = \frac{\text{Lateral strain}}{\text{Longitudinal strain}}$$

$$v = \frac{\varepsilon_y}{\varepsilon_x}$$

$$\varepsilon_y = v\,\varepsilon_x \tag{2.3}$$

For engineering metals the value of Poisson's ratio lies between 0.25 and 0.35. You should note that although the value of Poisson's ratio is always positive, it is usual in advanced work to adopt the sign convention that tensile stress and strain are positive and compressive stress and strain are negative. Although there is no compressive force acting in the y-direction in Figure 2.1, there is a reduction in the thickness of the material and the lateral strain ε_y would be given a negative sign.

You will recall that the modulus of elasticity of a material is given by:

$$E = \frac{\sigma_x}{\varepsilon_x} \quad \text{or} \quad \varepsilon_x = \frac{\sigma_x}{E}$$

Substituting for ε_x in equation (2.3) gives a formula for finding the lateral strain ε_y, directly from the longitudinal stress:

$$\text{i.e.} \quad \boldsymbol{\varepsilon_y = \frac{\nu\sigma_x}{E}} \qquad (2.4)$$

Example 2.1

A tubular steel strut of length 1.5 m, inner diameter 150 mm and outer diameter 200 mm carries a compressive load of 4 MN. If Poisson's ratio for the material is 0.3 and the modulus of elasticity is 200 GPa, determine the change of length and outer diameter of the column.

Finding cross-sectional area of column:

$$A = \frac{\pi}{4}(D^2 - d^2) = \frac{\pi}{4}(0.2^2 - 0.15^2)$$

$$\mathbf{A = 0.0137\,m^2}$$

Finding compressive stress in column (*x*-direction):

$$\sigma_x = \frac{F_x}{A} = \frac{-4 \times 10^6}{0.0137}$$

$$\boldsymbol{\sigma_x = -292 \times 10^6\,\text{Pa} \quad \text{or} \quad -292\,\text{MPa}}$$

Finding compressive longitudinal strain:

$$E = \frac{\sigma_x}{\varepsilon_x}$$

$$\varepsilon_x = \frac{\sigma_x}{E} = \frac{-292 \times 10^6}{200 \times 10^9}$$

$$\boldsymbol{\varepsilon_x = -1.46 \times 10^{-3}}$$

Finding change in length:

$$\varepsilon_x = \frac{\Delta x}{x}$$

$$\Delta x = x\,\varepsilon_x = 1.5(-1.46 \times 10^{-3})$$

$$\boldsymbol{\Delta x = -2.19 \times 10^{-3}\,\text{m} \quad \text{or} \quad -2.19\,\text{mm}} \quad \text{(i.e. decrease in length)}$$

Finding lateral strain (*y*-direction):

$$\nu = \frac{\varepsilon_y}{\varepsilon_x}$$

$$\varepsilon_y = \nu\varepsilon_x = 0.3(1.46 \times 10^{-3}) \quad \text{(Leave out the -ve sign)}$$

$$\boldsymbol{\varepsilon_y = +438 \times 10^{-6}}$$

Finding change in diameter:

$$\varepsilon_y = \frac{\Delta y}{y}$$

$$\Delta y = y\,\varepsilon_y = 0.2(438 \times 10^{-6})$$

$\Delta x = +0.0876 \times 10^{-3}\,\text{m}$ **or** $+0.0876\,\text{mm}$ (i.e. increase in diameter)

Two-dimensional loading

Two dimensional or bi-axial loading induces stress in both the x- and y-directions simultaneously as shown in Figure 2.2.

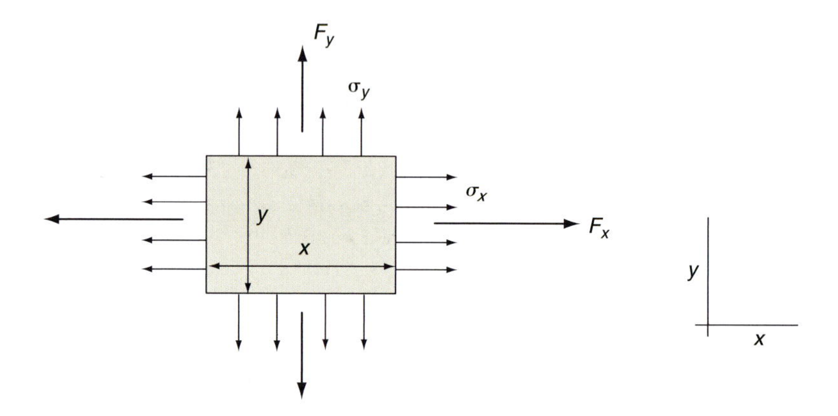

Figure 2.2 Bi-axial loading

The stress σ_x tends to cause tensile strain in the x-direction, making the component longer, i.e. positive strain. It also tries to produce compressive strain in the y-direction, making the component thinner, i.e. negative strain. In a similar way, the stress σ_y tends to cause tensile strain in the y-direction and compressive strain in the x-direction.

The strains in the x-direction will be:

$$\text{Strain due to } \sigma_x = +\frac{\sigma_x}{E}$$

$$\text{Strain due to } \sigma_y = -\frac{\nu\sigma_y}{E}$$

The resultant strain ε_x in the x-direction will be the sum of these:

$$\varepsilon_x = +\frac{\sigma_x}{E} - \frac{\nu\sigma_y}{E}$$

or $$\boldsymbol{\varepsilon_x = \frac{\sigma_x - \nu\sigma_y}{E}} \qquad (2.5)$$

The strains in the y-direction will be:

$$\text{Strain due to } \sigma_y = +\frac{\sigma_y}{E}$$

$$\text{Strain due to } \sigma_x = -\frac{\nu\sigma_x}{E}$$

The resultant strain ε_y in the y-direction will be the sum of these:

$$\varepsilon_y = +\frac{\sigma_y}{E} - \frac{\nu\sigma_x}{E}$$

$$\text{or} \quad \boldsymbol{\varepsilon_y = \frac{\sigma_y - \nu\sigma_x}{E}} \tag{2.6}$$

Having calculated these resultant strains, the changes in the x- and y-dimensions can be found from:

$$\boldsymbol{\Delta x = \varepsilon_x x} \tag{2.7}$$

$$\boldsymbol{\Delta y = \varepsilon_y y} \tag{2.8}$$

KEY POINT

Although the longitudinal and lateral strains may have different signs, Poisson's ratio is always positive.

Example 2.2

The plate shown in Figure 2.3 is subjected to tensile and compressive loading in perpendicular directions. The modulus of elasticity of the material is 150 GPa and Poisson's ratio is 0.35. Determine the dimensional changes that occur.

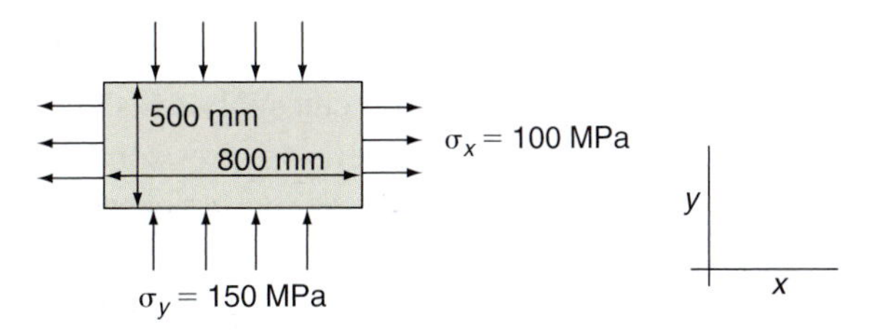

Figure 2.3

Finding resultant strain in the x-direction (note the stress σ_y is compressive, i.e. –ve):

$$\varepsilon_x = \frac{\sigma_x - \nu\sigma_y}{E}$$

$$\varepsilon_x = \frac{(100 \times 10^6) - (0.35 \times -150 \times 10^6)}{150 \times 10^9}$$

$$\boldsymbol{\varepsilon_x = 1.017 \times 10^{-3}}$$

Finding the change in the x-dimension:

$$\Delta x = \varepsilon_x x = 1.017 \times 10^{-3} \times 800$$

$$\boldsymbol{\Delta x = +0.813\,\text{mm}}$$

Finding resultant strain in the y-direction:

$$\varepsilon_y = \frac{\sigma_y - \nu\sigma_x}{E}$$

$$\varepsilon_y = \frac{(-150 \times 10^6) - (0.35 \times 100 \times 10^6)}{150 \times 10^9}$$

$$\boldsymbol{\varepsilon_y = -1.233 \times 10^{-3}}$$

Finding the change in the y-dimension:

$$\Delta y = \varepsilon_y y = -1.233 \times 10^{-3} \times 500$$

$$\mathbf{\Delta y = -0.617\,mm}$$

Three-dimensional loading

Three-dimensional or tri-axial loading induces stress in the x-, y- and z-directions simultaneously, are as shown in Figure 2.4.

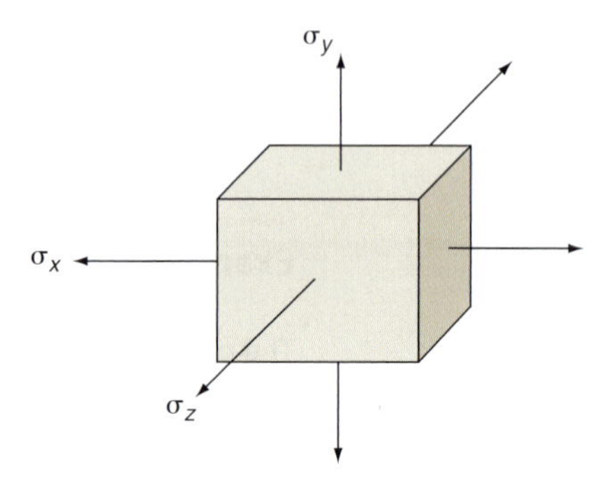

Figure 2.4 Tri-axial loading

You can apply the same reasoning as for two-dimensional loading to derive expressions for the strains ε_x, ε_y and ε_z in the x-, y- and z-directions, respectively.

Finding resultant strain in x-direction:

$$\varepsilon_x = \text{Strain due to } \sigma_x + \text{Strain due to } \sigma_y + \text{Strain due to } \sigma_z$$

$$\varepsilon_x = +\frac{\sigma_x}{E} - \frac{\nu\sigma_y}{E} - \frac{\nu\sigma_z}{E}$$

or $$\boldsymbol{\varepsilon_x = \frac{\sigma_x - \nu(\sigma_y + \sigma_z)}{E}} \tag{2.9}$$

Finding resultant strain in y-direction:

$$\varepsilon_y = \text{Strain due to } \sigma_y + \text{Strain due to } \sigma_x + \text{Strain due to } \sigma_z$$

$$\varepsilon_y = +\frac{\sigma_y}{E} - \frac{\nu\sigma_x}{E} - \frac{\nu\sigma_z}{E}$$

or $$\boldsymbol{\varepsilon_y = \frac{\sigma_y - \nu(\sigma_x + \sigma_z)}{E}} \tag{2.10}$$

Finding resultant strain in z-direction:

$$\varepsilon_z = \text{Strain due to } \sigma_z + \text{Strain due to } \sigma_x + \text{Strain due to } \sigma_y$$

$$\varepsilon_z = +\frac{\sigma_z}{E} - \frac{\nu\sigma_x}{E} - \frac{\nu\sigma_y}{E}$$

or $$\boldsymbol{\varepsilon_z = \frac{\sigma_z - \nu(\sigma_x + \sigma_y)}{E}} \tag{2.11}$$

KEY POINT

Compressive stresses must always be entered as negative quantities in the formulae for strain in the x-, y- and z-directions.

Volumetric strain

Consider the same block of material whose initial dimensions are x, y and z in the three perpendicular directions. Let us assume that the three tensile stresses σ_x, σ_y and σ_z cause the block to become larger in all three directions. This is quite possible as most solid material have a degree of volumetric elasticity (Figure 2.5)

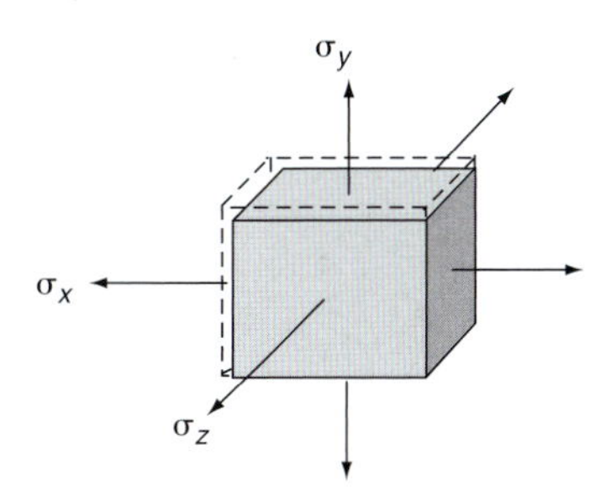

Figure 2.5 Effects of tri-axial loading

The final x-dimension is: $x + \Delta x = x + \varepsilon_x x = x(1 + \varepsilon_x)$
The final y-dimension is: $y + \Delta y = y + \varepsilon_y y = y(1 + \varepsilon_y)$
The final z-dimension is: $z + \Delta z = z + \varepsilon_z z = z(1 + \varepsilon_z)$

The initial volume V of the block is given by:

$$\boldsymbol{V = xyz} \tag{2.12}$$

The final volume $V + \Delta V$ of the block is given by:

$$V + \Delta V = x(1 + \varepsilon_x)y(1 + \varepsilon_y)z(1 + \varepsilon_z)$$
$$V + \Delta V = xyz(1 + \varepsilon_x)(1 + \varepsilon_y)(1 + \varepsilon_z)$$

Multiplying the bracketed terms together gives:

$$V + \Delta V = xyz(1 + \varepsilon_x + \varepsilon_y + \varepsilon_z + \varepsilon_x\varepsilon_y + \varepsilon_x\varepsilon_z + \varepsilon_y\varepsilon_z + \varepsilon_x\varepsilon_y\varepsilon_z)$$

Now the strains are very small and there products will be very small indeed. Neglecting these small products leaves:

$$\boldsymbol{V + \Delta V = xyz(1 + \varepsilon_x + \varepsilon_y + \varepsilon_z)} \tag{2.13}$$

The volumetric strain ε_v is given by:

$$\varepsilon_v = \frac{\text{Change in volume}}{\text{Original volume}} = \frac{\Delta V}{V}$$
$$\varepsilon_v = \frac{xyz(1 + \varepsilon_x + \varepsilon_y + \varepsilon_z) - xyz}{xyz}$$
$$\varepsilon_v = 1 + \varepsilon_x + \varepsilon_y + \varepsilon_z - 1$$
$$\boldsymbol{\varepsilon_v = \varepsilon_x + \varepsilon_y + \varepsilon_z} \tag{2.14}$$

The volumetric strain can thus be taken as the sum of the three perpendicular strains. The change in voume is given by:

$$\boldsymbol{\Delta V = \varepsilon_v V} \tag{2.15}$$

CHAPTER 2

Example 2.3

A cube whose sides are initially 200 mm is subjected to stresses of 100 MPa (tensile), 150 MPa (compressive) and 80 MPa (tensile) in the *x*-, *y*- and *z*-directions, respectively. Calculate the dimensional changes and the change in volume that results from the loading. The modulus of elasticity of the material is 50 GPa and Poisson's ratio is 0.3.

Finding strain in *x*-direction:

$$\varepsilon_x = \frac{\sigma_x - \nu(\sigma_y + \sigma_y)}{E}$$

$$\varepsilon_x = \frac{(100 \times 10^6) - [0.3(-150 + 80) \times 10^6]}{50 \times 10^9}$$

$$\boldsymbol{\varepsilon_x = +0.00242}$$

Finding change in *x*-dimension:

$$\Delta x = \varepsilon_x x = +0.00242 \times 200$$

$$\boldsymbol{\Delta x = +0.484\,\text{mm}}$$

Finding strain in *y*-direction:

$$\varepsilon_y = \frac{\sigma_y - \nu(\sigma_x + \sigma_z)}{E}$$

$$\varepsilon_y = \frac{(-150 \times 10^6) - [0.3(100 + 80) \times 10^6]}{50 \times 10^9}$$

$$\boldsymbol{\varepsilon_y = -0.00408}$$

Finding change in *y*-dimension:

$$\Delta y = \varepsilon_y y = -0.00408 \times 200$$

$$\boldsymbol{\Delta y = -0.816\,\text{mm}}$$

Finding strain in *z*-direction:

$$\varepsilon_z = \frac{\sigma_z - \nu(\sigma_x + \sigma_y)}{E}$$

$$\varepsilon_z = \frac{(80 \times 10^6) - [0.3(100 - 150) \times 10^6]}{50 \times 10^9}$$

$$\boldsymbol{\varepsilon_z = +0.0019}$$

Finding change in *z*-dimension:

$$\Delta z = \varepsilon_z z = +0.0019 \times 200$$

$$\boldsymbol{\Delta z = +0.38\,\text{mm}}$$

Finding volumetric strain:

$$\varepsilon_v = \varepsilon_x + \varepsilon_y + \varepsilon_z = +0.00242 - 0.00408 + 0.0019$$

$$\boldsymbol{\varepsilon_v = +0.00024}$$

Finding change in volume in cubic centimetres (each side is 20 cm):

$$\Delta V = \varepsilon_v V = +0.00024 \times 20^3$$

$$\boldsymbol{\Delta V = +1.92\,\text{cm}^3}$$

Test your knowledge 2.1

1. What is lateral strain?
2. What is Poisson's ratio and what is its range of values for common engineering metals?
3. What is the sign convention generally adopted for tensile and compressive loading?
4. What are the formulae for calculation of strain due to bi-axial loading?
5. What is volumetric strain and how is it calculated for a material subjected to tri-axial loading?

Activity 2.1

The rectangular prism shown in Figure 2.6 is acted upon by stresses of 300 MPa (compressive), 350 MPa (tensile) and 400 MPa (tensile) in the *x*-, *y*- and *z*-directions, respectively. Calculate the changes in the dimensions shown as a result of the loading and the change in volume. Poisson's ratio for the material is 0.29 and its modulus of elasticity is 85 GPa.

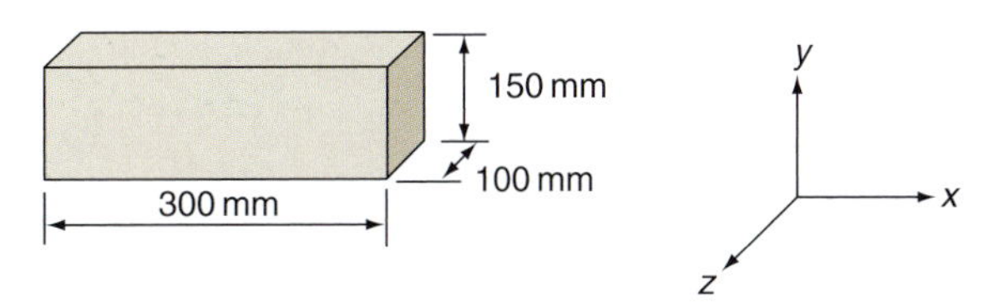

Figure 2.6

To check your understanding of the preceding section, you can solve Review questions 1–6 at the end of this chapter.

Bending in Beams

In Chapter 1 we covered the procedures for calculating the support reactions of a simply supported beam and plotting the distribution of shear force and bending moment along its length. We now need to examine the parameters that govern the stiffness and strength of a beam and the stresses that are caused by bending. A beam can be considered to be made up of an infinate number of layers, rather like the pages of a book. When bending occurs, some of the layers are in tension and some are in compression. In between these is a layer which, although bent like the others, is in neither tension or compression. It is called the *neutral layer* or *neutral axis* of the beam. In most cases it passes through or is very close to the centroid of its cross-section.

Second moment of area

There are two parameters which determine the stiffness of a beam and its resistance to bending. One of these will be quite familiar to you. It is the *modulus of elasticity* of the beam material. The greater the modulus of elasticity, the greater the stiffness and resistance to bending. Suppose that you have two beams with the same dimensions, one of mild steel whose modulus of elasticity is 200 GPa and one of an aluminium alloy whose modulus of elasticity is 100 GPa. These values indicate that the alloy beam has only half the stiffness of the steel beam. If the two beams carry the same load, it will be found that the deflection of the aluminium beam is double that of the steel beam.

The other parameter concerns the shape and orientation of the beam's cross-section. It is quite easy to bend a ruler when the bending forces are applied to its flat faces. When it is turned through 90° however, and the bending forces are applied to its edges it becomes very difficult to bend. The dimensional parameter which governs the stiffness of a beam is the *second moment of area* of its cross-section, taken about the neutral axis of bending.

The second moment of area is given the symbol I and its units are m^4. Consider a rectangular section beam of width b and depth d shown in Figure 2.7. The second moment of area δI, of the elemental strip about the neutral axis is its area δA multiplied by the square of its distance x from the neutral axis:

$$\text{i.e.} \quad \delta I = x^2 \delta A$$

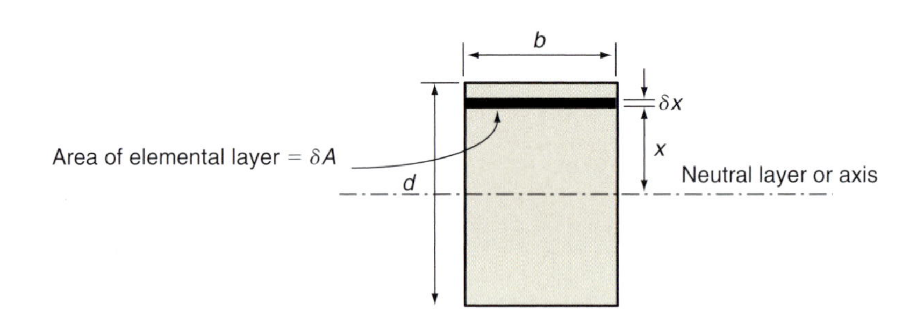

Figure 2.7 Rectangular cross-section

The total second moment of area I of the whole cross-section is the sum of all such elements:

$$\boldsymbol{I = \Sigma(x^2 \delta A)} \tag{2.16}$$

Now $\delta A = b\delta x$ and so:

$$I = \Sigma(x^2 b \delta x)$$

This sum can be found by integration between the limits of $x = +d/2$ and $x = -d/2$. As $\delta x \to 0$, the expression becomes:

$$I = \int_{-d/2}^{+d/2} x^2 b \, dx$$

$$I = b\left(\frac{x^3}{3}\right)_{-d/2}^{+d/2} = \frac{b}{3}\left(\left(\frac{+d}{2}\right)^3 - \left(\frac{-d}{2}\right)^3\right)$$

$$I = \frac{b}{3}\left(\frac{d^3}{8} + \frac{d^3}{8}\right) = \frac{b}{3}\left(\frac{d^3}{4}\right)$$

$$\boldsymbol{I = \frac{bd^3}{12}\ \mathrm{m}^4} \tag{2.17}$$

The same procedure can be applied to circular cross-sections, although the integration is a little more difficult. The formula which results for a circular section of diameter d is:

$$I = \frac{\pi d^4}{64}\ \text{m}^4 \qquad (2.18)$$

KEY POINT

The resistance to bending of a beam is directly proportional to the modulus of elasticity of the beam material and the second moment of area of its cross-section taken about an axis in the neutral layer.

Equation (2.17) shows that if you double the width b of a rectangular section beam you will double its stiffness. If however you double its depth d you will increase its stiffness by a factor of eight because of the d^3 term, i.e. $2^3 = 8$. This is why your ruler is so much more difficult to bend when the bending forces are applied to its edges. Equation (2.18) for a circular section beam shows that if you double the diameter, the stiffness will increase by a factor of sixteen because of the d^4 term, i.e. $2^4 = 16$.

Beams may of course be hollow rectangular or 'box' section and tubular section. In such cases equations (2.17) and (2.18) have to be modified as follows:

For a hollow rectangular section (Figure 2.8):

$$I = \frac{BD^3 - bd^3}{12} \qquad (2.19)$$

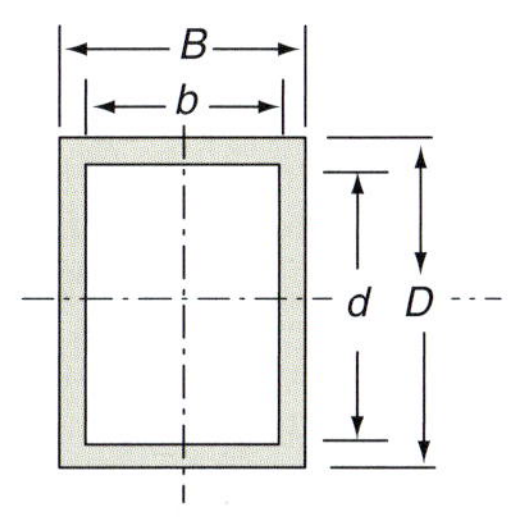

Figure 2.8

For a hollow rectangular section (Figure 2.9):

$$I = \frac{\pi(D^4 - d^4)}{64} \qquad (2.20)$$

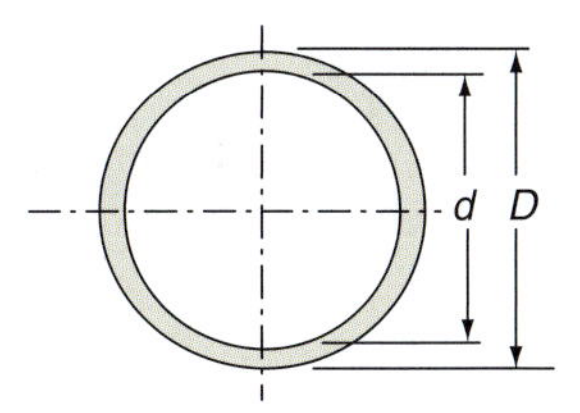

Figure 2.9

You may have noticed a similarity between the second moment of area and the moment of inertia of a rotating body that we covered in Chapter 1. The basic formulae used to calculate them are indeed very similar, i.e.

$$\text{Second moment of area} = \Sigma x^2\,\delta A$$
$$\text{Moment of inertia} = \Sigma x^2\,\delta m$$

Just like a rotating body, we sometimes talk about the radius of gyration of a beam section even though it is not rotating. You may recall that this is a root-mean square radius and it is used in the selection of standard section structural components such as columns and stanchions.

Example 2.4

Determine the second moment of area of a rectangular beam section of width 50 mm and depth 100 mm. What would be the diameter of a circular section of the same material and same stiffness? By how much would the depth of the rectangular section beam need to be increased in order to double its stiffness?

Finding second moment of area I_1 of the rectangular section about the neutral axis of bending for depth $d_1 = 100$ mm:

$$I_1 = \frac{bd_1^3}{12} = \frac{0.05 \times 0.1^3}{12}$$

$$\mathbf{I_1 = 4.17 \times 10^{-6}\ m^4}$$

Finding diameter of equivalent circular section beam:

$$I = \frac{\pi d^4}{64}$$

$$d = \sqrt[4]{\frac{64\,I}{\pi}} = \sqrt[4]{\frac{64 \times 4.17 \times 10^{-6}}{\pi}}$$

$$\mathbf{d = 0.096\ m \quad or \quad 96\ mm}$$

Finding second moment of area of rectangular section beam I_2 when stiffness is doubled, i.e. when second moment of area is doubled:

$$I_2 = 2I_1 = 2 \times 4.17 \times 10^{-6}$$

$$\mathbf{I_2 = 8.34 \times 10^{-6}\ m^4}$$

Finding new depth d_2 of the section:

$$I_2 = \frac{bd_2^3}{12}$$

$$d_2 = \sqrt[3]{\frac{12 I_2}{b}} = \sqrt[3]{\frac{12 \times 8.34 \times 10^{-6}}{0.05}}$$

$$\mathbf{d_2 = 0.0126\ m \quad or \quad 126\ mm}$$

The depth would need to be increased by **26 mm** to double the stiffness of the beam, i.e. a **26%** increase in depth.

KEY POINT

To double the stiffness of a solid rectangular section beam its depth should be increased by approximately 26% and to double the stiffness of a solid circular section beam its diameter should be increased by approximately 19%.

Test your knowledge 2.2

1. What is the *neutral layer* or *neutral axis* of a beam?
2. What are the two parameters which determine the stiffness of a beam?
3. What is the formula used to calculate the second moment of area of a rectangular section beam?
4. What is the formula used to calculate the second moment of area of a circular section beam?
5. What are the units of second moment of area?

Activity 2.2

A circular section beam of diameter 150 mm is to be replaced by a rectangular section of width 50 mm and whose material has double the modulus of elasticity value. What is the required depth of the rectangular section if the two beams are to have the same stiffness?

Stress due to bending

The bending moments that we examined in Chapter 1 give rise to tensile and compressive stresses in beams and cantilevers. These are found to be greatest at the upper and lower surfaces and zero at the neutral layer. The stress at any distance from the neutral layer can be calculated in either of two ways, depending on the information available:

1. From a knowledge of the modulus of elasticity of the beam material and the radius of curvature of bending.
2. From a knowledge of the bending moment and second moment of area of the beam cross-section about an axis across the neutral layer.

We make the following assumptions:

- The beam is initially straight and the cross-section is symmetrical about the plane of bending.
- Bending takes place in the plane of bending and plane transverse sections remain plane after bending.
- The stress at any layer is uniform across the width of a beam and each layer is assumed to be free of interference from adjacent layers.
- The beam material is homogenous and isotropic and has the same modulus of elasticity in tension and compression.
- The material obeys Hooke's law and the stresses do not exceed the elastic limit.

Stress in terms of curvature

Figure 2.10 shows a portion of a beam which is initially straight and is then bent by the application of a bending moment.

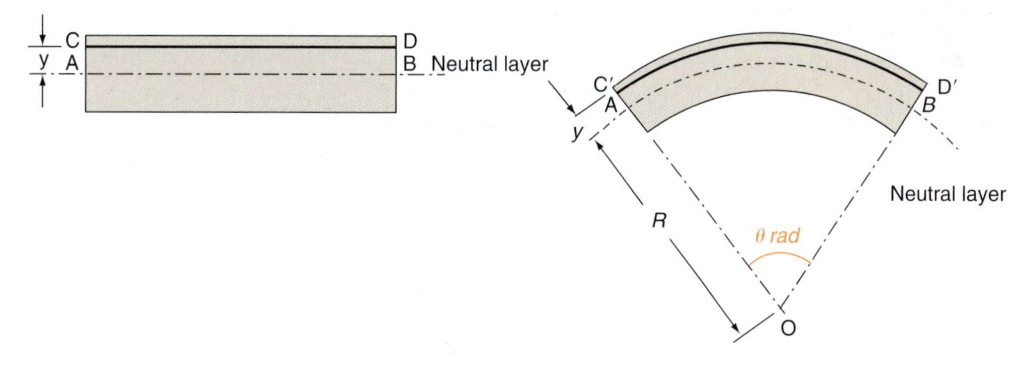

Figure 2.10 Beam element

A bending moment M bends the beam element into a circular arc of radius R measured to the neutral layer. As can be seen, the upper layers are in tension and the lower layers are in compression. The thin layer whose initial length is CD at distance y from the neutral layer AB is in tension and after bending this is extended to C′D′. Although it is bent, the length of the neutral layer AB is unchanged. The angle subtended by the element at the centre of curvature O is θ radians.

Consider now the extension of the layer CD:

Initial length of layer $= \text{CD} = \text{AB} = R\theta$
Final length of layer $= \text{C}'\text{D}' = (R + y)\theta$
Extension of layer $= (R + y)\theta - R\theta = R\theta + y\theta - R\theta$
Extension of layer $= y\theta$

The strain in the layer CD is given by:

$$\text{Strain} = \frac{\text{Extension}}{\text{Initial length}}$$

$$\varepsilon = \frac{y\theta}{R\theta}$$

$$\varepsilon = \frac{y}{R}$$

Now the stress in the layer can be found from the formula for modulus of elasticity:

KEY POINT

The formulae derived to calculate the stress in a beam assume that the material obeys Hooke's law and that the elastic limit stress has not been exceeded.

$$E = \frac{\sigma}{\varepsilon}$$

$$\sigma = E\varepsilon = E\frac{y}{R}$$

or $$\boldsymbol{\frac{\sigma}{y} = \frac{E}{R}} \quad (2.21)$$

Stress in terms of bending moment

Consider now the tensile force dF acting in the same layer, distance y from the neutral layer, width b and thickness dy. This will be the tensile stress σ multiplied by the cross-sectional area of the layer.

The moment of this force dM, about the neutral layer will be:

$$\mathrm{d}M = \mathrm{d}F\,y$$
$$\mathrm{d}M = \sigma b\,\mathrm{d}y\,y$$

Now from equation (2.21) the stress is given by $\sigma = Ey/R$. Substituting gives:

$$\mathrm{d}M = \frac{Eyb\,\mathrm{d}y\,y}{R}$$

$$\mathrm{d}M = \frac{Eby^2\,\mathrm{d}y}{R}$$

The total of all such elemental moments is called the *internal moment of resistance* of the beam which is equal to, and balances the external bending moment:

$$M = \int \frac{Eby^2\,\mathrm{d}y}{R}$$

$$M = \frac{E}{R}\int by^2\,\mathrm{d}y$$

Now $\int by^2 \mathrm{d}y = I$ the second moment of area of the beam section about the neutral layer and so:

$$M = \frac{EI}{R}$$

$$\text{or} \quad \frac{M}{I} = \frac{E}{R} \tag{2.22}$$

It is usual to combine equations (2.21) and (2.22) into a single three-part formula that is known as the *bending equation*:

$$\frac{\sigma}{y} = \frac{M}{I} = \frac{E}{R} \tag{2.23}$$

Position of neutral layer

Although tension and compression forces are both present, they will cancel each other out so that there is no net axial or horizontal load on the beam. Consider now the cross-section with the same layer whose width is b and thickness is $\mathrm{d}y$ as shown in Figure 2.11.

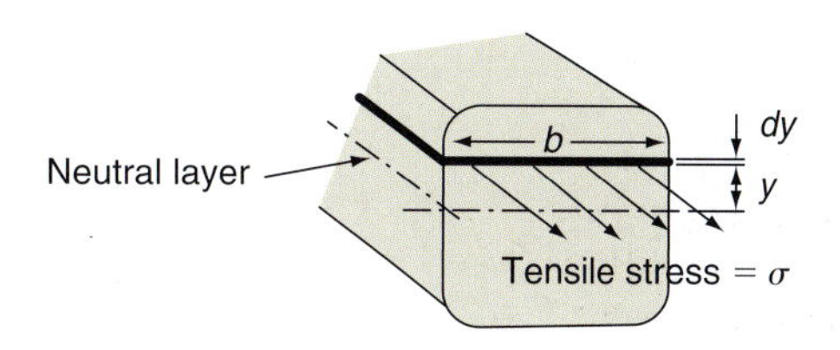

Figure 2.11 Stress in layer

The axial force on the layer is again given by:

$$\mathrm{d}F = \sigma b\,\mathrm{d}y$$

But the stress is given by $\sigma = Ey/R$ giving:

$$\mathrm{d}F = \frac{Eyb\,\mathrm{d}y}{R}$$

The total axial force is the sum of all such elements:

$$F = \int \frac{Eyb\,\mathrm{d}y}{R}$$

But we have already said that this is zero and so:

$$0 = \int \frac{Eyb\,\mathrm{d}y}{R}$$

Now E and R are constants, in which case:

$$0 = \int yb\,\mathrm{d}y$$

The product $yb\,\mathrm{d}y$ is the first moment of area of the layer about the neutral layer. The total of these, above and below the neutral layer, is the total first moment of the whole cross-section and this can only be zero if the neutral layer passes through the centroid of the section.

KEY POINT

Provided that the assumptions regarding the beam properties are valid, the neutral layer of bending will pass through the centroid of a section.

Example 2.5

A simply supported beam of hollow rectangular cross-section is 120 mm deep and 60 mm wide with a wall thickness of 10 mm. The beam has a span of 2 m and carries a concentrated load of 200 kN at its centre. Neglecting the weight of the beam itself, determine: (a) the maximum bending moment, (b) the maximum stress in the material, (c) the factor of safety in operation against an elastic limit stress of 300 MPa and (d) the minimum radius of curvature.

Take $E = 200\,\text{GPa}$.

(a) Finding maximum bending moment (Figure 2.12):

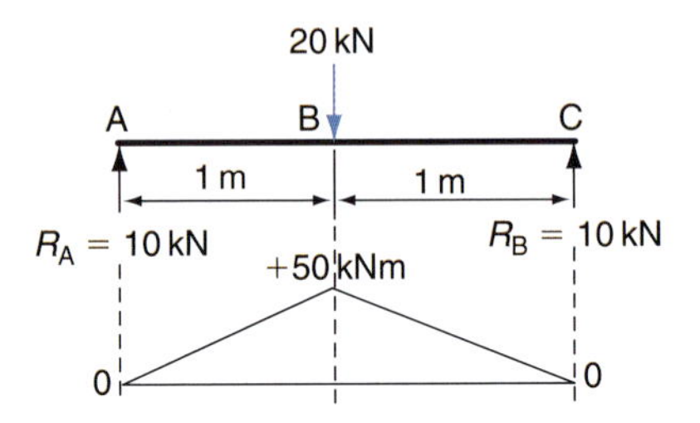

Figure 2.12

$$\text{BM at A} = 0$$
$$\text{BM at C} = +10 \times 1 = +10\,\text{kNm}$$
$$\text{BM at C} = 0$$
$$\boldsymbol{Max\ \text{BM} = +10\,\text{kNm at C}}$$

(b) Finding second moment of area about NA (Figure 2.13):

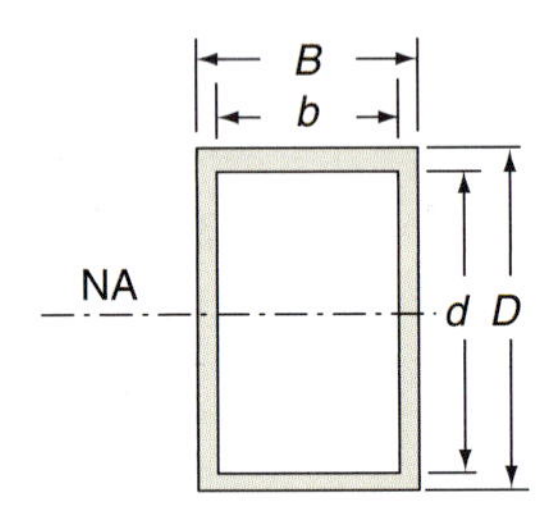

Figure 2.13

$$I = \frac{BD^3 - bd^3}{12} = \frac{(0.06 \times 0.12^3) - (0.05 \times 0.1^3)}{12}$$
$$\mathbf{I = 4.47 \times 10^{-6}\ m^4}$$

Finding maximum stress in material, i.e. at point of maximum bending moment where $y = 60\,\text{mm}$, at top and bottom faces of the beam:

$$\frac{\sigma}{y} = \frac{M}{I}$$
$$\sigma = \frac{My}{I} = \frac{10 \times 10^3 \times 0.06}{4.47 \times 10^{-6}}$$
$$\boldsymbol{\sigma = 134 \times 10^6\ \text{Pa} \quad \text{or} \quad 134\ \text{MPa}}$$

The stress distribution is as in Figure 2.14.

You should note that the stress is proportional to distance from the neutral layer irrespective of whether the section is hollow or solid (Figure 2.14).

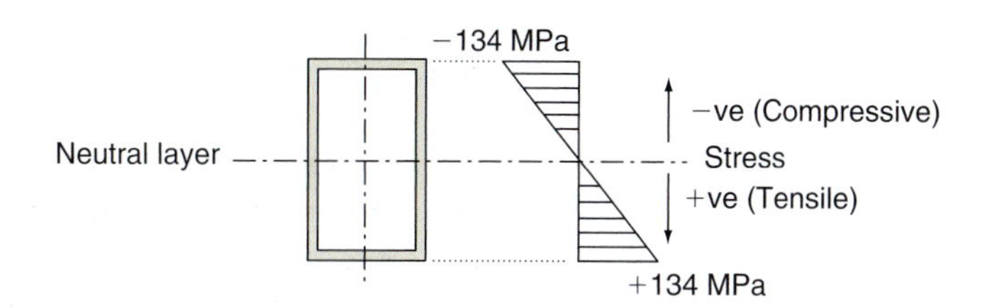

Figure 2.14

(c) Finding factor of safety in operation:

$$\text{FOS} = \frac{\text{Elastic limit stress}}{\text{Max stress in beam}} = \frac{300}{134}$$

$$\mathbf{FOS = 2.24}$$

(d) Finding minimum radius of curvature, i.e. at centre where bending moment is greatest:

$$\frac{M}{I} = \frac{E}{R}$$

$$R = \frac{EI}{M} = \frac{200 \times 10^9 \times 4.47 \times 10^{-6}}{10 \times 10^3}$$

$$\mathbf{R = 89.4\,m}$$

Test your knowledge 2.3

1. What is the internal moment of resistance of a beam?
2. How do you calculate stress in terms of modulus of elasticity and radius of curvature?
3. How do you calculate stress in terms of bending moment and second moment of area of cross-section?
4. How is the stress due to bending distributed over a beam cross-section?
5. Where along the span of a loaded beam would you expect to find the maximum value of stress due to bending?

Activity 2.3

A steel tube of outer diameter 200 mm and inner diameter 180 mm is used as a simply supported beam over a span of 3 m, carrying a uniformly distributed load of 4 kN m^{-1}. Neglecting the weight of the beam itself, determine: (a) the maximum bending moment, (b) the maximum stress in the material, (c) the factor of safety in operation against an allowed maximum of 250 MPa for the material and (d) the minimum radius of curvature. The modulus of elasticity of the steel is 205 GPa.

To check your understanding of the preceding section, you can solve Review questions 7–12 at the end of this chapter.

Experiment to determine the modulus of elasticity of a beam material

Apparatus

Simply supported beam apparatus with knife-edge or roller supports, metre rule, dial-test indicator and stand, micrometer, hangers and weights, rectangular-section beam to be tested (Figure 2.15).

Sketch

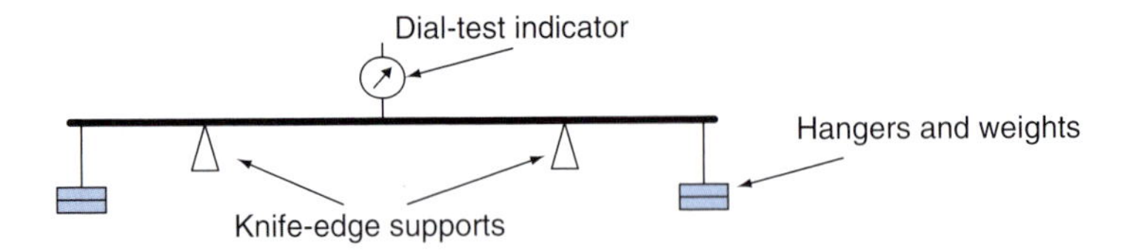

Figure 2.15

Procedure

1. Use the micrometer to measure the breadth b and depth d of the beam section at three different points along its length and calculate the average values and second moment of area about the neutral axis.
2. Set up the apparatus with the beam unloaded and the dial test indicator centrally positioned and set to read zero.
3. Record the distance between the supports $2l$, and the equal overhang distance a, from the supports to the weight hangers.
4. Carefully apply equal loads W to the hangers and record the central deflection y that they produce.
5. Calculate the radius of curvature between the supports and the modulus of elasticity of the beam material using the formulae derived in the following theory section.

Theory

The shear force and bending moment diagrams for the beam are as shown in Figure 2.16.

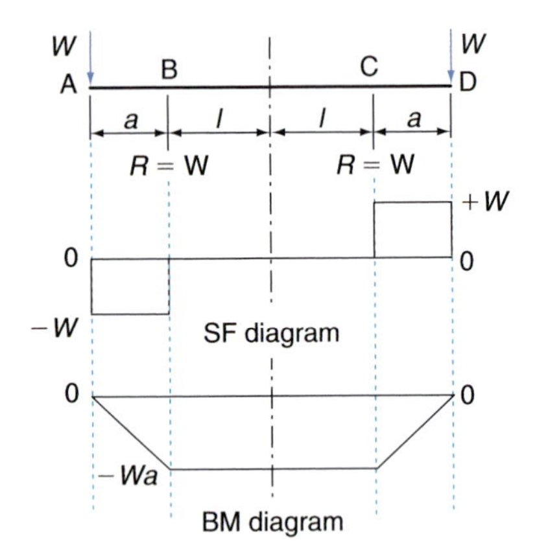

Figure 2.16

Because the beam is loaded symmetrically the reactions are equal to the loads:

SF from A to B $= -W$
SF from B to C $= 0$
SF from C to D $= +W$

BM at A $= 0$
BM from B to C $= -Wa$
BM at D $= 0$

The bending moment is constant between B and C indicating that the radius of curvature is also constant, i.e. the beam bends into a circular arc of radius R between B and C.

Consider the deflected shape of the beam between the supports B and C (Figure 2.17).

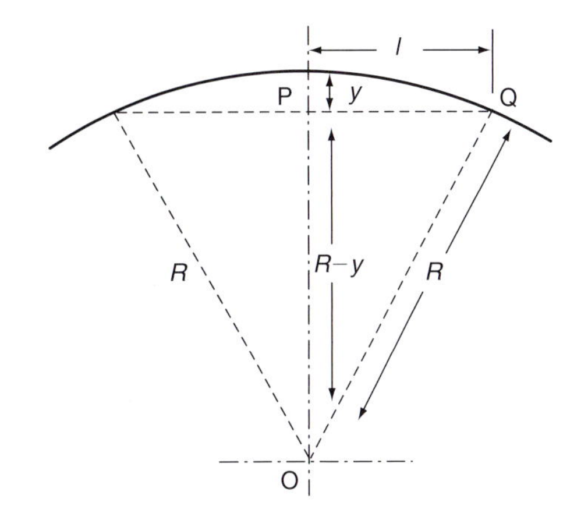

Figure 2.17

Applying Pythagoras' theorem to length OPQ gives:

$$R^2 = (R - y)^2 + l^2$$
$$\cancel{R}^2 = \cancel{R}^2 - 2Ry + y^2 + l^2$$
$$0 = -2Ry + y^2 + l^2$$
$$2Ry = y^2 + l^2$$
$$\boldsymbol{R = \frac{y^2 + l^2}{2y}} \qquad \text{(i)}$$

The bending moment between the supports is Wa and second moment of area of the section is given by:

$$\boldsymbol{I = \frac{bd^3}{12}} \qquad \text{(ii)}$$

Having calculated the radius of curvature and second moment of area the modulus of elasticity can be found from the bending equation, i.e.:

$$\frac{M}{I} = \frac{E}{R}$$
$$\boldsymbol{E = \frac{MR}{I}} \qquad \text{(iii)}$$

Torsion in Power Transmission Shafts

In Chapter 1 we covered the procedures for calculating shear stress and strain in static structural components. Shear stress also occurs in power transmission shafts such as those found in motor vehicles, ships and process machinery. As with bending in beams, we need to examine the parameters that govern the stiffness and strength of a shaft when

subjected to twisting. A circular section shaft can be considered to be made up of an infinate number of annular cylinders or tubes, packed tightly together. When twisting or torsion occurs, shearing forces are set up. The shear stress that they induce rises uniformly from zero at the central axis to a maximum at the outer surface. The central axis can thus be considered as the *neutral* axis rather like the neutral layer that we have encountered in beams.

Polar second moment of area

As with beams, there are two parameters which determine the stiffness of a shaft and its resistance to torsion. One of these is the *shear modulus* of the beam material which is sometimes also called the *modulus of rigidity*. The greater the modulus of rigidity, the greater the stiffness and resistance to twisting.

The other parameter concerns the dimensions of the shaft cross-section but it is not its cross-sectional area however. A solid shaft and a hollow tubular shaft may be made from the same amount of material and have the same length and cross-sectional area. The hollow shaft will however have the greater resistance to twisting. The dimensional parameter to which the stiffness of a shaft is proportional is the *polar second moment of area* of its cross-section, taken about its central axis.

The polar second moment of area is given the symbol J and its units are again m^4. Consider the polar second moment of area δJ of the elemental ring shown in Figure 2.18 about the centre O.

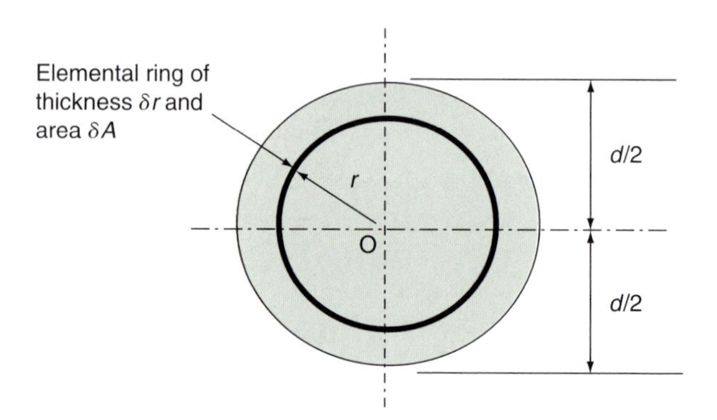

Figure 2.18 Elemental ring

$$\delta J = \delta A\, r^2$$

The total second moment of area J, of the whole cross-section is the sum of all such elements

$$\boldsymbol{J = \Sigma(r^2 \delta A)} \qquad (2.24)$$

Now $\delta A = 2\pi r\,\delta r$ and so

$$J = \Sigma(r^2 2\pi r \delta r)$$
$$J = \Sigma(2\pi r^3 \delta r)$$

This sum can be found by integration between the limits of $r = d/2$ and $r = 0$ as $\delta r \to 0$, the expression becomes:

$$J = \int_0^{d/2} 2\pi r^3 \, \mathrm{d}r$$

$$J = 2\pi \left(\frac{r^4}{4} \right)_0^{d/2} = \frac{\pi}{2} \left(\left(\frac{+d}{2} \right)^4 - 0 \right)$$

$$J = \frac{\pi}{2} \left(\frac{d^4}{16} \right)$$

$$\boldsymbol{J = \frac{\pi d^4}{32} \ \mathrm{m}^4} \qquad (2.25)$$

KEY POINT

The resistance to twisting in a shaft is directly proportional to the shear modulus of the shaft material and the second moment of area of its cross-section taken about the central polar axis.

In the case of a hollow shaft of outer diameter D and inner diameter d this is modified to:

$$\boldsymbol{J = \frac{\pi(D^4 - d^4)}{32} \ \mathrm{m}^4} \qquad (2.26)$$

Example 2.6

Compare the weight per metre length and torsional strength of a solid shaft 60 mm diameter with that of a hollow shaft of the same outer diameter and inner diameter 42.4 mm. Both shafts are made from the same material.

Finding cross-sectional area of solid shaft:

$$A_1 = \frac{\pi D^2}{4} = \frac{\pi \times 0.06^2}{4}$$

$$\boldsymbol{A_1 = 2.83 \times 10^{-3}\ \mathrm{m}^2}$$

Finding cross-sectional area of hollow shaft:

$$A_2 = \frac{\pi(D^2 - d^2)}{4} = \frac{\pi(0.06^2 - 0.0424^2)}{4}$$

$$\boldsymbol{A_2 = 1.42 \times 10^{-3}\ \mathrm{m}^2}$$

Finding percentage saving in weight (weight is directly proportional to cross-sectional area):

$$\%\text{ saving in weight} = \frac{(A_1 - A_2)}{A_1} \times 100 = \frac{(2.83 - 1.42)}{2.83} \times 100$$

$$\boldsymbol{\%\text{ saving in weight} = 49.8\%}$$

Finding polar second moment of area of solid shaft:

$$J_1 = \frac{\pi D^4}{32} = \frac{\pi \times 0.06^4}{32}$$

$$\boldsymbol{J_1 = 1272 \times 10^{-6}\ \mathrm{m}^2}$$

Finding second moment of area of hollow shaft:

$$J_2 = \frac{\pi(D^4 - d^4)}{32} = \frac{\pi(0.06^4 - 0.0424^4)}{32}$$

$$\boldsymbol{J_2 = 955 \times 10^{-9}\ \mathrm{m}^2}$$

CHAPTER 2

Finding percentage loss in torsional strength (torsional strength is directly proportional to polar second moment of area):

$$\% \text{ loss in torsional strength} = \frac{(J_1 - J_2)}{J_1} \times 100 = \frac{(1272 - 955)}{1272} \times 100$$

% loss in torsional strength = 24.9%

That is for a hollow shaft with this internal diameter the percentage saving in weight is double the percentage loss in torsional strength.

Activity 2.4

Taking Example 2.6 as a model, compare the effect of increasing the inner diameter of a hollow shaft on the percentage saving in weight and percentage loss on torsional strength. Plot graphs of percentage saving in weight and percentage loss of strength against inner diameter size and comment on the variations.

Shear stress due to torsion

In Chapter 1 we investigated shear stress and strain in structural components. These also occur in power transmission shafts which are subjected to twisting or torsion. The shear stress is found to be greatest at the outer surface and zero at the central polar axis. The shear stress at any distance from the central axis can be calculated in either of two ways, depending on the information available:

1. From a knowledge of the shear modulus of the beam material and the angle of twist when transmitting power.
2. From a knowledge of the torque transmitted and the second moment of area of the beam cross-section about the central polar axis.

We make the following assumptions:

- The shaft is initially straight and of solid or hollow circular cross-section.
- Twisting takes place along the axis of the shaft and radial lines remain radial after twisting.
- The shaft can be assumed to be made up of an infinite number of concentric annular tubes which do not interfere with each other and around which the shear stress is uniform.
- The shaft material is homogenous and isotropic.
- The shaft material obeys Hooke's law and the shear stress does not exceed the elastic limit in shear.

Shear stress in terms of angle of twist and shear modulus

Consider one of the annular tubes of which a shaft is assumed to be composed (Figure 2.19).

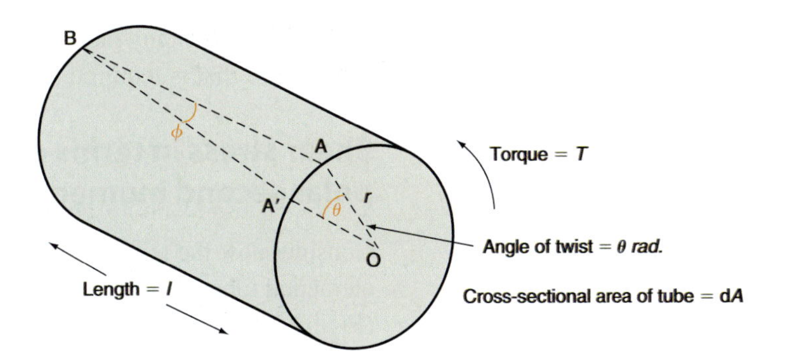

Figure 2.19 Elemental tube in torsion

During twisting the point A on the tube moves to A' because the angles θ and ϕ are small, it can be written that the distance $A'A$ is given by:

$$A'A = r\theta$$

Now if the tube were opened out flat, the material would be seen to be distorted from a rectangle to a parallelogram as shown in Figure 2.20.

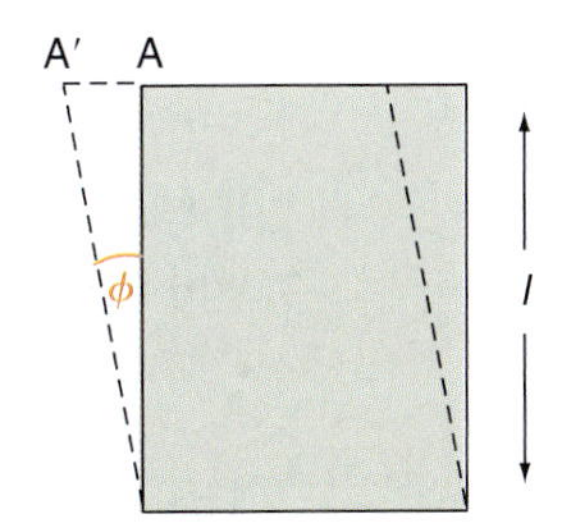

Figure 2.20 Effect of shear

The shear strain in the material is given by:

$$\gamma = \frac{A'A}{l}$$

$$\gamma = \frac{r\theta}{l}$$

The formula for shear modulus G can now by used to find the shear stress τ in the tube:

$$\text{Shear modulus} = \frac{\text{Shear stress}}{\text{Shear strain}}$$

$$G = \frac{\tau}{\gamma}$$

$$\tau = G\gamma$$

$$\tau = \frac{Gr\theta}{l}$$

This is more often written in the form:

$$\frac{\tau}{r} = \frac{G\theta}{l} \qquad (2.27)$$

KEY POINT

The shear stress in a transmission shaft increases uniformly from zero at the central axis to a maximum at the outer surface.

Because a solid shaft is considered to be made up of an infinite number of such elemental tubes, the above expression can be used to calculate the shear stress at any radius r. Generally it is the maximum value of shear stress that is required, which occurs at the outer radius.

Shear stress in terms of applied torque and polar second moment of area

Consider now the tangential shearing force acting on one of the elemental tubes that are assumed to make up a power transmission shaft (Figure 2.21).

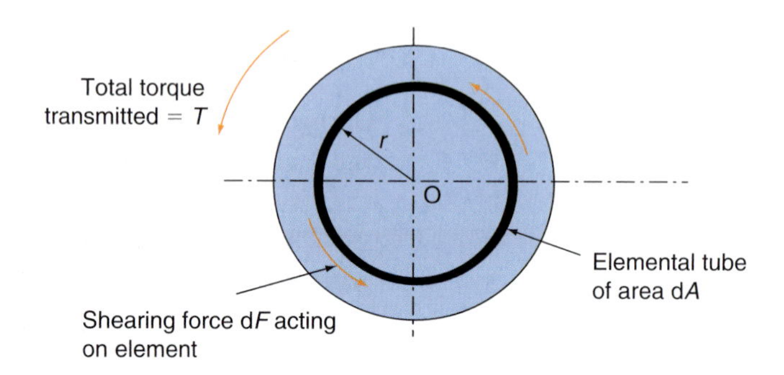

Figure 2.21 Shaft cross-section

If τ is the shear stress acting on the elemental tube, the shearing force dF will be:

$$dF = \tau dA$$

The elemental torque dT, transmitted by the tube is:

$$dT = r\,dF$$
$$dT = r\tau\,dA$$

The total torque transmitted is the sum of all such elements:

$$T = \int r\tau\,dA$$

Now from equation (2.27) the shear stress is given by $\tau = Gr\theta/l$

$$T = \int \frac{rGr\theta\,dA}{l}$$
$$T = \frac{G\theta}{l}\int r^2\,dA$$

But r^2 dA = J the polar second moment of area of the shaft, and so:

$$T = \frac{G\theta J}{l}$$
$$\frac{T}{J} = \frac{G\theta}{l} \qquad (2.28)$$

Equations (2.27) and (2.28) are generally combined together in a three-part formula that is known as the *torsion equation*:

$$\frac{\tau}{r} = \frac{T}{J} = \frac{G\theta}{l} \qquad (2.29)$$

KEY POINT

The conditions that must generally be satisfied in the design of transmission shafts are that the shear stress must not exceed a specified value and the angle of twist or *wind-up* must not exceed a specified value.

Power transmitted

As was stated in Chapter 1, the power transmitted in rotational motion is given by:

$$\textbf{Power} = T\omega \tag{2.30}$$

where ω rad s^{-1} is the angular velocity of rotation. Provided the power and transmission speed are known, the torque T can be calculated from equation (2.30) for use in the torsion equation.

Example 2.7

A hollow power transmission shaft of outer diameter 60 mm and inner diameter 30 mm is 3 m in length. The shaft is required to transmit a power of 150 kW at a speed of 1200 rpm. The shear modulus of the shaft material is 85 GPa. Determine (a) the maximum value of the shear stress induced and (b) the angle of twist along the shaft length measured in degrees.

(a) Finding angular velocity of shaft:

$$\omega = N\frac{2\pi}{60} = 1200 \times \frac{2\pi}{60}$$

$$\boldsymbol{\omega = 126\,\text{rads}^{-1}}$$

Finding torque transmitted:

$$\text{Power} = T\omega$$

$$T = \frac{\text{Power}}{\omega} = \frac{150 \times 10^3}{126}$$

$$\boldsymbol{T = 1190\,\text{Nm}}$$

Finding polar second moment of area of shaft:

$$J = \frac{\pi(D^4 - d^4)}{32} = \frac{\pi(0.06^4 - 0.03^4)}{32}$$

$$\boldsymbol{J = 1.19 \times 10^{-6}\,\text{m}^4}$$

Finding maximum value of shear stress, i.e. at outer surface where $r = 30$ mm:

$$\frac{\tau}{r} = \frac{T}{J}$$

$$\tau = \frac{Tr}{J} = \frac{1190 \times 0.03}{1.19 \times 10^{-6}}$$

$$\boldsymbol{\tau = 30 \times 10^6\,\text{Pa} \quad \text{or} \quad 30\,\text{MPa}}$$

(b) Finding angle of twist:

$$\frac{T}{J} = \frac{G\theta}{l}$$

$$\theta = \frac{Tl}{JG} = \frac{1190 \times 3}{(1.19 \times 10^{-6})(85 \times 10^9)}$$

$$\boldsymbol{\theta = 0.0353\,\text{rad}}$$

Change to degrees:

$$\theta^\circ = \theta\,\text{rad} \times \frac{360}{2\pi} = 0.0353 \times \frac{360}{2\pi}$$

$$\boldsymbol{\theta = 2.02^\circ}$$

Test your knowledge 2.4

1. What is the expression for polar second moment of area of a shaft?
2. Where does the maximum shear stress occur in a power transmission shaft?
3. What is the torsion equation?
4. What would be the advantage of replacing a solid shaft with a hollow one of the same mass and material?
5. What are the two conditions that may need to be satisfied when designing transmission shafts?

Activity 2.5

A solid power transmission shaft of length 2 m is required to transmit 600 kW at a speed of 2500 rpm. Determine the diameter of the shaft if the ultimate shear stress in the material is 350 MPa and a factor of safety of 10 is to apply. Determine also the angle of twist in degrees along the length of the shaft when transmitting this power. The shear modulus of the shaft material is 85 GPa.

To check your understanding of the preceding section, you can solve Review questions 13–18 at the end of this chapter.

Experiment to find the shear modulus of a shaft material

Apparatus

Torsion test apparatus, metre rule, micrometer, cord, hanger and weights (Figure 2.22).

Sketch

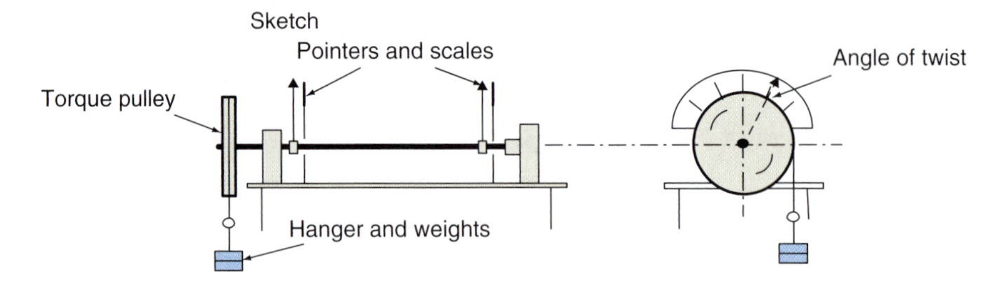

Figure 2.22

Procedure

1. Measure and record the distance l between the pointers and the radius of the torque pulley.
2. Measure the diameter of the shaft at three different linear and angular positions between the pointers and record the average value of its diameter d and radius r.
3. Calculate the polar second moment of area J of the shaft.
4. Set the pointers to read zero on the angular deflection scales.
5. Carefully add increasing weights to the hanger and observe the difference between the pointer readings for each increment.
6. Calculate and tabulate the torque T and the angle of twist θ rad, for each set of readings.

7. Plot a graph of torque against angle of twist and calculate its gradient.
8. Calculate the value of the shear modulus using the expression derived in the following theory.

Theory

The torsion equation gives:

$$\frac{G\theta}{l} = \frac{T}{J}$$

$$\text{or} \quad \frac{T}{\theta} = \frac{GJ}{l} \qquad \text{(i)}$$

Plotting torque T, against angle of twist θ should give a straight line graph of gradient T/θ, i.e.:

$$\text{Gradient of graph} = \frac{GJ}{l}$$

$$\mathbf{G = Gradient\ of\ graph} \times \frac{l}{J} \qquad \text{(ii)}$$

Resultant and Relative Velocity

Resultant velocity

A body may have two or more different velocities at the same time. If you have worked on a centre lathe you may have traversed the carriage and cross-slides at the same time in order to move the cutting tool directly to a new position. A boat steering directly across a river will have a velocity at right angles to the bank and also a velocity in the direction of flow due to the current. As a result the boat will not arrive directly across on the opposite bank, but somewhere downstream. The same will happen to an aircraft flying in a cross-wind. A steersman or pilot will of course be aware of this and alter course to compensate for wind or current.

Velocity is a vector quantity. It has magnitude, which we call speed measured in m s^{-1} or km h^{-1}, and also direction. A velocity can be represented by a line on a velocity vector diagram. The length of the line, drawn to a suitable scale, represents the speed and its angular orientation represents its direction. An arrow-head can be added to show the sense of the direction but very often this is not necessary. When a body has two or more velocities at the same time, their vectors can be added together to find the resultant velocity, as shown in Figure 2.23. The process is very similar to the one you have used to find the resultant force acting on a body.

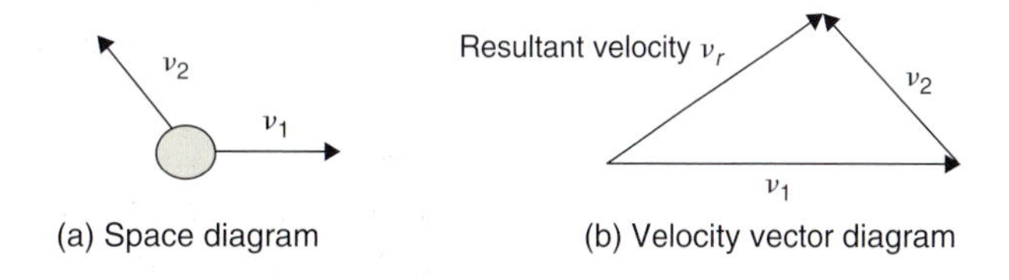

Figure 2.23 Space and vector diagrams

As a more accurate alternative to scale drawing, you can of course use trigonometry to find the resultant. If the velocities v_1 and v_2

KEY POINT

The resultant velocity of a body which has velocities in different directions can be found by velocity vector addition.

CHAPTER 2

are perpendicular, the velocity vector diagram will be a right angle triangle and Pythagoras' theorem can be used to calculate the resultant. Application of the cosine rule will however be required for examples such as that shown in Figure 2.23.

Example 2.8

A boat that can travel at 6 km h^{-1} in still water is steered directly across a river of width 50 m flowing at 4 km h^{-1}. (a) What will be the resultant velocity of the boat and how for down stream will it arrive at the opposite bank? (b) In what direction should the boat be steered in order to cross at right angles and how long would the crossing take?

(a) Finding resultant velocity when steered directly across river (Figure 2.24):

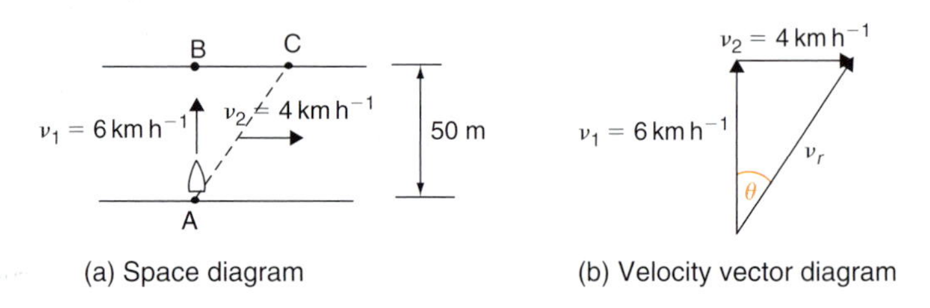

Figure 2.24

By Pythagoras,

$$v_r = \sqrt{v_1^2 + v_2^2} = \sqrt{6^2 + 4^2}$$

$$\mathbf{v_r = 7.21 kmh^{-1}}$$

Finding angle θ of path, the boat follows:

$$\text{Tan } \theta = \frac{v_2}{v_1} = \frac{4}{6} = 0.667$$

$$\boldsymbol{\theta = 33.7°}$$

Finding distance BC, boat is swept down stream:

$$BC = AB \text{ Tan } \theta = 50 \times 0.667$$

$$\mathbf{BC = 33.4 m}$$

(b) Finding resultant velocity in direction directly across river (Figure 2.25):

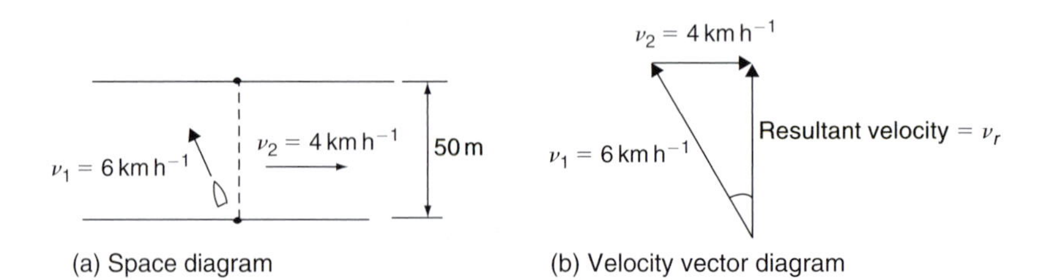

Figure 2.25

By Pythagoras,

$$v_r = \sqrt{v_1^2 - v_2^2} = \sqrt{6^2 - 4^2}$$

$$\mathbf{v_r = 4.47 kmh^{-1}} \quad \text{or} \quad \mathbf{1.24 ms^{-1}}$$

Finding angle θ

$$\text{Sin}\,\theta = \frac{v_2}{v_1} = \frac{4}{6} = 0.667$$
$$\theta = 41.8°$$

Finding time taken to cross:

$$t = \frac{s}{v_r} = \frac{50}{1.24} = 40.3\,\text{s}$$
$$\mathbf{t = 40.3\,s}$$

Relative velocity

The linear velocities of cars, ships aircraft, etc., are usually measured relative to the earth's surface that is assumed to be stationary. The earth is itself of course moving but we need a point of reference and for purposes other than space travel we generally assume it to be fixed. Very often we measure the velocity of a body relative to another which is itself moving. Figure 2.26 shows two bodies moving in the same direction. These could be cars on a motorway. When viewed from the rear car, the car in front appears to be leaving it behind not at its true velocity, i.e. relative to the earth, but at a velocity which is the difference between those of the two cars.

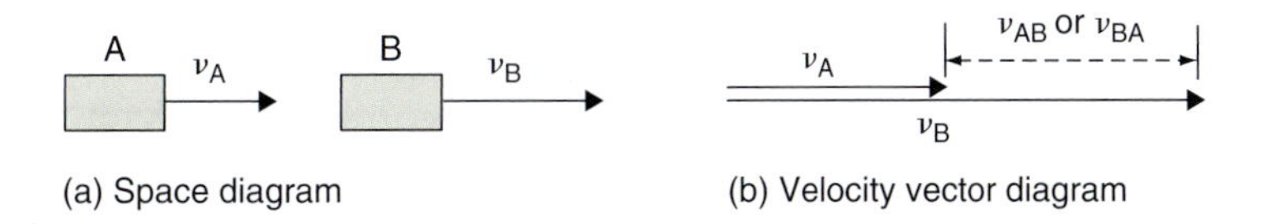

Figure 2.26 Space and vector diagrams

This is an example of vector subtraction. The velocity at which B is seen to be moving ahead from A, i.e. the velocity of B relative to A, is written as v_{BA}. In the same way the velocity at which A is seen to be falling behind from B, i.e. the velocity of A relative to B, is written as v_{AB}. Both have the same value but will be opposite in direction.

We can use the same vector subtraction method when the bodies are moving in different directions as shown in Figure 2.27.

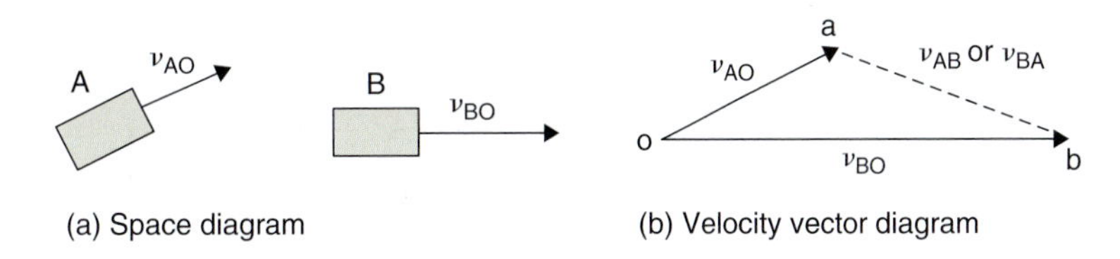

Figure 2.27 Space and vector diagrams

As you will see, the velocities have now been called v_{AO} and v_{BO}. The second letter of the subscript O, indicates the point relative to which these velocities are measured. Because v_{AO} and v_{BO} are true velocities, you can think of O, as being any point on the earth's surface. Also you will notice that the vectors for v_{AO} and v_{BO} have got the lower case letter o, at the start and the lower case letters a and b at their other ends.

These letters oa and ob give the direction of the velocities on the vector diagram, so there really is no need for the arrow heads.

The velocity of B as seen from A and the velocity of A as seen from B are again given by v_{AB} and v_{BA}. That is, the vector ab or ba depending which way round it is considered.

KEY POINT

The relative velocity between two bodies moving at different speeds can be found by velocity vector subtraction.

Example 2.9

An aircraft travelling northwest at 600 km h^{-1} passes behind another travelling due east at 500 km h^{-1}. What is the velocity of the second aircraft relative to the first?

Finding v_{BA} using cosine rule (Figure 2.28):

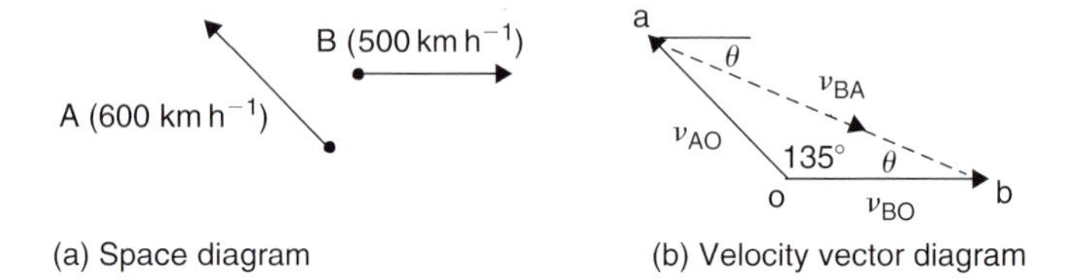

Figure 2.28

$$V_{BA}^2 = v_{AO}^2 + v_{BO}^2 - (2v_{AO}v_{BO}\ Cos\ 135°)$$
$$V_{BA}^2 = 600^2 + 500^2 - (2 \times 600 \times 500 \times Cos\ 135°)$$
$$V_{BA}^2 = 1.013 \times 10^6$$
$$\mathbf{V_{BA} = 1017\,km h^{-1}}$$

Finding angle θ using sine rule:

$$\frac{v_{AO}}{\sin\theta} = \frac{v_{BA}}{\sin 135°}$$
$$\sin\theta = \frac{v_{AO}\sin 135°}{v_{BA}}$$
$$\sin\theta = \frac{600 \times \sin 135°}{1017} = 0.4172$$
$$\mathbf{\theta = 24.7°}$$

That is the velocity of aircraft B as seen from A is 1017 km h^{-1} in a direction 24.7° south of east.

Example 2.10

Two ships are steaming towards each other. Ship A is travelling at 20 km h^{-1} and ship B is travelling at 15 km h^{-1}. When they are 1 km apart, ship B alters course by 30°. Determine the closest distance between the ships as they pass each other and the time taken to reach this position.

Finding v_{BA} using cosine rule (Figure 2.29):

$$v_{BA}^2 = v_{AO}^2 + v_{BO}^2 - (2v_{AO}v_{BO}\ \text{Cos}\ 150°)$$
$$v_{BA}^2 = 20^2 + 15^2 - (2 \times 20 \times 15 \times \text{Cos}\ 150°)$$
$$v_{BA}^2 = 1.145 \times 10^3$$
$$\mathbf{v_{BA} = 33.8\,km h^{-1}}$$

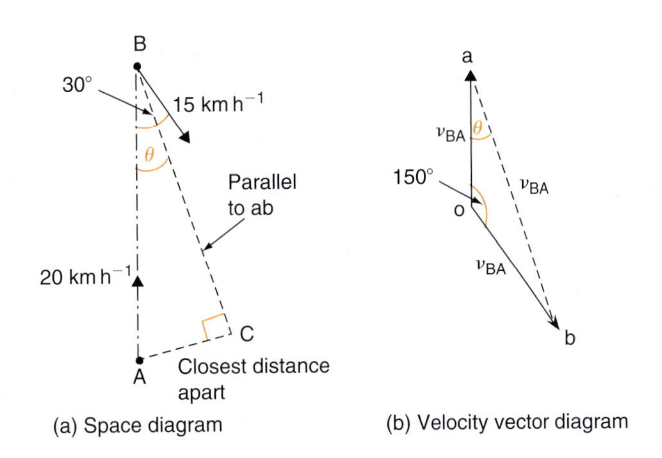

Figure 2.29

Finding angle θ using sine rule:

$$\frac{v_{AO}}{\sin\theta} = \frac{v_{BA}}{\sin 150°}$$

$$\sin\theta = \frac{v_{AO}\sin 135°}{v_{BA}}$$

$$\sin\theta = \frac{20 \times \sin 150°}{33.8} = 0.2959$$

$$\boldsymbol{\theta = 17.2°}$$

Finding closest distance of approach, i.e. distance AC:

$$AC = AB\sin\theta = 1000 \times 0.2959$$

$$\mathbf{AC = 296\ m}$$

Finding distance BC:

$$BC = AB\cos\theta = 1000 \times 0.955$$

$$\mathbf{BC = 955\ m \quad or \quad 0.955\,km}$$

Finding time t, to reach closest distance of approach, i.e. time to travel distance BC at the relative velocity v_{BA}:

$$t = \frac{BC}{v_{BA}} = \frac{0.955}{33.8}$$

$$\mathbf{t = 0.0283\,h \quad or \quad 1\,min\ 42\,s}$$

Test your knowledge 2.5

1. What is a vector quantity?
2. How are velocity vectors added?
3. How do you find the resultant velocity when a body has two velocities in two directions at the same time?
4. How are velocity vectors subtracted?
5. What is relative velocity between bodies moving in different directions?
6. How is the direction of a velocity vector indicated by the notation on a velocity vector diagram?

Activity 2.6

Two airports lie on a north–south line 100 km apart. An aircraft *A*, takes off from the southerly airport and flies due north at 200 km h^{-1}. At the same time a second aircraft *B*, takes off from the northerly airport and steers a course 30° east of south flying at 300 km h^{-1}. Determine (a) the closest approach distance of the two aircraft, (b) the time taken to reach this position.

To check your understanding of the preceding section, you can solve Review questions 19–25 at the end of this chapter.

Plane Linkage Mechanisms

Machines are devices in which the input work, or energy, is converted into a more useful form in order to do a particular job of work. The screw jacks, pulley systems and gear winches, which we have already examined in Chapter 1, are examples of simple machines. With these the input power, which may be manual or from a motor, is converted into a lifting force.

Machines may contain a number of parts. Levers, transmission shafts, pulleys, lead screws, gears, belts and chains are all typical machine components. Machines may also contain rigid links. The crank, connecting rods and pistons in an internal combustion engine are links. Car suspension units are made up of links and they are also to be found in machine tools, photocopiers and printers. A *linkage mechanism* may be defined as a device which transmits or transfers motion from one point to another.

Two of the most common plane linkage mechanisms are the slider-crank and the four-bar chain shown on Figure 2.30.

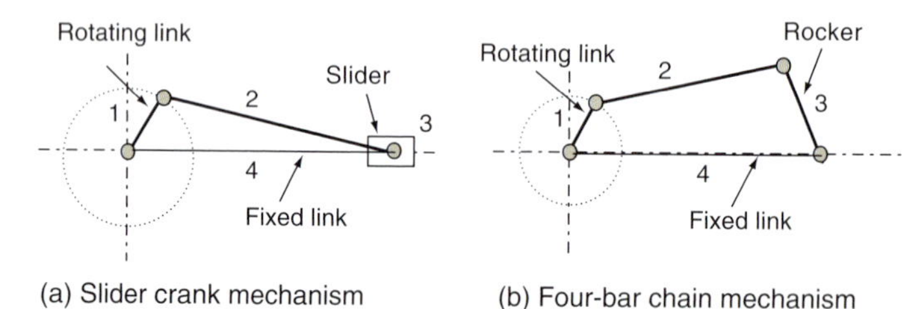

Figure 2.30 Linkage mechanisms

With the slider-crank mechanism, the crank (1) rotates at a uniform speed and imparts a reciprocating translational motion to the piston (3). It is said to have an input rotation and output translation. The connecting link (2) has both a rotational and translational motion and link (4) is formed by the stationary machine frame or cylinder block.

With the four-bar linkage the crank (1) rotates at a uniform speed and imparts a rocking rotational motion to the output link (3). It has input rotation and output rotation. Here again, the connecting link (2) has both a rotational and translational motion and link (4) is formed by the stationary machine frame.

KEY POINT

A link may have linear motion, rotational motion, or a combination of the two. Translational motion is another name for linear motion.

The nature of the output rotation can be altered by changing the lengths of the links. Different output characteristics can also be obtained by fixing a different link in the chain. Such mechanisms are called *inversions* of the slider-crank and four-bar chain. Whatever the arrangement, there are three types of link in the above mechanisms.

Links which have translational motion

These links, such as the piston (3) in the slider crank. Here, all points on the link have the same linear velocity. The velocity vector diagram for such a link is shown in Figure 2.31.

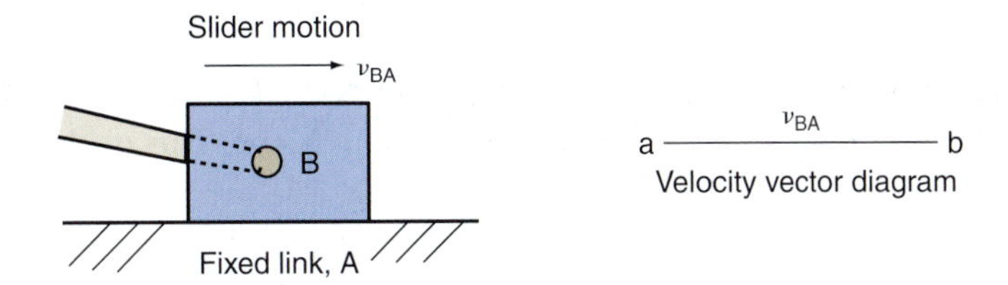

Figure 2.31 Slider velocity vector

The velocity vector ab gives the magnitude and direction of the piston velocity. No arrow is necessary. The sequence of the letters, a to b, gives the direction. The notation v_{BA}, indicates that this is the velocity of the piston B, relative to a point A on the fixed link or machine frame. The notation is similar to that which we have already used on vector diagrams, where upper case letters are used on the space diagram and lower case letters on the vector diagram.

Links with rotational motion

These are links where one end rotates about a fixed axis through a point on the link. Link (1) in the slider crank and links (1) and (4) in the four-bar chain are of this type. The velocity vector diagram for such a link is shown in Figure 2.32.

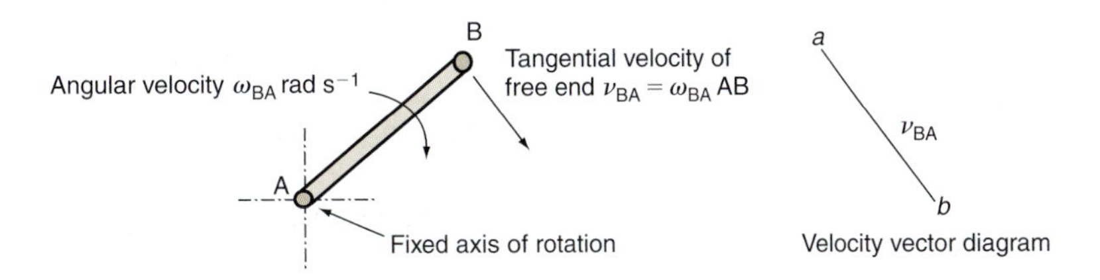

Figure 2.32 Rotating link velocity vector

Here the velocity vector ab, gives the magnitude and direction of the velocity of the free end B, as it rotates about the fixed axis through the end A, i.e. in the direction a to b. The notation v_{BA} again indicates that this is the velocity of B relative to A.

Links which have a combined translational and rotational motion

Links such as the connecting link (2) in both the slider-crank and the four-bar chain are of this type. The velocity vector diagram for such a link is shown in Figure 2.33.

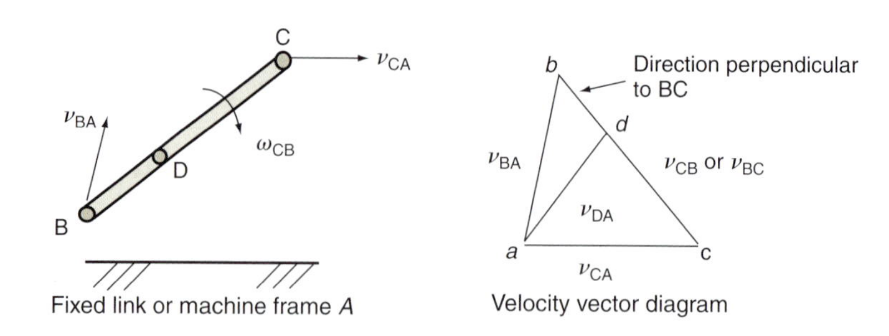

Figure 2.33 Velocity vector diagram for link with translational and rotational motion

Here the two ends of the link BC are moving in different directions at the instant shown. This gives both a rotational and translational motion to the link. The two velocity vectors ab and ac are drawn from the same point and the vector bc is the velocity of C relative to B. In other words, if you were sat on the end B, looking at C, that is what its velocity v_{CB}, would appear to be. You should note that the direction of the vector bc is always perpendicular to the link BC.

If you were sat on the end B looking at C, it would also appear to be rotating about you with angular velocity ω_{CB}. This angular velocity can be calculated by dividing v_{CB} by the radius of rotation which is the length BC, i.e.:

$$\omega_{CB} = \frac{v_{CB}}{CB} \tag{2.31}$$

Alternatively, if you were sat on the end C, looking at B, it would appear to be moving with velocity v_{BC} which is given by the vector cb. The velocity vector diagram also enables the velocity of any point D, on the link to be measured. Suppose that the point D is one-third of the way along the link from the end B. The point d, on the vector diagram is similarly located at one-third of the distance bc measured from b. The vector ad then gives the velocity v_{DA} of the point D relative to the fixed point A.

KEY POINT

When the two ends of a link are travelling in different directions, the velocity of one end relative to the other is always at right angles to the link.

Example 2.11

In the slider-crank mechanism shown in Figure 2.34, the crank AB is of length 200 mm and the connecting rod BC is of length 800 mm. The crank rotates clockwise at a steady speed of 240 rpm. At the instant shown determine (a) the velocity of the crank pin at B, (b) the velocity of the piston, (c) the velocity of the point D which is at the centre of the connecting rod and (d) the angular velocity of the connecting rod.

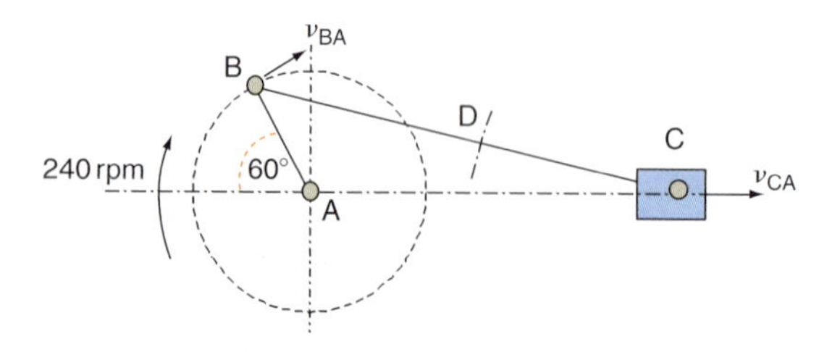

Figure 2.34

(a) Finding angular velocity of crank:

$$\omega_{BA} = N\frac{2\pi}{60} = 240 \times \frac{2\pi}{60}$$

$$\boldsymbol{\omega_{BA} = 25.1\,\text{rad}\,\text{s}^{-1}}$$

Finding velocity of crank pin at B:

$$v_{BA} = \omega_{BA} \times AB = 25.1 \times 0.2$$

$$\boldsymbol{v_{BA} = 5.03\,\text{m}\,\text{s}^{-1}}$$

Drawing velocity vector diagram (Figure 2.35):

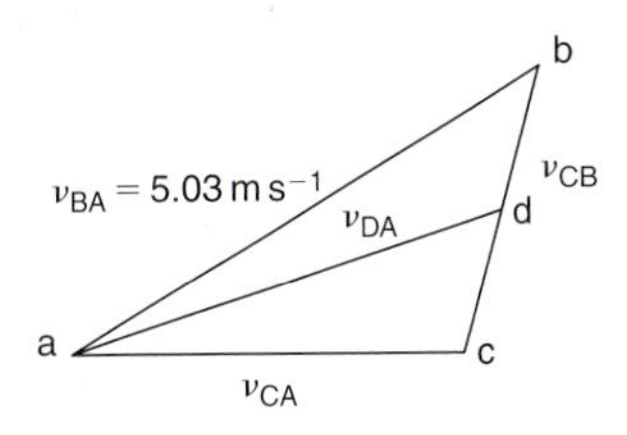

Figure 2.35

1. Draw vector ab for v_{BA} to a suitable scale.
2. From the point a draw in a horizontal construction line for the direction of vector ac.
3. From the point b draw a construction line perpendicular to the connecting rod BC for the vector bc. The intersection of the two construction lines fixes the point c.
4. Locate the point d which is the mid-point of vector bc and draw in the vector ad.

(b) Finding velocity of piston v_{CA}:

From diagram:

$$\boldsymbol{v_{CA} = ac = 3.8\,\text{m}\,\text{s}^{-1}}$$

(c) Finding v_{DA}, which is the velocity of D, the mid-point of the connecting rod:

From diagram:

$$\boldsymbol{v_{DA} = ad = 4.3\,\text{m}\,\text{s}^{-1}}$$

(d) Finding v_{CB}, which is the velocity of C relative to B:

From diagram:

$$\boldsymbol{v_{CB} = bc = 2.6\,\text{m}\,\text{s}^{-1}}$$

Finding ω_{CB}, the angular velocity of the connecting rod:

$$\omega_{CB} = \frac{v_{CB}}{BC} = \frac{2.6}{0.8}$$

$$\boldsymbol{\omega_{CB} = 3.25\,\text{m}\,\text{s}^{-1}}$$

Example 2.12

In the four-bar linkage shown in Figure 2.36, AB = 200 mm, BC = 250 mm, CD = 200 mm and the distance between the axes of rotation at A and D is

500 mm. At the instant shown, the crank AB is rotating clockwise at a speed of 300 rpm. Determine (a) the angular velocity of the crank AB and the velocity of the point B, (b) the velocity of the point C and the angular velocity of the rocker CD, (c) the velocity of the point E on the connecting rod BC which is 100 mm from B and (d) the angular velocity of the connecting rod.

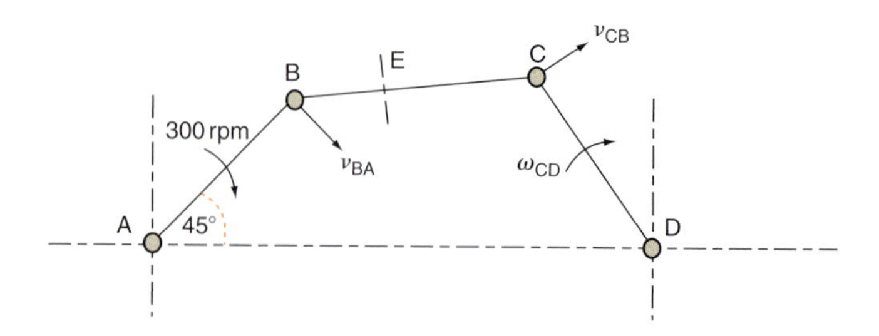

Figure 2.36

(a) Finding angular velocity of crank:

$$\omega_{BA} = N\frac{2}{60} = 300 \times \frac{2}{60}$$

$$\boldsymbol{\omega_{BA} = 31.4\ \text{rad s}^{-1}}$$

Finding velocity of B:

$$v_{BA} = \omega_{BA} \times AB = 31.4 \times 0.2$$

$$\boldsymbol{v_{BA} = 6.28\ \text{m s}^{-1}}$$

Drawing vector diagram (Figure 2.37):

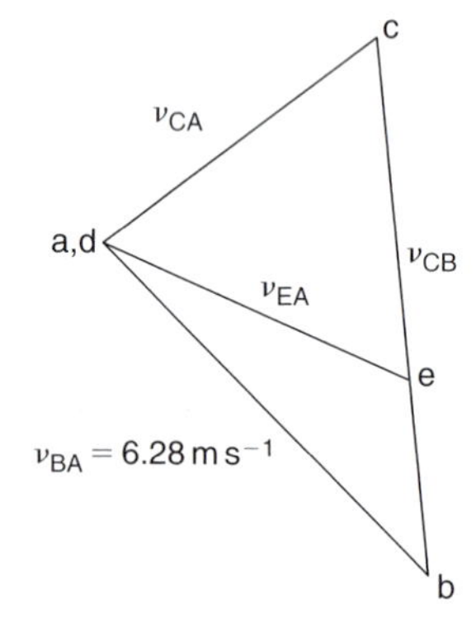

Figure 2.37

1. Draw vector ab for v_{BA} to a suitable scale.
2. From the point a, draw in a construction line in a direction at right angles to AC for the vector ac.
3. From the point b draw a construction line perpendicular to the connecting rod BC for the vector bc. The intersection of the two construction lines fixes the point c.
4. Locate the point e, on vector bc such that:

$$\frac{BE}{EC} = \frac{be}{ec} = \frac{2}{3}$$

5. Draw in the vector ae which gives the velocity of the point E.

(b) Finding velocity v_{CD}, of the point C:

From diagram:

$$v_{AD} = ac = 4.5\,m\,s^{-1}$$

Finding angular velocity of rocker CD:

$$\omega_{CD} = \frac{v_{CD}}{CD} = \frac{4.5}{0.2}$$

$$\omega_{CD} = 22.5\ rad\,s^{-1}$$

(c) Finding v_{EA}, which is the velocity of E, on the connecting rod:

From diagram:

$$v_{EA} = ae = 4.4\ m\,s^{-1}$$

(d) Finding v_{CB}, which is the velocity of C relative to B:

From diagram:

$$v_{CB} = bc = 7.1\,m\,s^{-1}$$

Finding ω_{CB}, the angular velocity of the connecting rod:

$$\omega_{CB} = \frac{v_{CB}}{BC} = \frac{7.1}{0.25}$$

$$\omega_{CB} = 28.4\,rad\,s^{-1}$$

Test your knowledge 2.6

1. What is the function of a linkage mechanism?
2. What constitutes the fixed link of the slider-crank mechanism in an internal combustion engine?
3. What is meant by an *inversion* of a slider-crank or four-bar chain mechanism?
4. Describe the lettering system used on space diagrams and velocity vector diagrams for plane mechanisms.

Activity 2.7

In the plane mechanism shown in Figure 2.38, the crank AB is rotating clockwise at 250 rpm at the instant shown. The links have the following dimensions.

AB = 165 mm
BC = 240 mm
CD = 200 mm
CE = 350 mm

Determine (a) the angular velocity of the crank AB and the velocity of the point B, (b) the velocity of the point C and the angular velocity of the rocker CD, (c) the angular velocity of the connecting rod BC and (d) the velocity of the piston E.

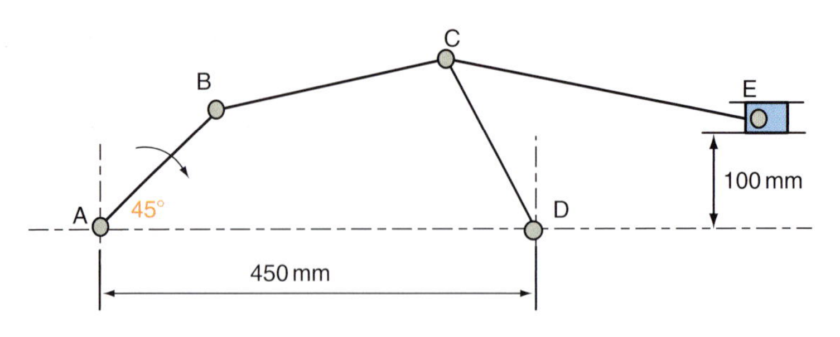

Figure 2.38

To check your understanding of the preceding section, you can solve Review questions 26–28 at the end of this chapter.

Natural Vibrations

There are certain mechanical systems which will vibrate if they are disturbed from their equilibrium position. A mass suspended or supported on a spring will vibrate if disturbed before finally settling back to its equilibrium position. This is a system with elasticity. Practical examples are the suspension systems in motor vehicles, motor cycles and some of the modern designs of bicycle. Another type of mechanical system which will vibrate is a mass suspended by a cord. This will swing from side to side if disturbed from its equilibrium position. The pendulum in a clock is an example of such a system.

KEY POINT

Free, or natural vibrations would experience no resistance to their motion and would never die away.

A *natuarally*, or *freely vibrating* system is one which would not experience any frictional resistance, air resistance or any other kind of resistance to its motion. If such a system could exist, the *natural vibrations* would never die away and there would be perpetual motion. This of course does not happen in practice, and all vibrations eventually die away due to *damping forces*. They are said to *attenuate*.

Simple harmonic motion

Simple harmonic motion (SHM) is a form oscillating motion which is often associated with freely vibrating mechanical systems. It is defined as follows:

> When a body moves in such a manner that its acceleration is directed towards, and is proportional to its distance from a fixed point in its path, it is said to move with simple harmonic motion.

The fixed point referred to is the point at the centre of the motion path. It is the equilibrium position to which all practical systems eventually return. The definition states that for a system which describes simple harmonic motion (SHM):

Acceleration = Constant × Displacement from centre of path

From this equation, it can be seen that the acceleration is continually changing. As a result, the equations that we have used for uniform acceleration cannot be applied. Some new expressions are required to find the displacement, velocity and acceleration at any instant in time for a body which moves with SHM. These may be derived by considering a body which moves in a circular path of radius r with uniform angular velocity ω rad s^{-1} as shown in Figure 2.39.

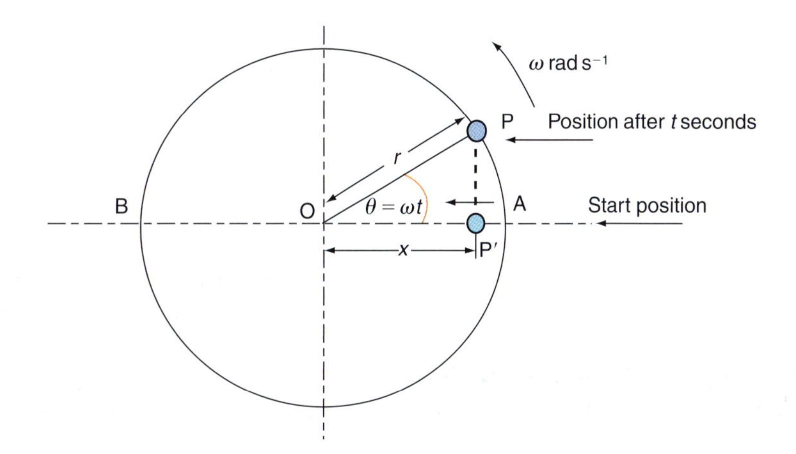

Figure 2.39 Uniform circular motion

As the body P travels with uniform angular velocity, its projection or shadow P′ on the diameter oscillates between A and B. At time t s after passing through the point A, the body P has travelled through an angle of θ radians and the projection P′ is distance x from the centre O. The angle θ radians is given by:

$$\boldsymbol{\theta = \omega t}$$

The distance x of P′ from the centre, O will be:

$$x = r \cos \theta$$

$$\text{or} \quad \boldsymbol{x = r \cos \omega t} \qquad (2.32)$$

The velocity v of P′ towards the centre, O is the rate of change of distance x with time. This can be obtained by differentiating equation (2.32) with respect to the time t.

$$v = \frac{dx}{dt} = \frac{d(r \cos \omega t)}{dt}$$

$$\boldsymbol{v = -\omega r \sin \omega t} \qquad (2.33)$$

Consider now the triangle OPP′ (Figure 2.40)

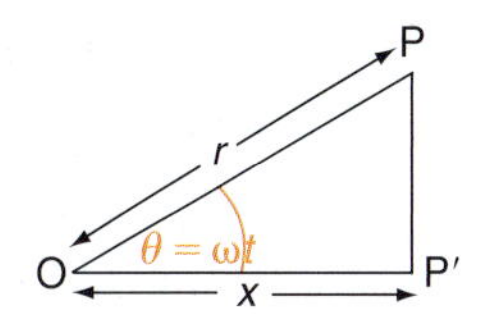

Figure 2.40 Triangle OPP′

By Pythagoras' theorem:

$$PP' = \sqrt{r^2 - x^2} \quad \text{and} \quad \sin \omega t = \frac{\sqrt{r^2 - x^2}}{r}$$

Sustituting for sin ωt in equation (2.33) gives:

$$v = -\omega \cancel{r} \frac{\sqrt{r^2 - x^2}}{\cancel{r}}$$

$$\mathbf{v = -\omega\sqrt{r^2 - x^2}} \qquad (2.34)$$

Equation (2.34) enables the velocity of P′ to be calculated at any distance x from the centre O. The negative sign denotes that when x is increasing, the velocity is decreasing. In many applications it may be disregarded. The acceleration a of P′ is the rate of change of velocity with time. This can be obtained by differentiating equation (2.33) with respect to the time t:

$$a = \frac{dv}{dt} = \frac{d}{dt}(-\omega r \sin \omega t)$$

$$a = -\omega^2 r \cos \omega t$$

But from equation (2.32), $x = r \cos \omega t$

$$\mathbf{a = -\omega^2 x} \qquad (2.35)$$

This is of the form:

Acceleration = Constant × Displacement from centre of path

By definition, the projection P′ thus describes SHM along the diameter AB of the circle. The negative sign in equation (2.35) denotes that when x is increasing, the projection P′ is retarding. In many applications it may be disregarded.

The time taken for one complete oscillation of P′ is known as the *periodic time* of the motion. It is also the time taken for the body P to complete one revolution of its circular path. The periodic time T is thus given by:

$$\mathbf{T = \frac{2\pi}{\omega}} \qquad (2.36)$$

For very rapid oscillations it is more convenient to consider the frequency measured in Hertz. This is the reciprocal of the periodic time.

$$\mathbf{f = \frac{\omega}{2\pi}\ Hz} \qquad (2.37)$$

KEY POINT

The circular frequency of a system is the angular velocity of the circular motion which would generate SHM with the same amplitude and periodic time.

In all of the above equations the term ω rad s^{-1} is the angular velocity of the circular motion which is generating the SHM. It is sometimes called the *circular frequency* of the motion and should not be confused with the frequency of vibration, given by equation (2.37). The projection P′ has maximum displacement from the central position at A and B where $x = r$. The distance r is called the *amplitude* of the motion.

Examination of equation (2.34) shows that the projection P′ is travelling fastest as it passes through the central position, i.e. when $x = 0$.

$$v_{max} = -\omega\sqrt{r^2 - 0}$$

$$\mathbf{v_{max} = -\omega r} \qquad (2.38)$$

KEY POINT

The maximum velocity of a system which describes SHM occurs at the central position whilst the maximum acceleration occurs at the amplitude of the motion.

Examination of equation (2.35) shows that the projection P′ has maximum acceleration at the maximum value of x, i.e. at the amplitude of the motion, where $x = r$.

$$a_{max} = -\omega r^2 \qquad (2.39)$$

Once you have calculated the acceleration of a body which moves with SHM, the force acting on the body at that instant in time can be found by applying Newton's second law of motion, i.e.:

$$F = ma \qquad (2.40)$$

Example 2.13

A body of mass 2.5 kg describes SHM of amplitude 500 mm and periodic time of 2.5 s. Determine (a) its velocity and acceleration at a displacement of 250 mm from the centre, (b) its maximum velocity and acceleration and (c) the maximum force acting on the body.

(a) Finding value of ω, the circular frequency of the motion (Figure 2.41):

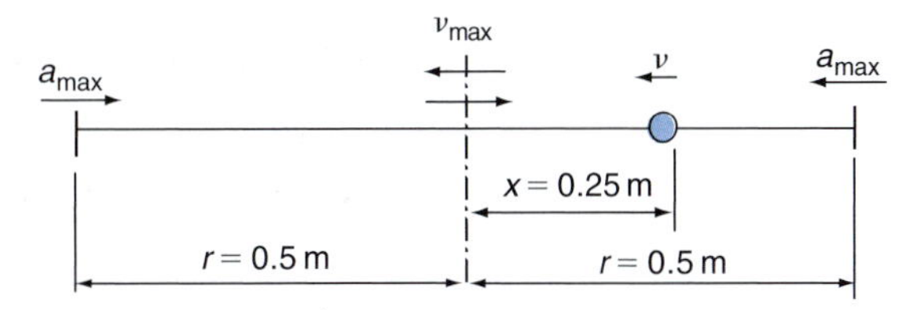

Figure 2.41 Simple harmonic motion

$$T = \frac{2\pi}{\omega}$$

$$\omega = \frac{2\pi}{T} = \frac{2\pi}{2.5}$$

$$\boldsymbol{\omega = 2.51\,\mathrm{rad\,s^{-1}}}$$

Finding velocity when $x = 0.25$ m

$$v = \omega\sqrt{r^2 - x^2} = 2.51\sqrt{0.5^2 - 0.25^2}$$

$$\boldsymbol{v = 1.09\,\mathrm{m\,s^{-1}}}$$

Finding acceleration when $x = 0.25$ m

$$a = \omega^2 x = 2.51^2 \times 0.25$$

$$\boldsymbol{a = 1.58\,\mathrm{m\,s^{-2}}}$$

(b) Finding maximum velocity:

$$v_{max} = \omega r = 2.51 \times 0.5$$

$$\boldsymbol{v_{max} = 1.26\,\mathrm{m\,s^{-1}}}$$

Finding maximum acceleration:

$$v_{max} = \omega^2 r = 2.51^2 \times 0.5$$

$$\boldsymbol{a_{max} = 3.15\,\mathrm{m\,s^{-2}}}$$

(c) Finding maximum force acting on body:

$$F_{max} = ma_{max}$$
$$F_{max} = 2.5 \times 3.15$$
$$\mathbf{F_{max} = 7.88\ N}$$

Test your knowledge 2.7

1. What is the definition of SHM?
2. What is meant by the *amplitude* of vibration?
3. What is meant by the *circular frequency* of a vibrating system?
4. What is meant by the *periodic time* of a vibrating system?
5. Where in its path do maximum velocity and acceleration occur when a body moves with SHM?

Activity 2.8

A body of mass 5 kg moves with SHM whose amplitude is 100 mm and frequency of vibration is 5 Hz. Determine (a) the circular frequency of the motion, (b) the maximum values of velocity and acceleration, (c) the displacements at which the velocity and acceleration are half of the maximum values and (d) the maximum force which acts on the body.

To check your understanding of the preceding section, you can solve Review questions 29–33 at the end of this chapter.

Mass-spring systems

A mass, suspended from the free end of a helical spring, will oscillate if displaced from its equilibrium position. To show that the oscillations are simple harmonic it is required to obtain an expression for the acceleration of the mass. If this is proportional to displacement from the equilibrium position, the definition is satisfied. Consider now a mass of m kg suspended from a spring of stiffness $S\,\text{N}\,\text{m}^{-1}$.

When the mass m is gently placed on the spring it produces a static deflection d and rests in the equilibrium position as shown in Figure 2.42(b). The forces acting on the mass are its weight and the spring tension which are equal and opposite as shown in Figure 2.43(a).

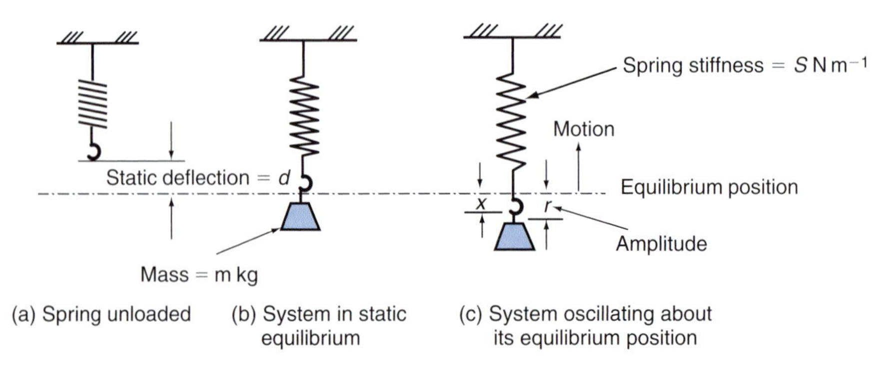

Figure 2.42 Mass-spring system

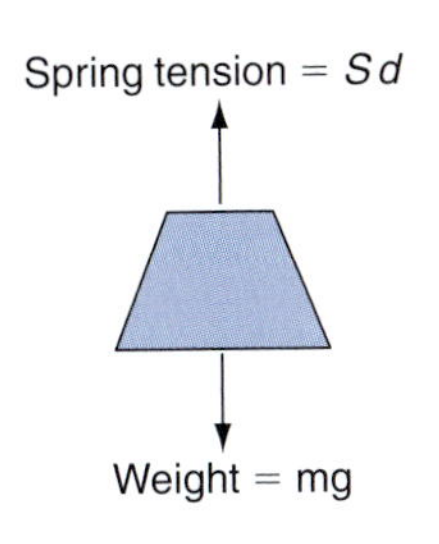

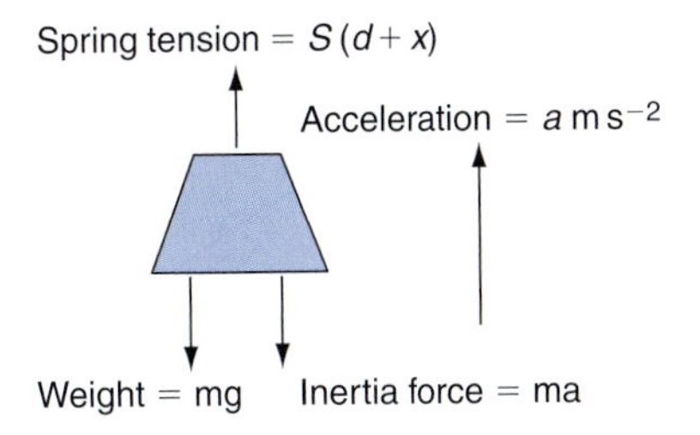

Figure 2.43 Forces acting on mass

When the system is in static equilibrium:

$$\text{Tension in spring} = \text{Weight of mass}$$

$$\boldsymbol{Sd = mg} \qquad \text{(i)}$$

$$\text{and} \quad S = \frac{mg}{d}$$

$$\text{i.e.} \quad \textbf{Spring stiffness} = \frac{\textbf{Static load}}{\textbf{Static deflection}} \qquad (2.41)$$

If the mass is given a downwards displacement r below its equilibrium position and then released, it will oscillate about the equilibrium position with amplitude r as shown in Figure 2.42(c). At some instant when the mass is distance x below its equilibrium position and accelerating upwards, the forces acting on it are as shown in Figure 2.43(b). It is assumed that the mass oscillates with *free* or *natural vibrations*, i.e. that the air resistance is negligible and that there are no energy losses in the spring material. Equating forces gives:

$$\text{Tension in spring} = \text{Weight} + \text{Inertia force}$$

$$S(d + x) = mg + ma$$

$$Sd + Sx = mg + ma$$

But from equation (i), $Sd = mg$, and so these terms cancel. This leaves:

$$Sx = ma$$

$$\boldsymbol{a = \frac{Sx}{m}} \qquad \text{(ii)}$$

The equation is of the form:

Acceleration = Constant × Displacement

This proves that the mass-spring system describes SHM. The acceleration is also given by the general formula:

$$a = \omega^2 x$$

The natural circular frequency of the system ω rad s^{-1} is thus given by:

$$\omega^2 = \frac{S}{m}$$

$$\boldsymbol{\omega = \sqrt{\frac{S}{m}}} \qquad (2.42)$$

The periodic time of the natural vibrations is given by:

$$T = \frac{2\pi}{\omega}$$

$$T = 2\pi\sqrt{\frac{m}{S}} \quad (2.43)$$

The frequency of the vibrations known as the *natural frequency* of the system. It is the reciprocal of periodic time, i.e.:

$$f = \frac{\omega}{2\pi}$$

$$f = \frac{1}{2\pi}\sqrt{\frac{S}{m}} \quad (2.44)$$

In practical mass-spring systems there is always some air resistance. There are also energy losses in the spring material as it is twisted and untwisted. They are known as *hysteresis losses*. The combined effect is known as *damping* which eventually causes the vibrations to die away.

Example 2.14

A close-coiled helical spring undergoes a static deflection of 30 mm when a mass of 2.5 kg is placed on its lower end. The mass is then pulled downwards through a further distance of 20 mm and released so that it oscillates about the static equilibrium position. Neglecting air resistance, energy losses in the spring material and the mass of the spring determine (a) the periodic time and natural frequency of vibration and (b) the maximum velocity and acceleration of the mass.

(a) Finding stiffness of spring:

$$\text{Spring stiffness} = \frac{\text{Static load}}{\text{Static deflection}}$$

$$S = \frac{mg}{d} = \frac{2.5 \times 9.81}{0.03}$$

$$\mathbf{S = 818\,Nm^{-1}}$$

Finding natural circular frequency of system:

$$\omega = \sqrt{\frac{S}{m}} = \sqrt{\frac{818}{2.5}}$$

$$\boldsymbol{\omega = 18.2\,rad\,s^{-1}}$$

Finding periodic time:

$$T = \frac{2\pi}{\omega} = \frac{2\pi}{18.2}$$

$$\mathbf{T = 0.345\,s}$$

Finding natural frequency of vibration:

$$f = \frac{1}{T} = \frac{1}{0.345}$$

$$\mathbf{f = 2.90\,Hz}$$

(b) Finding maximum velocity of mass:

$$v_{max} = \omega r = 18.2 \times 0.02$$

$$\mathbf{v_{max} = 0.364\,m\,s^{-1}}$$

Finding *maximum acceleration of mass:*

$$a_{max} = \omega^2 r = 18.2^2 \times 0.02$$

$$\mathbf{a_{max} = 6.62\,ms^{-2}}$$

Simple pendulum

A simple pendulum consits of a concentrated mass, known as the pendulum 'bob', which is free to swing from side to side at the end of a light cord or a light rod. When it is given a small angular displacement and released, the pendulum bob appears to describe SHM. It does not travel along a perfectly straight line, but for small angular displacements it is approximately straight, with a central equilibrium position.

If the mass is given a small displacement r from its equilibrium position and then released, it will oscillate about the equilibrium position with amplitude r as shown in Figure 2.44(a). At some instant when the bob is distance x from its equilibrium position, and accelerating towards it, the forces acting on it will be as shown in Figure 2.44(b). It is assumed that the bob oscillates with *free* or *natural vibrations*, i.e. that the air resistance is negligible and that there are no energy losses at the point of suspension. Equating forces in the direction of motion at the instant shown gives:

$$\text{Inertia force} = \text{Component of weigh in direction of motion}$$

$$\not{m}a = \not{m}g \sin\theta$$

$$a = g \sin\theta$$

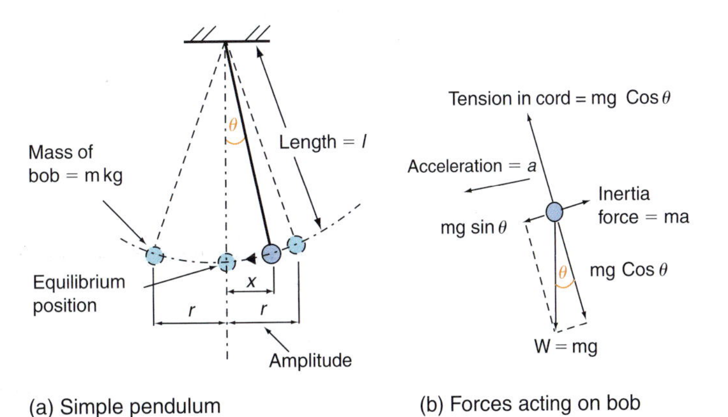

Figure 2.44 Simple pendulum

But

$$\sin\theta = \frac{x}{l}$$

$$a = \frac{gx}{l}$$

$$\text{or} \quad \mathbf{a = \frac{gx}{l}}$$

This is of the form:

Acceleration = Constant × Displacement

It shows that for small displacements, the simple pendulem describes SHM. You will recall that the acceleration is also given by the general formula:

$$a = \omega^2 x$$

The natural circular frequency of the system $\omega\,\text{rad}\,\text{s}^{-1}$ is thus given by:

$$\omega^2 = \frac{g}{l}$$

$$\boldsymbol{\omega = \sqrt{\frac{g}{l}}} \quad (2.45)$$

The periodic time of the natural vibrations is given by:

$$T = \frac{2\pi}{\omega}$$

$$\boldsymbol{T = 2\pi\sqrt{\frac{l}{g}}} \quad (2.46)$$

The natural frequency of the vibrations is the reciprocal of periodic time, i.e.:

$$f = \frac{\omega}{2\pi}$$

$$\boldsymbol{f = \frac{1}{2\pi}\sqrt{\frac{g}{l}}} \quad (2.47)$$

As with the mass-spring systems there is always some air resistance. There is also some energy loss at the point of suspension and the combined effect eventually causes the vibrations to die away.

Example 2.15

A simple pendulum describes 45 complete oscillations of amplitude 30 mm in a time of 1 min. Assuming that the pendulum is swinging freely, calculate (a) the length of the supporting cord and (b) the maximum velocity and acceleration of the bob.

(a) Finding the periodic time:

$$T = \frac{\text{Recorded time}}{\text{Number of oscillations}} = \frac{60}{45}$$

$$\boldsymbol{T = 1.33\,\text{s}}$$

Finding length of pendulum:

$$T = 2\pi\sqrt{\frac{l}{g}}$$

$$T^2 = \frac{4\pi^2 l}{g}$$

$$l = \frac{T^2 g}{4\pi^2} = \frac{1.33^2 \times 9.81}{4\pi^2}$$

$$\boldsymbol{l = 0.440\,\text{m}}$$

(b) Finding the natural circular frequency of the system:

$$T = \frac{2\pi}{\omega}$$

$$\omega = \frac{2\pi}{T} = \frac{2\pi}{1.33}$$

$$\boldsymbol{\omega = 4.72\,\mathrm{rad\,s^{-1}}}$$

Finding maximum velocity of pendulum bob:

$$v = \omega r = 4.72 \times 0.03$$

$$\boldsymbol{v = 0.142\,\mathrm{m\,s^{-1}}}$$

Finding maximum acceleration of pendulum bob:

$$a = \omega^2 r = 4.72^2 \times 0.03$$

$$\boldsymbol{a = 0.668\,\mathrm{m\,s^{-2}}}$$

Experiment to verify that a mass-spring system describes SHM

Apparatus

A light close-coiled helical spring, retort stand and clamps, hanger and slotted masses, metre rule, stop watch (Figure 2.45).

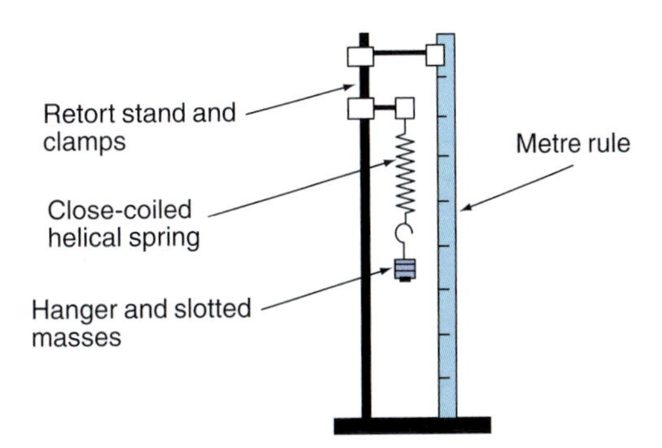

Figure 2.45 Arrangement of apparatus

Procedure

1. Hang the spring on one of the the retort stand clamps and position the metre rule alongside it, secured by the other clamp.
2. Place the hanger and a slotted mass on the lower end of the spring and note the static displacement.
3. Displace the mass from its equilibrium position and release it so that it oscillates freely.
4. Note the number of complete oscillations that the mass makes in a time of 1 min and calculate the periodic time of the motion.
5. Repeat the procedure with increasing values of the slotted mass until at least six sets of readings have been taken.
6. Calculate the square of the periodic time for each set of readings and tabulate the results.
7. Plot a graph of load against static deflection and from it determine the spring stiffness.
8. Plot a graph of mass against the square of periodic time, observe its shape and calculate its gradient.

Theory

Ploting static load W against static deflection d should give a straight line whose gradient is the stiffness S of the spring (Figure 2.46).

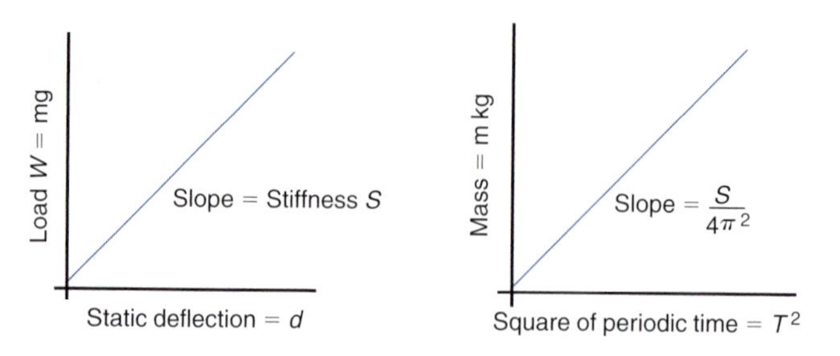

Figure 2.46 Characteristic graphs for mass-spring system

The periodic time of a mass-spring system which oscillates with SHM is given by the expression:

$$T = 2\pi\sqrt{\frac{m}{S}}$$

$$T^2 = \frac{4\pi^2 m}{S}$$

or $$\boldsymbol{m = \frac{ST^2}{4\pi^2}}$$

If the mass-spring system is decribing SHM, the graph of mass against the square of the periodic time will be a straight line whose gradient is $S/4\pi^2$.

Experiment to verify that a simple pendulum describes SHM

Apparatus

Length of light cord, pendulum bob, retort stand and clamp, metre rule, stopwatch (Figure 2.47).

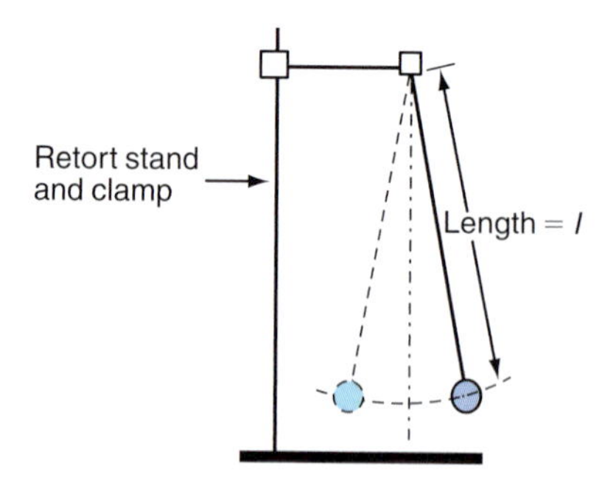

Figure 2.47 Arrangement of apparatus

Procedure

1. Attach one end of the cord to the bob and the other to the retort stand clamp.
2. Measure and record the length of the cord from the support to the centre of the bob.

3. Displace the bob from its equilibrium position and release it so that it oscillates freely.
4. Note the number of complete oscillations made by the bob in a time of 1 min and calculate the periodic time of the motion.
5. Repeat the procedure with different lengths of the cord until at least six sets of readings have been taken.
6. Calculate the square of the periodic time for each set of readings and tabulate the results.
7. Plot a graph of cord length against the square of the periodic time, observe its shape and calculate its gradient.

Theory

The periodic time of a simple pendulum which oscillates with SHM is given by the expression:

$$T = 2\pi\sqrt{\frac{l}{g}}$$

$$T^2 = \frac{4\pi^2 l}{g}$$

$$\text{or} \quad l = \frac{gT^2}{4\pi^2}$$

If the pendulum is decribing SHM, the graph of mass against the square of the periodic time will be a straight line whose gradient is $g/4\pi^2$ (Figure 2.48).

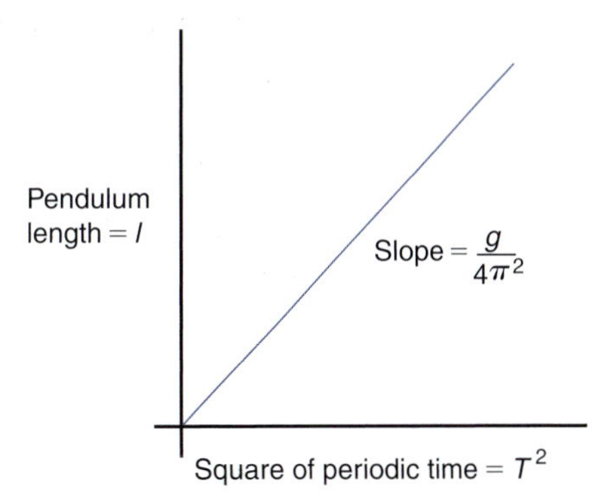

Figure 2.48 Characteristic graphs for simple pendulum

Test your knowledge 2.8

1. How can the stiffness of a spring be determined?
2. How can the circular frequency of a mass-spring system be determined?
3. What is the effect on the periodic time of a mass-spring system of (a) increasing the mass and (b) increasing the spring stiffness?
4. How can the circular frequency of a simple pendulum be calculated?
5. What is the effect on the periodic time of a simple pendulum of (a) increasing the mass of the bob and (b) increasing the pendulum length?

Activity 2.9

The length of a close-coiled helical spring extends by a distance of 25 mm when a mass of 0.5 kg is placed on its lower end. An additional mass of 1.5 kg is then added and displaced so that the system oscillates with an amplitude of 50 mm. Determine (a) the periodic time and frequency of the oscillations, (b) the maximum velocity and acceleration of the mass and (c) the length of a simple pendulum which would have the same periodic time and frequency.

To check your understanding of the preceding section, you can solve Review questions 34–38 at the end of this chapter.

Review questions

1. A steel tie bar in a structure is initially 2 m long and 25 mm diameter. What will be the change in length and diameter when carrying a tensile load of 100 kN? The modulus of elasticity is 200 GPa and Poisson's ratio is 0.33.

2. A steel bar is initially 250 mm in length with a square cross-section of 50 × 50 mm. The modulus of elasticity is 200 GPa and Poisson's ratio is 0.32. Determine the change in the dimensions of the bar when carrying a tensile load of 600 kN.

3. The plate shown in Figure 2.49 initially measures 75 × 50 mm. Find the changes to these dimensions when subjected to the stresses shown. Take $E = 207$ GPa and $\nu = 0.3$.

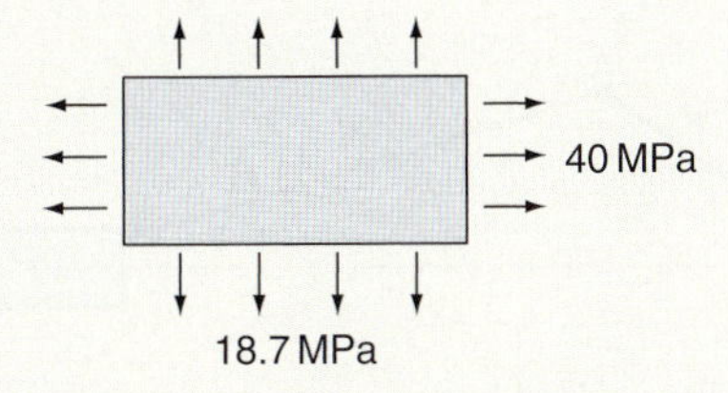

Figure 2.49

4. The plate shown in Figure 2.50 is made from brass whose modulus of elasticity is 105 Gpa and Poisson's ratio 0.3. The plate is initially 160 mm long and 70 mm high. Determine the dimensions after loading.

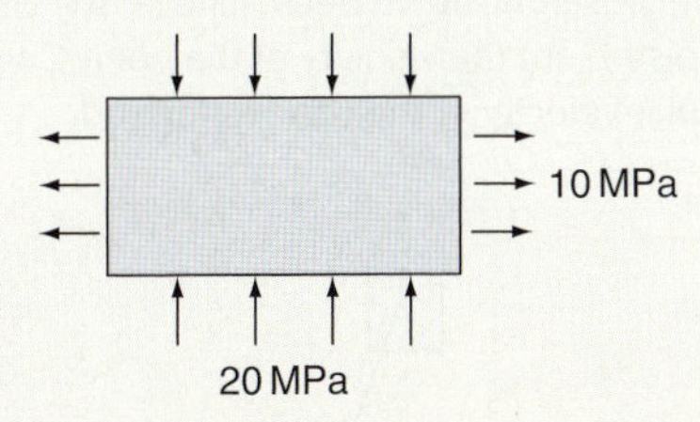

Figure 2.50

5. A cube has sides that are initially 100 mm and is subjected to stresses of 60 MPa (tensile), 40 MPa (compressive) and 80 MPa (tensile) in the *x*-, *y*- and *z*-directions, respectively. Calculate (a) the strains in these directions, (b) the changes in dimensions, (c) the volumetric strain and (d) the change in volume. Take $E = 150$ GPa and $\nu = 0.3$.

6. A rectangular prism of sided 100, 80 and 50 mm before being subjected to a uniform fluid pressure of 80 MPa on all faces. Determine the changes to these dimensions and the change in volume which occurs. Take $E = 75$ GPa and $\nu = 0.3$.

7. A solid square section beam 100 × 100 mm is to be replaced by a hollow section of the same depth and wall thickness 15 mm. If the replacement is to have the same stiffness, calculate its required width and the percentage saving in weight that it will bring.

8. A tubular steel cantilever of outer diameter 100 mm and inner diameter 75 mm is 3 m long and carries a concentrated load of 2 kN at its free end. Calculate the maximum stress due to bending and the radius of curvature at its fixed end. The modulus of elasticity is 200 GPa.

9. A rectangular timber beam of width 60 mm and depth 600 mm has a span of 12 m and carries a uniformly distributed load of 2.5 kN m^{-1} including its own weight. Calculate the maximum stress due to bending and the radius of curvature where this occurs. The modulus of elasticity of the timber is 15 GPa.

10. A length of flat steel strip is to be coiled around a drum 2 m diameter. The elastic limit stress of the steel is 216 MPa and its modulus of elasticity is 216 Gpa. What is the maximum thickness that the strip may have if there is to be no permanent distortion when it is uncoiled?

11. A solid section axle for a heavy vehicle is 2.25 m long and 175 mm diameter. The axle is suspended from the body at points 1.5 m apart and each wheel carries a load of 140 kN. The modulus of elasticity of the material is 208 GPa. Plot the shear force and bending moment diagrams and determine the maximum stress due to bending, the factor of safety in operation if the UTS of the material is 500 MPa, and the value radius of curvature between the suspension points.

12. A horizontal lever 1.2 m long is made of hollow rectangular section steel tube the depth of which in the plane of bending is twice the width. The UTS of the steel is 450 MPa and a factor of safety of 6 is to apply. The modulus of elasticity is 100 GPa. Determine the minimum dimensions of the cross-section and the radius of curvature at the point of maximum stress.

13. The drive shaft to a boat's propeller is 50 mm in diameter. If a 25 mm diameter hole is bored along its length calculate (a) the percentage saving in weight and (b) the percentage reduction in torsional strength.

14. A solid shaft transmits 1.6 MW at a speed of 75 rpm. The shaft length is 6 m and its diameter is 300 mm. The shear modulus of the shaft material is 80 GPa. Calculate (a) the maximum shear stress in the shaft and (b) the angle of twist.

15. A propeller shaft of outer diameter 75 mm and inner diameter 60 mm is made from steel whose shear modulus is 85 GPa and ultimate shear stress 140 MPa. The shaft is to transmit 150 kW at a speed of 1650 rpm. Determine (a) the factor of safety

in operation and (b) the angle of twist per metre length.

16. The following test data relates to a power transmission shaft:

Operating speed 280 rpm

Shear modulus of material 80 GPa

Shaft diameter 75 mm

Shaft length 15 m

Measured angle of twist 20°

Determine (a) the maximum shear stress in the shaft material and (b) the power transmitted.

17. The design specification for a transmission shaft states that the angle of twist must be no more than 1° over a length 20 times the diameter. Using this specification determine the diameter of a solid steel shaft to transmit 150 kW at a speed of 300 rpm. Take $G = 80$ GPa.

18. A solid transmission shaft is connected to a pump by a flanged coupling containing eight 12 mm diameter bolts on a pitch circle of 225 mm diameter. The power transmitted is 110 kW at a speed of 500 rpm and the maximum shear stress in the shaft is to be equal to that in the bolts. It can be assumed that the bolts are loaded in shear only and carry all of the driving torque. Calculate (a) the shaft diameter and (b) the maximum stress in the shaft and bolts.

19. An aircraft that can travel at 300 km h^{-1} attempts to fly due south in a westerly cross-wind of 50 km h^{-1}. What is the resultant velocity of the aircraft?

20. A shaft rotates in a bearing with a tangential speed of 1.5 m s^{-1} whilst moving axially at a speed of 0.8 m s^{-1}. What is the resultant sliding velocity of the shaft?

21. An aircraft that can fly at 400 km h^{-1} wishes to travel in a north easterly direction. If there is a wind of 120 km h^{-1} blowing from the north, in what direction should the aircraft be piloted and what will its speed be?

22. A ship A is travelling at a speed of 30 km h^{-1} in a direction 60° south of east. A second ship B is sighted which is travelling at 20 km h^{-1} in a direction 30° east of north. What is the apparent velocity of ship B when viewed from ship A?

23. At an instant in time, two ships A and B are separated by a distance of 15 km with ship B south-east of A. Ship A is travelling east at 30 km h^{-1} and ship B is travelling north at 50 km h^{-1}. Determine (a) the velocity of B relative to A, (b) the closest distance between the two ships and (c) the time taken to reach this position.

24. Two straight railway lines running north south and east–west intersect. A train running south approaches the crossing at a speed of 80 km h^{-1} whilst a train travelling east approaches it at 40 km h^{-1}. When train A is 6 km south of the crossing bridge, train B is 8 km east. Determine the shortest distance between the two trains and the time from these initial positions that they take to reach it.

25. To a ship A, travelling due west at 30 km h^{-1} a second ship B, appears to be travelling north-west at 24 km h^{-1}. What are the two possible directions in which ship B, might actually be heading.

26. In the slider crank mechanism shown in Figure 2.51 the crank OA is 150 mm long and the connecting rod AB is 300 mm long. The crank rotates clockwise at a speed of 3600 rpm and C is the mid-point of the connecting rod. At the instant shown, determine (a) the velocity of the piston, (b) the velocity of the point C and (c) the angular velocity of the connecting rod.

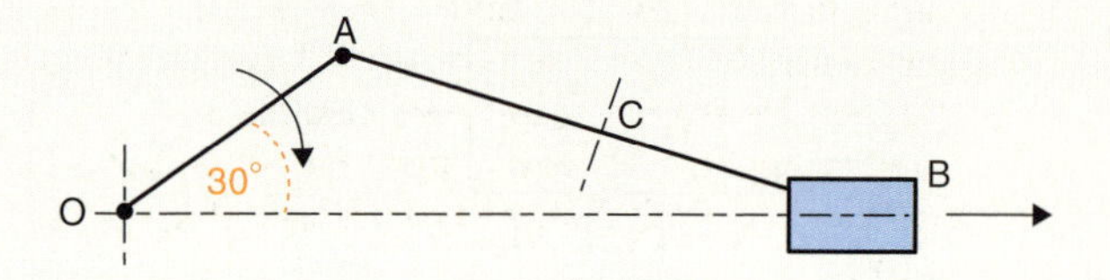

Figure 2.51

27. The crank OA of the engine mechanism shown in Figure 2.52 rotates clockwise at a speed of 3000 rpm. The length of the crank is 200 mm and the length of the connecting rod AB is 500 mm. The point C on the connecting rod is 200 mm from the crankpin at A. At the position shown determine (a) the velocity of the piston, (b) the velocity of the point C and (c) the angular velocity of the connecting rod.

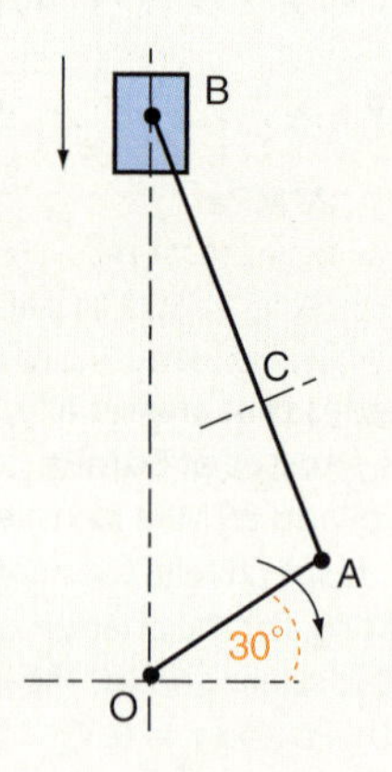

Figure 2.52

28. The four-bar linkage shown in Figure 2.53. The crank OA, the connecting rod AB and the rocker BC are 20 mm, 120 mm and 60 mm long, respectively. The crank rotates at a speed of 60 rpm and the centre distance OC is 90 mm. At the instant shown determine (a) the velocity of C, (b) the angular velocity of the rocker BC and (c) the angular velocity of the connecting rod AB.

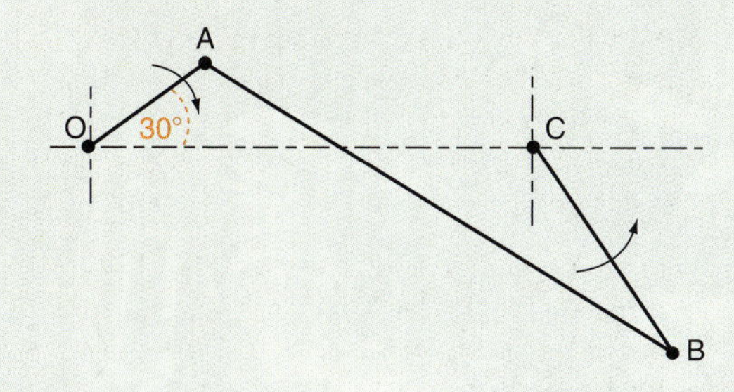

Figure 2.53

29. A machine component of mass 2 kg moves with SHM of amplitude 250 mm and periodic time 3 s. Determine (a) the maximum velocity, (b) the maximum acceleration and (c) the acceleration and the force acting on the component when it is 50 mm from the central position.

30. A body moves with SHM in a straight line and has velocities of 24 m s^{-1} and 12 m s^{-1} when it is 250 mm and 400 mm, respectively from its central position. Determine (a) the amplitude of the motion, (b) the maximum velocity and acceleration and (c) the periodic time and frequency of the motion.

31. The length of the path of a body which moves with SHM is 1 m and the frequency of the oscillations is 4 Hz. Determine (a) the periodic time of the motion, (b) the maximum velocity and acceleration and (c) the velocity and acceleration when the body is 300 mm from its amplitude.

32. A body of mass 50 kg describes SHM along a straight line. The time for one complete oscillation is 10 s and the amplitude of the motion is 1.3 m. Determine (a) the maximum force acting on the body, (b) the velocity of the body when it is 1 m from the amplitude position and (c) the time taken for the body to travel 0.3 m from the amplitude position.

33. A machine component of mass 25 kg describes SHM along a straight line path. The periodic time is 0.6 s and the maximum velocity is 5 ms^{-1}. Calculate (a) the amplitude of the motion, (b) the distance from the central position when the force acting on the component is 750 N and (c) the time taken for the component to move from its amplitude to a point mid-way from the central position.

34. A mass of 25 kg is suspended from an elastic coiled spring which undergoes a 50 mm increase in length. The mass is then pulled downwards through a further distance of 25 mm and released. Determine (a) the periodic time of the resulting SHM, (b) the maximum velocity of the mass and (c) the maximum acceleration of the mass.

35. A helical spring is seen to undergo a change in length of 10 mm when a mass of 1 kg is gently suspended from its lower end. A further 3 kg are then added, displaced from the equilibrium through a distance of 30 mm and released. Determine (a) the frequency of the oscillations, (b) the maximum velocity of the mass and (c) the maximum acceleration of the mass.

36. A mass of 6 kg is suspended from a helical spring and produces a static deflection of 6 mm. The load on the spring is then increased to 18 kg and settles at a new equilibrium position. It is then displaced through a further 10 mm and released so that system oscillates freely. Determine (a) the frequency of the oscillations, (b) the maximum velocity and acceleration and (c) the maximum tension in the spring.

37. An elastic spring of stiffness 0.4 kN m^{-1} is suspended vertically with a load attached to its lower end. When displaced, the load is seen to oscillate with a periodic time of 1.27 s. Determine (a) the magnitude of the load, (b) the acceleration of the load when it is 25 mm from the equilibrium position and (c) the tension in the spring when the load is 25 mm from the equilibrium position.

38. A simple pendulum is made from a light cord 900 mm long and a concentrated mass of 2.25 kg. What will be the periodic time of the oscillations when the pendulum is given a small displacement and allowed to swing freely? What will be the stiffness of the spring which will have the same periodic time when carrying the same mass?

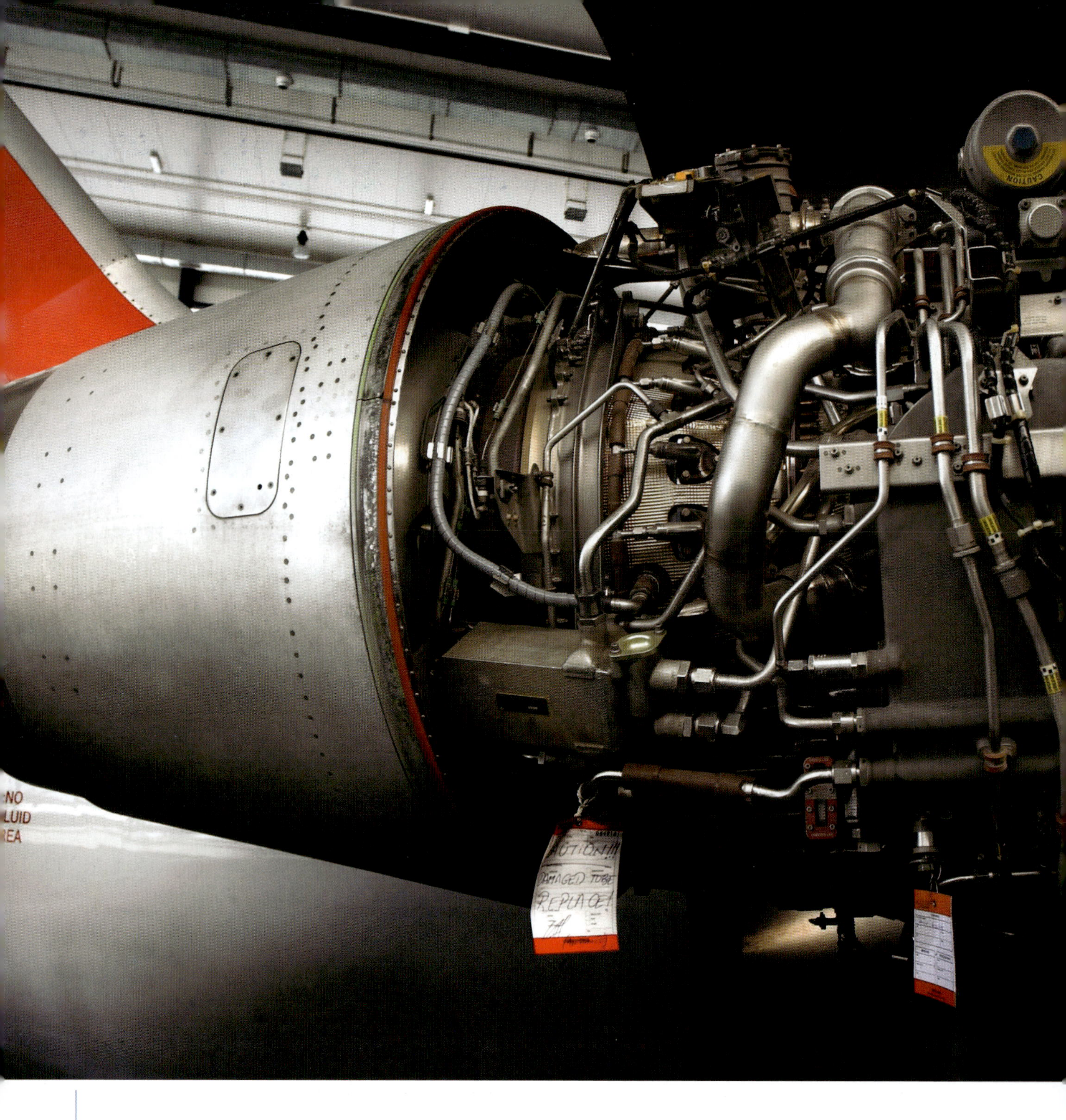

A gas turbine contains a multitude of mechanical components. Bearings, seals, drives and linkage mechanisms are all to be found in a modern aero-engine. Components rotating at high speed require lubrication with the appropriate choice of lubricant and lubrication system. A variety of screwed fastenings are used to assemble the components and to provide for replacement and maintenance.

Photo courtesy of iStockphoto, Tomas Bercic, Image # 4056469

Chapter 3

Applications of Mechanical Systems and Technology

Mechanical engineering embraces a wide field of activity. Its range includes power generation, land, sea and air transport, manufacturing plant and machinery and products used in the home and office such as the photocopier, computer printer and washing machine. The term ***mechatronics*** is often used to describe systems that incorporate mechanical devices, electrical and electronic circuits and elements of information technology. These are to be found in all of the above areas and it is the aim of this unit to investigate some of the more common mechanical systems and components.

Moving parts generally require lubrication and the first part of this chapter examines lubricant types and lubrication systems. Pressurised systems require seals and gaskets to prevent the escape of lubricants and other working fluids. Rotating parts require bearings and all mechanical systems incorporate fixing devices to hold the various components in position. These will also be examined in this chapter.

A prime purpose of mechanical systems is to transmit power and motion and the various ways in which this can be achieved is investigated. The chapter also provides an overview of hydraulic and pneumatic systems, steam plant, refrigeration and air-conditioning systems and mechanical handling equipment.

Lubricants and Lubrication Systems

In your studies you may have calculated the forces needed to overcome static and dynamic frictional resistance between surfaces in sliding contact. It is essential that this should be kept to a minimum in most mechanical systems. The exceptions are of course belt drives, clutches and braking systems that depend on frictional resistance for their operation. Generally however friction causes wear, generates heat and wastes power. Efficient lubrication will not eliminate these effects but it can minimise them.

Lubricant purposes and types

A lubricant separates the sliding and rolling surfaces so that there is little on no metal to metal contact. In effect, it forms a cushion between the surfaces and if properly chosen it will keep the surfaces apart even when squeezed by high contact pressures. As the surfaces move over each other the lubricant is subjected to shearing and offers some resistance to the motion. This however is far less than that when the surfaces are in dry contact. In the case of liquid lubricants such as oils, the resistance depends on a property known as *dynamic viscosity.* It is a measure of the internal resistance of the liquid to being stirred, poured or, in the case of a lubricant, to being sheared. The SI units of dynamic viscosity are N s m^{-2} and some typical values for different grades of lubricating oil are given in Table 3.1.

KEY POINT

Dynamic viscosity is a measure of the resistance of a liquid to being stirred or poured.

A lubricant fulfils purposes other than reducing frictional resistance and wear. A steady flow of lubricant can carry away heat energy and solid particles from the contact area. This is an important function of the cutting lubricants used in the machining of metal components. They reduce the friction between the cutting tool face and the metal being removed and they also carry away the heat energy and the metal swarf which is produced by the cutting tool.

Another function that a lubricant performs is to prevent corrosion which might occur from the presence of moisture or steam. It is important that a lubricant has good adhesion with the contact surfaces and does not drain away when they are at rest. It is then able to protect them from corrosion and maintain surface separation in readiness for start-up. A lubricant must not react with any other substances in the working environment. Its function is also to prevent contamination of the contact surfaces and prevent the ingress of solid particles that would damage them.

Lubricant types and applications

Lubricants may be broadly divided into mineral oils, synthetic oils, vegetable oils, greases, solid lubricants and compressed gas. Mineral oils and greases are perhaps the most widely used but each has its own particular characteristics and field of application. We will now examine some of these.

Mineral oils

Mineral oils are hydrocarbons. That is to say that they are principally made up of complex molecules composed of hydrogen and carbon. The chemists who specialise in this field refer to the hydrocarbon types as *paraffins*, *naphthenes* and *aromatics*. Mineral oils for different applications contain different proportions of these compounds. This affects their viscosity and the way that it changes under the effects of temperature and pressure.

Traditionally, lubricating oils have been grouped according to their different uses and viscosity values as shown in Table 3.1.

Table 3.1 Lubricating oils

Group	Applications	Dynamic viscosity
Spindle oils	Lubrication of high speed bearings	Below 0.01 Ns m^{-2} at 60°C
Light machine oils	Lubrication of machinery running at moderate speeds	0.010 to 0.02 Ns m^{-2} at 60°C
Heavy machine oils	Lubrication of slow moving machinery	0.02 to 0.1 Ns m^{-2} at 60°C
Cylinder oils	Lubrication of steam plant components	0.1 to 0.3 Ns m^{-2} at 60°C

KEY POINT

The viscosity index of a lubricating oil gives an indication of how its viscosity changes with temperature rise.

Within the different types shown in Table 3.2, lubricating oils are graded according to the change in viscosity which takes place with a rise in temperature. This is indicated by the *viscosity index* number that has been allocated to the oil. It can range from zero to over one hundred as shown in Table 3.2. The higher the value, the less is the effect of temperature rise. A value of 100+ indicates that the viscosity of the oil is not much affected by temperature change.

Table 3.2 Viscosity index groupings

Viscosity index group	Viscosity index
Low viscosity index (LVI)	Below 35
Medium viscosity index (MVI)	35 to 80
High viscosity index (HVI)	80 to 110
Very high viscosity index (VHVI)	Above 110

Table 3.3 shows the typical viscosity and viscosity index values for the different types of lubricating together with the likely percentage composition of paraffins, naphthenes and aromatic hydrocarbon molecules which might be expected.

You might notice that within each group, the high viscosity index oils, which are least affected by temperature change, have the highest percentages of paraffin compounds. Also within each group, the oils

Table 3.3 Group analysis of mineral lubricating oils

Group	Viscosity ($Ns\,m^{-2}$)	Viscosity index (VI)	% paraffins	% naphthenes	% aromatics
Spindle oils					
Low VI	2.7×10^{-3}	Below 35	46	32	22
Light machine oils					
Medium VI	3.9×10^{-3}	35 to 80	59	37	4
High VI	4.3×10^{-3}	80 to 110	68	26	6
Heavy machine oils					
Low VI	7.4×10^{-3}	Below 35	51	26	23
Medium VI	7.5×10^{-3}	35 to 80	54	37	8
High VI	9.1×10^{-3}	80 to 110	70	23	7
Cylinder oils					
High VI	26.8×10^{-3}	80 to 100	70	22	8

with the lowest viscosity have the highest percentages of aromatic compounds. It might thus be stated as a general rule, that aromatic compounds lower the viscosity of an oil whilst paraffin compounds help it to retain its viscosity as the temperature rises.

There are other ways of measuring viscosity and if you go to buy oil for a motor car or motor cycle, the service manual might recommend SAE 10W-40 lubricating oil for the engine and gearbox. The letters stand for the Society of Automotive Engineers of America and the numbers indicate the viscosity and a measure of its variation with temperature. The system is widely used by the motor industry but is not entirely suitable for other industrial applications where wider ranges of working temperatures, pressures, running speeds and service environments are to be found.

Additives and synthetic oils

Plain mineral oils are suitable for applications such as the lubrication of bearings, gears and slide-ways in machine tools and reciprocating pumps and compressors. These are applications where the operating temperature is relatively low and the service environment does not contain substances that will readily contaminate or degrade the lubricant. Typical contaminants are air, (which is unavoidable) ammonia, water, oil of another grade, soot, dust and wear particles from the lubricated components. Special chemicals are often added to plain mineral oils to improve their properties and prolong their life. Table 3.4 lists some of the more common additive types.

Polymers are sometimes added to mineral oils to enhance their properties. Theses are often referred to as synthetic oils although it is the mineral oil that forms the bulk of the lubricant. Various types of ester, silicone and other polymers are added, mainly to increase the viscosity index and wear resistance of the oil.

Some of the more expensive multi-grade motor oils contain polymers. They appear to be much thinner than the less expensive engine oils at normal temperature but undergo a much smaller change of viscosity

Table 3.4 Additive types

Additive types	Purpose
Acid neutralizers	To neutralise contaminating acids such as those formed by the combustion of sulphur in solid and liquid fuels.
Anti-foam	To reduce the formation of surface foam where the lubricant is subjected to aggressive agitation.
Anti-oxidants	To reduce the build-up of sludge and acidic products which result from oxygen from the air reacting with the oil.
Anti-corrosion agents	To reduce the corrosion of ferrous metals, copper alloys and bearing metals.
Anti-wear agents	To minimise surface contact and reduce wear under heavy loading conditions.
Detergents	To clean away or reduce surface deposits such as those which occur in internal combustion engine cylinders.

KEY POINT

Modern multi-grade motor oils with synthetic additives give better cold starting and offer less resistance to the moving parts during engine warm-up.

as the temperature rises. This makes cold starting easier and gives improved circulation of the lubricant during the warming-up period. At running temperatures their viscosity is similar to that of the less expensive multi-grades but they are purported to have better wear resistant properties.

Vegetable oils

Soluble vegetable oils are used as coolants and cutting tool lubricants in metal machining. The vegetable oils used have a low flash point. This is essential since high temperatures can be generated at the tip of a cutting tool. When mixed with water the oil takes on a low viscosity milky appearance. It can be delivered under pressure to the cutting tool and assist in the clearance of metal particles. This is of particular importance in deep-hole drilling operations.

Castor oil has been a popular additive to the lubricating oil used for the highly tuned engines in motor cycle and formula car racing. Indeed, it gives its name to a popular brand of motor oil. When used the exhaust gases have a distinctive smell that immediately identifies the additive.

Greases

A grease consists of a lubricating fluid which contains a thickening agent. It may be defined as a semi-solid lubricant. The lubricating fluid may be a mineral oil or a polymer liquid and the thickening agent may be a soap or a clay. Table 3.5 lists some of the more common thickeners.

The main advantage of a grease is that it stays where it is applied and is less likely to be displaced by pressure or centrifugal action than oil. It can act as both a lubricant and a seal, preventing the entry of water, abrasive grit and other contaminants to the lubricated surfaces.

Table 3.5 Greases

Lubricating fluid	Soap or thickener	Temperature range	Uses
Mineral oil	Calcium (Lime)	−20°C to 80°C	General purpose, ball and roller bearings
	Sodium	0°C to 175°C	Glands, seals, low to medium speed ball and roller bearings
	Mixed sodium and calcium	−40°C to 150°C	High speed ball and roller bearings
	Lithium	−40°C to 150°C	General purpose, ball and roller bearings
	Clay	−30°C to 200°C	Slideways
Ester polymer	Lithium	−75°C to 120°C	General purpose, ball and roller bearings
Silicone polymer	Lithium	−55°C to 205°C	General purpose, ball and roller bearings
	Silicone soap	−55°C to 260°C	Miniature bearings

KEY POINT

Greases do not dissipate heat energy like a lubricating oil but they have good adherence and are not easily displaced by gravity or centrifugal action.

Bearings such as the wheel bearings of motor vehicles can be pre-packed with grease and function for long periods of time without attention. The main disadvantages of grease are that it does not dissipate heat energy so well as a lubricating oil and because of its higher viscosity, it offers greater resistance to motion.

Solid lubricants

There are many applications, such as at very low and very high working temperatures, where lubricating oils and greases are unsuitable. In such cases the use of a solid lubricant can reduce wear without adversely affecting the contact surfaces or the working environment. Three of the most common solid lubricants are graphite, molybdenum disulphide and polytetrafluoroethylene, which is better known as PTFE. You will be quite familiar with non-stick cooking utensils that are coated with PTFE. Lead, gold and silver are also used as dry lubricants for aerospace applications but their use is expensive.

Solid lubricants may be applied to the contact surfaces as a dry powder and you might recall that cast iron is self-lubricating because it already contains graphite flakes. They may be mixed with a resin and sprayed on the surfaces to form a bonded coating, or they may be used as an additive to oils and greases. Molybdenum disulphide, whose chemical formula is MoS_2, has been used in this way by car owners for a number of years. It can be purchased from motor accessory dealers in small tins as a suspension in mineral oil. Solid lubricants can also be added to molten metal in the forming process so that when solidified, the metal is impregnated with particles of the solid lubricant. Graphite is often added to phosphor-bronze in this way to improve its qualities as a bearing material.

KEY POINT

Solid lubricants are able to function outside the working temperature range of oils and greases.

Compressed gases

Compressed air and inert gases such as carbon dioxide have a very low viscosity compared to oils and greases. With compressed gas bearings the contact surfaces are separated by a thin cushion of gas which offers very little resistance to motion. The main disadvantage is the cost. A gas delivery system is required and the stationary outer surface of a journal bearing or lower stationary slide-way must be machined with evenly spaced exit orifices to provide a dry and clean gas cushion. The usual operating pressures are 2 to 5 bar with higher pressures up to 10 bar being used for heavy duty applications. The bearing surfaces may be coated with a solid lubricant to guard against dry running should the air supply fail.

Gas bearings are well suited to high speed applications. They can operate at speed up to 300 000 rpm and at temperatures well outside the range of oils and greases. A further advantage is that they do not allow dirt of moisture to enter the bearing and are surprisingly rigid. You might think that the air cushion could be easily be penetrated or displaced by high contact pressures and shock loads. This is not the case however as gas bearings have a self-correcting property. Shaft displacement causes the air gap to increase on one side and the pressure in that region to fall. At the same time, the air gap on the opposite side decreases and the pressure rises forcing the shaft back to the central position.

KEY POINT

Air bearings are rigid and self-correcting under load, offer very little resistance to motion and can operate at high and low temperatures.

Test your knowledge 3.1

1. What are the purposes and functions of a lubricant?
2. What does the viscosity index of a lubricating oil indicate?
3. What are the main constituents of a grease?
4. Name two solid lubricants.
5. What are the advantages of air bearings?

Activity 3.1

Using only the raw materials available in your science laboratory and workshop devise a method which can be used to compare the viscosity of different liquids and the effect of temperature rise on viscosity.

To check your understanding of the preceding section, you can solve Review questions 1–5 at the end of this chapter.

Lubrication systems and maintenance

Oil lubricating systems can be divided into three categories, total loss, self-contained and re-circulating. With total loss lubrication the oil is applied to the moving parts by means of an oil can or an aerosol spray. Over a period of time the lubricant evaporates, drips away or takes in dust and dirt to become semi-solidified. The chain of a bicycle is lubricated in this way and periodically, after appropriate cleaning, the lubricating oil needs to be re-applied. An alternative method, which is used on larger items of equipment, is to supply lubricating oil or oil mist periodically from a central reservoir by means of a hand operated lever or an automated pump.

KEY POINT

A total loss lubrication system is one in which the lubricant becomes free to drain away of evaporate.

With self-contained lubrication the oil is contained in a reservoir. The gearbox of a car or a lathe is generally lubricated in this way. The gears are partly submerged in the oil as shown in Figure 3.1 and the process is known as *splash lubrication*. Oil is carried up to the parts that are not submerged, and an oil mist is created inside the gearbox. This too has a lubricating effect.

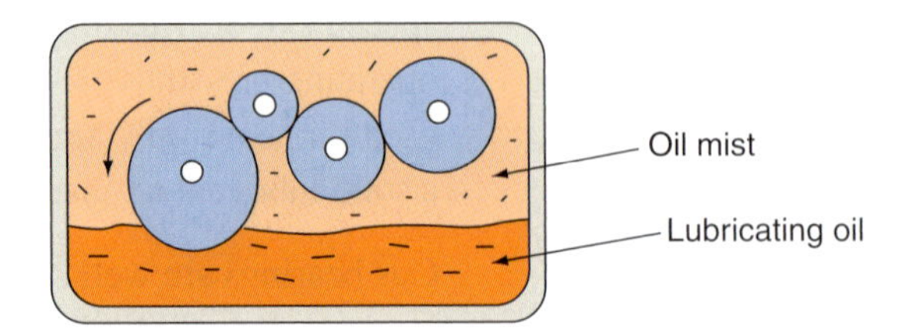

Figure 3.1 Splash lubrication

Ring oiling is another self-contained system in which oil from a reservoir is carried up to the rotating parts of a mechanism. This is shown in Figure 3.2 where a ring rotating with a shaft is partly submerged in the oil and carries it up to the shaft bearings.

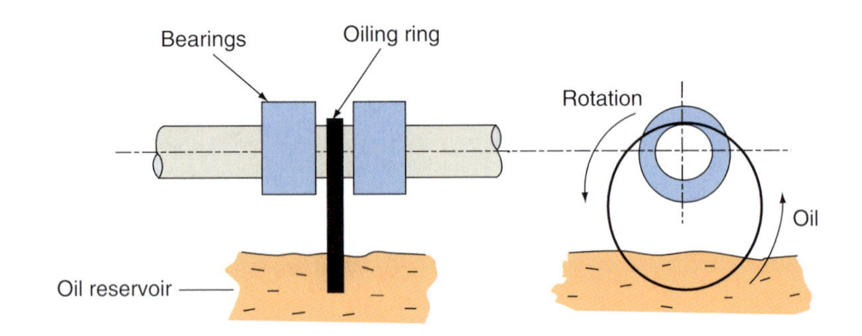

Figure 3.2 Ring oiling

Plane bearings and machine slides that are lightly loaded are sometimes lubricated by means of a pad or wick feed from a small reservoir as shown in Figure 3.3. This too is a self-contained method of lubrication. The oil travels down the wick by capillary action to the moving parts.

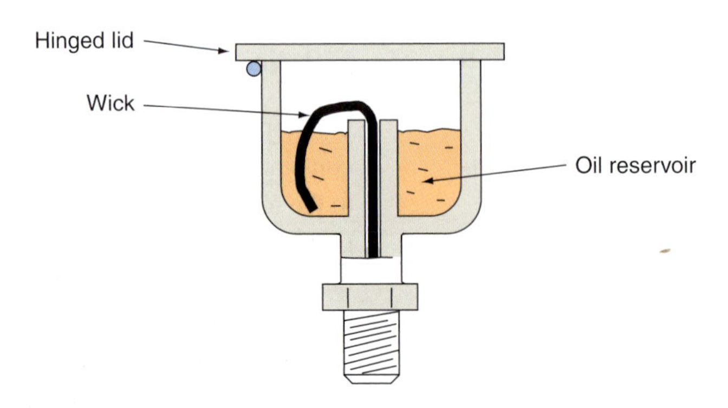

Figure 3.3 Wick feed lubrication

With re-circulating lubrication systems oil from a reservoir is fed under pressure direct to the moving parts or delivered as a spray. The flow is continuous and after passing over the contact surfaces the oil runs back into the reservoir under the effects of gravity. A single pump may deliver lubricant to several locations as shown in Figure 3.4.

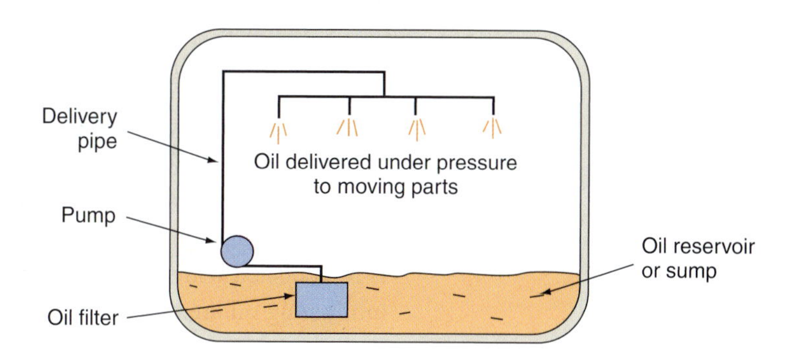

Figure 3.4 Re-circulating oil system

KEY POINT

Wheel bearings should not be packed too tightly with grease when they have been serviced or replaced. Over-packing can cause leakage onto the brake shoes and pads.

Grease lubrication systems fall into two categories, replenishable and non-replenishable. The bearings in some domestic appliances and power tools are examples of the non-replenishable type. They are packed with grease on assembly and this is judged to give sufficient lubrication for the life of the equipment. The wheel bearings on cars, caravans and trailers also fall into this category. They are packed with grease when new and only need to be repacked when stripped down after long periods of service.

Replenishable systems incorporate the grease nipples and screw-down grease cups as shown in Figure 3.5. Lubrication is carried out at recommended periods by means of a grease gun or by screwing down the grease cup.

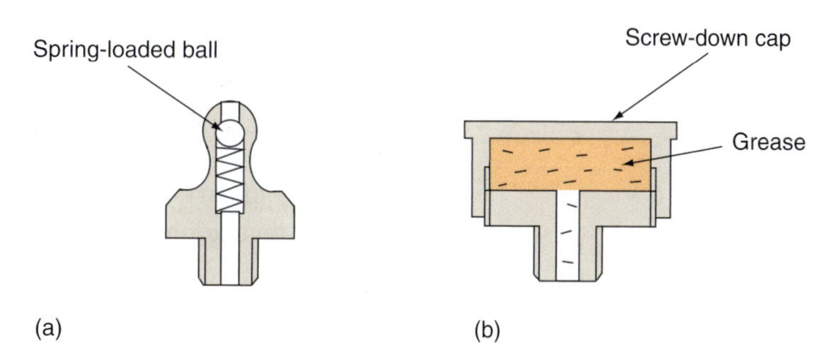

Figure 3.5 Grease lubricators: (a) grease nipple and (b) screw-down grease cup

As with lubricating oils, grease can be supplied under pressure from a central reservoir, by means of a hand operated or motor-driven pump, to a number of points in a mechanism. Some trucks and buses are fitted with automated chassis lubrication which supplies grease to the suspension and steering linkages in this way. Unlike oil lubrication systems, all greasing systems are total loss.

As a general rule, oil lubrication systems should be checked weekly. This can be carried out by plant operators or vehicle drivers and needs very little skill. If necessary, the tank or reservoir should be topped up taking care not to let any dirt into the system during the process. Systems should not be over-filled as this might cause increased resistance to splash-lubricated parts. It might also lead to overheating of the oil due to excessive churning. Dip-sticks and sight gauges are usually provided to indicate the correct depth or level in reservoirs.

It should be noted that the oil level whilst a machine or engine is running might be different to when it is stationary. Servicing instructions should indicate clearly whether the equipment is to be stopped for oil replenishment. It is usually only continuous process machinery which is replenished whilst in operation. In some establishments, samples of lubricating oil are taken for laboratory analysis at regular intervals as part of a condition monitoring procedure. Here they are examined for contamination and the presence of solid wear particles. This can provide useful information to plant managers and maintenance engineers as to the condition of the plant and machinery.

KEY POINT

Safety equipment and protective clothing should always be used when replenishing or disposing of lubricating oils.

All lubricating oils degrade over a period of time due to oxidation and contamination. They should be changed at the recommended intervals together with the filters in re-circulating systems. Lubricating oils contain chemicals that can cause skin irritation and in extreme cases, exposure can lead to skin, and other forms of cancer. Safety equipment should be used, and protective clothing worn, at all times when handling lubricating oils. Oil spillage should be dealt with immediately. Supervision should be informed and the spillage treated with an absorbing material and removed. Water and a suitable detergent can then be used to remove all traces from the work area.

Waste oil should be disposed of in the approved manner. In some cases this might be controlled incineration at temperatures in excess of 1500°C by specially licensed contractors. Some may be fit for re-cycling. After filtering it may be used as a lower grade lubricant or in the production of greases. New supplies should be stored in a safe manner away from substances where cross-contamination might occur. They should be clearly marked as to their grade and type in accordance with COSHH regulations, i.e. The Care of Substances Hazardous to Health.

Test your knowledge 3.2

1. What is total loss lubrication?
2. How does a ring oiler function?
3. How does a wick feed oiler operate?
4. What is provided to indicate the correct level of the oil in a supply tank or sump.
5. What do the initials COSHH stand for and how do they apply to lubricating oils?

Activity 3.2

Re-circulating lubrication systems require a pump to supply the oil under pressure to the different lubrication points. Describe the kind of pump that is most commonly used in car engines and the kind of filter that is used to remove solid particles from the oil.

To check your understanding of the preceding section, you can solve Review questions 6–10 at the end of this chapter.

Engineering Components

There are certain components that are common to wide range of engineered products. Nuts, bolts, screws, rivets and other fixing devices are used to connect and hold the parts together. Seals and gaskets are used to prevent the escape of fluid and the ingress of dirt, and bearings are used to support rotating and reciprocating parts. We will now examine some of these components and their applications.

Seals and packing

The seals used on engineered products may be divided into stationary seals and dynamic seals. Dynamic seals may be further sub-divided into those used with reciprocating components and those used with rotating components. Other relevant terms which you might encounter are *gaskets*, which are a form of stationary seal, and *glands* which are assemblies containing seals or packing to maintain the pressure inside a system at the entry or exit point of a reciprocating rod or a rotating shaft.

Stationary seals are used between the flanged joints of pipes, on the cylinder heads of engines pumps and compressors, between crank cases and oil sumps and under inspection and access covers. Sometimes a liquid sealant is used and there are many proprietary brands that are resistant to oil and water. Some of them solidify after assembly whilst others are intended to remain tacky. Separation of the surfaces for maintenance can sometimes be a problem after which they have to be carefully cleaned before re-assembly.

KEY POINT

Always fit new seals and gaskets after maintenance.

Where there are possible irregularities in the contact surfaces it is usual to employ a seal which can be compressed to accommodate them. A variety of materials are used. They include paper, cork, plastic, rubber, bonded fibre and copper. Laminated materials are also widely used. The cylinder head gaskets of a great many internal combustion engines is of this type. They consist a bonded fibre sandwiched between two thin sheets of copper. Rubber seals sometimes take the form of an O-ring which is seated in semi-circular recesses as shown in Figure 3.6.

Three of the most common dynamic seals are the rotational lip seal, the packed gland and the mechanical seal. They are used as a seal around the drive shafts to pumps and compressors, and around the input and output shafts of gearboxes.

Rotary lip seals are used to prevent the escape of fluids from systems where the pressure difference across the system boundary is relatively

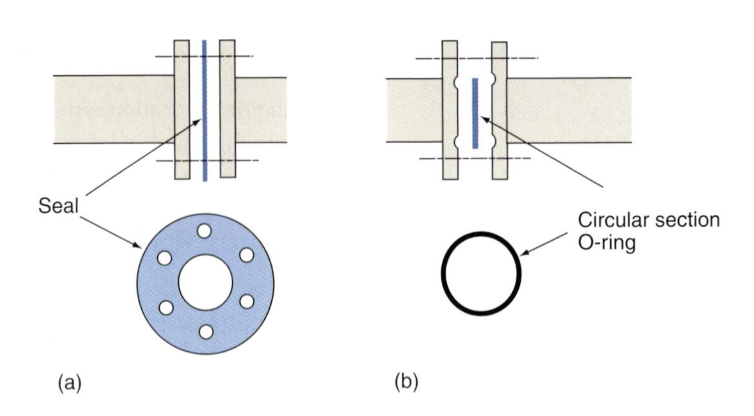

Figure 3.6 Pipe joints: (a) joint seal and (b) rubber O-ring

small. They are widely used on the input and output shafts to gearboxes to prevent the escape of lubricating oil and are suitable for moderate to high running speeds. A section through a lip seal is shown in Figure 3.7. A rubber garter surrounds the rotating shaft and is held in contact by a spring around its circumference.

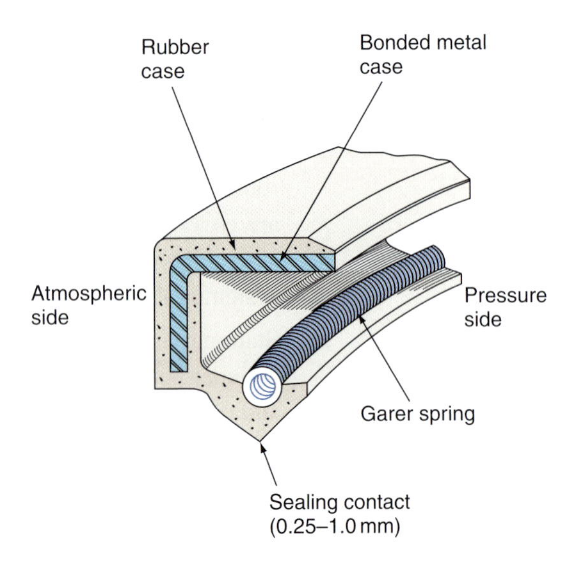

Figure 3.7 Rotary lip seal (Redrawn from *Drives and Seals*, M J Neale, Butterworth-Heinemann, p. 98, Fig. 19.1, 1993, with permission from Elsevier.)

Additional contact force is provided by the internal pressure which pushes down on the garter. During assembly the seal is pressed into a machined recess in the system casing, usually alongside the shaft bearing, and covered by a retaining plate. Different grades of rubber and flexible plastic are used for different operating temperatures and contained fluids.

Packed glands, as shown in Figure 3.8, are able to withstand higher internal pressures then lip seals. A compression collar is tightened so that packing material forms a tight seal around the shaft. A wide variety of packing materials is used. It includes compressed mineral and vegetable fibres such as asbestos, cotton, flax and jute. Compressed synthetic fibres such as nylon and PTFE are also used and sometimes they are impregnated with graphite to assist lubrication. Packed glands may be used as a seal for rotating and reciprocating shafts.

KEY POINT

Packed glands need to be inspected regularly. If there is sign of excess leakage, the compression collar should then be tightened just enough to prevent it.

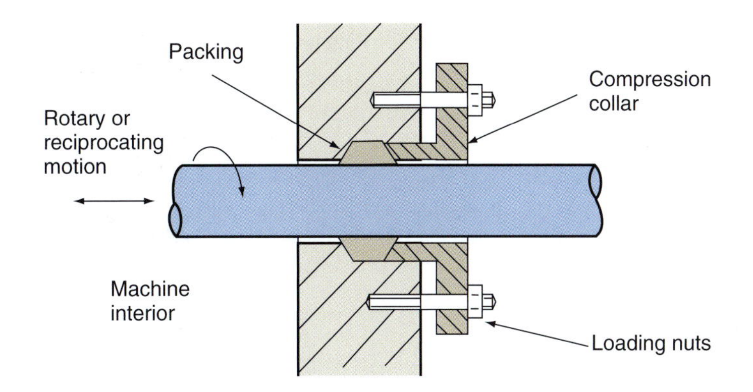

Figure 3.8 Packed gland

KEY POINT

Wherever possible use a press and dedicated centralising tools when fitting new lip seals and mechanical seals.

Packed glands need periodic adjustment and replacement of the packing material. They are still widely used with reciprocating shafts but to reduce the need for maintenance, the mechanical seal, such as that shown in Figure 3.9, have been developed for use with rotating shafts. There are a variety of designs but the operating principle is the same. The seal is formed between the stationary and rotating rings. The loading collar, spring, flexible sheath and rotating ring rotate with the shaft. The stationary ring is pressed into a recess in the machine casing. The loading collar is a tight fit on the shaft and the spring applies a contact force between the rings. Lubrication is from the interior of the machine.

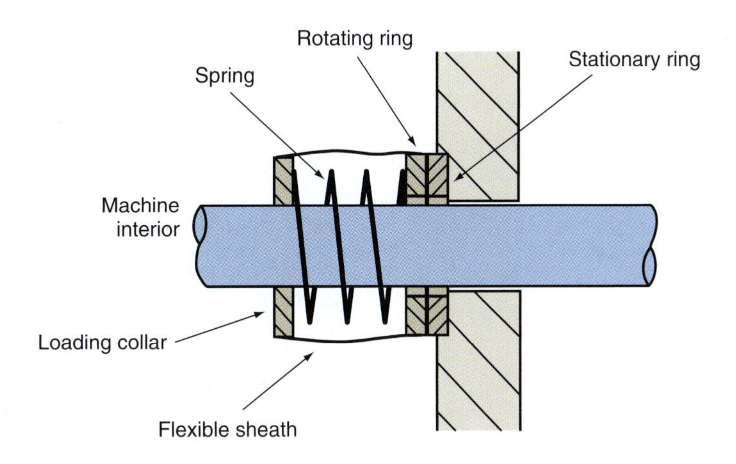

Figure 3.9 Mechanical seal

The pistons in engines and compressors are fitted with piston rings whose purpose is to prevent the leakage of high pressure gas around the piston and into the crankcase. Piston rings are made from a springy metal such as steel, cast iron and bronze. A typical arrangement is shown in Figure 3.10.

When fitted into the cylinder, the ends of the rings are sprung almost together to leave only a very small gap and provide firm contact with the cylinder walls. This ensures that the amount of compressed gas that is able to bypass the piston is negligible. The upper of the two rings provide the seal whilst the lower controls lubrication. As can be

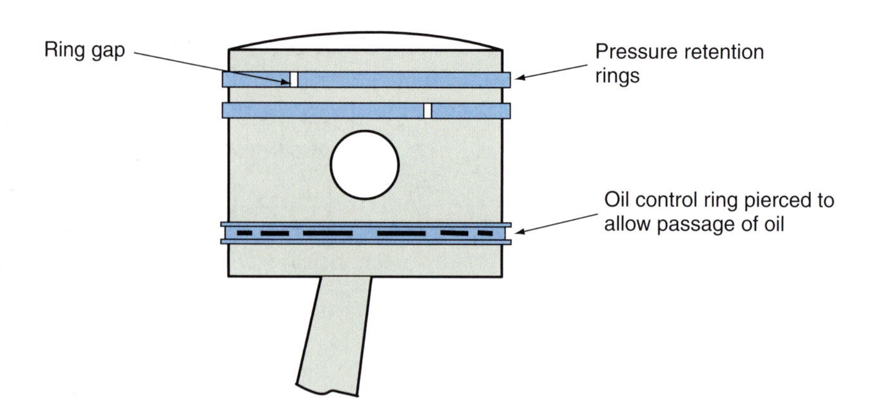

Figure 3.10 Piston rings

seen from Figure 3.10, the oil control ring has a channel section that is pierced with elongated slots. Its function is to control the lubrication of the cylinder walls and scrape away carbon deposits.

Care needs to be taken when fitting new lip seals, mechanical seals and piston rings. Lip seals and the fixed rings of mechanical seals need to be fitted using a press and special centralising tools to ensure that they are properly aligned with the machined recesses. Care also needs to be taken when fitting new piston rings and reassembling the pistons in the cylinders of an engine or compressor. Cast iron rings in particular are brittle and they should be opened only sufficiently for them to pass over the piston and be positioned in the turned grooves. When replacing the pistons in the cylinders, a compression sleeve should be placed around the rings. This can be tightened by means of a screw to compress the rings. The piston can then be pushed through the sleeve and into the cylinder without damage.

Test your knowledge 3.3

1. What do you understand by the terms *gasket* and *gland*?
2. What are the limitations of rotary lip seals?
3. How does a mechanical seal function?
4. What is the purpose of the different kinds of piston ring used in internal combustion engines?
5. What are the materials most commonly used for piston rings?

Activity 3.3

In addition to the oils seal which we have considered there is a type known as a labyrinth seal. There is also a related type that incorporates a screw thread. It is known as a viscoseal, or sometimes a screw seal or wind-back seal. Describe with the aid of sketches how these oil seals function and applications to which they are suited.

Bearings

Rotating shafts need to be supported. Depending on the application, the support is provided by plain bearings, ball bearings or roller bearings.

In addition to providing radial support, the bearings sometimes also need to accommodate axial forces or *end thrust* along a shaft. The plain bearings for steel shafts are very often made from phosphor-bronze. This has good load carrying properties which can be enhanced by impregnating it with graphite, as will be described in Chapter 4. Nylon is sometimes used for light duty low speed applications. Figure 3.11 shows a plane bearing which is designed to give both radial and axial support.

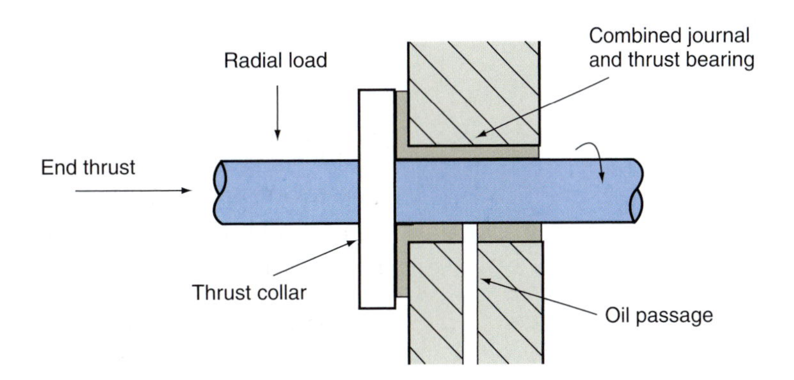

Figure 3.11 Journal and thrust bearing

Plain bearings are best suited to low rotational speeds and steady loads. It is essential that they are adequately lubricated with a relatively low viscosity oil. This is usually delivered under pressure in a re-circulating oil system. The oil also carries away heat and maintains the bearing at steady temperature. Very often a spiral groove, leading from the oil passage, is cut in the internal bearing surface to ensure even distribution of the oil.

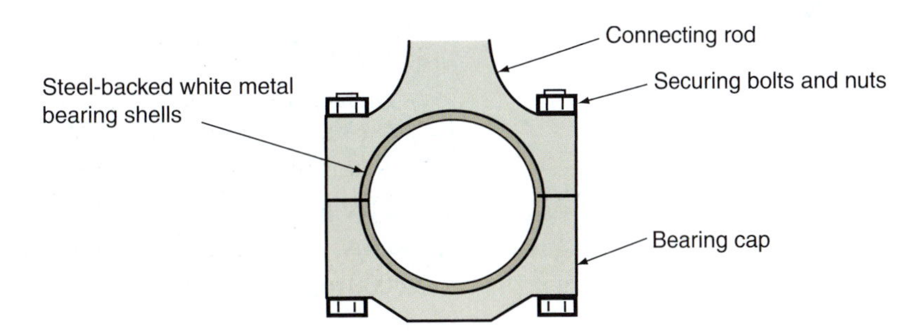

Figure 3.12 A split big-end bearing

Another type of plain bearing that is widely used in internal combustion engines and reciprocating compressors is the split bearing with steel-backed white metal bearing shells. They are used for the crankshaft main bearings and the connecting rod big-end bearings. Figure 3.12 shows a connecting rod big-end bearing. The shells are made from mild steel coated with a soft tin-lead alloy which is able to absorb the shock loading of the power strokes.

As with other plain bearings, adequate lubrication is essential and oil is delivered under pressure through oil passages in the crankshaft. White bearing metal has a very low melting point and overheating can be disastrous. Failure of the white metal layer is indicated by vibration and an unmistakable knocking sound. Great expense will then incurred in acquiring and fitting a replacement engine.

Ball and roller bearings are classed as rolling element bearings. They offer less resistance to motion than plain bearings and are able to operate at higher speeds and carry greater loads. They are of course more expensive but when correctly fitted and lubricated they give trouble-free service for long periods. Ball bearings may be contained in a cage that runs between an inner and outer ring or race, as shown in Figure 3.13. The balls and the rings are made from hardened and toughened steel which have been precision ground to size. The balls are not always contained in a cage, and for some low speed applications they are packed freely in grease between the rings. This is known as a *crowded assembly* and is to be found in the wheel and steering head bearings of some bicycles.

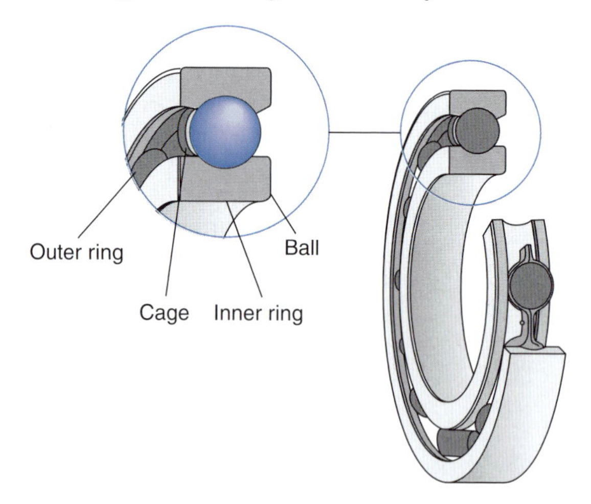

Figure 3.13 Caged ball bearing (Redrawn from *Mechanical Engineer's Reference Book*, 12th edition, E H Smith, Butterworth-Heinemann, p. 9/50, Fig. 9.35, 1998, with permission from Elsevier.)

Because of their low friction rolling action, grease lubrication is often sufficient for ball bearings and in recent years a range has been produced which are greased and sealed for life. As an example of their speed capability, a ball race with 8 mm diameter balls is capable of speeds up to 32 000 rpm with grease lubrication, and this can be exceeded with a re-circulating oil system.

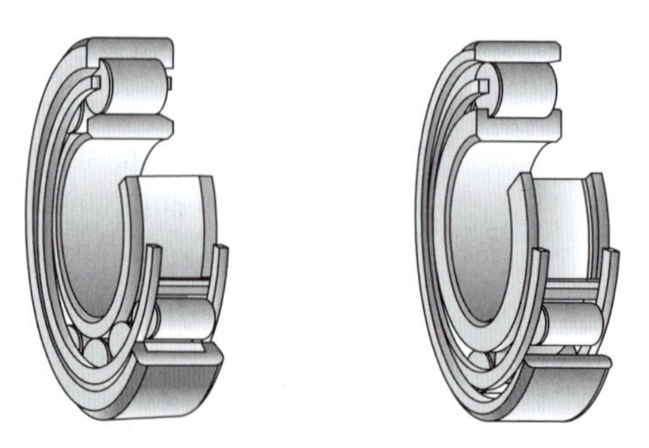

Figure 3.14 Caged roller bearings (Redrawn from *Mechanical Engineer's Reference Book*, 12th edition, E H Smith, Butterworth-Heinemann, p. 9/51, Fig. 9.39, 1998, with permission from Elsevier.)

Ball bearings make point contact with the rings whereas cylindrical roller bearings make line contact. As might be expected, they are able to carry greater loads than ball bearings of the same size. A selection of caged roller bearing is shown in Figure 3.14. It will be noted that some of the rings are fitted with flanges. These enable the bearings to carry a limited amount of axial load. Where the inner ring has no flanges or one flange only, it can be removed enabling the two parts of the bearing to be fitted separately. The cages for some ball and roller bearings are made from reinforced plastic material to reduce friction. These perform well provided that a low running temperature is maintained. Pressed steel cages should be used for running temperatures above 100°C.

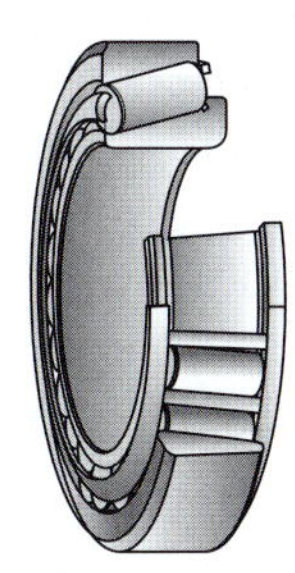

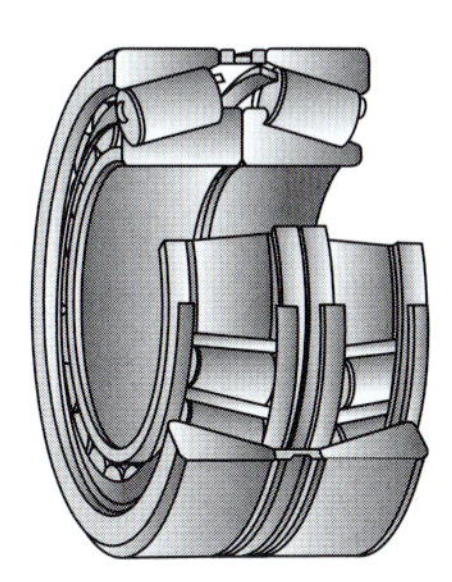

Figure 3.15 Tapered roller bearings (Redrawn from *Mechanical Engineer's Reference Book*, 12th edition, E H Smith, Butterworth-Heinemann, p. 9/51, Fig. 9.40, 1998, with permission from Elsevier.)

For applications where high radial and axial loads are to be carried it is advisable to use tapered roller bearings. These are shown in Figure 3.15. Single row tapered roller bearings can carry axial loads in one direction only. If axial loads are present in both directions the bearings may be paired back to back as shown. The front-wheel bearings in motor vehicles, which have to carry considerable radial loads and axial loads in both directions when cornering, contain tapered roller bearings that are paired in a similar way.

Needle bearings are roller bearings with long, thin rollers. They are designed for applications where there is limited radial space. Very often the inner race is dispensed with so that the needle rollers run directly on the rotating shaft. They are intended to carry light loads at relatively low or intermittent rotational speeds.

Care needs to be taken when assembling bearing races in position. Very often the rings needs to be pressed onto a shaft or into a housing and as with oil seals, a special centralising tool should be used to ensure correct alignment and seating. A possible arrangement on an input or output shaft from a machine shown in Figure 3.16.

The slide-ways on machine beds and worktables may be classed as linear bearing surfaces. The hardened cast iron slide-ways on lathe beds are V-shaped as are the mating surfaces in the tailstock and saddle assemblies. Tool slides and worktable slides are generally of dovetail section. The contact materials are mostly steel and cast iron. Because the movement on these slides is relatively slow and intermittent, no special bearing materials are required and regular lubrication by hand

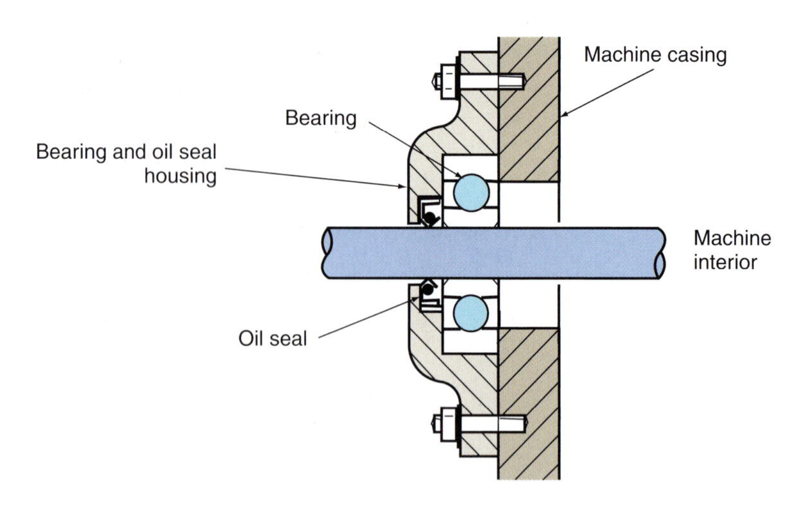

Figure 3.16 Bearing and oil seal arrangement

is all that is required for smooth operation. Where the movement on a slide is rapid and continuous, as with the ram of a shaping machine, the slides are often faced with a bearing material such as phosphor-bronze. Continuous lubrication is also required either by a wick feed or forced feed from a pump.

Test your knowledge 3.4

1. What is the usual material used for plain journal bearings?
2. Describe the bearing shells used in the main and big-end bearings of internal combustion engines?
3. What is meant by a crowded assembly of ball bearings?
4. What types of bearing are used to accommodate axial thrust?
5. What are needle bearings?

Activity 3.4

Use your resource centre or public library to find out about the components which make up the front-wheel hub of a motor car. Make a labelled sketch which shows how the major components are arranged. What is the method of lubrication and the recommended lubricant in the particular model which you have researched?

To check your understanding of the preceding section, you can solve Review questions 11–15 at the end of this chapter.

Fastenings

Screw fastening and rivets are the most widely used semi-permanent fastenings in engineered products. Screwed fastenings on covers and internal components allow access and removal for maintenance and repair. Components joined by rivets are not intended to be separated, but if the need arises this can be achieved by grinding off the rivet heads or drilling out the rivets. Most countries throughout the world now use the International Organisation for Standardisation (ISO) metric screw thread. The USA is an exception where feet and inches are still in widespread

use. Small sized screw fastenings, particularly those used in electrical and IT equipment, employ the British Association (BA) screw thread. Although this is British in origin, it is used internationally in these applications. The form of these screw threads is shown in Figure 3.17.

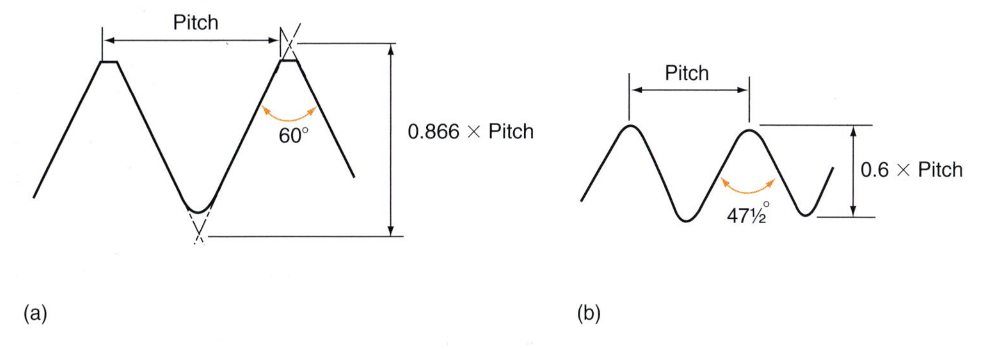

Figure 3.17 Screw thread forms: (a) ISO metric thread and (b) British Association (BA) thread

The most commonly used screwed fastenings are nuts and bolts, setscrews, studs and self-tapping screws. Metric nuts and bolts are specified in a way that describes their shape of head and nut, nominal diameter, pitch and bolt length. Figure 3.18 shows a metric hexagonal head bolt and nut.

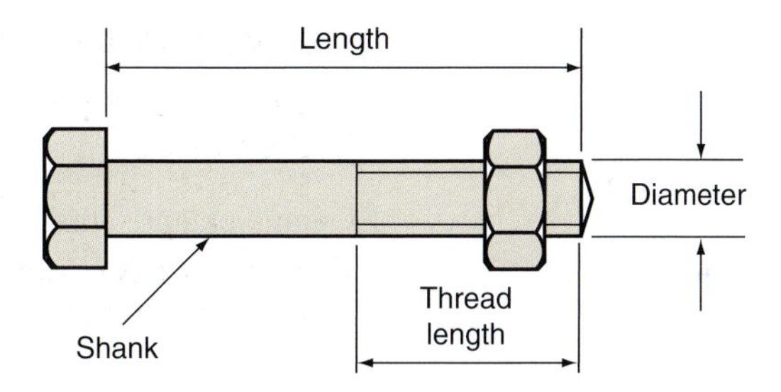

Figure 3.18 Hexagonal head nut and bolt

A hexagonal head nut and bolt might be specified as follows:

$$\text{Steel, hex hd bolt} - \text{M}12 \times 1.25 \times 100$$
$$\text{Steel, hex hd nut} - \text{M}12 \times 1.25$$

The M, specifies that it is metric, the 12 specifies the diameter in millimetres, the 1.25 specifies the pitch in millimetres and the 100 specifies the length in millimetres.

Bolts can be obtained with different lengths of thread for different applications. Bolts for general use are forged to shape and are generally known as *black bolts*. For applications where precision is required, fitted bolts can be obtained that have been accurately machined to size. Bolts made from high tensile steel are also supplied for heavy duty applications. Nuts and bolts are made from steel, brass, aluminium alloys and plastics. Steel nuts and bolts are sometimes cadmium plated to improve their resistance to corrosion.

Figure 3.19(a) shows two components that are joined by a nut and bolt. A bolt should always be selected such that the plain unthreaded part of

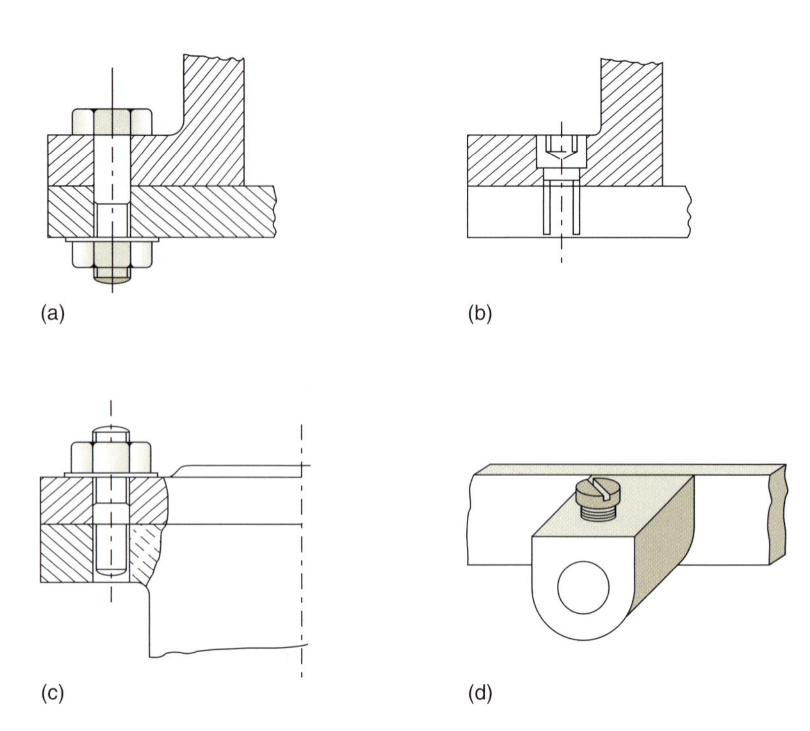

Figure 3.19 Typical screwed fastenings: (a) section through a bolted joint (plain shank extends beyond joint face), (b) cap head socket screw (head recessed into counterbore to provide flush surface), (c) stud and nut fixing for inspection cover (used where joint has to be regularly dismantled) and (d) cheese head brass screw (for clamping electrical conductor into terminal) (Redrawn from *Engineering GNVQ: Intermediate*, Second edition, M Tooley, Newnes, p. 135, Fig. 3.10, 2000, with permission from Elsevier.)

its shank extends through the joint interface. This is to ensure that there will be no shearing force on the weaker threaded part. It will be seen that a washer has been placed under the nut. This is good practice. It prevents damage to the component face and has the effect of spreading the load.

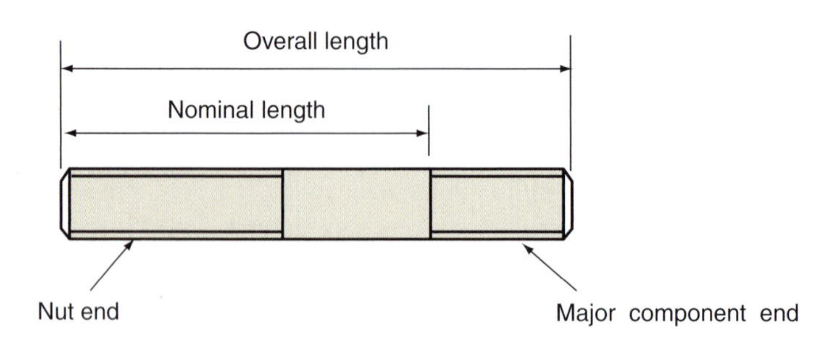

Figure 3.20 Stud

Figure 3.19(c) shows an alternative type of fastening using a stud and nut. A stud is a length of bar which has been threaded at each end as shown in Figure 3.20. The shorter threaded end is screwed into the major component, sometimes using a locking fluid to prevent it from easily becoming unscrewed. It is good practice to use studs where a component such as a machine inspection cover needs to be removed regularly for maintenance purposes. Because the studs are permanently in position in the machine casing, wear to the threaded holes is prevented. Any excessive wear will be at the nut end of the stud, which can be replaced.

The unthreaded portion, in the middle of a stud should be a little shorter than the component through which it passes and the threaded nut end should protrude through by a distance greater than the thickness of the nut. As with bolted joints, it is good practice to place a washer under the nut (Figure 3.21).

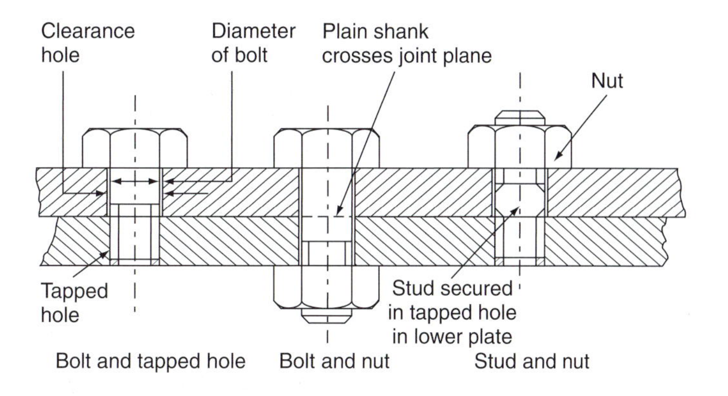

Figure 3.21 Use of bolts and studs ((Redrawn from *Engineering GNVQ: Intermediate*, Second edition, M Tooley, Newnes, p. 206, Fig. 3.87, 2000, with permission from Elsevier.)

Setscrews, which are also called machine screws, do not use a nut. Figure 3.22 shows how a thin plate might be joined to a larger component using setscrews. When joining sheet metal or thin plate components to a bulky object it is usual to use screws that are threaded over the whole of the shank.

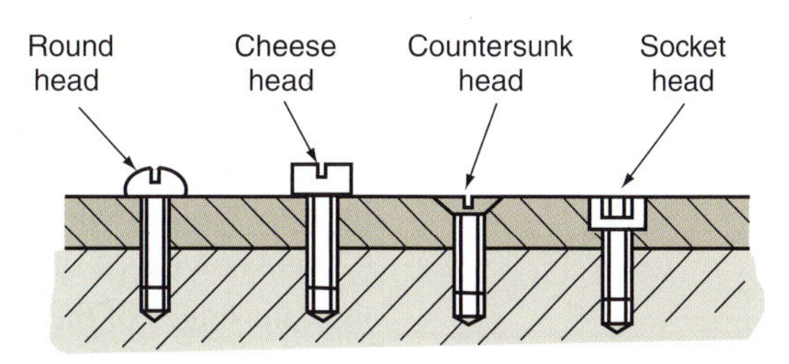

Figure 3.22 Set screw types

Screwed fastenings that are subject to vibration require a locking device to prevent them working loose. There are two basic categories of locking device. Those which have a positive locking action and those that depend on friction. Figure 3.23 shows a selection of both types. The positive locking category includes castle nuts, secured with a split-pin. The split-pin passes through a hole which is drilled in the bolt and its ends are opened to ensure that it stays in position. Split-pins should only be used once. After removal for maintenance they should be discarded and new ones used when the components are reassembled. Tab washers are also positive locking devices. After tightening the nut, one side is bent up against the nut face and the opposite side is bent over the

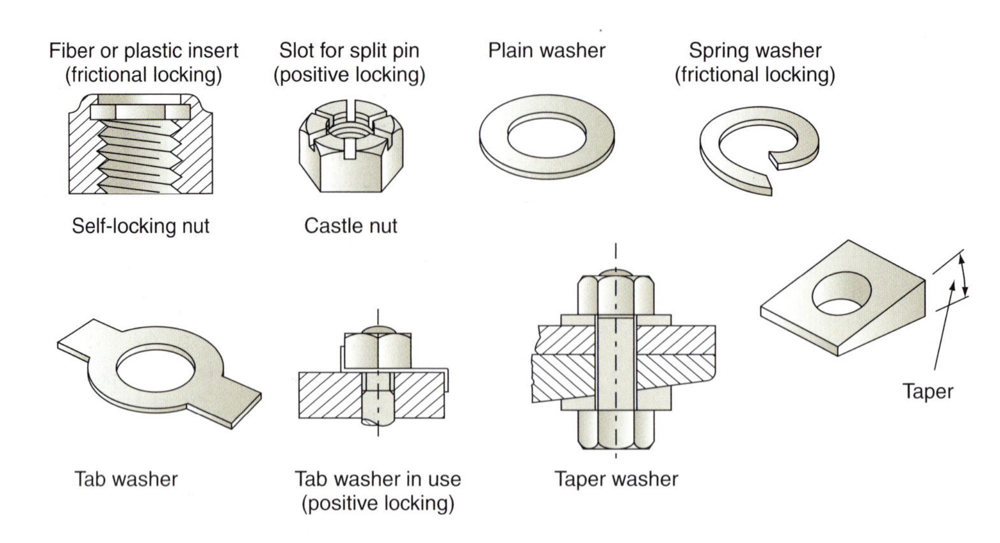

Figure 3.23 Locking devices (Redrawn from *Engineering GNVQ: Intermediate*, Second edition, M Tooley, Newnes, p. 207, Fig. 3.88, 2000, with permission from Elsevier.)

component as shown in Figure 3.23. A positive locking method which is used with setscrews is to drill a small hole across opposite faces of the head and secure two or more together with wire.

Among the locking devices that depend on friction are lock nuts, stiff nuts and spring washers. A lock nut is a secondary nut of smaller thickness that is tightened down on top of a plain nut. The combined frictional resistance is generally sufficient to withstand the effects of vibration. There are a variety of types of stiff nut. The type shown in Figure 3.23 can have a compressed fibre or a nylon insert. Those with a fibre insert are traditionally called Simmonds nuts and those with the nylon insert are often referred to as 'nyloc' nuts. Friction between these materials and the screw thread is sufficient to prevent them vibrating loose. As with split-pins it is good practice to use new friction nuts when reassembling components after maintenance.

Spring washers come in a variety of shapes. The one shown in Figure 3.23 is known as a split spring washer. When the nut is tightened down on it, the edges of the split resist anticlockwise turning. Other kinds of spring washer are stamped from spring steel to the shape of a many pointed star. The edges of the star are turned slightly so that like the split spring washer, they oppose anticlockwise turning of the nut. Figure 3.23 also shows how a wedge-shaped washer should be used when one of the joined components, such as a rolled steel section, has a tapering face.

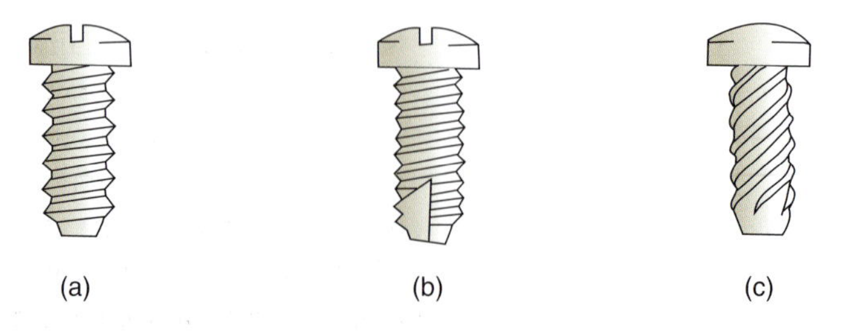

Figure 3.24 Self-tapping screws: (a) thread forming, (b) thread tapping and (c) drive in type

Self-tapping screws are widely used with sheet metal and plastic components (Figure 3.24). The type used depends on the hardness of the receiving material. For each type, a pilot hole is drilled whose diameter is the same, or slightly less than the root diameter of the thread. The thread forming type is used on soft materials such as thermoplastics. As it is screwed home it displaces the material to form its own mating thread. The thread cutting type is used on harder materials. It has a cutting edge at the start of its shank. This cuts a thread in the receiving material, rather like a screw cutting tap. The third type has a multi-start thread and is intended to be driven in by a hammer. It is used for applications where removal is not expected.

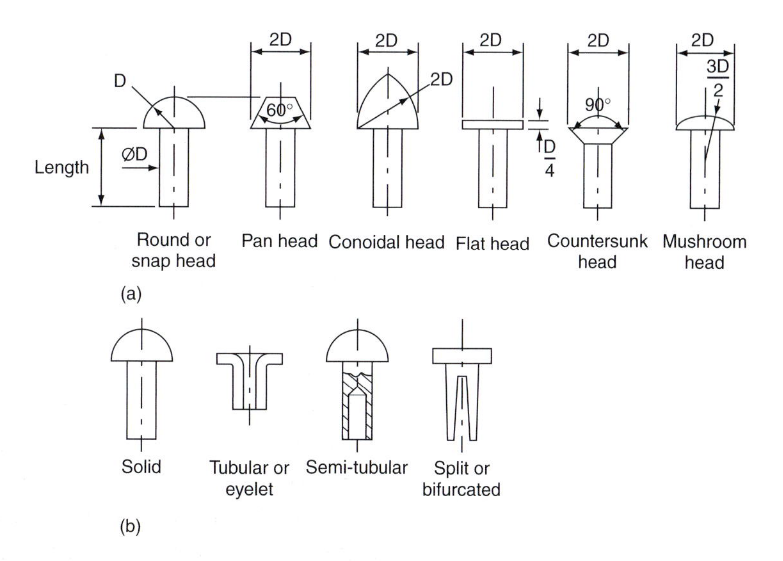

Figure 3.25 Rivets (a) Some type of rivet head and approximate proportions (b) Types of rivets (Redrawn from *Engineering GNVQ: Intermediate*, Second edition, M Tooley, Newnes, p. 137, Fig. 3.13, 2000, with permission from Elsevier.)

Rivets are often preferred to screw fastenings for components which are expected to be joined permanently. Many of their former uses have been taken over by welding but they are still widely used in aircraft production, particularly for joining light metal alloys which require special welding skills. They produce very strong joints and have the advantage that they are not so rigid as welded joints. They are able to flex a little under load and this is often desirable particularly with shock loads.

The composition of rivet material should always be as close as possible to that of the materials being joined. Otherwise, electrolytic corrosion of the type described in Chapter 4 might occur in the presence of moisture. Figure 3.25 shows a variety of rivet types for different applications. Rivets should not be loaded in tension. They should only be subjected to the single or double shearing loads described in Chapter 1. Some tensile stress will inevitably be present from the setting operation, particularly after hot riveting as the rivet cools. Round, conoidal and pan heads have the greatest strength whilst the countersunk and flat heads are used where a flush joint surface is required. Figure 3.26 shows the correct proportions for a riveted joint and the correct way in which rivets should be closed.

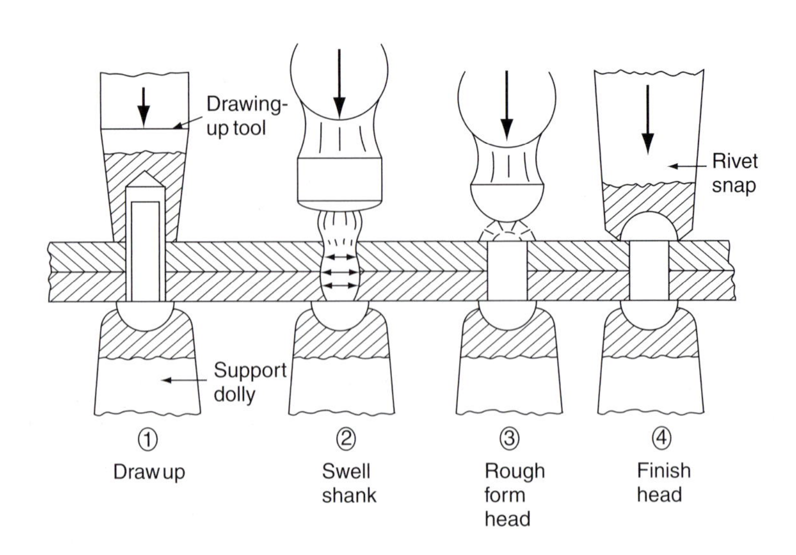

Figure 3.26 Correct riveting procedure (Redrawn from *Engineering GNVQ: Intermediate*, Second edition, M Tooley, Newnes, p. 208, Fig. 3.90, 2000, with permission from Elsevier.)

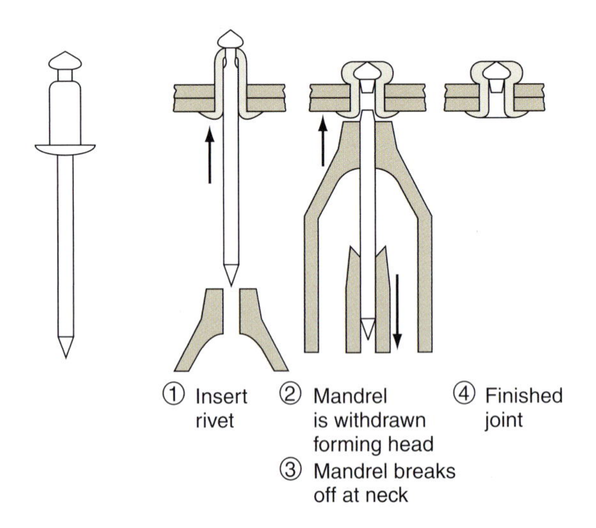

Figure 3.27 Pop riveting procedure (Redrawn from *Engineering GNVQ: Intermediate*, Second edition, M Tooley, Newnes, p. 208, Fig. 3.90, 2000, with permission from Elsevier.)

The pop rivets shown in Figure 3.27 are widely used with sheet metal fabrications where access is only possible from one side of the joint. After inserting the rivet in the joint, the head on the blind side is formed by pulling the central mandrel through the rivet using a special tool. As the rivet is formed, the mandrel breaks off at its head and is discarded.

Test your knowledge 3.5

1. What does the following specification indicate?

   Steel, hex hd M8 × 1 × 50

2. What are the two main categories of locking device used with screwed fastenings?
3. What kind of application are studs used for?
4. Under what circumstances might a riveted joint be preferred to welding?
5. For what application are pop rivets useful?

Activity 3.5

a) There are types of stiff nut other than the ones that we have considered. Describe two of these with the aid of sketches and the way that they function.

b) What are shims, and how they are used in the assembly of engineered products?

To check your understanding of the preceding section, you can solve Review questions 16–20 at the end of this chapter.

Mechanical Power Transmission Systems

The purpose of mechanical power transmission systems is to transmit force and motion from one point in a machine to another. Rotational power is transmitted by means of drive shafts, belt drives and chain drives. In some cases it is required to convert rotational motion to linear motion of a particular type. Cams, slider-crank mechanisms and the rack and pinion are used for this purpose. The drive from a motor or engine may need to be connected and disconnected to a machine as part of a working cycle. A clutch is used for this purpose and if it is required to bring the moving parts quickly to rest, some braking device is needed.

Cams

A *cam* is a rotating or oscillating body which imparts reciprocating or oscillatory motion to a second body which is in contact with it. The second body is called a *follower*. The most common types of cam are the radial or plate cam and the cylindrical cam and the face cam as shown in Figure 3.28.

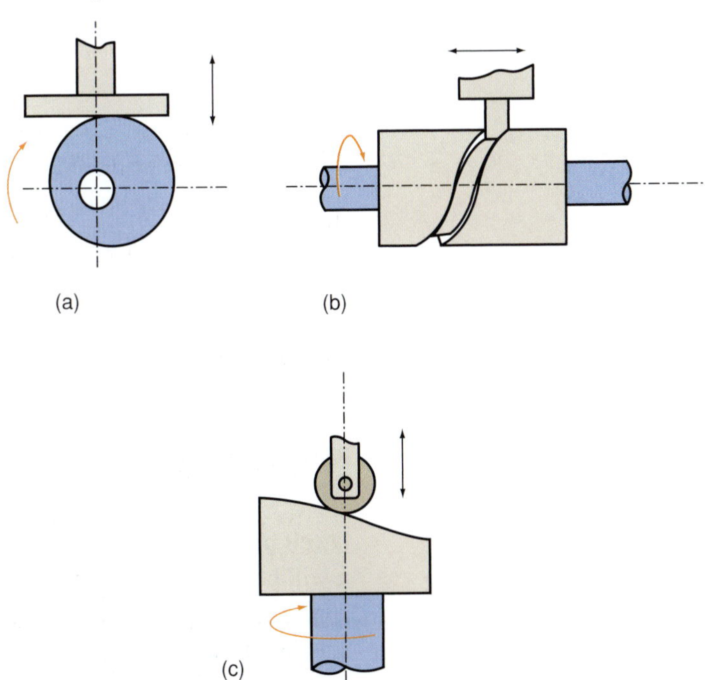

Figure 3.28 Types of cam: (a) eccentric circular plate cam with flat-faced follower, (b) cylindrical cam and (c) face cam with roller follower

KEY POINT

Dwell angle is the angle turned by a cam whilst the follower is stationary.

Knife edge, flat-faced and roller followers are all used with plate and face cams. The knife edge type is only suitable for low speed applications where there is light contact force. The roller type is more wear resistant and is widely used in automated machinery. In automobile engines where the space is confined the flat-faced follower is generally used, offset as shown in Figure 3.29(b). This causes it to rotate about its own avis so that wear is evenly distributed over the contact face. If a follower is stationary for part of the rotation cycle, it is said to *dwell* and the angle through which the cam turns during the dwell period is known as the *dwell angle*. Figure 3.29(b) shows a dwell angle which will occur when the follower at its lowest position. Here the cam profile is circular with its centre on axis of rotation.

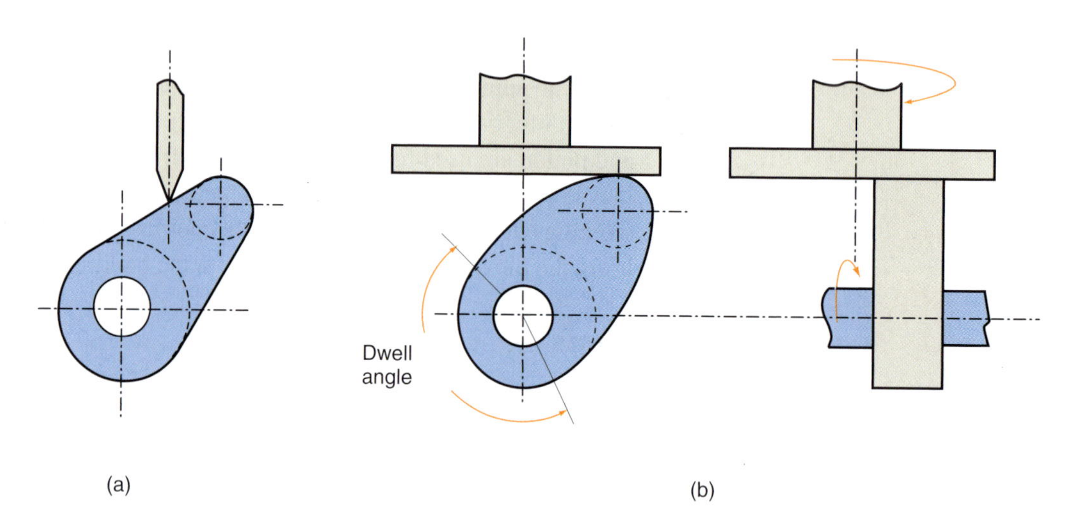

Figure 3.29 Cams with offset followers: (a) plate cam made up of circular arcs and tangents with knife-edged follower offset from axis of rotation and (b) plate cam made up of circular arcs with side offset flat-faced follower

KEY POINT

There are two categories of cam. Those where the chosen shape determines the motion of the follower and those where the specified follower motion determines the shape of the cam.

Cam profiles can be divided into two categories. Cams for which the cam profile is made up of circular arcs and tangents, as shown in Figure 3.29 and cams for which the required nature of the follower motion determines the cam profile, e.g. rise and fall with uniform velocity, rise and fall with simple harmonic motion and rise and fall with uniform acceleration and retardation. This latter category are more expensive to produce and confined to special purpose applications. Performance graphs showing the rise and fall of a knife edged follower against angular position of the cam for these motions are shown in Figure 3.30.

Dwell periods have been omitted from the cams in Figure 3.30 and it will be noted that the cams have a characteristic heart-shaped appearance. In certain applications, two or more of the different motions may be required together with dwell periods when the follower is at its extreme positions. Additionally, the cam may have a roller or flat-faced follower with offset. The projection construction process then becomes more complex and is beyond the range of this unit.

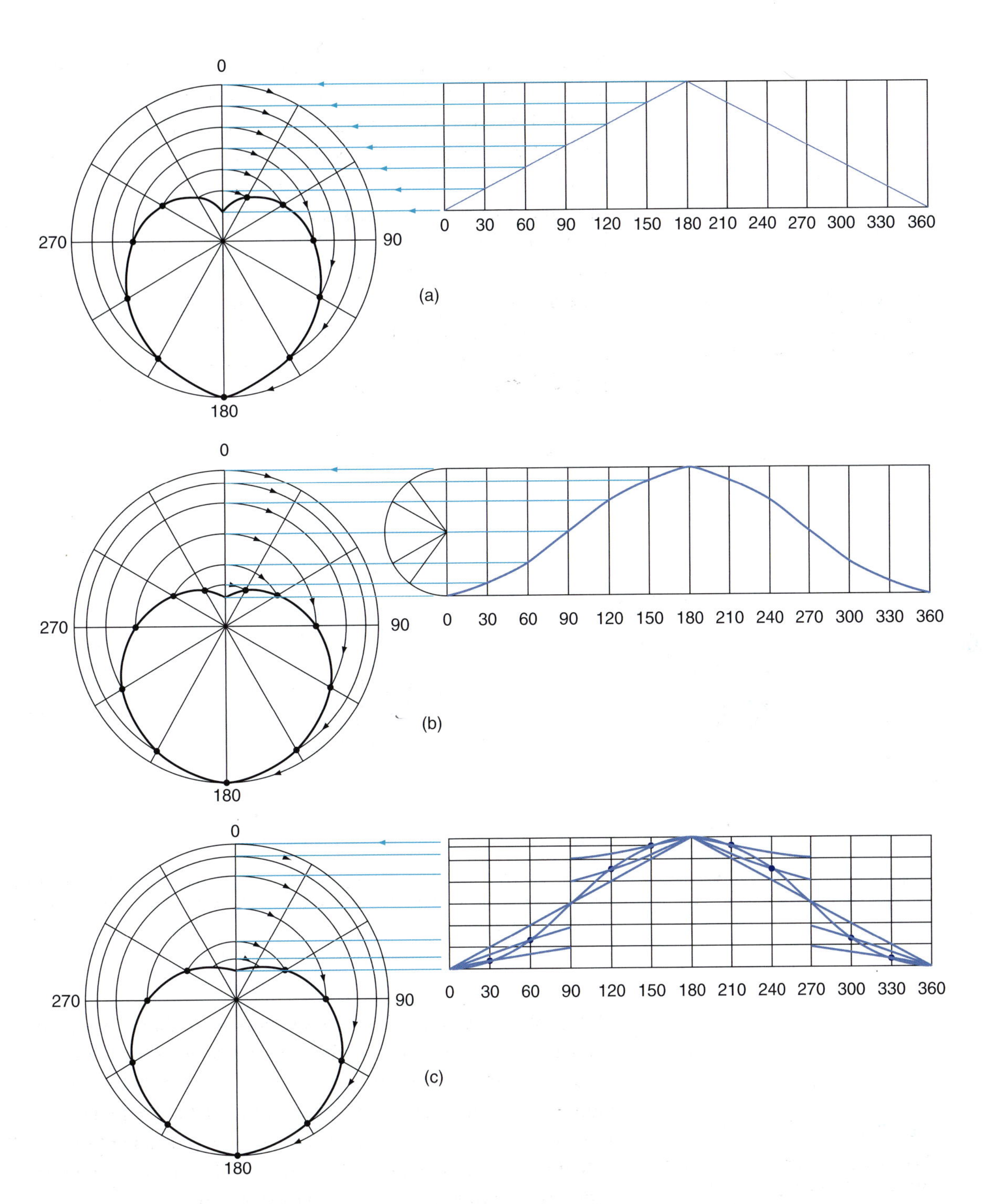

Figure 3.30 Performance graphs and cam profiles

Test your knowledge 3.6

1. What is a face cam?
2. How can flat-faced followers be made to rotate as they rise and fall?
3. What is a dwell angle?
4. What are the three main types of cam follower?

Activity 3.6

Plot the performance graph and cam profile for a plate cam with a knife edged follower to the following specifications:

Base circle diameter 25 mm.

Rise 75 mm with simple harmonic motion whilst rotating through an angle of 120°

Dwell for a further 60° of rotation.

Fall 75 mm with uniform acceleration and retardation whilst rotating through an angle of 150°.

Dwell for final 30° of rotation.

Linkage mechanisms

A mechanism may be defined as a *kinematic chain* in which one element is fixed for the purpose of transmitting or transforming motion. A kinematic chain is made up of a number of linked elements. Two elements which are linked together constitute a *kinematic pair*. If they are hinged, they are called a *turning pair*. If one element is constrained to slide through or around another, they are called a *sliding pair*. If one element rotates inside another in a screw thread, they are called a *screwed pair*.

KEY POINT

Linked elements in a kinematic chain may constitute a turning pair, a sliding pair or a screwed pair.

Two of the most common mechanisms are the slider-crank and the four-bar chain and we have given these some consideration in Chapter 2. Closely related to these are the slotted link and Whitworth quick return motion and Watt's parallel motion. When a mechanism is required to transmit power, the various links and joints have to be designed, with an appropriate factor of safety, to carry the forces to which they will be subjected. The mechanism, or a series of linked mechanisms, is then classed as a *machine*.

Slider-crank mechanisms

The slider-crank mechanism finds widespread use in reciprocating engines, pumps and compressors. It is made up of three links and a slider as shown in Figure 3.31. Link AB is the crank which in Figure 3.31(a) rotates at a steady speed. Link BC is the connecting rod or coupler, which causes the piston C to slide in the cylinder. The third link is the cylinder block itself, AD which is stationary and on which the crank rotates at A. As stated above, a mechanism is a chain of links, one of which is fixed. In this case it is the cylinder block AD.

KEY POINT

An inversion of a mechanism is obtained by changing the link which is fixed.

In Chapter 2 we used a graphical method to find the speed of the piston at a given instant in time, when the crank was at a particular angular position and rotating at a given speed. We now need to consider *inversions* of the mechanism. An inversion is obtained by holding another link in the fixed position. The inversion shown in Figure 3.31(b) is called an oscillating cylinder mechanism. It is obtained by holding the link BC in the fixed position so that the crank AB rotates about B. This imparts a combined rocking and oscillating motion to the cylinder which pivots about the stationary piston.

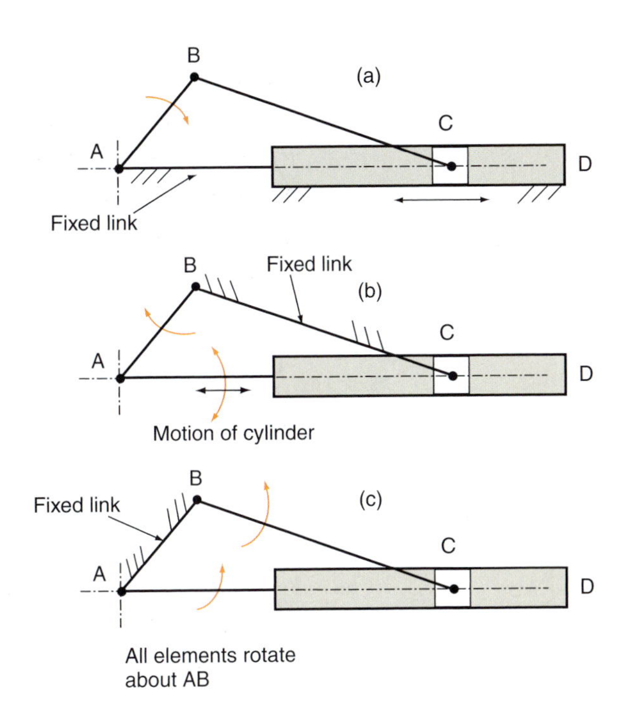

Figure 3.31 Slider-crank mechanisms and inversions

The inversion shown in Figure 3.31(c) is called a rotating cylinder mechanism. It is obtained by holding the link AB in the fixed position. Both the link BC and the cylinder AD rotate about the fixed link. The arrangement was used in the early days of aviation for rotating radial cylinder aero-engines. It also form the basis for the Whitworth quick return motion which we shall shortly be considering.

Four-bar linkage mechanisms

The four-bar linkage is another mechanism which finds widespread use. It is to be found in applications such as windscreen wiper drives, vehicle suspension units and everyday uses such as the hinges on kitchen cupboard doors and squeeze-mop mechanisms. Two of the links rotate about fixed centres and are joined by a coupler link. The fourth link is formed by the frame or bed plate that contains the fixed centres of rotation.

It should be noted that the number of inversion of a mechanism is equal to the number of links, which in this case is four. In Figure 3.32 the links are all of different length and chosen so that the sum of the longest and shortest is less that the sum of the other two.

$$\text{i.e.} \quad AB + AD < BC + C$$

When this is the case, three distinct types of mechanism can be obtained from the inversions. The inversions shown in Figure 3.32(a) and (c) are both crank-rocker mechanisms. It is the longest two links which are fixed whilst the input crank AB rotates at a steady speed, first about end A and then about end B. This imparts a rocking action to the output link CD.

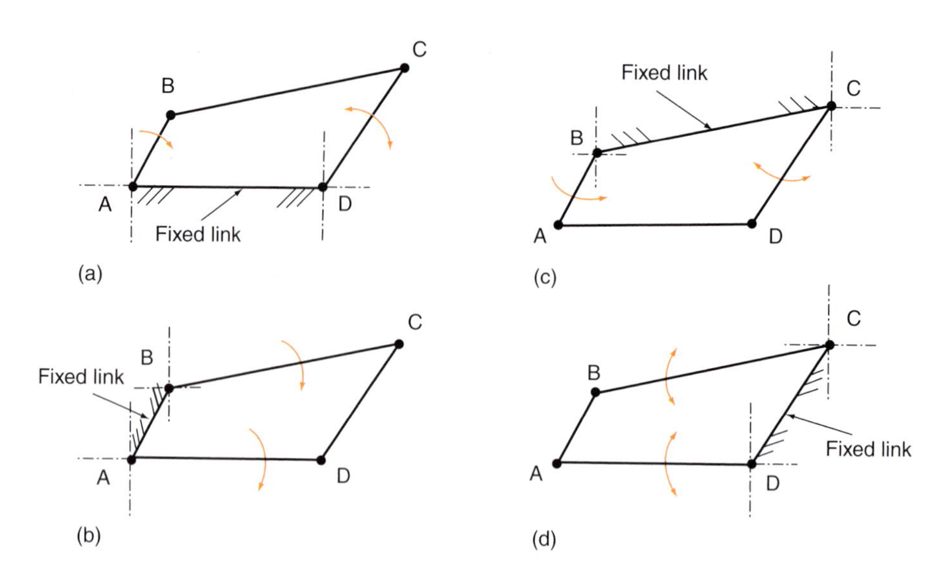

Figure 3.32 Four-bar linkage mechanism inversions: (a) crank-rocker nechanism, (b) double crank mechanism, (c) crank rocker and (d) double rocker

KEY POINT

The number of possible inversions of a mechanism is equal to the number of links.

With the inversion shown in Figure 3.32(b), the shortest link AB is fixed and this produces a double crank mechanism. Link BC rotates about B and link AD rotates about A. If BC is the input crank, rotating at a steady speed, the link AD will rotate at a varying speed and the links must be able to cross over certain times. The link CD will have in a complicated motion. It will rotate about the fixed link AB and also rotate about its own centre as the links BC and AD cross over each other.

With the inversion shown in Figure 3.32(d), the second shortest link CD is fixed and this produces a double rocker mechanism. The links BC and AD rotate about their fixed ends, C and D but are unable to describe a full revolution.

Watt's parallel motion

This is an application of the four-bar linkage which gives approximate straight line motion to a particular point, *P* on the coupler link. The arrangement is shown in Figure 3.33.

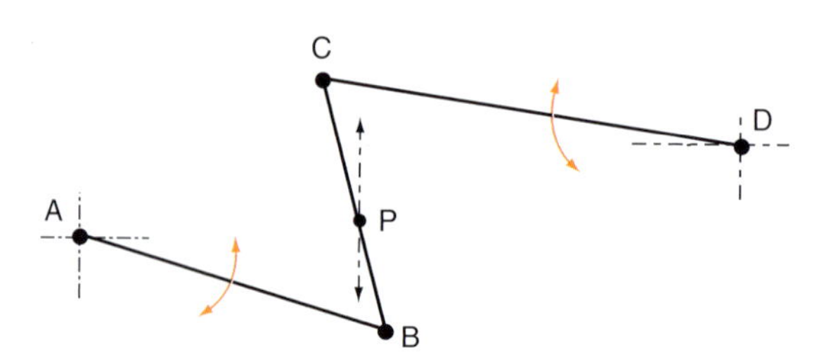

Figure 3.33 Watt's parallel motion

The point *P* travels in a line that is approximately straight between the positions where AB is horizontal and CD is horizontal. The position of the point *P* on the coupler link should be chosen so that:

$$\frac{AB}{CD} = \frac{BP}{PC}$$

Quick return mechanisms

Slow forward and quick return mechanisms are to be found on some metal cutting machines, printers and scanners. The output gives a slow forward linear motion followed by a quick return to the start position along the same path. Two of the most common which are found in shaping, planning and slotting machines for metal cutting are the slotted link mechanism and the Whitworth quick return motion.

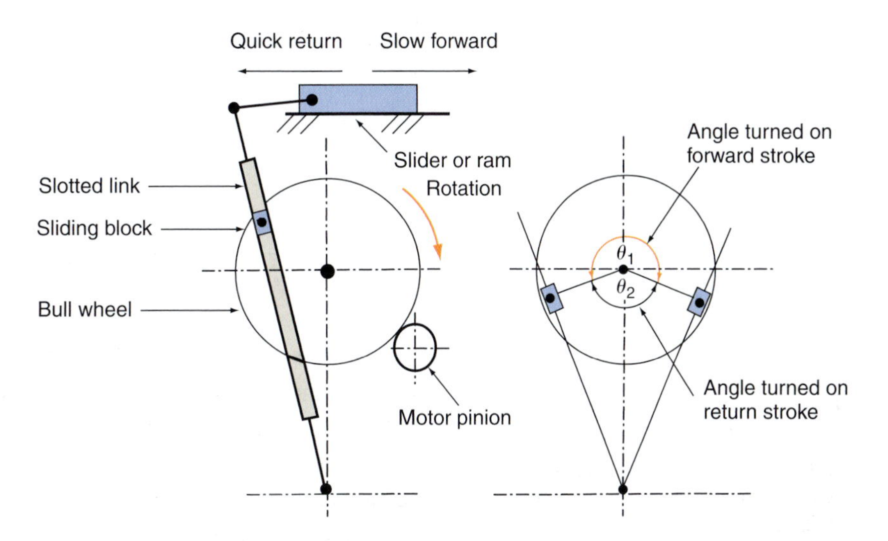

Figure 3.34 Slotted link quick return mechanism

The slotted link mechanism is used on shaping machines where single point cutting tool is mounted on the front of the slider or ram, in a hinged tool post. The tool cuts on the slow forward stroke and lifts over the workpiece on the quick return stroke. The slotted link rocks from side to side, driven by the sliding block on the bull wheel.

The bull wheel rotates at a constant speed and as can be seen from Figure 3.34, the angle through which it rotates on the forward stroke is greater than the angle through which it rotates on the return stroke. This imparts the slow forward and quick return motion to the slotted link and slider. The distance of the sliding block from the centre of the bull wheel can be altered to vary the length of stroke of the slider.

The Whitworth quick return motion also employs a slotted link and sliding block as shown in Figure 3.35. The mechanism is used on planing machines, which are quite large, and on slotting machines which are small and compact. With slotting machines a single point tool is fixed to the front of the slider and is used for cutting fine grooves and key-ways. With planing machines the slider is the worktable on which the workpiece is secured. This moves with slow forward and quick return motion beneath a stationary single point cutting tool.

The driving gear which contains the sliding block, rotates at a constant speed. The sliding block causes the slotted link to rotate but because it has a different centre of rotation, its speed is not constant. As can be

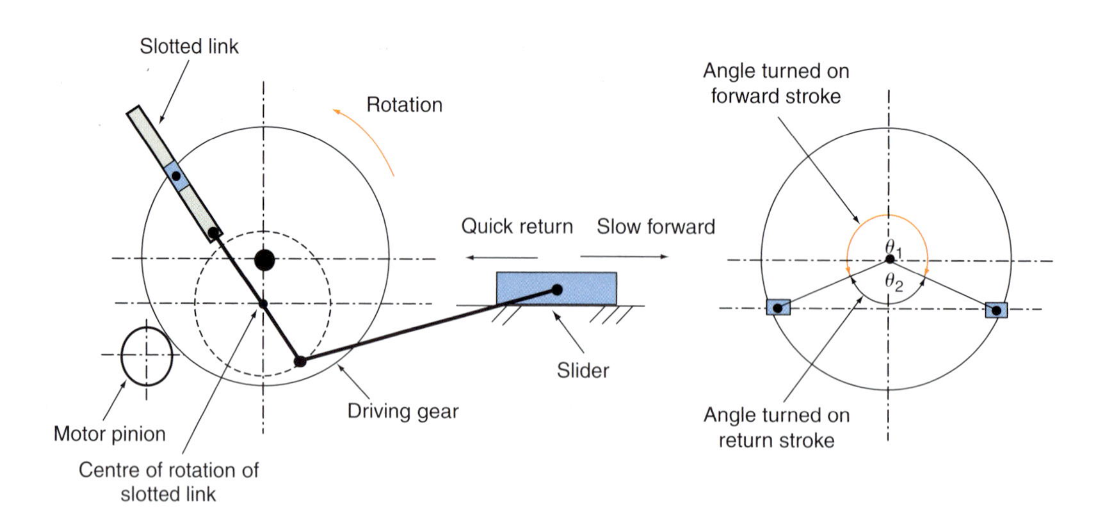

Figure 3.35 Whitworth quick return mechanism

KEY POINT

The slotted link and Whitworth quick return mechanisms are practical applications of the slider-crank inversion shown in Figure 3.31(c).

seen from Figure 3.35, the angle through which the driving gear and slotted link rotate on the forward stroke is greater than the angle through which they rotate on the return stroke. This imparts the slow forward and quick return motion to the slider.

***Test your knowledge* 3.7**

1. What constitutes a kinematic pair?
2. What is an inversion of a mechanism?
3. Which link of a slider crank mechanism is fixed to produce the oscillating cylinder inversion?
4. What are the conditions which must apply for the inversions of a four-bar linkage to produce three distinct mechanisms?
5. Which of the slider-crank inversions is used in the Whitworth quick return motion?

Activity 3.7

When turning a corner in a car, the front wheel on the inside of the curve has to run through a larger angle than the wheel on the outside of the curve. Find out the name and sketch the layout of the steering linkage used. Which of the mechanisms that we have described is it based upon?

To check your understanding of the preceding section, you can solve Review questions 21–25 at the end of this chapter.

Shafts, clutches and brakes

Transmission shafts are widely used to transmit power from prime movers such as electric motors and internal combustion engines. They are used to drive machinery, pumps, compressors and road vehicles. Transmission shafts are generally made from steel and may be of solid or tubular section.

Where the drive extends over long distances, it is often necessary to couple sections of shaft together and support them in bearings.

Several different kinds of coupling are available and a clutch is often incorporated in the system to connect and disconnect the drive. Good alignment is desirable between the driving motor and driven machinery but where this is not possible, universal and constant velocity joints can be used turn the line of drive into the required direction.

Joints and couplings

When two shafts are joined directly together, the joint is called a coupling. The coupling may be rigid or flexible. A rigid coupling is shown in Figure 3.36. It should only be used where the centres of rotation of the shafts are concentric. It consists of two coupling flanges keyed to the ends of the separate shafts and bolted together.

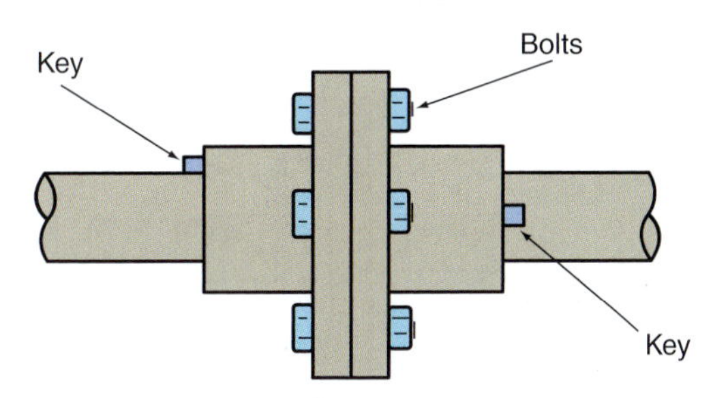

Figure 3.36 Rigid coupling

If the shafts are in line but there is likely to be shock loading or excessive vibration, a degree of flexibility is required. The same applies if one of the machines is likely to undergo a temperature change, causing a change in shaft height due to expansion. This is known as *thermal growth* and also requires a degree of flexibility. This can be achieved by the fitting of rubber or plastic bushes around the bolts or by separating the flange faces with flexible disc, as shown in Figure 3.37(a) and 3.37(b). There are many variations of these designs where rubber bushes and pads are used to provide flexibility. They are generally known as *elastomeric* couplings.

KEY POINT

Elastomeric couplings employ rubber or flexible plastic bushes and pads to absorb shock loads and vibration.

The type shown in Figure 3.37(c) employs a metal disc to which the forked ends of the shafts are connected. In addition to allowing some flexibility, this can also function as a safety device which will fail in the event of an overload. The type shown in Figure 3.37(d) allows for a small degree of misalignment. Hubs with external gear teeth are keyed on the ends of the shafts. These locate in the flanged parts which have internally cut teeth. The length of the teeth on the hubs is relatively short which allows limited angular deflection of the shafts as they rotate.

The degree of accuracy required when aligning shafts that are to be coupled depends on the operating conditions and the type of coupling that is to be used. In the case where two shafts have only a small gap between them and a flexible coupling is to be used it may be sufficient to align them by means of a straight edge as shown in Figure 3.38.

Assuming the machine carrying shaft A to be fixed, the position of shaft B is adjusted until the straight edge makes even contact with both shafts. The straight edge is then moved through 90° to the position shown by the dotted line and if required, further adjustments are made.

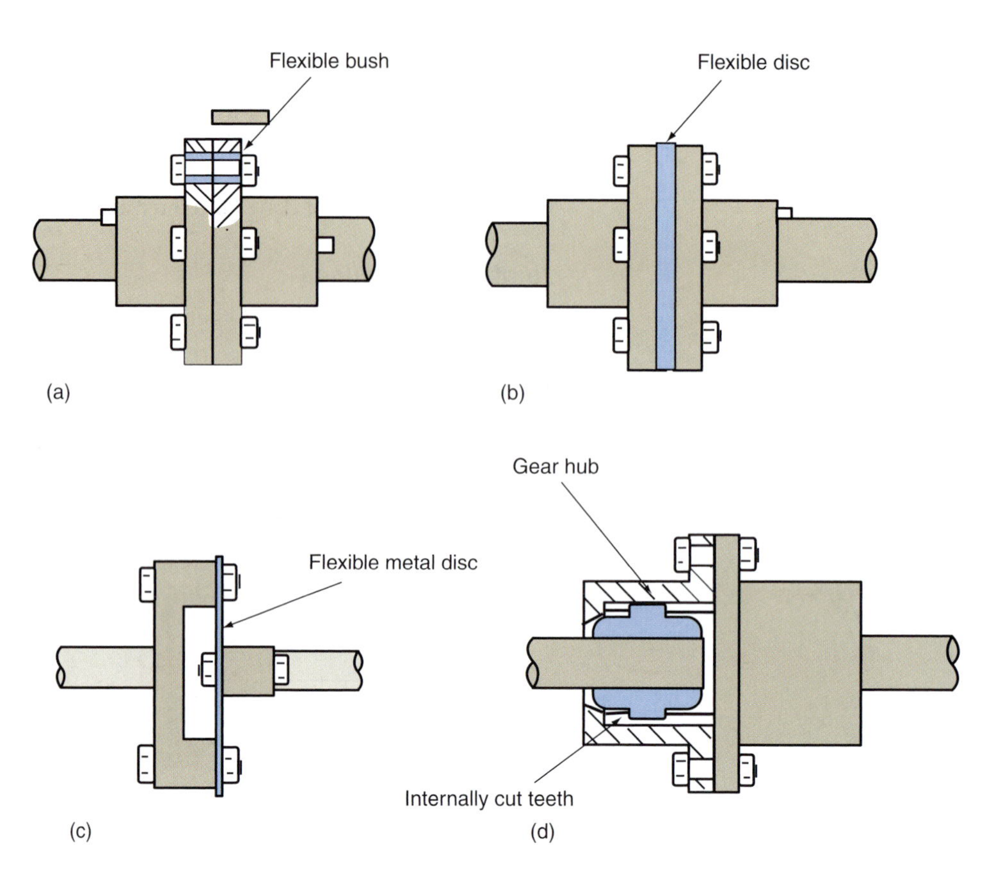

Figure 3.37 Flexible couplings: (a) flexible bush, (b) flexible disc, (c) flexible disk and (d) gear hub

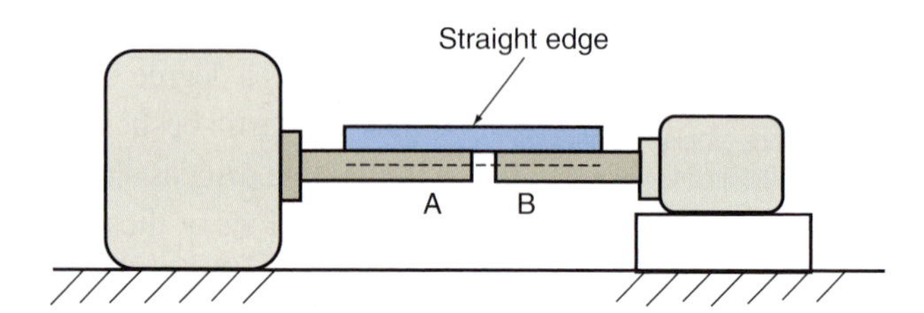

Figure 3.38 Alignment using a straight edge

KEY POINT

Misalignment is present if the shaft axes are parallel but not coincidental, if their axes intersect at some point there is a combination of these two faults.

There are two kinds of misalignment. Parallel misalignment is when the shaft axes are parallel but not in line. Angular misalignment is when the axes intersect at some point. There may of course be a combination of the two requiring linear and angular adjustment of the movable shaft.

If the shafts are to be joined by a rigid coupling, a more precise degree of alignment is required and if the gap between the shafts is small, a dial test indicator can be used as shown in Figure 3.39. The dial test indicator is mounted on a Vee-block and positioned as shown. The Vee-block is then moved along shaft A and deflection of the pointer denotes angular misalignment in that plane which can be corrected by turning the machine of B. When there is no movement of the pointer, the Vee-block is moved through 90° to the position shown by the dotted line. Once again it is moved along shaft B and further angular adjustments are made until there is no movement of the pointer.

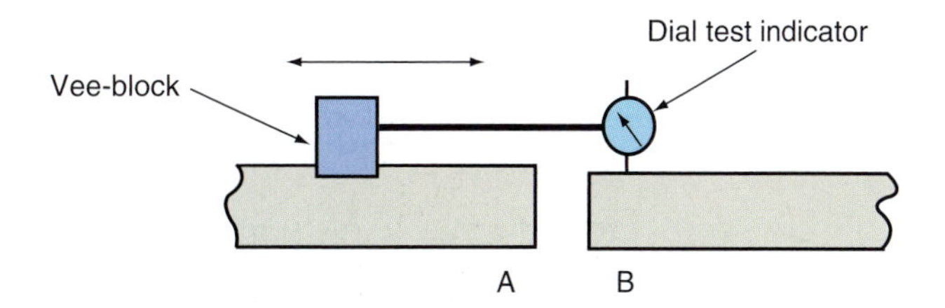

Figure 3.39 Alignment using a dial test indicator

There should then be no angular misalignment but parallel misalignment may still be present. To check for this, the Vee-block is clamped to shaft *A* which is then rotated with shaft *B* stationary. Deflection of the pointer denotes parallel misalignment and the machine of shaft *B* is moved over or raised until there is no pointer movement. The Vee-block is then moved to shaft *B* with the dial test indicator in contact with shaft *A*, and the procedure is repeated.

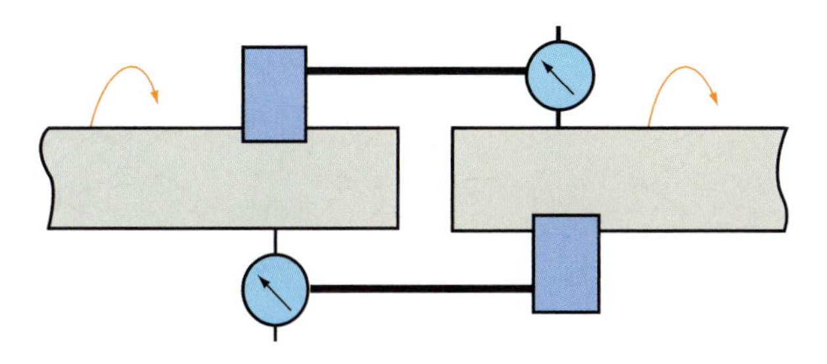

Figure 3.40 Alignment using two dial test indicators

An alternative method is to use two dial test indicators positioned on opposite sides of the two shafts as shown in Figure 3.40. The Vee-blocks are clamped to the shafts which are rotated together. The deflection of the two pointers at different angular positions enables judgements to be made as to the kind of misalignment present. Adjustments are then made until the shafts rotate without any movement of the pointers. The advantage of this method is that there is no sliding of the Vee-blocks or the dial test indicator plungers. Errors due to surface roughness and ovality of the shafts are thus eliminated.

KEY POINT

The method which uses two dial test indicators eliminates errors due to surface finish and ovality.

In cases where alignment is required between machines that are separated by longer distances it is modern practice to use laser equipment. The laser beam is used to ensure that the drive lines from the two machines lie in the same horizontal and vertical planes. One method is to mount the laser to one side of the drive line at the required shaft height, and set so that it can be swung from one machine to the other in a true horizontal plane. After the necessary adjustments, the process is repeated with the laser mounted above the machines and set so that it can be swung from one to the other in a true vertical plane. The laser can also be mounted to act along the drive line to enable the accurate positioning of bearings.

A *Hooke joint*, which is also called a *universal coupling*, can be used where it is required to turn the drive line of a shaft through an angle. Angles of up to 30° are possible with this kind of joint which is shown

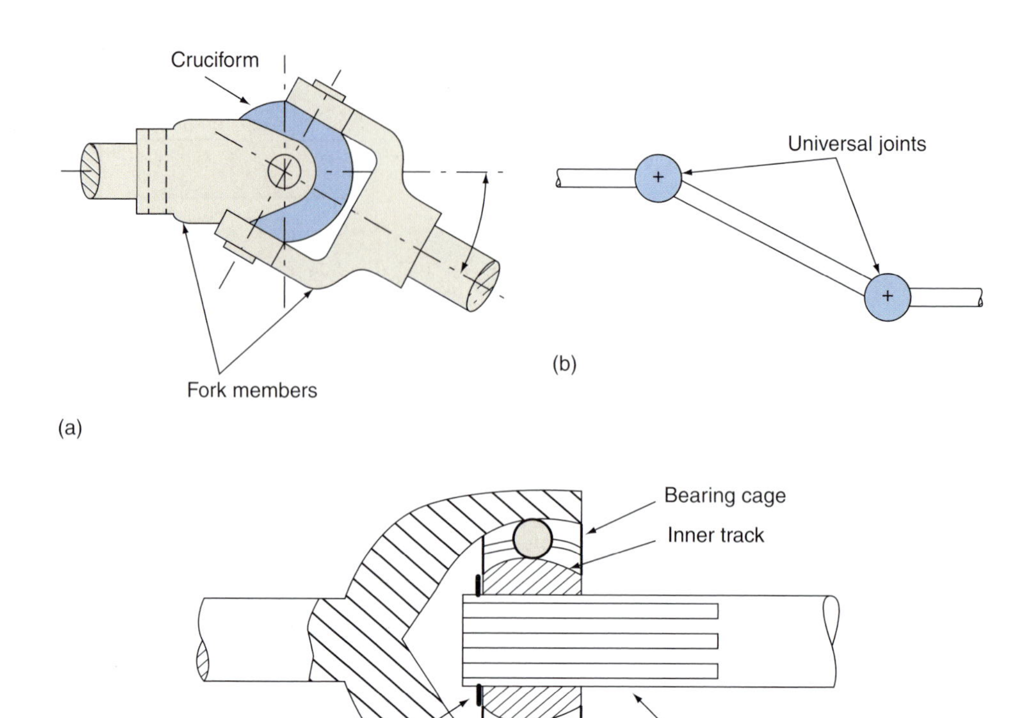

Figure 3.41 Hooke coupling and constant velocity joint: (a) Hooke joint, (b) arrangement of Hooke joints and (c) constant velocity joint

in Figure 3.41. It is widely used in cars and commercial vehicles at each end of the propeller shaft that connects the gearbox to the rear axle. The joint consists of two forked members that are attached to the ends of the shafts to be connected. The connection is made through a cross-shaped or cruciform member on which the fork ends pivot. Needle roller bearings, packed with grease, are generally used on the pivot.

The main disadvantage of the joint is that the output speed fluctuates and the effect increases with the angle between the connected shafts. For this reason, it is usual to use two Hooke joints as shown in Figure 3.41(c). The fork ends at each end of the central shaft are positioned in the same plane and in this way, the speed fluctuations are cancelled out. The speed of the central connecting shaft will still fluctuate but the input and output shafts will rotate at a steady speed.

KEY POINT

The condition for a universal coupling to transmit constant velocity is that the joint driving contacts must always be in a plane which bisects the angle between the connected shafts.

The condition for a universal coupling to transmit constant velocity is that the joint driving contacts must always be in a plane which bisects the angle between the connected shafts. In the case of the Hooke joint, the driving contacts are the arms of the cruciform member. As can be seen from Figure 3.41(a), the condition is not met. At the instant shown, the plane of the cruciform is at right angles to centre line of the shaft on the right. The *constant velocity joint* shown in Figure 3.41(b) overcomes the problem. The drive is transmitted through ball bearings which can move in tracks on the joint members. The geometry of the design is such that the plane in which the balls lie exactly bisects the angle between the shafts. The constant velocity joint is widely used in the drive between the gearbox to the wheels in front-wheel drive cars.

Test your knowledge 3.8

1. What is *thermal growth*?
2. What is an *elastomeric* coupling?
3. What is the advantage of the method of shaft alignment that uses two dial test indicators?
4. What is the main disadvantage of a Hooke joint?
5. What are the conditions required for a universal coupling to give a constant velocity ratio?

KEY POINT

Dog clutches should only be engaged and disengaged when the shafts are stationary.

Clutches and brakes

Clutches are used to connect and disconnect the drive in mechanical power trains. The dog clutch is one of the most basic types which is sometimes regarded more as a coupling than a clutch. The two shafts are joined by engaging interlocking teeth on the input and output shafts. Three different kind are shown in Figure 3.42. These are positive engagement clutches which should only be operated when the shafts are at rest.

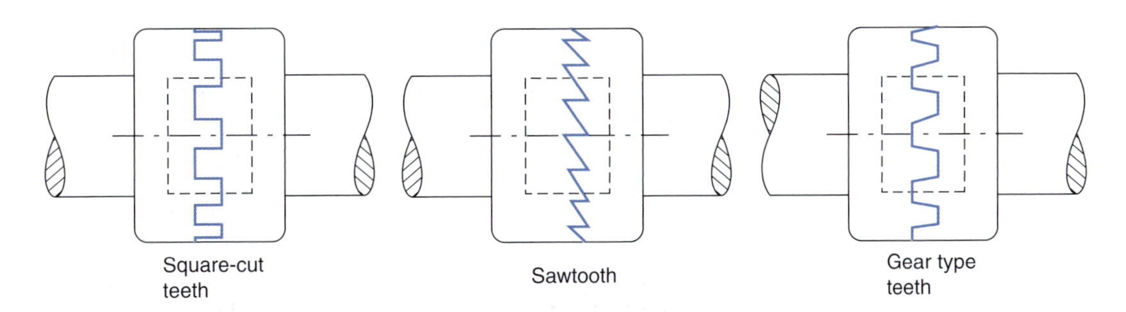

Figure 3.42 Dog clutches (Redrawn from *Mechanical Engineer's Reference Book*, 12th edition, E H Smith, Butterworth-Heinemann, p. 10/45, Fig. 10.99, 1998, with permission from Elsevier.)

Flat plate clutches, similar to that shown in Figure 3.43, are widely used in cars and commercial vehicles. The single plate type is shown but multi-plate clutches are also widely used, particularly in motor cycles. The driven plate is lined with friction material and has a splined hole at its centre. This engages with the splines on the output shaft to which it transmits the drive. When the clutch is engaged, the driven plate in sandwiched between the driving member and the pressure plate. Coiled springs or a spring diaphragm provide the clamping pressure. To disengage the clutch the pressure plate is pulled back against the springs by the release collar and thrust bearing which slides on the splines of the output shaft.

The conical clutch shown in Figure 3.44 is used in machine tools and also for some heavy duty applications in contractors plant. In a slightly different form it is used in synchromesh manual gearboxes and overdrive units. The clutch has a wedge action which reduces the spring force needed to hold the driving member and the driven cone in engagement.

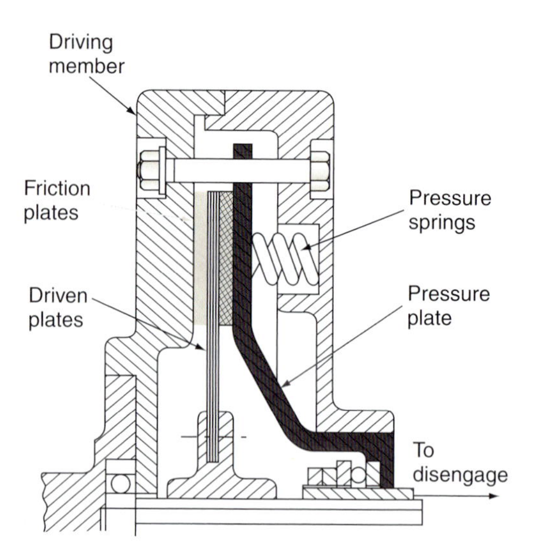

Figure 3.43 Single plate friction clutch (Redrawn from *Mechanical Engineer's Reference Book*, 12th edition, E H Smith, Butterworth-Heinemann, p. 10/46, Fig. 10.104, 1998, with permission from Elsevier.)

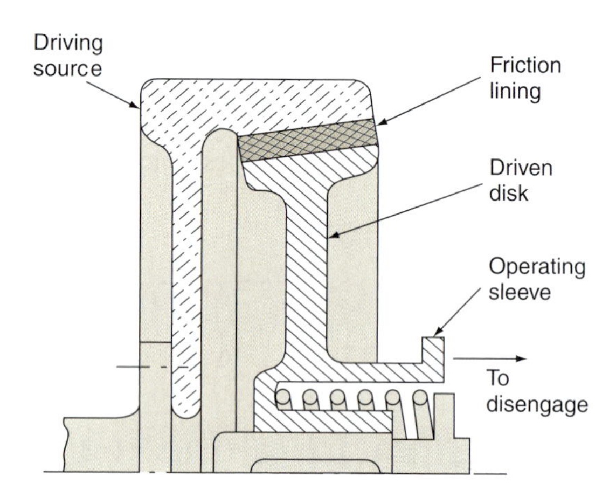

Figure 3.44 Conical friction clutch (Redrawn from *Mechanical Engineer's Reference Book*, 12th edition, E H Smith, Butterworth-Heinemann, p. 10/46, Fig. 10.104, 1998, with permission from Elsevier.)

KEY POINT

Centrifugal clutches engage and disengage automatically at a pre-determined speed.

The operation of centrifugal clutches was described in Chapter 1 (Figure 3.45). Their advantage is that they are automatic in operation and enable motors with a low starting torque to commence engagement gradually without shock. The strength of the control springs determines the speed at which engagement begins.

KEY POINT

There is no mechanical connection with fluid clutches and the take-up is smooth and gradual.

Fluid couplings of the type shown in Figure 3.46 have many advantages. There is no mechanical connection between the input and output shafts, the take-up is smooth and gradual and the operation is fully automatic. There are two basic elements, the runner which is driven by the input shaft and the impeller which is connected to the output shaft. The assembly is filled with oil which is made to circulate as shown, by the vanes on the runner and impeller. As the runner speed increases, the oil transmits the drive to the impeller. The slippage between the runner and impeller speeds becomes less and less as the speed increases until eventually the oil transmits an almost solid drive to the output shaft. A similar form of fluid couplings is used in road vehicles with automatic gearboxes.

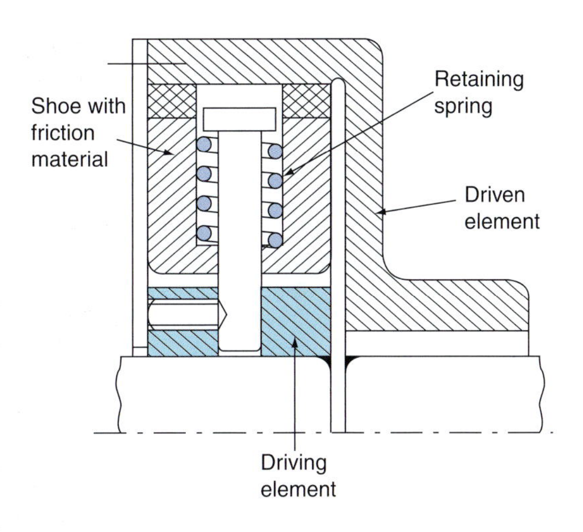

Figure 3.45 Centrifugal clutch (Redrawn from *Drives and Seals*, M J Neale, Butterworth-Heinemann, p. 39, Fig. 7.9, 1993, with permission from Elsevier.)

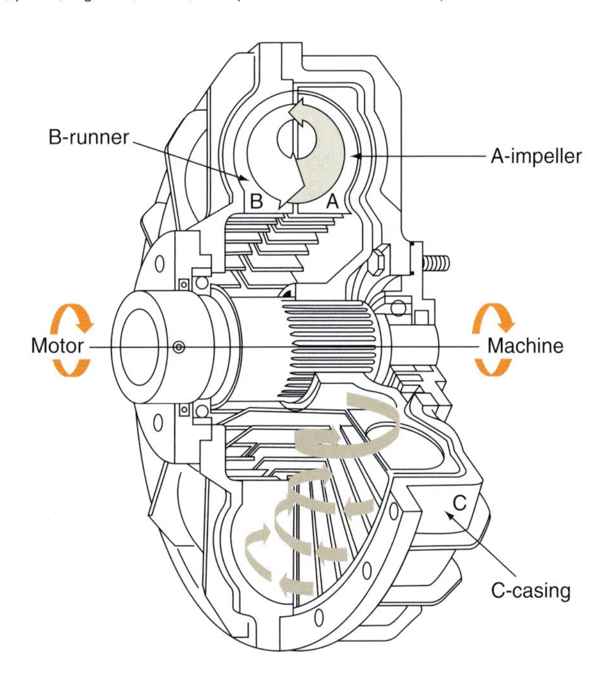

Figure 3.46 Fluid coupling (Redrawn from *Mechanical Engineer's Reference Book*, 12th edition, E H Smith, Butterworth-Heinemann, p. 10/48, Fig. 10.111, 1998, with permission from Elsevier.)

The purpose of friction brakes is the opposite to that of friction clutches. They are intended to slow down the rotating parts of a machine or a road vehicle. Figure 3.47 shows the operating principle of external band type brakes. Figure 3.48 shows the principle of internal expanding brakes and disc brakes of the type used on road vehicles. Brakes may be applied through a mechanical linkage but it is more usual for them to be activated by a pneumatic or hydraulic actuator.

A dynamometer is a particular type of brake that is used to measure the output torque from electric motors, internal combustion engines and gas turbines. The basic type of friction dynamometer is the rope brake shown in Figure 3.49. These are used only on low speed oil and gas engines, mainly for educational purposes. The brake drum is coupled to the engine output shaft and has an internal channel section. Whilst

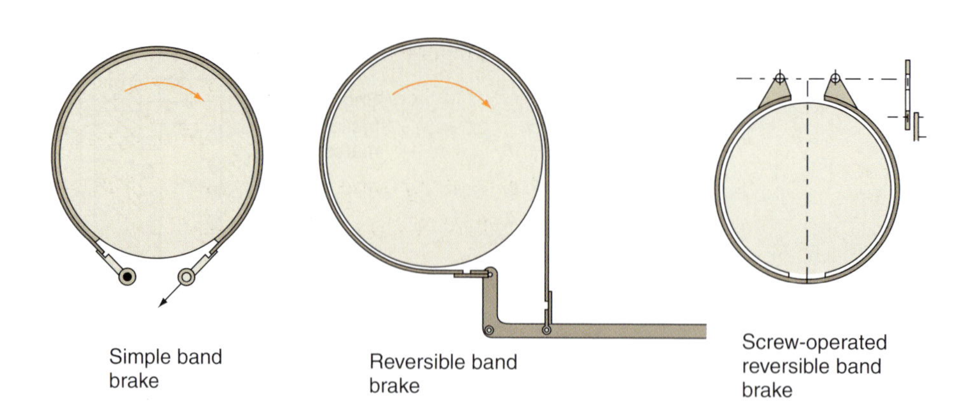

Figure 3.47 External friction brakes (Redrawn from *Mechanical Engineer's Reference Book*, 12th edition, E H Smith, Butterworth-Heinemann, p. 10/49, Fig. 10.113, 1998, with permission from Elsevier.)

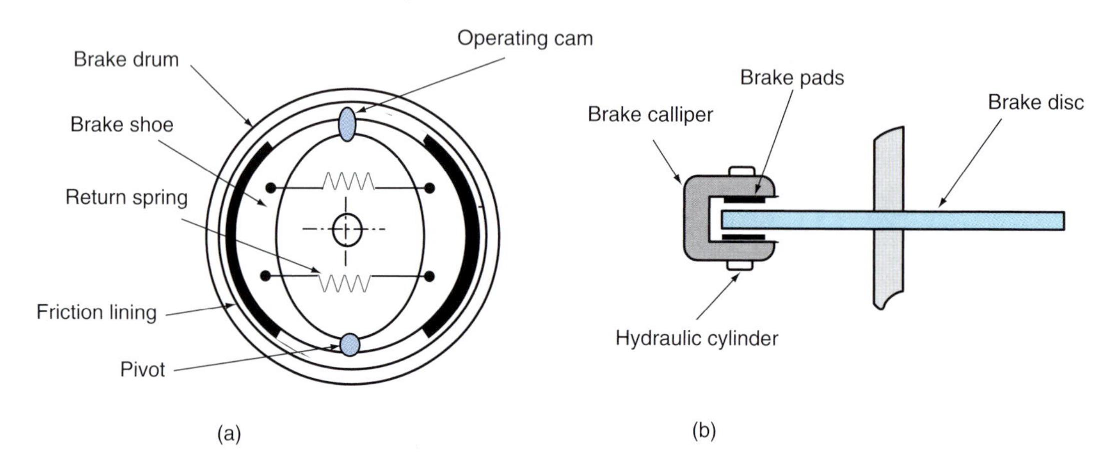

Figure 3.48 (a) Internal expanding and (b) disc brakes

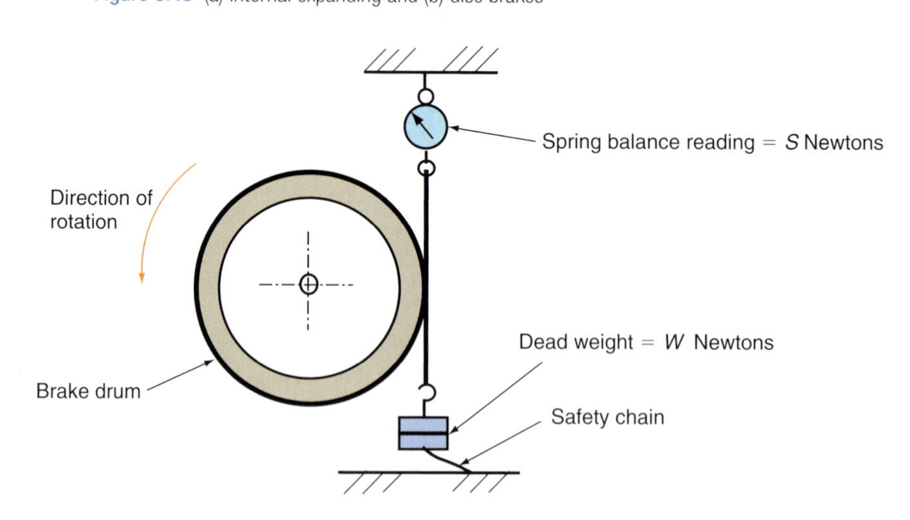

Figure 3.49 Rope brake

rotating, it is filled with water which is kept evenly distributed around the internal circumference by centrifugal force. This absorbs the heat generated by friction between the rope and brake drum. A dead weight is hung on the lower end of the rope and the upper end is attached to a spring balance. A safety chain to the dead weight hanger prevents the weight being flung off should the engine stall or back-fire.

The tangential force F acting on the brake drum is the difference between the dead weight and the spring balance reading.

$$\text{i.e.} \qquad F = W - S \text{ Newtons}$$

The braking torque T is the product of the tangential force and the effective radius r of the brake. i.e.:

$$T = Fr$$

$$\boldsymbol{T = (W - S)r}$$

The hydraulic and electrical dynamometers shown in Figure 3.50 are the widely used by test and research departments in industry. The hydraulic dynamometer, through which there is a steady flow of water, functions in the same way as a fluid coupling. Its casing is mounted on trunnions but is prevented from rotating by the dead weight and spring balance attached to the torque arm. The rotor, which is driven by the engine or motor under test, contains vanes that rotate alongside similar stationary vanes in the casing. The water is made to swirl in such a way as to transmit force from the moving vanes to the casing.

KEY POINT

The effective radius of a rope brake is measured from the centre of the brake drum to the centre of the rope.

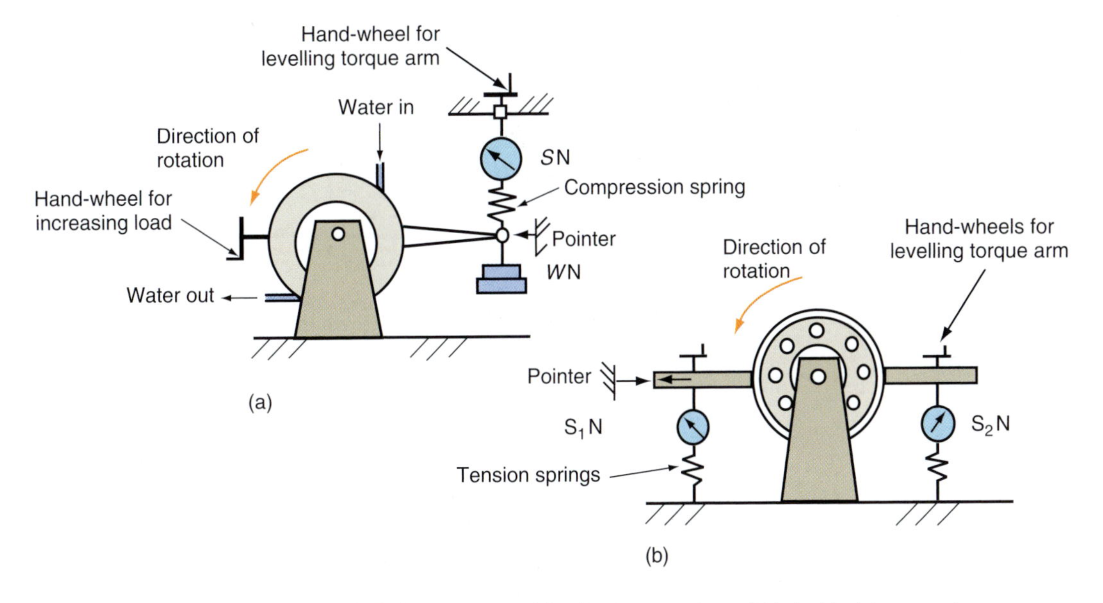

Figure 3.50 Dynamometers: (a) water dynamometer and (b) electrical dynamometer

The clearance between the fixed and moving vanes can be adjusted by means of a hand-wheel. Reducing the clearance increases the load on the engine or motor. The torque arm is adjusted to the horizontal position, indicated by the fixed pointer, using the upper hand-wheel. If W and S are the dead weight and spring balance reading and r is the torque arm radius, the braking torque T is given by:

KEY POINT

The torque arm should always be adjusted to the horizontal position before any readings are taken.

$$\text{Torque} = \text{Force on torque arm} \times \text{Torque arm radius}$$

$$\boldsymbol{T = (W - S)r}$$

The electrical dynamometer is in fact a generator. It functions in a similar way to the hydraulic dynamometer except that the reaction torque is magnetic. The electrical power generated is usually dissipated as heat through banks of electrical resistors. The load is decreased and increased by switching additional resistors in and out of the circuit. The type shown in Figure 3.50(b) has two spring balances reading S_1 and S_2 Newtons. The hand-wheels above the spring balances are adjusted so that for a particular load, the torque arm is horizontal. If r, is the torque arm radius to the spring balances, the braking torque is given by:

$$\text{Torque} = \text{Force on torque arm} \times \text{Torque arm radius}$$

$$T = (S_2 - S_1)r$$

Routine maintenance for couplings, clutches, brakes and dynamometers chiefly involves checking coupling bolts for tightness, cleaning, lubricating and checking brake bands and friction linings and pads for wear. Fluid couplings should be checked for signs of leakage and the hydraulic fluid replaced at the recommended intervals.

Test your knowledge 3.9

1. What is a *dog clutch*?
2. Which of the clutches described are automatic in operation?
3. Why does a conical clutch require less spring force than the equivalent flat plate clutch?
4. What is the purpose of a dynamometer?
5. To what position should the torque arm be adjusted when using fluid and electrical dynamometers?

Activity 3.8

Two single plate clutches as shown in Figure 3.39, have the same pressure springs and the same outer diameter. The inner diameters of the friction liner are however different. The clutch with the smaller contact surface is found to transmit more power before slipping. Explain why should this be so?

To check your understanding of the preceding section, you can solve Review questions 26–30 at the end of this chapter.

Belt and chain drives

Belt drives are used two connect shafts which may be some considerable distance apart. The belts may be flat, V-section or toothed and the pulleys on which they run can be selected to give a particular velocity ratio. Flat and V-section belts rely on friction between the belt and pulley to transmit power. There is a limit to the power that they can transmit before slipping occurs. Toothed or *synchronous* belts, running on toothed pulleys, give a more positive drive. If the belt is in good condition and correctly tensioned, slipping should not occur. The limiting factor on the power that can be transmitted is the allowable stress in the belt material. Figure 3.51 shows some typical belt sections and profiles.

KEY POINT

The maximum power that a flat or Vee-section belt can transmit before slipping depends on the initial tension setting. This should not be so high that maximum allowable stress in the belt material is exceeded before slipping occurs.

Belts are generally made from synthetic neoprene rubber with nylon or sometimes metal, reinforcing fibres. Flat belts are used mainly for transmitting light loads. Because they are flexible, this makes them

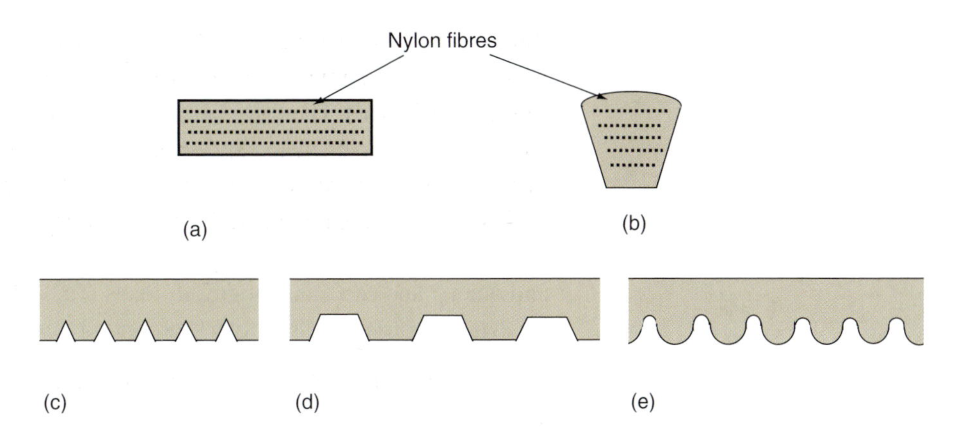

Figure 3.51 Belt sections and profiles: (a) flat belt section, (b) V-belt section, (c) ribbed V-belt, (d) trapezoidal synchronous belt and (e) HRD synchronous belt

suitable for applications where there is some misalignment between shafts and as can be seen in Figure 3.52, they may be crossed to give opposite directions of rotation to the pulleys. They can also be twisted to connect shaft which are not in the same plane.

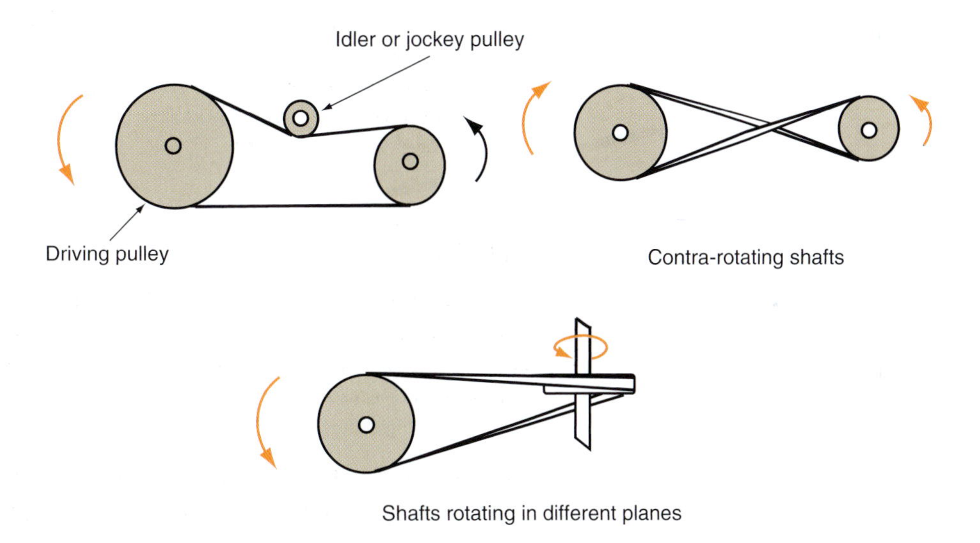

Figure 3.52 Belt drive configurations

The pulleys should be slightly wider than the belt with a convex crown on which the belt will ride. High speeds are possible but centrifugal effects as the belt passes round the pulleys can result in loss of grip at belt speeds in excess of around 18–20 ms^{-1}. In applications where the shaft centre distance is fixed, idler pulleys positioned on the slack side, are the usual means of tensioning flat belts.

The V-section belt is the standard choice for a great many power transmission systems. It is widely used in machine tools, automobiles, washing machines and tumble dryers. Multiple belts are used for large power transmissions. The wedge action of the belt in its pulley gives approximately three times the grip of a flat belt made from the same materials. The ribbed V-belt shown in Figure 3.51 has been developed to combine the grip of a V-belt with the flexibility of a flat belt.

Synchronous belts have a similar power capacity to V-belts. They give a positive drive and are used where slipping would be detrimental to the driven components. The drive belt to the camshaft on automobile engines is a typical example where slippage can alter the valve timing and possibly result in the pistons striking the valves. The HRD belt, which stands for *high torque drive*, has been developed from the trapezoidal synchronous belt and runs on toothed pulleys in the same way. It is said to be smoother and quieter and can transmit greater loads. The recommended speed limit for synchronous and HRD belts is around to 60 ms^{-1}.

Routine maintenance on belt drives involves checking for wear, checking the tension setting and checking tensioning devices. The failure of synchronous belts can have serious consequences, particularly in automobile engines, and they should be changed at the recommended service intervals.

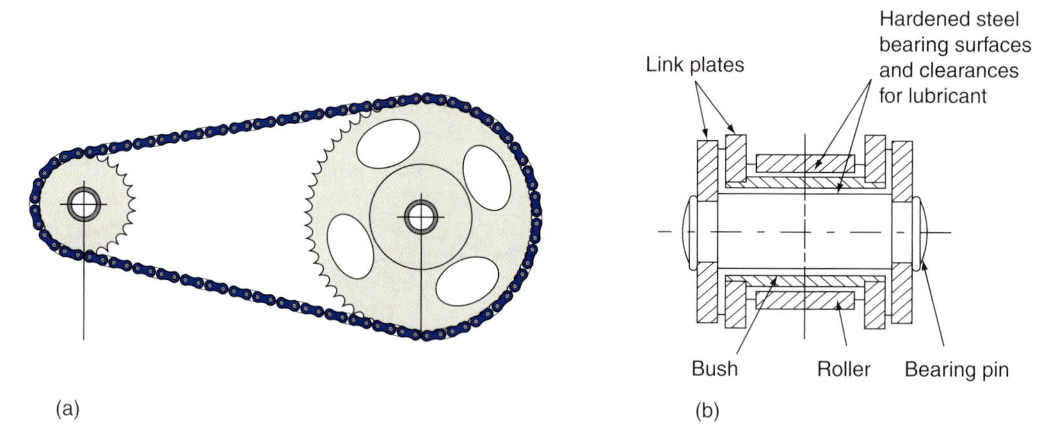

Figure 3.53 (a) Roller chain and sprockets (Redrawn from *Drives and Seals*, M J Neale, Butterworth-Heinemann, p. 11, no fig. no., 1993, with permission from Elsevier.) and (b) section through a roller chain (Redrawn from *Mechanical Engineer's Reference Book*, 12th edition, E H Smith, Butterworth-Heinemann, p. 10/36, Fig. 10.84, 1998, with permission from Elsevier.)

KEY POINT

Good alignment of the shafts and sprockets is essential with chain drives to prevent excessive strain on the links.

Like toothed belts, chain drives provide a positive means of transmitting power between parallel shafts. The standard type is the bushed roller chain of the kind used on bicycles (Figure 3.53). Other types, intended to provide a better rolling action, have been designed but the roller chain is by far the most widely used. Belts can tolerate some misalignment between the shafts and pulleys but a higher degree of alignment is required for chain drives. Misalignment can produce sideways bending which will strain the links and cause rapid wear. Chain drives can transmit larger loads than belts without the possibility of slipping. Multiple chains are used for very large load requirements. Until the introduction of toothed belts, roller chains were used universally used for the camshaft drive on automobile engines.

Chain drives need to be adequately lubricated as shown in Figure 3.54, to slow down the rate of wear. The method of lubrication depends on the chain speed and the power which it is required to transmit. Tensioning devices are often incorporated, consisting of an idler sprocket or a spring loaded friction pad acting on the slack side of the chain.

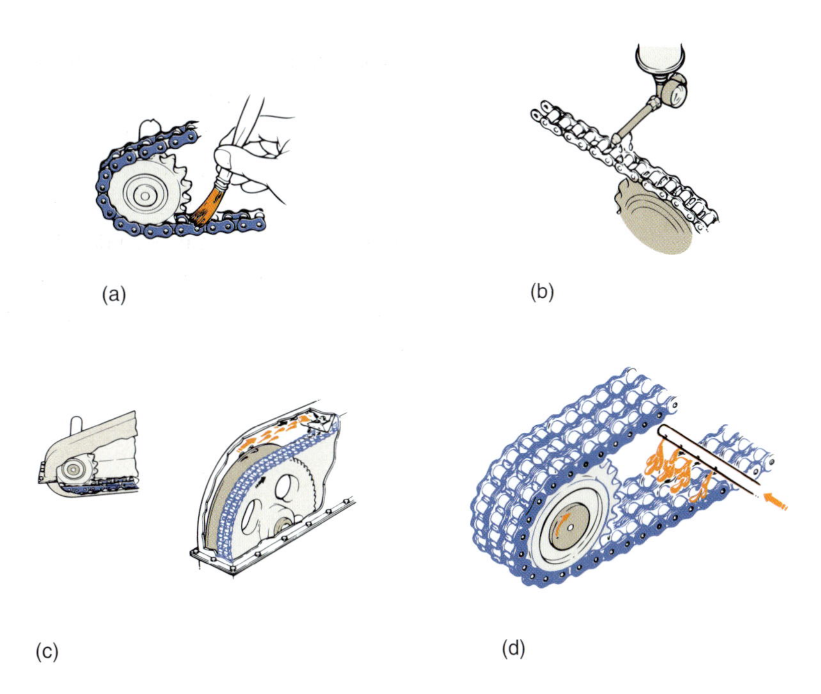

Figure 3.54 Chain drive lubrication: (a) manual lubrication for low power and speed, (b) drip lubrication for low power and medium speed, (c) oil bath for medium power and speed and (d) continuous circulation from pump for high power and speed (Reprinted from *Drives and Seals*, M J Neale, Butterworth-Heinemann, p. 15, no Fig. nos, 1993, with permission from Elsevier.)

Routine maintenance involves cleaning exposed chains, lubricating or replenishing lubricants, checking the operation of tensioning devices and checking the teeth on the sprockets for wear. The chain length should also be checked periodically. A 2% increase in length due to wear in the rollers indicates that replacement is due.

Test your knowledge 3.10

1. Why can a V-belt transmit more power than a flat belt of the same material and tension setting?
2. What is the advantage of a ribbed v-section belt?
3. What are synchronous belts used for?
4. Why is accurate alignment more important for chain drives than belt drives?
5. How should a chain drive that transmits high power at high speed be lubricated?

Activity 3.9

Two parallel shafts whose centre distance is not to exceed 400 mm and are to be joined by a roller chain. The sprockets on the shafts have effective diameters of 200 mm and 100 mm and the chain has links of pitch 20 mm. What will be the length of the chain required to give the greatest possible centre distance between the shafts?

***Note*: A roller chain can only be lengthened or shortened by adding or taking off two links.**

Gear trains

Gear trains form an essential part of a great many power transmission systems. By far the most common type used in engineering is gears with an *involute* tooth profile. An involute curve is generated by the end of a cord as it is unwound from around the surface of a cylinder as shown in Figure 3.55.

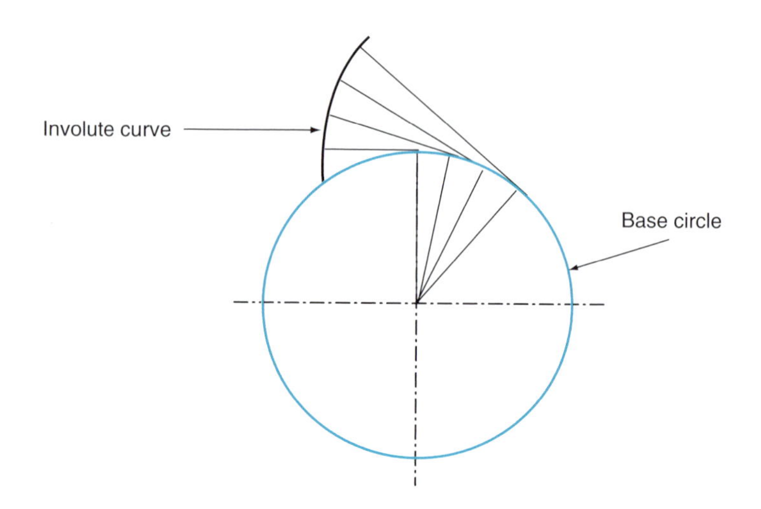

Figure 3.55 Involute to a circle

The reason for adopting the involute tooth form is that the teeth are strong, the velocity ration between mating gears is constant and the teeth can be accurately machine with modern gear cutting machinery. A pure rolling action between meshing teeth would be the ideal situation but no such tooth form has yet been discovered. All gear teeth mesh with a rolling and sliding action which is why good lubrication is essential.

An alternative tooth form based on the Russian Novokov gear has aroused some interest in recent years. It has been developed by Westland helicopters under the name of the *conformal* gear profile and is claimed to be superior for some applications. However it has still to find widespread approval and we will concentrate only on the involute form.

A term commonly used with gears is the *pitch circle diameter*. This is the effective diameter of a gear. The pitch circle diameters of two mating gears are the diameters of the discs which would transmit the same velocity ratio by frictional contact alone. They are shown in Figure 3.56.

The *base circles*, from which the involute teeth are generated, are smaller than the pitch circles. Their common tangent passes through the point where the pitch circles touch and makes an angle ψ (Psi) with the common centre line. This is known as the *pressure angle* of the gears. It is along part of this line that the point of contact between the gear teeth passes. In modern gears the pressure angle has been standardized at $\psi = 20°$. For a gear of pitch circle diameter D, the base circle diameter is given by:

$$\textbf{Base circle dia} = \boldsymbol{D} \cos \Psi \qquad \text{(i)}$$

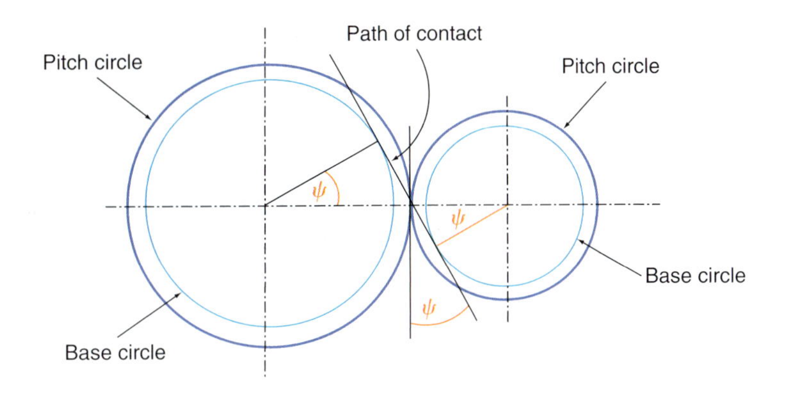

Figure 3.56 Pitch and base circles

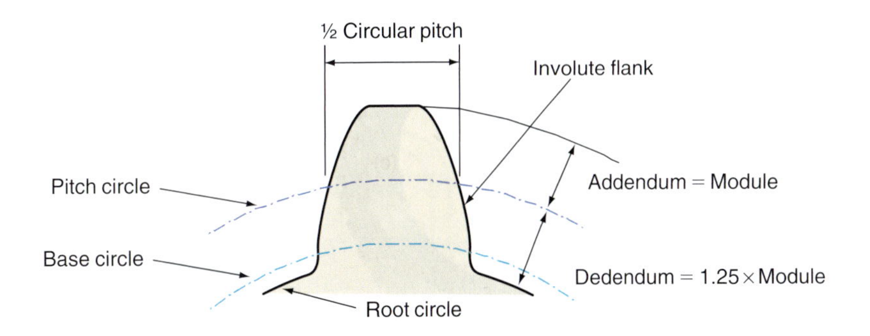

Figure 3.57 Gear tooth geometry

The *circumferential* or *circular pitch* of a gear is the length of the arc of the pitch circle between the same point on successive teeth. If t is the number of teeth, the circular pitch p is given by the formula:

$$p = \frac{\pi D}{t} \qquad \text{(ii)}$$

A parameter which is of great importance is the *module* of a gear. The module m is obtained by dividing the pitch circle diameter by the number of teeth. It is this quantity which determines the size of the teeth. It is essential for mating gears to have the same module, otherwise they will not mesh

$$m = \frac{D}{t} \qquad \text{(iii)}$$

KEY POINT

Gears can only mesh if they have the same module as this fixes the size and pitch of the teeth.

The *addendum* of a gear tooth is its height above the pitch circle and this is equal to the module. The *dedendum* is the depth of the tooth below the pitch circle to the root circle. This is generally 1.25 times the module to give sufficient clearance between the root and the tip of mating teeth (Figure 3.57).

Some of the different types of gear are shown in Figure 3.58. External spur gears have straight teeth which are cut or moulded parallel to the gear axis. Amongst other materials they are made from hardened steel, cast iron, phosphor-bronze and nylon. The meshing gears rotate in opposite directions. Spur gears give good results at moderate speeds but tend to be noisy at high speeds.

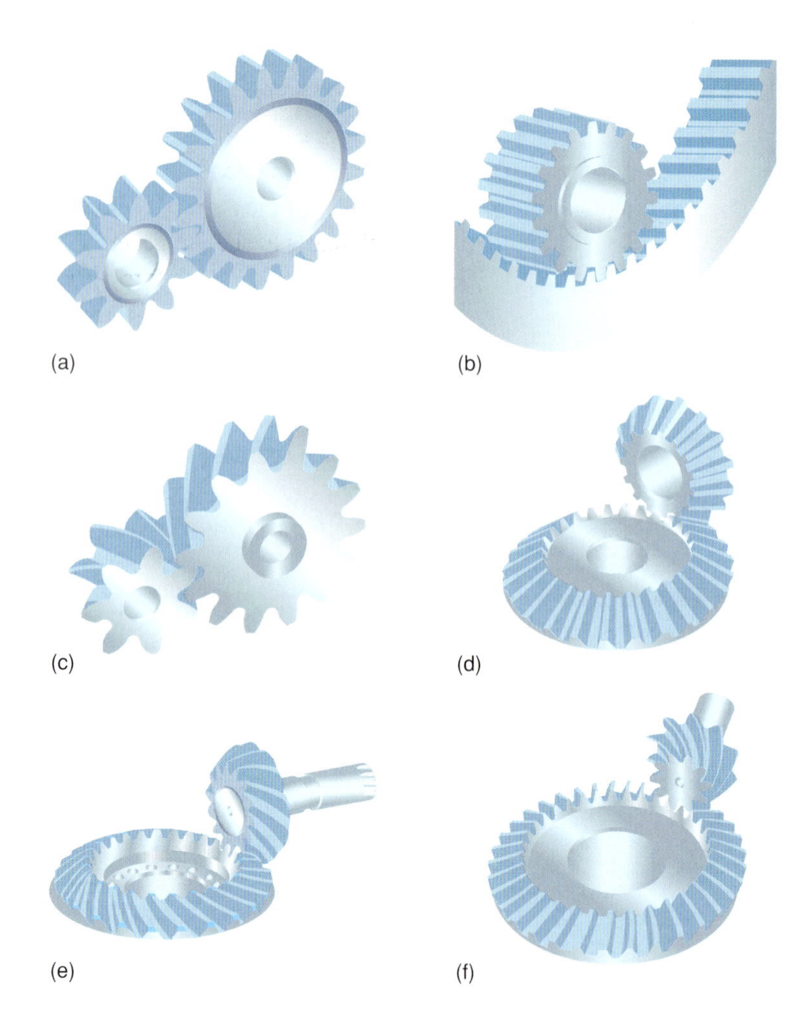

Figure 3.58 Gear types: (a) external spur gears, (b) internal spur gears, (c) helical gears, (d) straight bevel gears, (e) spiral bevel gears and (f) hypoid bevel gears

KEY POINT

Meshing spur gears with external teeth rotate in opposite directions but meshing spur and annulus gears rotate in the same direction.

Spur gears with internally cut teeth are known as annulus gears. They are used in epicyclic gear trains which we will shortly be describing. When meshing with an external spur gear as shown in Figure 3.58, both gears rotate in the same direction. Automatic gearboxes in cars contain epicyclic gear trains.

KEY POINT

Helical gears are quieter and smoother running than spur gears but produce end thrust along the axis of the shafts.

Helical gears are use to connect parallel shafts in the same way as spur gears but have a superior load carrying capacity. With helical teeth, the point of contact moves across the tooth and it is possible for more than one pair of teeth to be simultaneously in mesh. This makes for smoother and quieter running, particularly at high speeds. The one disadvantage with helical teeth is that they produce end thrust. To eliminate this, double helical or herring-bone teeth are sometimes used but these are expensive to manufacture.

Straight bevel gears are used to connect shafts whose axes intersect at some angle. The gear teeth radiate outwards from the point of intersection of the axes. Considerable end thrust is developed by bevel gears under load and this tends to push the teeth apart. Spiral bevel gears fulfil the same purpose but have a better load carrying capacity. As with helical gears, the teeth mesh gradually to give smoother and quieter running. Hypoid bevel gears are closely related spiral bevels but are used where the axes of the two shafts do not intersect, as shown in Figure 3.58. Cars with rear wheel drive generally use hypoid bevel gears to transmit power from the propeller shaft to the road wheels. The velocity ratio of all of these gear combinations is given by the formula:

$$\text{Velocity ratio} = \frac{\text{Number of teeth on output gear}}{\text{Number of teeth on input pinion}}$$

Figure 3.59 Worm gears

Worm gears consist of a worm shaft and worm wheel (Figure 3.59). Their axes do not intersect and are usually at right angles. The worm shaft may have one single helical tooth, rather like a screw thread, or multi-start teeth. The worm wheel teeth are specially machined to mesh with the worm. Very high velocity ratios are possible with single start worms since one complete revolution will only move the worm wheel through an angle subtended by the circumferential pitch, i.e. it will only move it forward by one tooth. Its velocity ratio is given by the formula:

$$\text{Velocity ratio} = \frac{\text{Number of teeth on worm wheel}}{\text{Number of starts on worm}}$$

It is usual to make the two parts from different materials. The worm shaft is often made from hardened steel whilst the outer ring of the worm wheel is made from phosphor-bronze. Worm gears are sometimes used in preference to hypoid bevel gears in the driving axles of heavy slow moving vehicles.

In Chapter 1 we discussed the use of simple and compound gear trains in gear winches. We will now consider two different kinds of gearbox which might be used in a power transmission system. They might be used to reduce or increase the input speed depending on the numbers of teeth on the gears.

The conventional gearbox shown in Figure 3.60 might contain spur or helical gears. The cluster ACE can be moved to the left and right along

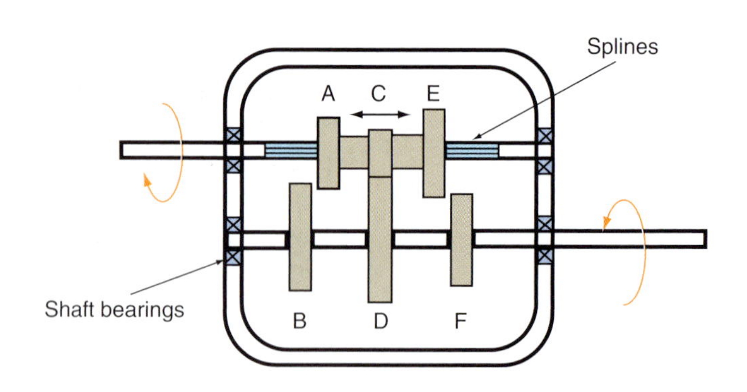

Figure 3.60 Conventional gearbox

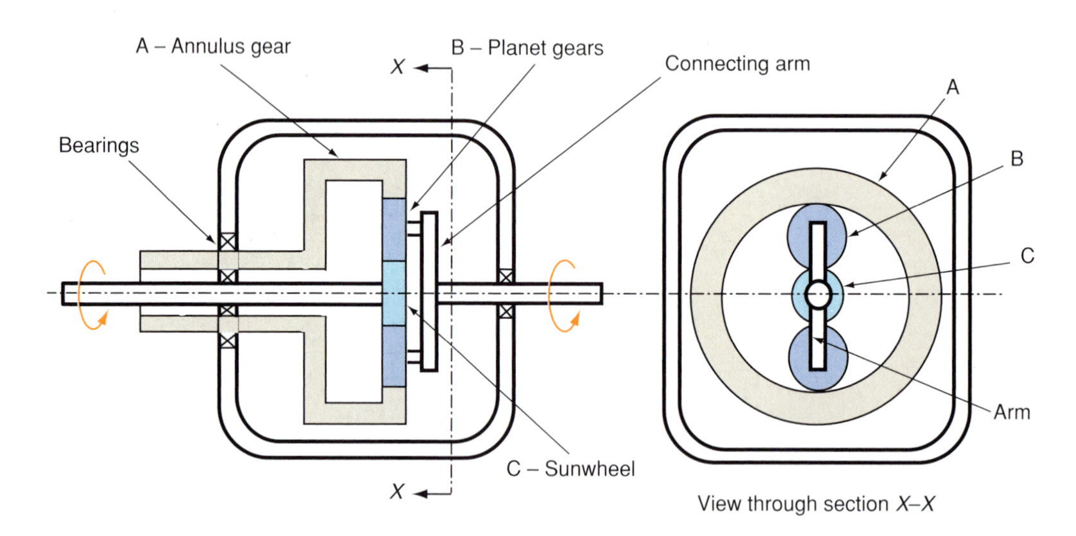

Figure 3.61 Epicyclic gearbox.

splines so that three velocity ratios can be obtained, i.e. A driving B, C driving D and E driving F. Consider now the epicyclic gearbox shown in Figure 3.61.

This type of gear train and gearbox gets its name from the curve traced out by a point on a circle as it rolls inside a circle of larger diameter. It is called an *epicycloid*. The gear A in Figure 3.61 has internally cut teeth and is the annulus gear. The gear C is known as the sunwheel and the gears B, which can rotate about the sunwheel, are called planet gears. The two planet gears are connected to the arm which can rotate around the sunwheel with them. The gears are in mesh all of the time and six different velocity ratios can be obtained by holding different gears stationary as follows:

1. With the annulus fixed, input to the sunwheel C, and output from the shaft connected to the planet arm or vise-versa.
2. With the sunwheel C fixed, input to the annulus A, and output from the shaft connected to the planet arm or vise-versa.
3. With the planet arm fixed, input to the sunwheel C, and output from the annulus A or vise-versa.

The velocity ratios will of course depend on the number of teeth on each gear. The values are not so easy to calculate as with conventional gear trains and the method is best left for study at a higher level. Epicyclic gear trains are to be found in automatic gearboxes, hub-reduction gears on heavy commercial vehicles, three-speed hubs on bicycles and speed reduction gearing for turbines. Epicyclic gear trains have the advantage of being compact and there is no radial force on the bearings other than that exerted by the weight of the gears. Their main disadvantage is cost. A greater degree of precision is required in producing the gears and the gear change mechanism, which is required to clamp the different gears in the stationary position, is more complex than that on a conventional gearbox.

KEY POINT

Epicyclic gearboxes are more compact and do not exert radial forces on the bearings.

The routine maintenance of geared systems generally involves the checking and replenishment of lubrication levels and the cleaning or replacement of oil filters. Excessive noise, vibration and overheating should be reported for expert attention.

Test your knowledge 3.11

1. What is the name of the curve that is most commonly used for the profile of gear teeth?
2. What is the pitch circle of a gear?
3. What is the module of a gear?
4. What is the advantage of using helical gears rather than spur gears?
5. What are the advantages of using epicyclic gears in a gearbox?

Activity 3.10

Cut two circles of diameters 100 mm and 50 mm from a piece of stiff cardboard. Fix the larger circle to a flat surface using blue-tack and place the smaller one touching it. Mark the point of contact on both circles. Now roll the smaller circle around the circumference of the larger stationary circle in the same way that a planet gear rotates around a stationary sunwheel. Count the number of turns that the planet circle rotates on its own axis whilst making one orbit of the sunwheel circle. Discuss, and give a reason for your findings.

To check your understanding of the preceding section, you can solve Review questions 31–35 at the end of this chapter.

Plant Equipment and Systems

A great many engineering and manufacturing processes make use of pneumatic and hydraulic devices for mechanical handling and positioning, material forming and process control. Steam is widely used for processes that require sustained high temperatures. It is also used for space heating and the bulk of our electricity is generated using steam. Refrigeration systems are used where sustained low temperatures are required for storage and processing. Refrigeration is also used in air-conditioning systems to provide comfortable working conditions. We will now describe how some of these systems operate, their major components, safety aspects and the routine maintenance duties required.

Hydraulic and pneumatic systems

Hydraulics and pneumatics are widely used in engineering processes and servo-control to convey energy from one location to another. The enclosed fluids most commonly used are specially formulated hydraulic oils and compressed air. Both kinds of system are used to transmit force and produce linear and rotational motion.

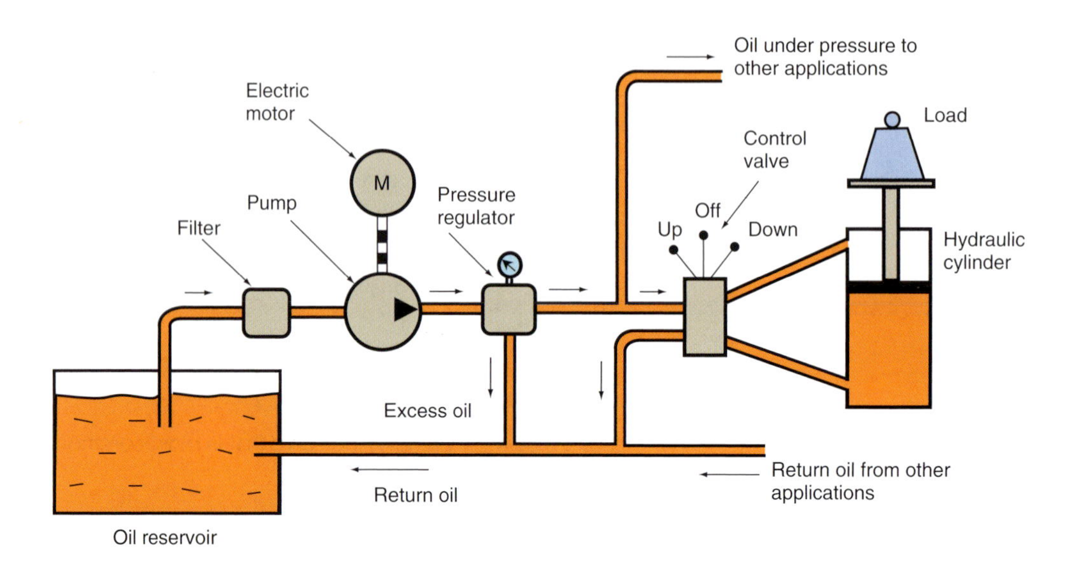

Figure 3.62 Hydraulic system

Air has a low density and is compressible whilst hydraulic oil has a much higher density and is almost incompressible. As a result, hydraulic systems are able to operate at much higher pressure and deliver the very large positive forces which are required in applications such as hydraulic presses and lifts. The major components of a hydraulic actuation system are shown in Figure 3.62.

KEY POINT

Hydraulic systems are able to deliver higher and more positively controlled forces than pneumatic systems.

The heart of the system is the motor-driven pump which draws filtered oil from the reservoir and delivers it via a pressure regulator to the points where it is required. The pump runs continuously and the excess oil which is not required for operations is diverted back to the reservoir by the pressure regulator. It should be noted that the system usually serves a relatively small work area in the locallity of the pump and reservoir. It is not practicable to supply oil under pressure over large distances because of pressure drop and the need for a return pipe. A manual or automatic control valve supplies oil to the actuation cylinder and directs return oil to the reservoir.

KEY POINT

Hydraulic systems usually serve a small localised area whereas compressed air can be delivered as a plant service over much larger distances.

Pneumatic systems have a softer action and are not able to deliver such large forces. They do however have certain advantages. Compressed air is readily available in many industrial installations, being supplied as a sevice to the operational areas. Furthermore it can be supplied over greater distances and is vented to the atmosphere after use. The major components of a typical pneumatic system is shown in Figure 3.63.

The compressor takes in filtered air and delivers it via an after-cooler to the compressed air receiving vessel. Compressors come in a variety of

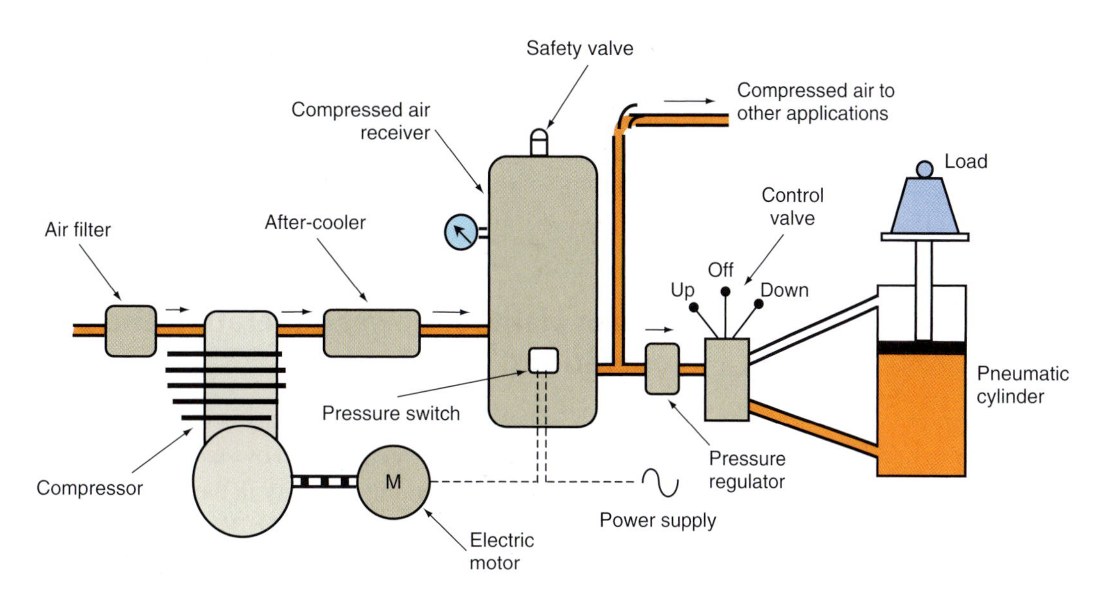

Figure 3.63 Pneumatic system

sizes. They may be single stage as shown, or two stage. With two-stage compressors the air undergoes initial compression in the larger diameter low pressure cylinder. It then passes through an inter-cooler to the smaller diameter high pressure cylinder for further compression before delivery through the after-cooler. For some applications where the air must be perfectly dry, the system also contains a moisture separator.

The receiving vessel is equipped with a pressure switch which cuts out the compressor motor when the supply pressure is reached and restarts it when the pressure falls. In addition it is fitted with a safety valve which will open should the pressure switch fail to operate. A pressure regulator adjusts the supply pressure to that required at the point of application. As with hydraulic systems, a manual or automatic control valve directs compressed air to the actuation cylinder from which it is exhausted to the atmosphere.

KEY POINT

Pneumatic systems exhaust to the atmosphere. They do not require a return line.

Leaks from hydraulic systems can be both messy and dangerous whilst those from compressed air systems pose less of a problem for maintenance engineers. Routine maintenance involves checking the systems for leaks, the replenishment or replacement of hydraulic fluid and the cleaning or replacement of filters. In compessed air systems the lubricant in the compressor should be checked periodically and replenished. Also, the safety valve in the air receiver should be checked periodically for correct operation by over-riding the pressure switch until the blow-off pressure is reached.

Test your knowledge 3.12

1. What are the advantages and disadvantages of hydraulic systems?
2. Which system is able to exert the greater controlled force?
3. Why is it not practical to supply hydraulic pressure over long distances?
4. What is the purpose of the pressure switch in a compressed air receiver?
5. What kind routine maintenance activities should the operators of pneumatic and hydraulic equipment carry out?

CHAPTER 3

Activity 3.11

The brakes on all makes of cars are hydraulically operated whilst those on heavy commercial vehicles are operated by compressed air. What is the reason for this?

Steam plant for power generation and process operations

By far the greater part of our electricity is produced by power stations in which the generators are powered by steam turbines. An approximate breakdown of the generating capacity in the United Kingdom is 37% coal fired, 31% gas fired, 25% nuclear, 2% oil fired and 5% from renewable sources such as hydro-electric and wind power. Whatever the heat source, steam generating plants have the same major components. A typical arrangement is shown in Figure 3.64.

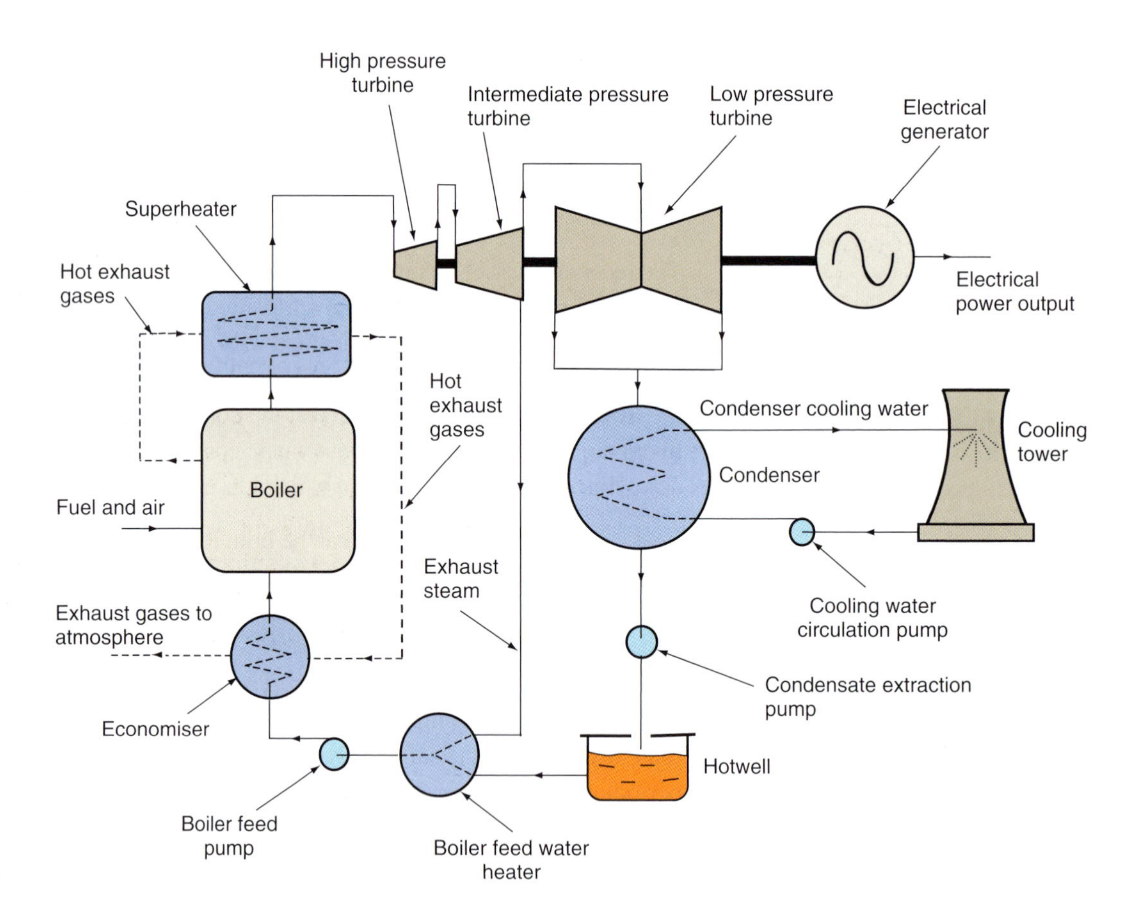

Figure 3.64 Steam plant circuit

Steam is also widely used for space heating and industrial processes. Integrated systems are to be found in many parts of the world where condenser cooling water and exhaust steam from industrial processes provide heating for adjacent office and apartment blocks. Industrial processes that make use of steam as a heat source include food processing, bleaching and dying of textiles and chemical processing. Hospitals also make use for steam for cleaning and sterilising medical equipment. The advantages of using steam as a heat source are that it is clean and transportable. You might also be aware that the pressure of wet steam that has not received all of its latent heat of vaporisation is directly related to its temperature. By controlling its pressure, its temperature can thus be tightly controlled. This is an important requirement in food and chemical processing.

Boilers and superheaters

In coal gas and oil fired systems, the fuel and air enter the boiler where the hot gases from combustion heat the feed water to produce wet steam. There are two basic kinds of boiler, the *fire tube* type and the *water tube* type. In fire tube boilers the hot gases from combustion pass through a system of tubes around which water is circulating. These are usually to be found in small installations where hot water or low pressure steam is required for industrial processes and space heating. In water tube boilers the hot gases from combustion circulate around a system of tubes containing water. This is the type used in power stations for producing large quantities of high pressure steam.

KEY POINT

Steam boilers for power generation are usually of the water tube variety.

The wet steam passes through a system of tubes in the *superheater* where additional heat energy is supplied from the combustion gases. This dries it out and raises its temperature to produce superheated steam. Every possible unit of heat energy is extracted from the exhaust gases and before escaping to the atmosphere, they are used to heat up the boiler feed water in the *economiser*. They are also used to pre-heat the incoming air, but this is not shown in Figure 3.64.

KEY POINT

Superheated steam is steam that has been heated to a temperature above that at which it was generated in the boiler.

In nuclear installations, the heat source is enriched uranium. This is bombarded with neutrons in the reactor vessel causing some of the uranium atoms to split and release heat energy. The process is called *nuclear fission*. There are many different designs of reactor in operation throughout the world. In Britain pressurised carbon dioxide is used to transfer heat energy from the reactor to the boilers, superheaters and economisers whilst in American designs, pressurised water is preferred.

Turbines

The superheated steam passes to the *high pressure turbine* where it expands and does work on the rotor blades. It then passes to the *intermediate pressure turbine* where the blades have a larger diameter. Here it continues to expand and do work. You will note that some of the exhaust steam from the intermediate pressure turbine is fed to the boiler feed water heater where it is injected into the feed water from the hotwell. The remainder of the steam from the intermediate pressure turbine passes to the *low pressure turbine*. Here the blades are of a still larger diameter and arranged so that the steam enters centrally, as shown in Figure 3.64, and expands outwards through the two sets of low pressure blades. All three turbines are connected by a common shaft which drives the electrical generator.

KEY POINT

The blade diameter increases to allow for expansion of the steam as it passes through the various turbine stages.

Condensers and feed water heaters

The exhaust steam from the low pressure turbine passes to the condenser as low pressure wet steam. There are two basic types of condenser. In the *spray type*, cooling water is injected into the steam causing it to condense. In the *surface type*, the steam condenses on the surface of a system of pipes through which there is a flow of cooling water. The surface type is used in all large power stations. Sea water and river water are used for cooling wherever practical and a great many of our nuclear and gas fired power stations are sited on the coast. Coal fired power stations tend to be sited inland near the remaining coal fields, to reduce transportation costs. Here the cooling water for the condensers is generally re-circulated through cooling towers as shown in Figure 3.64.

KEY POINT

The pressure inside a condenser is below atmospheric pressure and so a pump is required to extract the condensate.

When the steam condenses it occupies a much smaller volume and the pressure in the condenser falls to below atmospheric pressure. This is beneficial because it creates as large a pressure drop as is possible across the low pressure turbine, allowing steam to expand freely and do the maximum possible amount of work. The condensed steam must however be extracted from the condenser by the *condensate extraction pump*.

The condensate passes to a reservoir called the *hotwell* where make-up water is added for evaporation losses. Depending on the local water supply, the make-up feed water may need to be treated before use. This is especially the case in 'hard water' areas where there is a high calcium content. It is this that causes 'furing' and 'scaling' in kettles and it can have the same effect on boiler tubes. Chemical treatment in an ancillary water softening plant reduces the problem. The feed water from the hotwell is heated first in the *feed water heater* by exhaust steam, and then in the economiser by the exhaust gases from the boiler. The *boiler-feed pump* delivers the feed water through the economiser to the boiler. The objective of the feed water heaters is to raise the temperature of the water to as close to its boiling point as possible before it enters the boiler.

KEY POINT

A feed heater takes heat from exhaust steam whilst an economiser takes heat from the boiler exhaust gases.

There are a great many operational and maintenance duties on a steam plant. Routine maintenance includes the lubrication of moving parts, checking for leaks, checking that valves and steam traps are working properly and checking that pressure gauges and temperature measuring instruments are giving accurate readings. Steam traps are automatic devices used to drain off the water which sometimes collects in steam pipes, without allowing the steam itself to escape. The presence of high temperatures and pressures can make this work hazardous and permits to work are generally required in some parts of a plant. Maintenance should only be carried out by qualified personnel working to set procedures. Some parts of a steam plant such as the boilers and turbines, require specialised maintenance and this is usually sub-contracted to outside firms.

Test your knowledge 3.13

1. What are the two basic types of boiler used in steam plant?
2. What is the purpose of a superheater?
3. What are the two basic types of condenser used in steam plant?
4. What is an economiser and how does it function?
5. What is the purpose of a cooling tower?

Activity 3.12

The steam generated in a boiler always contains water droplets. It is known as wet steam and when it is used for industrial processes water tends to collect in the steam pipes. Steam traps are used to drain off water without allowing the steam to escape. The three main types are thermostatic traps, float traps and impulse traps. Select any particular one and with the aid of a diagram, describe how it functions.

To check your understanding of the preceding section, you can solve Review questions 36–40 at the end of this chapter.

Refrigeration systems

There are two basic types of refrigeration system. They are the vapour compression system and the vapour-absorption system. Both types are used in commercial applications and domestic refrigerators and both work on the principle that when a liquid evaporates it takes in latent heat from its surroundings. The liquids used in refrigerators and freezers are called *refrigerants*. They are made to evaporate at a temperature below 0°C and in doing so, they take in latent heat and maintain the cold space at a sub-zero temperature.

A refrigerant must have a low freezing point so that it does not solidify or form slush in the low temperature part of the refrigeration cycle. Also it should have a high value for its latent heat of vaporisation to maximise the transfer of heat energy during the cycle. Until recently, refrigerants were only judged on how well they performed when taking in and giving out heat. The range of CFC (chlorofluorocarbon) refrigerants developed during the last century were thought to be quite suitable and safety was only considered in relation to the danger of poisonous leaks, fires and explosions. All this changed however when it was discovered that CFC's were damaging the earth's ozone layer and their use was banned under the Montreal Protocol of 1987.

One of the oldest, and still widely used refrigerants is ammonia, NH_3. Luckily, it is ozone friendly and is not affected by the protocol. Its main disadvantage is its toxicity but the installation of leak detection equipment makes it reasonably safe to use. Ammonia is used mostly in industrial refrigeration systems and in the small absorption refrigerators used in caravans and boats. Refrigerant R12, diclorodifluoromethane, CCl_2F_2, better known as Freon, was one of the most widely used CFC's in domestic and commercial refrigerators. This has now been banned and refrigerants such as R134a tetrafluoroethane, are used in its place. The basic circuit for a vapour compression refrigerator is shown in Figure 3.65.

KEY POINT

When they are no longer required, the disposal of old refrigerators containing CFC's must be carried out in the approved manner under local authority control.

The main components of a vapour-compression system are the compressor, the throttling valve, the condenser and the evaporator. You will find the condenser grid at the rear of your domestic refrigerator. When you touch it, it feels quite warm. The evaporator grid is inside, or surrounds the cold chamber where you can store ice cubes. The high pressure liquid refrigerant passes through the throttling valve where it undergoes a rapid fall in pressure and temperature. Temperatures in the range −10°C to −15°C are quite common. This is well below that of the

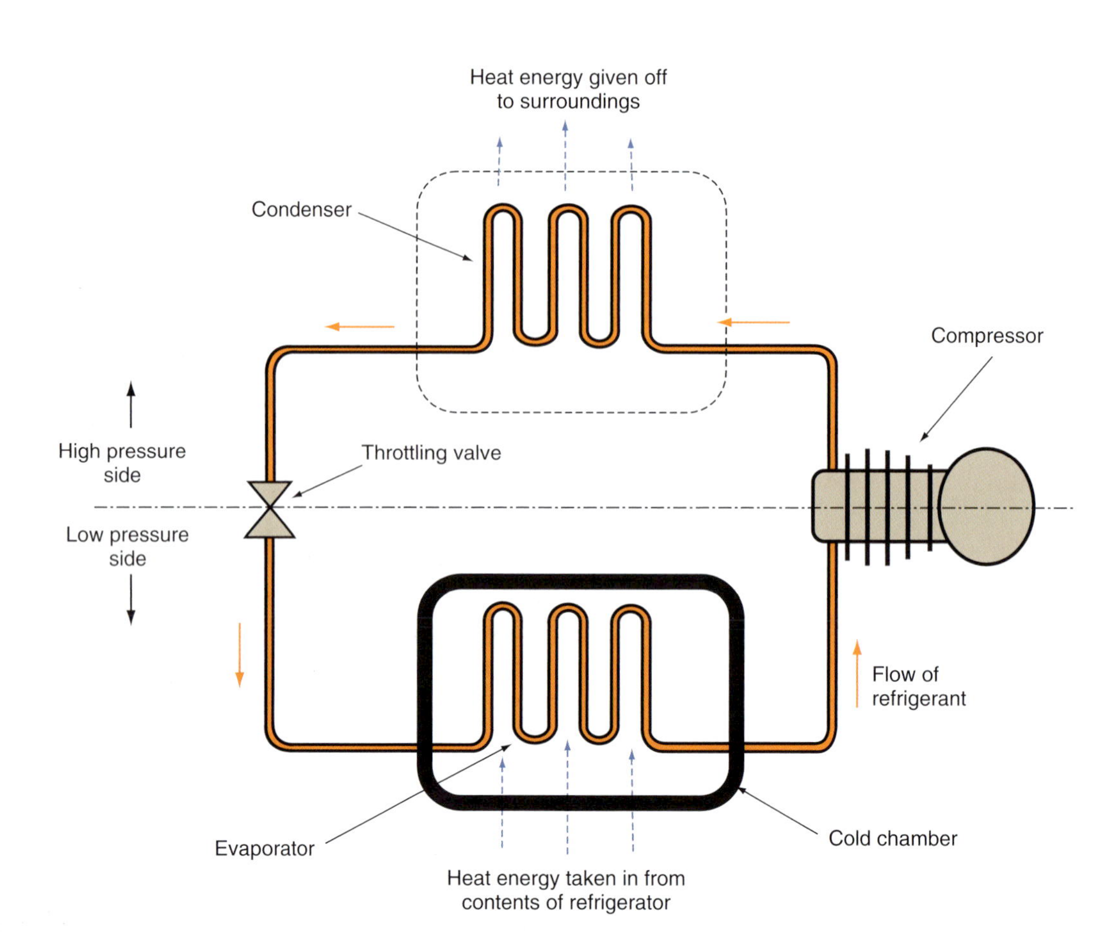

Figure 3.65 Vapour-compression refrigeration circuit

KEY POINT

Adequate ventilation space should provided around the condenser matrix for a refrigerator to work efficiently.

refigerator contents and as it passes through the evaporator grid it takes in latent heat and evaporates.

The refrigerant vapour passes from the evaporator to the compressor. During the compression process its pressure increases and its temperature rises to above that of the surroundings. Temperatures of 30°C to 35°C are common. This is well above the normal temperature in countries with a temperate climate, such as the United Kingdom. As it passes through the condenser grid the vapour condenses, giving up its latent heat to the atmosphere. Large industrial installations sometimes have two-stage compressors with an inter-cooler between the stages. They might also be equipped with an oil separator to filter out any lubricating oil which has become mixed with the refrigerant during the compression process.

KEY POINT

The waste heat from industrial processes can be used as the energy source for vapour-absorption refrigerators.

Vapour-absorption refigerators have been in use for some time in large industrial installations and in the small refrigerators used in caravans and boats. The refrigerants that they use are ozone friendly and they can be operated using waste heat from hot water or exhaust steam from manufacturing processes. The basic circuit for a vapour-absorption refrigerator is shown in Figure 3.66.

A vapour-absorption refrigerator functions with a condenser, throttling valve and evaporator in the same way as a vapour-compression refrigerator. The difference between the two systems is that the

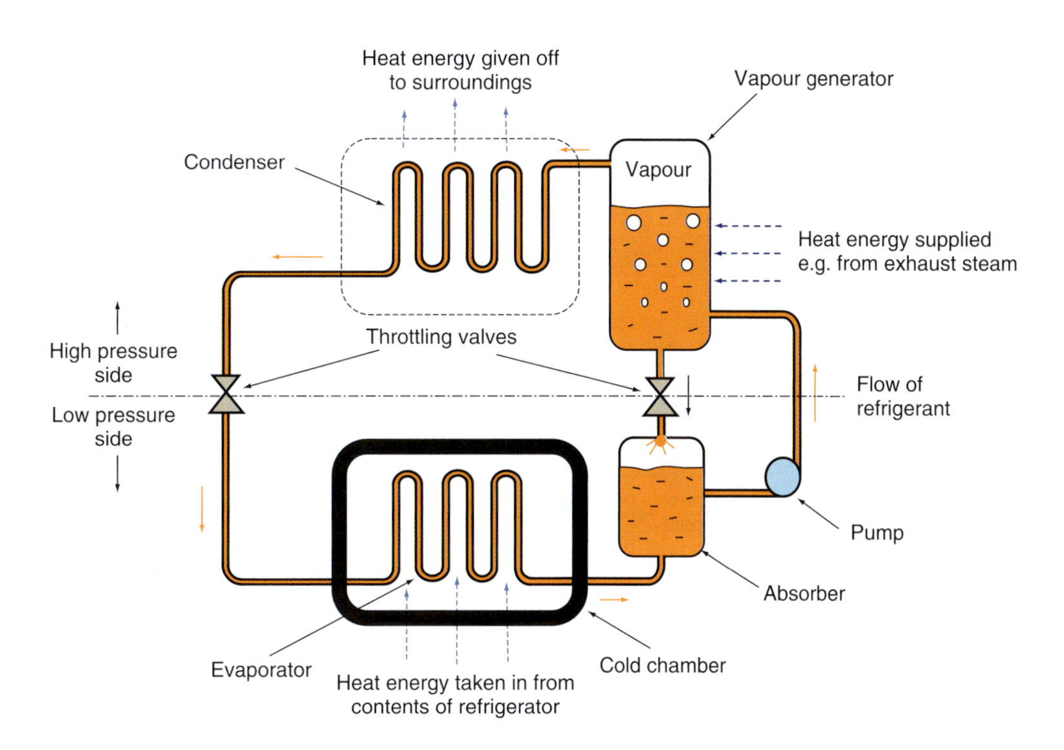

Figure 3.66 Vapour-absorption refrigeration circuit

compressor is replaced by a vapour absorber, pump and vapour generator. These contain a secondary liquid which readily absorbs the refrigerant at low temperature and pressure and releases it at a later stage when the temperature and pressure have risen. The most common combination of liquids for industrial installations is water and lithium bromide. Water is the refrigerant and lithium bromide the absorber. The combination used for the small caravan refrigerators is ammonia and water, where ammonia is the refrigerant and water the absorber.

After taking in latent heat in the evaporator the refrigerant vapour enters the absorber at low temperature and pressure where it is readily absorbed by the secondary liquid. The refrigerant-absorber solution is then pumped up to the vapour generator which is at a higher pressure. Here it receives heat energy from an external source, its temperature rises and the refrigerant vaporises out of solution. The absorber liquid passes back to the absorber vessel through the throttling valve between the two vessels, and the refrigerant vapour passes through the condenser. Here it gives up its latent heat of vaporisation to the atmosphere as it condenses before passing through the throttling valve to begin the cycle again. The pump is dispensed with in caravan and boat refrigerators. Here, gravity alone provides the circulation in the form of a convection current.

The routine maintenance of refrigeration systems involves inspection for leaks and signs of corrosion in the pipework, replenishment of the lubricant in the compressor and the periodic replacement of valves and seals.

Air-conditioning

Air-conditioning is the full mechanical control of the indoor environment to maintain comfortable and healthy conditions.

Is objective is to provide clean fresh air at a temperature and humidity level that is comfortable to the occupants. The general arrangement of an air-conditioning system is shown in Figure 3.67.

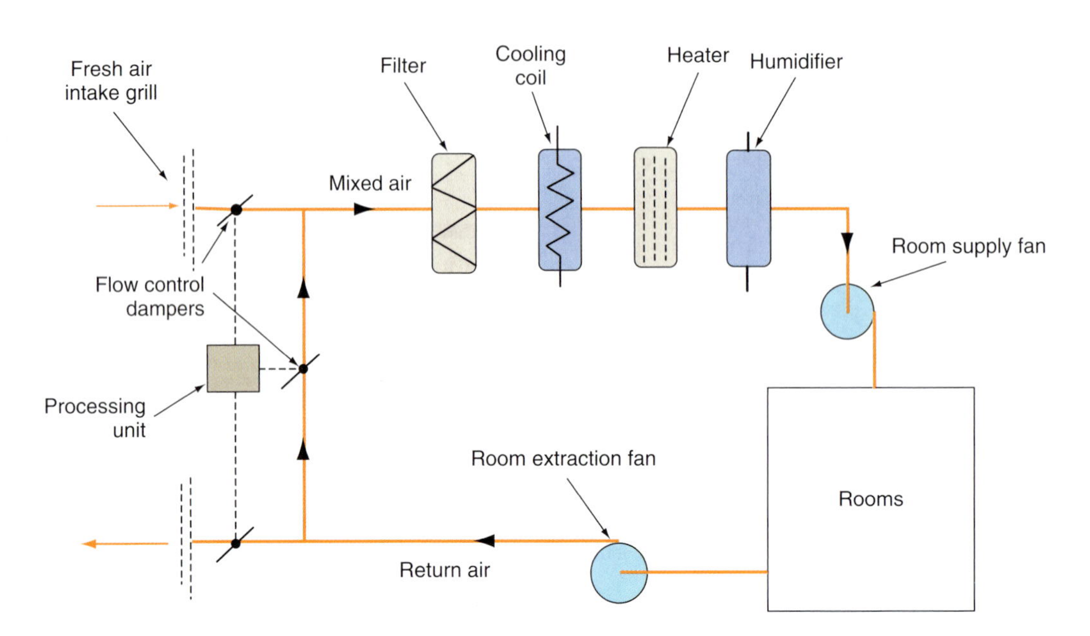

Figure 3.67 Air-conditioning system

KEY POINT

A full air-conditioning system controls the temperature, humidity cleanliness and freshness of the air.

The major components of a modern air conditioning system are an air filter, cooling coil, heater, humidifier, circulation fans, flow control dampers, room monitoring equipment and a processing unit. The processing unit controls the mixing of fresh air and return air. It is programmed to maintain the required standard of freshness, humidity, temperature and flow rate whilst keeping the energy input to a minimum.

After mixing, the air is filtered and then heated or cooled to the required temperature. It is heated if the fresh air temperature is lower than that required, and the system is then said to be working on a winter cycle. If the fresh air temperature is higher than that required, the mixture is cooled and the system is then said to be working on a summer cycle. In both cases the humidity, or moisture content of the air, may need to be adjusted to the required comfort level. If the air is too moist, the occupants of the room will tend to perspire, especially if they are doing physical work. If the air is too dry, the skin will feel dry and they will tend to feel thirsty.

The temperature of the air may be raised in the heater by the flow of hot water or steam or it may be heated by an electrical element. The temperature of the cooling coil is maintained by the flow of refrigerated water. Routine maintenance involves the cleaning or replacement of filters, lubrication of the moving parts and inspection for leaks.

Test your knowledge 3.14

1. What is the purpose of the throttling valve in a vapour-compression refrigerator?
2. What is the purpose of the compressor in a vapour-compression refrigerator?
3. What form does the external energy supply take in a vapour-absorption refrigerator?
4. Why is return air mixed with fresh air in an air-conditioning system?
5. What is the difference between the summer and winter cycles of an air-conditioning system?

Activity 3.13

(a) When you use an aerosol spray the container starts to feel cold. Why is this, and how does it relate to the working of refrigerators?

(b) The 'relative humidity' of the air in a building gives a measure of its moisture content. How is it defined and measured?

Mechanical handling and positioning equipment

The range of mechanical handling, lifting and positioning equipment used in the engineering and manufacturing industries is very wide and in some cases it is highly specialised. The transfer of materials, components and assemblies through the production stages often takes place on roller or belt conveyors. The roller conveyer is probably the simplest form where products are passed between work stations along a track containing rollers. The track is usually set on a slight incline so that transfer is effected by the force of gravity. Bulk materials are often transferred by means of a motor-driven belt conveyer. The belt is sometimes supported on concave rollers so that it sags in the centre. This enables it to carry loose materials in granular form. Roller and belt conveyers are shown in Figure 3.68.

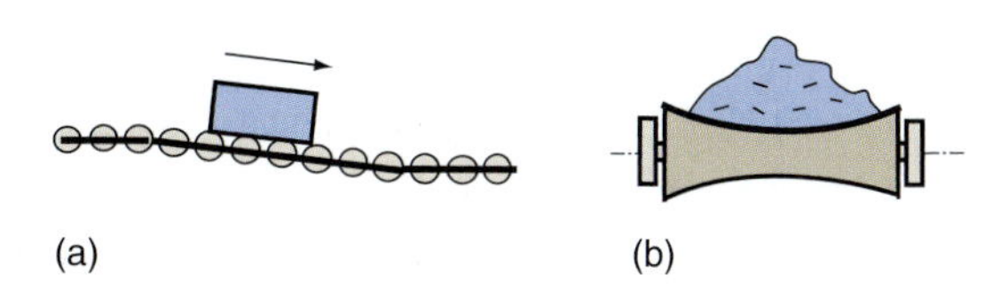

Figure 3.68 (a) Roller conveyer and (b) section through a belt conveyers

The assembly of cars and other mass production items is generally carried out on slow moving track conveyers. At some stages of assembly the track may be set at ground level whist at others it is raised so that parts can be fitted from below. Raised tracks, on which the components can be hung, are used for the transfer of components to the assembly stages and also for the transfer of components through washing, de-greasing and paint spraying booths.

A wide range of lifts, hoists and cranes are used for the vertical movement of products and materials. Passenger lifts and the lifts used to tranport materials between the floors of industrial installations are generally raised and lowered by means of a cable system driven by an electric motor. A counterbalance weight is sometimes included to reduce the power requirement. Fail-safe devices are always included in the design in case of a cable or power failure. Hydraulic lifts are used where heavy loads need to be raised through a comparatively small distance. The vehicle lifts in motor repair workshops are often of this type in which pump supplies oil under pressure to the ram in a hydraulic cylinder. The principle of operation is similar to that of the hydraulic jack.

Heavy loads need to be lifted by a crane. The types used in engineering and manufacturing workshops are mainly jib and gantry cranes. Gantry cranes bridge the workshop or workshop bay. They run on overhead rails at either side of the service area as shown in Figure 3.69.

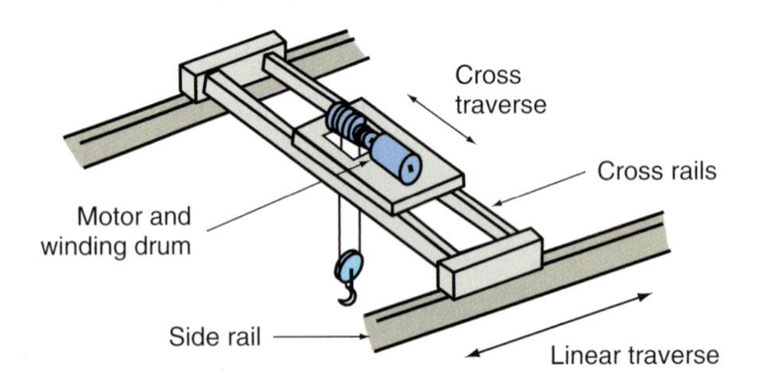

Figure 3.69 General arrangement of a gantry crane

KEY POINT

A gantry crane can lift and re-position loads from any part of its workshop or workshop bay.

The carriage which carries the driving motor and lifting drum runs on cross rails and is able to pick up and re-position loads from any part of the service area. In some installations the gantry rails extend outside the building so that the crane can be used for loading and unloading operations. Gantry cranes with a large lifting capacity are usually controlled by an operator seated on the crane. Lighter versions may be controlled from ground level by means of a handset.

Mobile jib cranes are often used in outdoor storage areas for loading, unloading and re-positioning materials. Static jib cranes such as that shown in Figure 3.70 are sometimes installed in indoor work areas for lifting heavy components at a work station. All types of crane and lifting device should be clearly marked with the safe working load, e.g. SWL 500 kg. This must never be exceeded. Routine maintenance on cranes is generally limited to cleaning and lubricating the moving

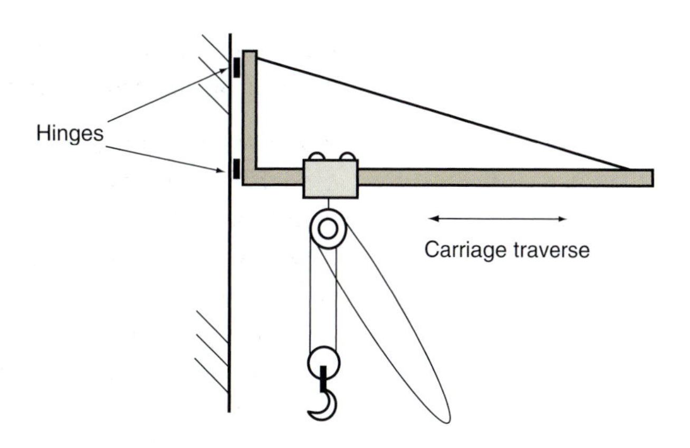

Figure 3.70 Work station jib crane

KEY POINT

Cranes and lifts are marked with the maximum load or number of passengers which can be safely carried. This should not be exceeded.

parts and inspecting the wire ropes and electrical cables for wear or damage.

Fork-lift trucks find widespread use for transporting products and materials loaded on pallets. They should be clearly marked with the loading capacity when the forks are extended at maximum lift. Cranes and fork-lift trucks should be used only by operators who have received the appropriate training. Large storage warehouses with high rise storage racks often have automated positioning and retrieval lifts which work on the fork-lift principle. They run on tracks between the storage racks and can be directed to any position within the matrix from a remote control centre.

The automated manipulation, positioning and securing of components has seen many advances in recent years. CNC machining and automated assembly and packaging operations make widespread use of pneumatic and hydraulic systems and stepper motors. Sometimes these are incorporated in robot arms that enable components to be securely gripped, rotated, lifted and placed with great accuracy during manufacturing processes.

Stepper motors are a form of electric motor that rotates through a precise angle on receipt of a digital signal. They are widely used in CNC machining processes and robotic systems. Automated systems are generally well guarded with safety interlocks that ensure the system will not operate unless the guards are in place. Guard rails should also be provided around the working envelope of robot arms. Routine maintenance usually involves no more than cleaning and the replenishing of lubricants and coolants.

Test your knowledge 3.15

1. What is a gantry crane?
2. What do the initials SWL stand for on cranes and hoists?
3. What is the purpose of a counterbalance weight as used in some lift systems?
4. Why are concave rollers sometimes used to support conveyer belts?
5. How does a stepper or stepping motor function?

Activity 3.14

Automated handling and positioning equipment often contains proximity sensors to detect whether or not a component is present on the work station. Describe two such sensors which confirm the presence of a component by sending an electrical signal to a control unit so that the processing cycle can begin.

To check your understanding of the preceding section, you can solve Review questions 41–45 at the end of this chapter.

Review questions

1. A measure of the resistance of a liquid to shearing forces is given by its
   (a) density
   (b) solubility
   (c) viscosity
   (d) resistivity

2. The viscosity index of a lubricating oil is an indication of its
   (a) change of viscosity with pressure
   (b) resistance to shearing forces
   (c) change of viscosity with temperature
   (d) resistance to contamination

3. The purpose of an anti-oxidant additive is to
   (a) reduce the formation of surface foam
   (b) give increased corrosion resistance
   (c) neutralise the effects contaminating acids
   (d) reduce the effects of reaction with atmospheric oxygen

4. Polytetrafluoroethylene is widely used as
   (a) a solid lubricant
   (b) an acid neutraliser
   (c) a detergent additive
   (d) an anti-foam agent

5. Lithium is used as
   (a) an anti-oxidant additive to lubricating oil
   (b) a thickener in greases
   (c) an additive to increase the viscosity index of lubricating oil
   (d) a non-stick surface coating

6. A lubrication system in which the lubricating oil is free to drain away is known as a
   (a) self-contained system
   (b) non-replenishable system
   (c) re-circulating system
   (d) total loss system

7. COSHH regulations apply to the
   (a) specifications of lubricating oils
   (b) handling and storage of lubricating oils
   (c) periodic replenishment of lubricating oils
   (d) recommended uses of lubricating oils

8. Motor vehicle gearboxes generally incorporate
   (a) splash lubrication
   (b) total loss lubrication
   (c) screw-down grease cup lubrication
   (d) wick feed lubrication

9. A re-circulating oil lubrication system is incorporated in
   (a) washing machines
   (b) diesel engines
   (c) electric motors
   (d) power tools

10. Wick feed lubrication utilises
    (a) a circulation pump
    (b) the force of gravity
    (c) capillary action
    (d) grease nipples

11. The cylinder head seal that is widely used on automobile engines is a
    (a) composite metal gasket
    (b) mechanical seal
    (c) rotary lip seal
    (d) packed gland

12. Rotary lip seals are used
    (a) between pipe flanges
    (b) around pistons
    (c) around rotating shafts
    (d) around reciprocating shafts

13. The type of bearing best able to carry axial thrust is the
    (a) plane journal bearing
    (b) needle bearing
    (c) caged ball bearing
    (d) tapered roller bearing

14. The crankshaft and big-end bearings of a multi-cylinder internal combustion engine are generally
    (a) Phosphor-bronze
    (b) steel-backed white metal
    (c) hardened steel
    (d) cast iron

15. Needle roller bearings that run directly on a rotating shaft are mainly used for applications where there is
    (a) high speed and high axial thrust
    (b) low speed and light radial loading
    (c) high speed and high radial loading
    (d) low speed and high axial thrust

16. The British Association screw thread is widely used for
    (a) self-tapping screws
    (b) electrical and IT applications

(c) accurately machined fitted bolts

(d) heavy engineering applications

17. The screwed fastening made from a short length of bar and threaded at both ends is called a

(a) setscrew

(b) black bolt

(c) stud

(d) machine screw

18. Which of the following is classed as a positive locking device?

(a) a tab washer

(b) a lock nut

(c) a spring washer

(d) a friction nut

19. For applications where rivet heads must be flush with the joint surface it is usual to use

(a) conoidal headed rivets

(b) round headed rivets

(c) pan headed rivets

(d) countersunk headed rivets

20. For applications where access is only possible from one side of a riveted joint it is usual to use

(a) flat headed rivets

(b) pan headed rivets

(c) pop rivets

(d) round headed rivets

21. The angle turned through by a cam whilst its follower is stationary is called the

(a) pressure angle

(b) angle of lap

(c) dwell angle

(d) angle of friction

22. An inversion of a plane mechanism can be obtained by

(a) changing the input speed

(b) changing the lengths of the links

(c) reversing the input motion

(d) changing the link that is fixed

23. Two links in a mechanism that are hinged or pin-jointed together constitute a

(a) turning pair

(b) sliding pair

(c) screwed pair

(d) fixed pair

24. A practical application of a four-bar linkage is the

(a) Whitworth quick return motion

(b) oscillating cylinder mechanism

(c) slotted link quick return motion

(d) Watt's parallel motion

25. A flat follower is sometimes offset to the side of the cam in order to

(a) rotate the follower and equalise wear

(b) increase the dwell angle

(c) reduce the travel of the follower

(d) increase the speed of the follower

26. Alignment for a connecting shaft between machines that are some distance apart is best carried out using

(a) a straight edge

(b) a single dial test indicator

(c) optical equipment

(d) two dial test indicators

27. That the driving contacts of a universal coupling must be in a plane that bisects the angle between the coupled shafts is a condition for

(a) maximum power transmission

(b) a constant velocity ratio

(c) accurate alignment

(d) a secure joint

28. Automatic engagement at a pre-determined speed is a characteristic of a

(a) dog clutch

(b) flat plate clutch

(c) conical clutch

(d) centrifugal clutch

29. A dynamometer is a form of brake that is used

(a) to measure output torque

(b) on motor vehicles

(c) to measure angular velocity

(d) on power transmission shafts

30. Fluid clutches are widely used in motor vehicles in conjunction with

(a) manually operated gearboxes

(b) power steering systems

(c) automatic gearboxes

(d) anti-lock braking systems

31. The effective diameter of a gear is called the

(a) base circle diameter

(b) root circle diameter

(c) pitch circle diameter

(d) outer diameter

32. The pitch circle diameter of a gear divided by the number of teeth is its

(a) module

(b) circumferencial pitch

(c) pressure angle
(d) dedendum

33. The maximum power that a belt drive can transmit before slipping depends primarily on
(a) the pulley diameters
(b) the centre distance between the pulleys
(c) the initial tension setting
(d) the mass of the belt

34. Hypoid bevel gears are used where the axes of the connected shafts
(a) lie in the same plane
(b) do not intersect
(c) intersect at right angles
(d) are parallel

35. A synchronous belt is used in applications where
(a) a positive drive is required
(b) the pulleys must rotate at the same speed
(c) the belt is required to slip at a pre-determined speed
(d) the shaft axes intersect at right angles

36. An advantage of a hydraulic actuation system is that it
(a) can operate over long distances
(b) vents to the atmosphere
(c) can transmit large positive forces
(d) does not require a return pipe

37. The purpose of a pressure switch in a pneumatic system is to
(a) regulate the supply pressure
(b) control the pneumatic actuator
(c) stop and restart the compressor
(d) display the air pressure

38. The purpose of a superheater in a steam plant is to
(a) heat up the boiler feed water
(b) dry out the steam and raise its temperature
(c) pre-heat the air supply to the boiler
(d) control the fuel supply to the boiler

39. Cooling towers are installations that are used to
(a) cool the exhaust gases from the boiler
(b) condense the low pressure steam
(c) release exhaust steam to the atmosphere
(d) cool the water circulating through the condenser

40. An economiser heats up the boiler feed water using
(a) exhaust gases from the boiler and superheater
(b) low pressure exhaust steam
(c) electrical energy that is generated
(d) low cost fuel

41. The purpose of the compressor in a vapour-compression refrigerator is to
(a) take in latent heat
(b) release heat energy to the atmosphere
(c) raise the temperature and pressure of the refrigerant
(d) allow the refrigerant to condense

42. In a refrigeration cycle the high pressure refrigerant expands to a lower pressure and temperature on passing through the
(a) throttling valve
(b) evaporator matrix
(c) compressor
(d) condenser matrix

43. The mixing of return air and fresh air in a modern air-conditioning system is controlled by the
(a) humidifier
(b) air filter
(c) processing unit
(d) cooling coil

44. Loads can be re-positioned in any part of a workshop bay if it is served by a
(a) roller conveyer
(b) work station jib crane
(c) belt conveyer
(d) gantry crane

45. Stepper motors are able to rotate through a precise angle on receipt of a
(a) analogue signal
(b) pneumatic signal
(c) digital signal
(d) hydraulic signal

A modern aircraft such as the Boeing 747 is constructed from a variety of materials. Aluminium and titanium alloys are widely used for the fuselage and wings because they are both strong and light. A range of steel is used in the undercarriage components where high strength and toughness are required. The tyres are of course rubber and a range of thermoplastics and thermosetting plastics are used for the interior fittings.

Photo courtesy of iStockphoto, Dan Barnes, Image # 4941515

Chapter 4

Properties and Applications of Engineering Materials

Design and manufacturing engineers need to be aware of the range of materials available for use in engineered products. They need to know about the properties of the different materials, their cost, their availability and how they can be processed. This enables them to select materials which are fit for their purpose and which can be processed at a reasonable cost.

Maintenance engineers also need to have a knowledge of materials. They need to know how they are affected by service conditions and how they might be protected to prolong their service life. Replacement components must be made of the same material, or from a suitable alternative which has similar properties. When a component fails in service, engineers must be able to identify the cause. A design change or a change to a more suitable material can then be implemented to improve the quality of the product.

The aim of this chapter is to provide you with a basic knowledge of the most common engineering materials. This will include their structure, properties, the effects of processing, their relative costs and forms of supply. It will also direct you to sources of information which will enable you to find out more about the range of available materials.

Atomic Structure of Materials

There are 92 naturally occurring chemical elements. A chemical element cannot be split into other substances. It is made up of atoms that are all the same. Sometimes, groups of two or more atoms combine together to form molecules. These too are identical in a pure element. Atoms are made up of particles. The heaviest of these are protons and neutrons which are roughly the same size and form the central core, or nucleus, of an atom.

A cloud of orbiting electrons surrounds the nucleus. These have only about two-thousandths of the mass of protons and neutrons, and it is at the nucleus that the mass of an atom is concentrated. Protons carry a small positive electrical charge. The orbiting electrons, which are equal in number to the protons, carry a negative electrical charge. The combined effect is to make an atom electrically neutral.

The atoms of the different chemical elements have different numbers of protons, neutrons and electrons. Hydrogen has the smallest and lightest atoms. They have just one proton in the nucleus and one orbiting electron. Uranium has the largest and heaviest naturally occurring atoms with 92 protons and electrons and an even larger number of neutrons. The electrons in the different atoms are found to be orbiting in distinct shells. There are seven naturally occurring shells which are given the letters K, L, M, N, O, P and Q. The maximum number of electrons which shells can hold is 2, 8, 18, 32, 32, 9 and 2, respectively.

The lighter elements only have electrons in the inner shells. The heavier the element, the more shells are occupied. Hydrogen atoms, with one proton in the nucleus, have just one electron orbiting in the K shell. Iron atoms have 26 protons and 20 neutrons in the nucleus and 26 orbiting electrons. They are distributed 2, 8, 14 and 2 in the K, L, M and N shells. Atoms with eight electrons in their outer shell are very stable. Although the M and N shells can hold more than eight electrons, atoms with their outer electrons in these shells will readily shed or share electrons to empty the shell or to achieve the stable number of eight.

The forces that bind the atoms of a material together arise from the electrical charges carried by their protons and electrons. Atoms of an element with a small number of electrons in the outer shell will readily donate them to the atoms of another element whose outer shell needs them to give a stable number (Figure 4.1).

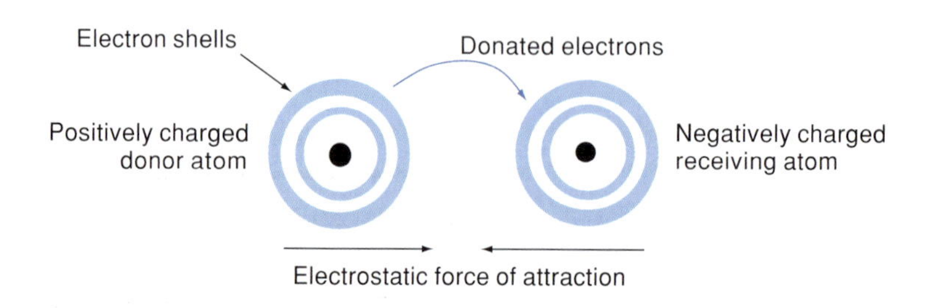

Figure 4.1 Ionic or electrovalent bonding

KEY POINT

Ionic or electrovalent bonds are formed when an atom donates electrons to another to give it a stable number in its outer electron shell. The donor becomes positively charged and the receiver becomes negatively charged. The atoms then become bonded together by the forces of electrostatic attraction.

The donor atom then becomes positively charged and the receiving atom becomes negatively charged. The result is that the atoms become joined together by electrostatic forces to form a compound molecule. The process is called ionic or electrovalent bonding and is one of the ways in which chemical compounds are formed. Sodium and chlorine bond together in this way. Sodium has only one electron in its outer shell whilst chlorine has seven. Their atoms combine to form molecules of sodium chloride, NaCl, which is better known as common salt.

Another way in which atoms can combine together to form molecules is by sharing their outer electrons to give a stable number in a common outer shell (Figure 4.2). These are called valence electrons. Sharing them produces strong bonds which hold the atoms together. The process is called covalent bonding.

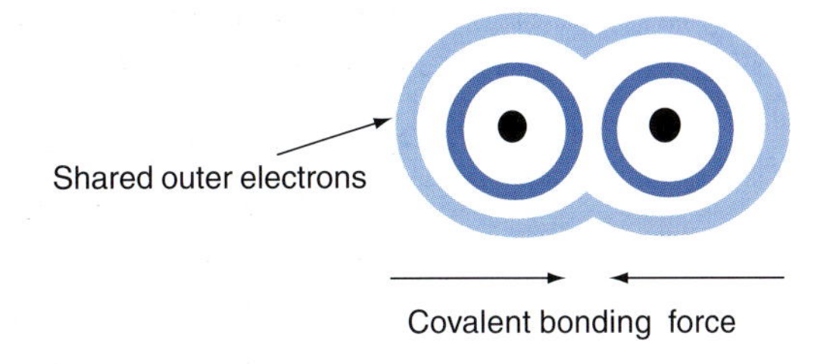

Figure 4.2 Covalent bonding

The atoms of hydrogen combine in this way to form H_2 molecules. Similarly, atoms of nitrogen form covalent bonds with three atoms of hydrogen to form molecules of ammonia, NH_3.

KEY POINT

Covalent bonds are formed when atoms share their outer electrons to give a common outer shell which contains a stable number. The outer shell binds the atoms together.

Metals have a small number of electrons in the outer shell which easily become detached and are shared between all the atoms in the material. They are known as *free electrons* which circulate between and around the atoms in a random fashion. The effect is to bond the atoms together with a form of covalent bond. It is sometimes referred to as *metallic bonding* which is quite strong (Figure 4.3).

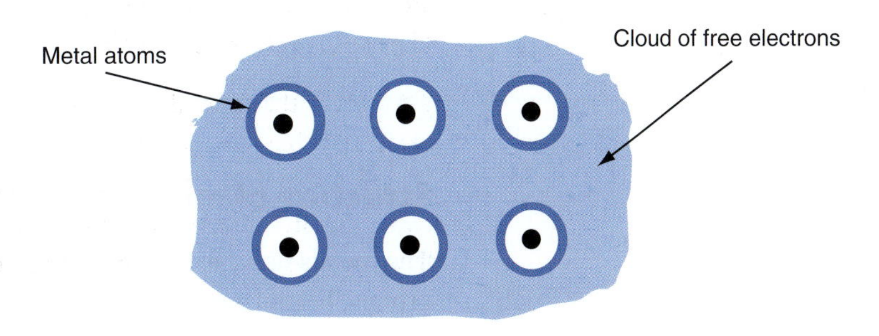

Figure 4.3 Metallic bonding

All of the materials used to make engineered products are thus made up of atoms and molecules which are held together by strong bonds. A closer examination reveals that in different materials, the atoms and molecules are arranged in different ways. The structure of a solid material may be *amorphous*, *crystalline* or *polymer*. In amorphous materials the atoms or molecules are not arranged in any particular pattern. They are disordered rather like those in a liquid. Glass is one of the few truly amorphous materials used in engineering.

All metals, and a great many ceramics, are crystalline. That is to say that when they solidify from being molten, crystals start to form in different parts of the melt. These multiply and grow until the material is completely solidified. Within the crystals or *grains* the atoms pack themselves into a regular geometric pattern. The size of the grains depends on the rate of cooling. Slow cooling usually gives large grains. Grain size affects the properties of a material. Large grains tend to increase the brittleness and reduce the strength of a metal.

Sometimes you can see the grains in a metal with the naked eye, such as the grains of zinc on galvanised steel. In other cases you will need to polish the surface of the metal and then etch it with chemicals to show up the grain boundaries. The grains can then be seen using a magnifying glass or a microscope.

Polymer materials include plastics and rubbers. Here the atoms are arranged in long molecular chains, made up mainly of carbon and hydrogen atoms, which are known as *polymers*. Each polymer can contain several thousand atoms. In plastics and rubbers the polymers are intertwined, rather like spaghetti. The strength of the covalent bonds that form between the polymers affects the properties of plastic materials and rubbers. As you know from experience, some are flexible and easy to stretch whilst others are quite stiff and brittle.

Test your knowledge 4.1

1. How are electrovalent and covalent bonds formed between atoms of different materials?
2. How are the atoms bonded together in metals?
3. What is the difference between an amorphous and a crystalline material?
4. What is a polymer?
5. How does that rate of cooling from the molten state affect the grain size in crystalline materials?

Structure of metals

KEY POINT

An alloy is a mixture of metals or a mixture of a metal and other substances which results in a material which displays metallic properties.

The metals used in engineering may be subdivided into two main groups. Ferrous metals, which contain iron as a major constituent, and non-ferrous metals, which do not contain iron, or in which iron is only present in small amounts. Some metals such as copper and lead are used as an almost pure form. Others are mixed to form *alloys*. An alloy can be a mixture of metals. It can also be a mixture of a metal and other substances so long as the resulting material displays metallic properties.

When a pure molten metal solidifies, latent heat is given off and crystals, or grains, start to form at different points in the fluid. These embryo grains are known as *dendrites*, which grow until they come into contact with each other (Figure 4.4). Solidification is then complete and within the grains, the atoms are found to have arranged themselves in a regular geometric pattern. The orientation of the grains is different and they come into contact at irregular angles. This is why some of them seem darker than others when you view them.

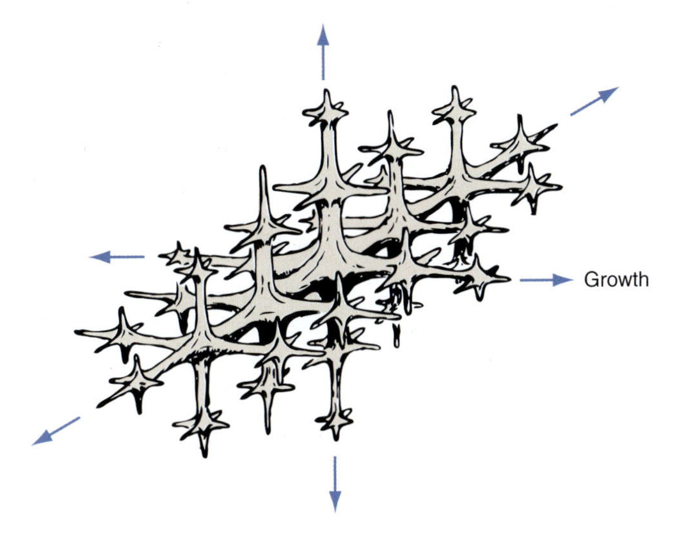

Figure 4.4 Growth of a dendrite (Redrawn from *Properties of Engineering Materials*, Second Edition, R Higgins, Butterworth-Heinemann, p. 48, Fig. 3.19, 1994, with permission from Elsevier.)

The pattern in which the atoms arrange themselves is known as the *crystal lattice structure*. It is held together by the metallic bond produced by the cloud of free electrons. The atoms take up positions of minimum potential energy and in most engineering metals they arrange themselves in one of the three possible lattice structures.

The most open packed is the *body-centred cubic* (BCC) formation (Figure 4.5). The pattern is continuous throughout the crystal but it is convenient to consider a unit cell where the atoms are arranged at the eight corners of a cube, surrounding another atom at the centre of the cube.

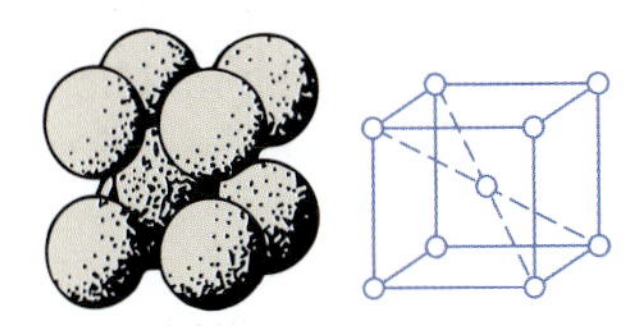

Figure 4.5 BCC structure (Redrawn from *Properties of Engineering Materials*, Second Edition, R Higgins, Butterworth-Heinemann, p. 4, Fig. 3.15, 1994, with permission from Elsevier.)

This structure is taken up by iron at normal temperatures and also by chromium, tungsten, niobium, molybdenum and vanadium. The open-packed planes of atoms which make up this structure do not easily move over each other when external forces are applied. This is the main reason why the metals listed above are hard to deform in comparison to some others.

Some of the softest and more easily deformed metals take up a *face-centred cubic* (FCC) formation (Figure 4.6). This is made up of planes

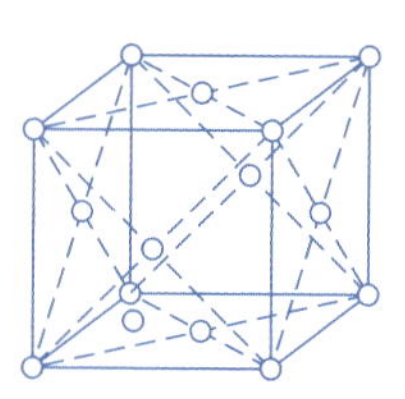

Figure 4.6 FCC structure (Redrawn from *Properties of Engineering Materials*, Second edition, R Higgins, Butterworth-Heinemann, p. 43, Fig. 3.14, 1994, with permission from Elsevier.)

of atoms that are more closely packed. The unit cell consists of a cube with atoms at the eight corners and atoms at the centre of each face.

This structure is taken up by aluminium, copper, silver, gold, platinum, lead and also iron when heated to a temperature of about 800°C. These are some of the softest and easily deformed metals. The closely packed planes of atoms move more readily over each other when external forces are applied. They do not actually slide over each other. It is thought to be more of a rippling action which occurs due to imperfections in the lattice, but the effect is the same.

KEY POINT

Metals with a body-centred cubic crystal lattice structure tend to be hard and difficult to deform whereas metals with a face-centred cubic crystal lattice structure tend to be softer and easier to deform.

Iron has a BCC structure at normal temperatures but at temperatures approaching 800°C, the atoms rearrange themselves in the solid material to take up an FCC structure. The iron then becomes much softer and easier to shape by forging and hot pressing. Materials such as iron, which can exist with different crystal lattice structures, are said to be *allotropic*.

The third type of crystal lattice formation that is to be found in engineering metals is the *close-packed hexagonal* (CPH) structure. This is shown in Figure 4.7. It is made up of layers of atoms which are just as tightly packed as those in an FCC structure. A good example of this packing is the way in which the red snooker balls are packed in the triangular frame at the start of a game. The unit cell of the CPH consists of a hexagonal prism running through three layers of atoms. It has atoms at the 12 corners of the prism and an atom at the centre of each hexagonal face. Between these are sandwiched another three atoms from the middle layer.

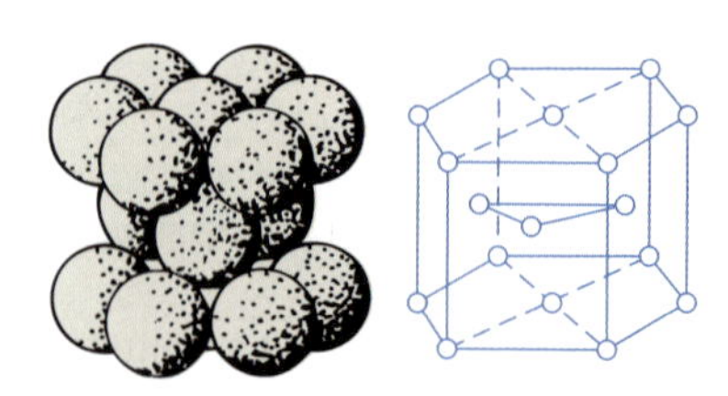

Figure 4.7 CPH structure (Redrawn from *Properties of Engineering Materials*, Second edition, R Higgins, Butterworth-Heinemann, p. 42, Fig. 3.13, 1994, with permission from Elsevier.)

The difference between CPH and FCC structures is that every third layer is displaced by half an atomic distance. This has the effect of making the metal a little more difficult to deform. Zinc, magnesium, beryllium and cadmium are some of the metals which take on this formation.

Alloys are formed when different metals, and sometimes also metals and non-metallic substances, are mixed together in the molten state. After cooling, the resulting *solid solution* may be an *interstitial alloy* (Figure 4.8), a *substitutional alloy* or an *intermetallic compound*. In an interstitial alloy, the atoms of one of the constituents are relatively small compared to those of the other. As the molten mixture solidifies and the crystals start to form, the smaller atoms are able to occupy the spaces between the larger atoms.

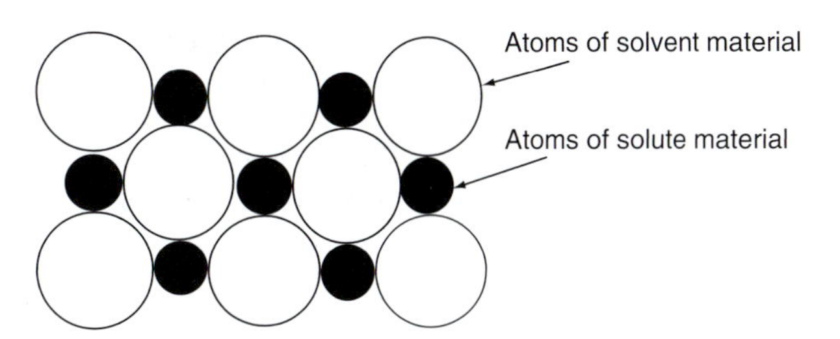

Figure 4.8 Interstitial alloy

The material with the larger atoms is called the *solvent* and that with the smaller atoms is called the *solute*. This is what occurs with some of the carbon atoms when carbon is mixed with iron to form steel. The effect is generally to enhance the strength and toughness of the parent metal.

In a substitutional alloy, two or more materials whose atoms are of roughly the same size are mixed together in the molten state. During the cooling process, atoms of the solute material replace atoms of the solvent in the crystal lattice structure. Brasses, bronzes and cupro–nickel are substitutional alloys (Figure 4.9).

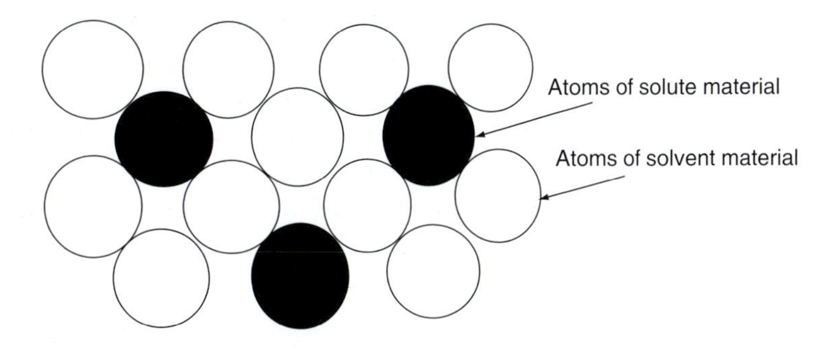

Figure 4.9 Substitutional alloy

KEY POINT

In an interstitial alloy the small solute atoms occupy the spaces between larger sized and loosely packed solvent atoms. In a substitutional alloy the solute atoms are of a similar size to the solvent atoms which they replace in the crystal lattice structure.

The difference in size of the atoms tends to distort the crystal lattice structure. This makes it more difficult for the planes of atoms to slip over each other when external forces are applied. As a result the alloy is generally stronger and tougher than its main constituent, the solvent metal.

Phase-equilibrium diagrams

Phase-equilibrium diagrams, which are also called thermal-equilibrium diagrams, enable materials experts to understand what happens during the alloying process. The materials are mixed together in the molten state and allowed to cool slowly. Temperature readings are taken and as with water when it solidifies to form ice, there is a slowing down or complete arrest of the temperature whilst solidification is taking place. This is of course due to latent heat being given off but because there is a mixture of materials, the process is a little more complicated than for water changing to ice. The process is repeated for different mixture percentages. Typical cooling curves are as shown in Figure 4.10.

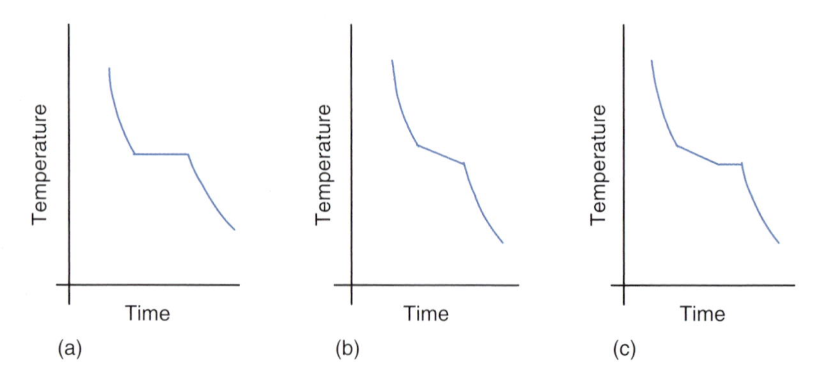

Figure 4.10 Cooling curves

The type of cooling curve that results depends on the materials and the percentage composition of the mixture. Some metals will mix together in all proportions and will remain completely mixed after cooling. These have cooling curves like Figure 4.10(b) and we say that they are completely soluble. They are usually substitutional alloys composed of metals with similar properties and whose atoms are roughly the same size. Copper and nickel behave like this when mixed together to give a range of cupro–nickel alloys. The phase-equilibrium diagram, shown in Figure 4.11, is obtained by plotting the temperature change points for mixtures ranging from 100% Cu and 0% Ni to 0%Cu and 100%Ni.

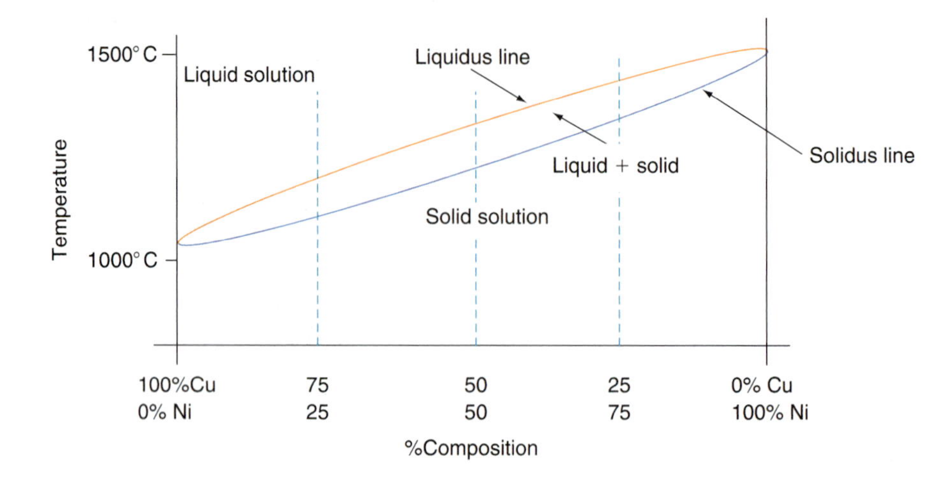

Figure 4.11 Copper–nickel phase-equilibrium diagram

At the other extreme we have materials which are completely soluble when molten but which separate as solidification takes place. Bismuth and cadmium fall into this category and their phase-equilibrium diagram is as shown in Figure 4.12.

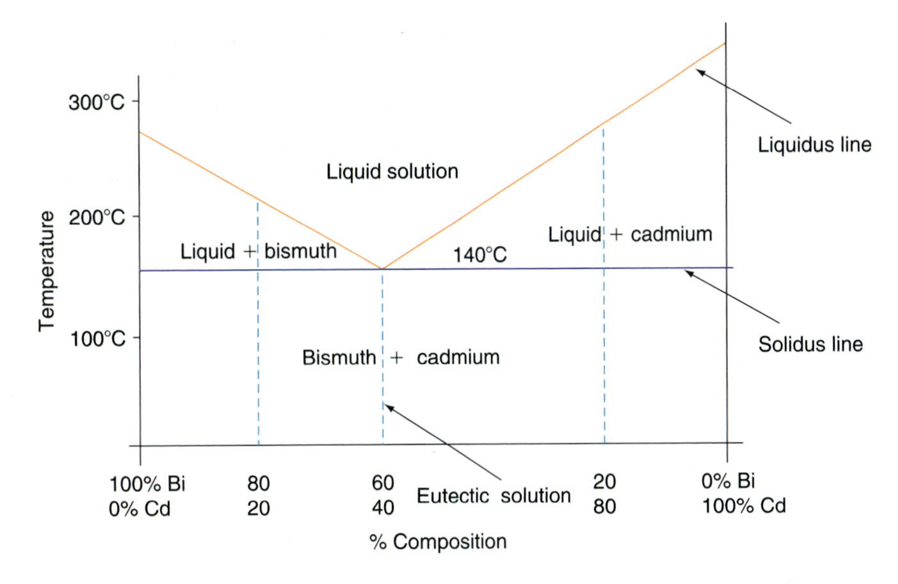

Figure 4.12 Bismuth–cadmium phase equilibrium diagram

The 40% Bi and 60% Cd solution is rather special. It is called the *eutectic solution* and as you can see, it has the lowest melting/solidification temperature. This is around 140°C and when viewed under a microscope the solid crystals or *grains* are made up of very thin alternate layers of bismuth and cadmium (Figure 4.13(b)). The cooling curve that it follows is shown in Figure 4.10(a).

To the left of the eutectic composition, bismuth crystals first start to solidify until the remaining liquid is of the eutectic composition at which point it solidifies into the layered eutectic crystals at a temperature of 140°C (Figure 4.13(a)). The cooling curve that it follows is shown in Figure 4.10(c). To the right of the eutectic composition it is cadmium crystals first start to solidify. This again continues until the remaining liquid is of the eutectic composition at which point it solidifies at 140°C into the same layered eutectic crystals (Figure 4.13(c)).

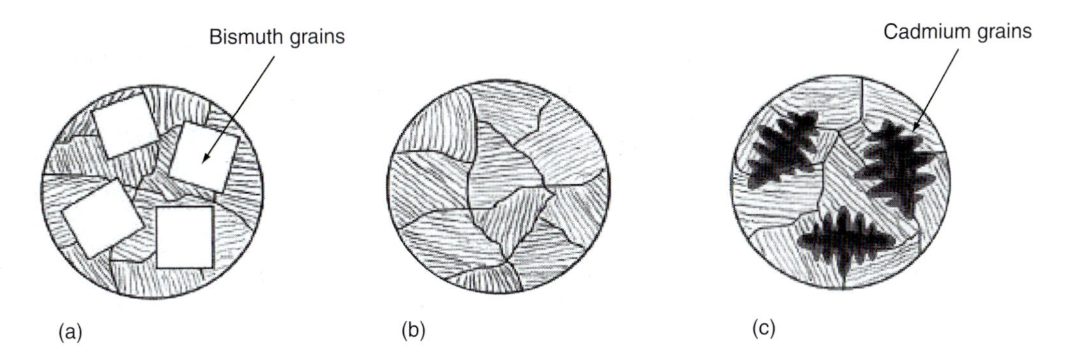

Figure 4.13 Bismuth–cadmium ally crystals (a) 80% Bi, 20% Cd (b) 60% Bi, 40% Cd (c) 20% Bi, 80% Cd

A phase-equilibrium diagram containing a eutectic is typical of alloys where the two constituents are not soluble, or only partly soluble in the solid state. The tin–lead alloy that we use for soft solder is an example, where the eutectic composition of 38% lead and 62% tin is used as electricians solder because it solidifies quickly. Plumber's solder contains more lead and has a longer solidifying time, which allows wiped joints to be shaped. A similar thing occurs in part of the iron–carbon phase-equilibrium diagram. This is relevant to the composition of plain carbon steels and we shall be looking at this later.

Sometimes a chemical reaction takes place between the solvent and solute materials. This results in the formation of an intermetallic compound. An example of this is the reaction which takes place between some of the carbon and iron atoms in steel. The iron can only take in a relatively small number of carbon atoms in the interstitial spaces. The remainder combine with some of the iron atoms to form iron carbide, Fe_3C.

Intermetallic compounds are usually quite different in their appearance and properties to the two parent materials. Iron carbide is hard and brittle but its presence in some of the grains of steel contributes to its hardness and toughness. Another example of an intermetallic compound is the combination of antimony and tin in some bearing metals to form tin-antimonide, SbSn. Its presence is found to improve the load-carrying properties of the alloy when it is used as a bearing material.

Test your knowledge 4.2

1. What are ferrous and non-ferrous metals?
2. What constitutes an alloy?
3. What is a dendrite?
4. How are the atoms arranged in body-centred cubic, face-centred cubic and close-packed hexagonal crystal lattice structures?
5. How do substitutional and interstitial alloys differ in the way in which solute atoms are accommodated?

Ferrous metals

As has been stated, these are metals in which iron is the main constituent. Pure iron is a relatively soft metal. It is not easy to machine it to a good surface finish because it tends to tear when being cut. When molten, it does not have good fluidity and as a result it is difficult to cast. In years gone by, wrought iron was used for engineering purposes. This is almost pure iron but containing strands of impurities which enhance its strength. Except for some decorative forge work its use has been superseded by mild steel. When small amounts of carbon are added to iron, this greatly improves its strength and machinability. The resulting alloys are called plain carbon steels. The different grades, over which the carbon content ranges from 0.1% to 1.4%, are shown in Table 4.1.

KEY POINT

Plain carbon steel is an alloy of iron and carbon with a carbon content between 0.1% and 1.4%. Increasing the carbon content increases the tensile strength and hardness of the material.

As can be seen from Table 4.1, the tensile strength of plain carbon steel increases with carbon content. Small amounts of other elements may be present from the iron smelting and steel production process and others may be added in small amounts to improve the properties of the metal.

Table 4.1 Plain carbon steels

Classification	Carbon (%)	Tensile strength (MPa)	Applications and uses
Dead mildsteel	0.1–0.15	400	Steel wire, nails, rivets, tube, rolled sheet for the production of pressings
Mild steel	0.15–0.3	500	Bar for machining, plate for pressure vessels, nuts, bolts, washers, girders and stanchions for building and construction purposes
Medium carbon steel	0.3–0.8	750	Crankshafts, axles, chains, gears, cold chisels, punches, hammer heads
High carbonsteel	0.8–1.4	900	Springs, screwcutting taps and dies, woodworking tools, craft knives

The main constituents of plain carbon steel are however iron and carbon and it is the carbon content which most affects on its properties.

When mild steel is viewed under a microscope it can be seen to have two distinct types of grain. One type appears white and is made up of iron atoms with carbon atoms absorbed in the interstitial spaces. These are known as *ferrite* grains. The other type has a layered mother-of-pearl appearance and is made up of alternate layers of ferrite and the intermetallic compound iron carbide, Fe_3C, which is known as *cementite*. These are known as *pearlite* grains. It is in the pearlite grains that most of the carbon content resides. The higher the carbon content of steel, the more pearlite grains are present, and as a result the steel is harder and tougher (Figure 4.14).

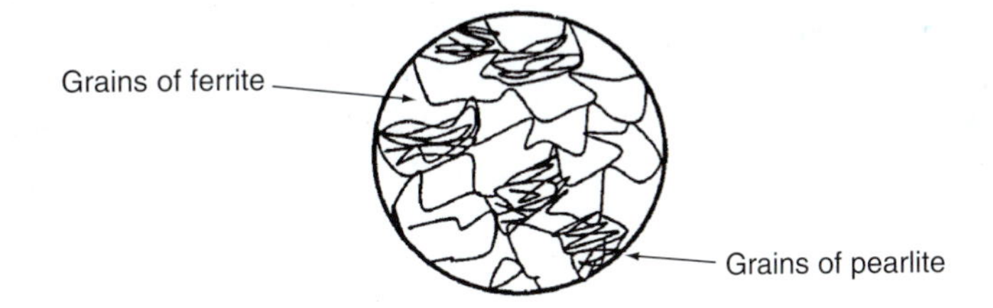

Figure 4.14 Microstructure of mild steel

Dead mild and mild steel are general purpose materials which are used for a host of components in the engineering and the building industries. The surface hardness of mild steel can be increased by heat treatment processes known as case carburising and case hardening. These give improved wear resistance and will be discussed later.

The increased carbon content of medium carbon steel gives it increased toughness and impact resistance, making it suitable for power transmission components and tools such as hammers, cold chisels and rivet snaps. The hardness and toughness of both medium and high carbon steel can be enhanced by heat treatment processes known as hardening and tempering and these will also be discussed later.

The carbon content in high carbon steel, together with appropriate heat treatment, results in a material which is very hard and wear resistant.

This makes it most suitable for cutting tools such as files, tin-snips, woodwork tools and craft knives where the cutting process takes place at normal temperatures and at a relatively slow speed. Alternative heat treatment can produce a material which is less hard but very tough and resilient. It is widely used for coil and leaf springs.

In addition to carbon, manganese is also present in steel. It is added during the steel making process to get rid of excess iron oxide. It combines with this to form manganese oxide which rises to joined the slag on the surface of the molten steel. The remaining manganese improves the harness and toughness of the steel. Steel also contains impurities such as silicon, sulphur and phosphorus and nitrogen. Silicon improves fluidity and is beneficial in steel that is to be used for casting in amounts up to 0.3%. Phosphorus up to around 0.05%, sulphur up to around 0.3% and nitrogen up to around 0.02% are all undesirable impurities which remain from the smelting process but cannot be removed entirely.

Steels in which elements other than carbon are present in relatively large amounts are referred to as *alloy steels*. The elements added are selected to give the steel special properties. Three common alloy steels that are produced for their tensile strength, corrosion resistance and for high-speed cutting operations are listed in Table 4.2.

Table 4.2 Alloy steels

Classification	Carbon and other elements (%)	Tensile strength	Applications and uses
Nickel–chrome–molybdenum steel	0.4% Carbon 1.5% nickel 1.2% chromium 0.55% manganese 0.3% molybdenum	1080–2010 MPa depending on degree of heat treatment	Highly stressed machine components where a high resistance to fatigue and shock loading is required
Stainless steel	Up to 1.0% carbon Up to 27.0% chromium Up to 0.8% manganese	500–1670 MPa depending on composition and degree of heat treatment	Food processing, kitchen and surgical equipment, pressings, decorative trim
High-speed steel	0.7% carbon 18.0% tungsten 4.0% chromium 1.0% vanadium	Not relevant	Twist drills, milling cutters, turning tools, hacksaw blades

Stainless steels can be classified as 'austenitic' (which accounts for 70% of the production), 'ferritic' and 'martensitic' depending on their crystal structure. The term 'ferrite' has already been mentioned. It refers to grains of BCC iron, containing a small amount of interstitial carbon. The term 'austenite' refers to FCC iron containing larger amounts of interstitial carbon. The term 'martensite' refers to a needle-like crystal structure that occurs when the material is quenched.

KEY POINT

Alloy steels contain alloying elements other than carbon. They are selected to give the steel special properties.

The austenitic range contains a maximum of 0.15% carbon and a minimum of 16% chromium. They are used for a wide range of domestic articles and decorative trim. Ferritic stainless steels contain between 10.5% and 27% chromium and very little nickel. They are more corrosion resistant but less durable in other ways than the austenitic range. Martensitic stainless steels contain 0.1–1% carbon, 12–14% chromium, 0–2% nickel and 0.2–1% molybdenum. They are not as corrosion resistant as the other two classes but compensate for this by being stronger, tougher and easily machinable.

When carbon is added to molten iron in quantities greater than 1.7%, all cannot be absorbed interstitially or as iron carbide. After the mixture has cooled down, the excess is seen to be present as flakes of graphite between the grains. The material is then known as *grey cast iron*. A carbon content of 3.2–3.5% and silicon of up to 2.5% is usual which gives the molten metal good fluidity. This enables it to be cast into intricate shapes. It is easy to machine without the use of a cutting fluid since the graphite flakes have a self-lubricating effect. Grey cast iron is strong in compression but tends to be weak and brittle in tension. The graphite flakes also act as vibration absorbers which make it an ideal material for machine beds. A low silicon content coupled with a fast rate of cooling does not allow the graphite flakes to separate out. The material then has a white appearance and is known as *white cast iron*. This tends to be weak and brittle but it can have its properties modified.

KEY POINT

Grey cast iron has good fluidity when molten, it is self-lubricating and easy to machine, it is strong in compression but weak in tension.

The properties of cast iron can be improved by heat treatment processes to produce what are known as *malleable cast irons*. The effect is to modify the graphite content so that the material is less brittle and better able to withstand shock and impact. The properties can also be improved by adding certain alloying elements, particularly small amounts of magnesium. This has the effect of producing small spheres of graphite instead of graphite flakes. The material is then known as *spheroidal graphite iron*. Spheroidal graphite or SG iron has the fluidity of the grey cast iron but is much tougher and stronger when solidified.

Cast irons are used for a variety of engineering components. They are used for internal combustion engine cylinder blocks and cylinder heads, brake drums, pump and turbine casings, water pipes, manhole covers and gratings. In recent years the popularity of cast iron as a decorative material has returned. In particular, there has been an increase in its use for lamp standards, garden furniture and decorative structural castings in buildings and public places.

Test your knowledge 4.3

1. What is a plain carbon steel?
2. What are high carbon steels used for?
3. How does the carbon content affect the properties of plain carbon steel?
4. Give an example of an alloy steel.
5. What are the properties of grey cast iron?
6. How is the carbon content distributed in grey cast iron?

Activity 4.1

Investigate modern methods of iron and steel production and prepare a report on your findings. Make particular reference to:

1. **The different kinds of iron ore and their suitability for processing.**
2. **The way in which iron ore is processed to extract the iron.**
3. **Modern methods of steel production.**

Non-ferrous metals

Non-ferrous metals do not contain iron except in relatively small amounts. In particular, it is included in some cupro–nickel alloys to improve their properties. The major non-ferrous base metals used in engineering are copper, zinc, tin, lead, aluminium and titanium. Their properties and uses are listed in Table 4.3.

Table 4.3 Non-ferrous metals

Metal	Tensile strength (MPa)	Density ($kg\,m^{-3}$)	Melting point (°C)	Applications and uses
Copper	220	8900	1083	Water pipes, heat exchangers, vehicle radiators, electrical wire and cable, major constituent of brasses, bronzes and cupro–nickel alloys.
Zinc	110	7100	420	Major constituent of brasses and some die-casting alloys. Used for galvanising, i.e. as a protective coating for steel products.
Tin	15	7300	232	Major constituent of tin–bronze alloys, soft solders and white bearing metals. Used as a protective coating for mild steel.
Lead	Very low	11300	327	Major constituent of soft solders and white bearing metals. Used for protective cladding on buildings and protective lining in tanks.
Aluminium	60	2700	660	Major constituent of aluminium alloys and aluminium–bronzes. Used for electrical cables and conductors and domestic utensils.
Titanium	216	4500	1725	Major constituent of titanium alloys used in aircraft production and for gas turbine compressors blades. Alloyed with nickel to produce 'smart' thermo-responsive materials.
Silver	Low	10500	960	Used for electrical contacts and as a constituent of silver solder with copper and zinc.
Gold	Low	19300	1063	Very few engineering uses. Some specialised aerospace applications.

Copper is an excellent conductor of heat and electricity and like most non-ferrous metals, it is corrosion resistant. Although it does not have a high tensile strength, it is malleable and ductile and easily drawn out into wire and tube. Zinc is used to give a protective coating to mild steel products by a process known as galvanising. Other than this, it is not widely used alone because it is rather brittle and has a relatively low tensile strength.

Tin has a very low tensile strength and melting point. It is very soft and malleable and highly corrosion resistant. It is used as a protection for mild steel sheet which is then known as tinplate. Its other main use is as a constituent of soft solder.

Lead is very soft and malleable. It has a low tensile strength and being relatively heavy, it can creep and fracture under its own weight. Properly supported, it is widely used for weather protection on buildings and also for lining tanks containing chemicals that are corrosive to iron and steel. Until fairly recently lead was used for domestic water pipes and these are still to be found in some older properties. This use has now been discontinued because of the danger from lead poisoning. Most of the lead piping has now been replaced by copper and plastic materials.

Like copper, aluminium is an excellent conductor of heat and electricity. It is relatively light in weight and has good corrosion resistance. In its pure form it has a low tensile strength but is very malleable and ductile. It is widely used for domestic utensils and as the core of electrical power transmission cables. Titanium is also light in weight, ductile, highly corrosion resistant and has a high melting point. It is however expensive and this has limited its use as an engineering material. It is most widely used in the aerospace industries.

The main non-ferrous alloys used for engineering components are brasses, tin–bronzes, cupro–nickels, aluminium–bronzes, aluminium alloys and white-bearing metals. There are of course many other special purpose alloys but these are the ones that you will encounter most frequently.

KEY POINT

Brass is an alloy of copper and zinc. Brasses with a high copper content are the more ductile and suitable for cold forming. Brasses with a high zinc content are less ductile and more suitable for hot forming operations.

Brasses are alloys in which the main constituents are copper and zinc. As a general rule, brasses with the higher copper content are the more ductile and suitable for cold forming operations such as pressing and drawing. Those with the higher zinc content are less ductile and more suitable for hot forming operations such as casting, forging, extrusion and hot stamping. Some of the more common brasses alloys are listed in Table 4.4.

Table 4.4 Brasses

Type of brass	Composition	Properties	Applications and uses
Cartridge brass	70% copper 30% zinc	Very ductile, suitable for cold forming operations such as deep drawing	Cartridge cases, condenser tubes
Standard brass	65% copper 35% zinc	Ductile, suitable for cold working and limited deep drawing	Pressings and general purposes
Naval brass	62% copper 37% zinc 1% tin	Strong and tough, malleable and ductile when heated, corrosion resistant	Hot formed components for marine and other structural purposes
Muntz metal	60% copper 40% zinc	Strong and tough, malleable and ductile when heated, good fluidity when molten	Hot rolled plate, castings extruded tubes and other sections

Tin–bronzes are alloys in which the main constituents are copper and tin. Phosphorous is sometimes added in small amounts to prevent the tin from oxidising when molten. This gives rise to the name phosphor–bronze which is very malleable and ductile and widely used for plane bearings. Tin–bronzes which contain zinc are known as gunmetal. They have good fluidity when molten and can be cast into intricate shapes. When cooled they are strong, tough and corrosion resistant. Small amounts of lead are also sometimes added to tin–bronzes to improve their machinability. Four of the more common tin–bronzes are listed in Table 4.5.

Table 4.5 Tin–bronzes

Types of tin–bronze	Composition	Properties	Applications and uses
Low tin–bronze	96% copper 3.9% tin 0.1% phosphorous	Malleable and ductile Good elasticity after cold forming	Electrical contacts, instrument parts, springs
Cast phosphor–bronze	90% copper 9.5% tin 0.5% phosphorous	Tough, good fluidity when molten, good anti-friction properties	Plane bush and thrust bearings, gears
Admiralty gunmetal	88% copper 10% tin 2% zinc	Tough, good fluidity when molten, good corrosion resistance	Pump and valve components, marine components, miscellaneous castings
Bell metal	78% copper 22% tin	Tough, good fluidity when molten, sonorous when struck	Church and ships bells, other miscellaneous castings

KEY POINT

The main constituents of tin–bronzes are copper and tin. The addition of zinc improves fluidity when molten and the addition of phosphorous gives good load-bearing properties.

Cupro–nickel alloys contain copper and nickel as the main constituents with small quantities of manganese and sometimes iron, added to improve their properties. Cupro–nickels are strong, tough and corrosion resistant. They find use in chemical plant and marine installations and are widely used throughout the world as 'silver' coinage. Two of the more common cupro–nickel alloys are listed in Table 4.6.

Table 4.6 Cupro–nickel alloys

Types of cupro–nickel	Composition	Properties	Applications and uses
Coinage metal	74.75% copper 25% nickel 0.25% manganese	Strong, tough, corrosion and wear resistant	'Silver' coins
Monel metal	29.5% copper 68% nickel 1.25% iron 1.25% manganese	Strong, tough and highly corrosion resistant	Chemical, marine and engineering plant components

Aluminium is also alloyed with copper to produce a range of metals known as aluminium–bronzes. They also contain nickel and manganese and are highly corrosion resistant. Two of the more common aluminium–bronzes are listed in Table 4.7.

Table 4.7 Aluminium–bronzes

Types of aluminium–bronze	Composition	Properties	Applications and uses
Wrought aluminium–bronze	91% copper 5% aluminium 2% nickel 2% manganese	Ductile and malleable for cold working Excellent corrosion resistance at high temperatures	Boiler and condenser tubes, chemical plant components
Cast aluminium–bronze	86% copper 9.5% aluminium 1% nickel 1% manganese	Strong, good fluidity when molten, corrosion resistant	Valve and pump parts, boat propellers, gears, miscellaneous sand and die cast components

KEY POINT

Pure aluminium is soft, malleable and ductile. It has a low tensile strength but is a good conductor of both heat and electricity. Some aluminium alloys are produced for cold working whilst others are produced for casting.

There is a wide range of alloys in which aluminium is the major constituent. Aluminium in its pure form is soft, malleable and ductile with a low tensile strength. The addition of copper, silicon, nickel, silver, manganese and magnesium in various small quantities greatly enhances its properties. Some of the resulting aluminium alloys are produced for cold working whilst others are produced for casting. They all retain the low weight characteristic of aluminium and some may be hardened by a heat treatment process known as precipitation hardening. Some of the most common types are listed in Table 4.8.

Table 4.8 Common aluminium alloys

Types of aluminium alloy	Composition	Properties	Applications and uses
Casting alloy BS 1490/LM4M	92% aluminium 5% silicon 3% copper	Good fluidity when molten, moderately strong	Miscellaneous sand and die cast components for light duty applications
Casting alloy BS 1490/LM6M	88% aluminium 12% silicon	Good fluidity and strength	Miscellaneous sand and die cast components for motor vehicle and marine applications
'Y' alloy	92% aluminium 2% nickel 1.5% manganese	Good fluidity when molten and able to be hardened by heat treatment	Internal combustion engine pistons and cylinder heads
Duralumin	94% aluminium 4% copper 0.8% magnesium 0.7% manganese 0.5% silicon	Ductile and malleable in the soft condition. Tough and strong when heat treated	Motor vehicle and aircraft structural components
Wrought alloy BS 1470/5:H30	97.3% aluminium 1% magnesium 1% silicon 0.7% manganese	Ductile in the soft condition, good strength when heat treated	Ladders, scaffold tubes, overhead electric power lines

Test your knowledge 4.4

1. What are the main uses and applications of zinc?
2. What are the main constituents of brass?
3. What is cast phosphor–bronze widely used for?
4. What is the non-ferrous alloy from which 'silver' coins are made?
5. What are the main constituents of aluminium–bronze?
6. Which aluminium alloy is widely used for aircraft and motor vehicle structural components?

Activity 4.2

In the early years of the last century £1 gold coins, known as sovereigns, were in general circulation and the accompanying silver coins were really made from silver. Nowadays, gold sovereigns are only minted for special occasions and have a value which is much greater than £1. Take a modern £1 coin, a 10p piece and a 1p piece and examine them.

What materials do you think the modern coins are made from? Why do you think these materials have been selected?

Which crystal lattice structure would you expect gold and silver to have? Give your reasons.

A little research in the chemistry or metallurgy section of your library or on the Internet will tell you if you have made the correct choices.

To check your understanding of the preceding section, you can solve Review questions 1–5 at the end of this chapter.

Polymers

Polymer materials include plastics and rubbers whose atoms are arranged in long molecular chains known as *polymers*. The use of natural rubber, obtained from the rubber tree, began in the nineteenth century. Unfortunately natural rubber degrades or 'perishes' over a relatively short period of time and is readily attacked by solvents. It is not used much in engineering today except when mixed with synthetic rubbers. Developments over the last hundred years have seen the introduction of many new synthetic rubbers and plastics with many different properties. The raw materials that are used to make them come from oil distillation, coal and from animal and vegetable substances.

The name 'plastic' is a little misleading. Plasticine, chewing gum and linseed oil putty are examples of materials which are truly plastic. They are easily deformed by tensile, compressive and shearing forces and retain whatever shape they are moulded into. Polymer materials are in fact quite elastic at normal temperatures and the name 'plastics' refers to their condition when they are being formed into shape. At this stage the raw material is often a resin which can be poured or injected into a mould. Alternatively the raw material may be a powder or granules which are heated into a liquid prior to moulding.

Polymers are made up mainly of carbon and hydrogen atoms but may also have oxygen, chlorine and fluorine atoms attached to them depending on the type of plastic. Polyethylene, which is better known as polythene, is one of the simplest examples of a polymer. It is produced from the chemical compound ethylene, C_2H_4, which is a by-product of oil distillation. This is known as the *monomer*, from which the polymer is formed. When suitably treated the bonds between the atoms are broken to make two units, each containing one carbon atom and two hydrogen atoms. These cannot exist on their own and so they immediately link up to form long chain-like molecules as shown in Figure 4.15.

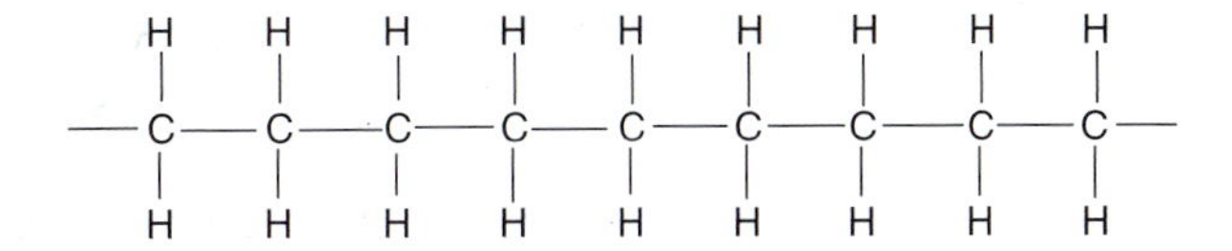

Figure 4.15 The polythene polymer

The process is called *polymerisation* and each polymer can contain several thousand atoms. If fully extended they can be anything up to a millimetre long and possibly more. The polymers are intertwined and held together by covalent bonds. The strength of the bonds affects the properties of plastic materials and rubbers. Some are very flexible and easy to stretch whilst others are hard and brittle. Some will melt and burn at relatively low temperatures whilst others are heat resistant. The polymers of two common plastics, polyethylene (polythene) and polychloroethene (PVC), are shown in Figures 4.15 and 4.16.

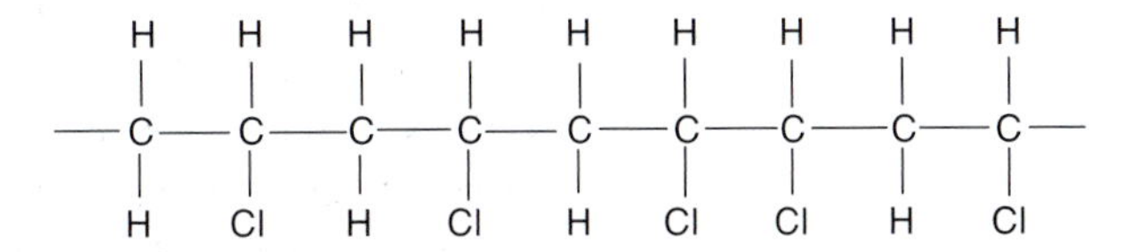

Figure 4.16 The PVC polymer

Plastics can be part amorphous, where the polymers are randomly entangled, and part crystalline where the polymers become arranged in a geometric pattern. These regions are called *crystallites* (Figure 4.17). The relative proportions of amorphous regions and crystallites vary

KEY POINT

Plastic and rubbers are made up of intertwined chains of molecules called polymers. Polymers are mainly composed of carbon and hydrogen atoms.

with different plastics and the way in which they have been processed. Forces of attraction are set up between the polymers as they form. They are known Van der Waal forces and it is these that hold plastics such as polythene and PVC together.

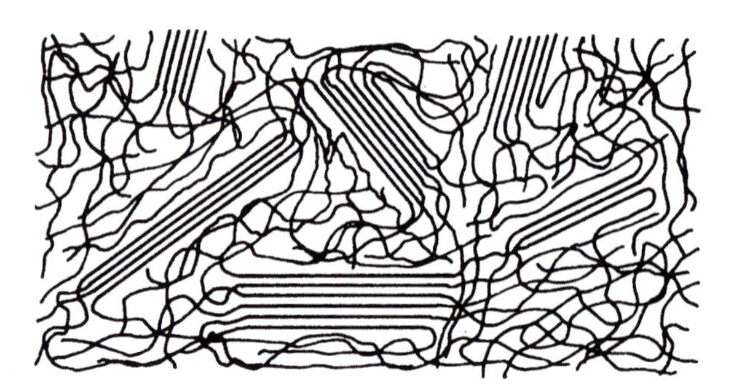

Figure 4.17 Crystalline and amorphous regions (Reprinted from *Properties of Engineering Materials*, Second edition, R Higgins, Butterworth-Heinemann, p. 259, Fig. 12.7, 1994, with permission from Elsevier.)

Plastics can be subdivided into *thermoplastics*, *thermosetting plastics* and *rubbers*. Thermoplastics can be softened and remoulded by heating whereas thermosetting plastics cannot. Rubbers are polymer materials known as *elastomers* which have the property of being able to return to their original shape after large amounts of deformation.

Thermoplastics

Some thermoplastic materials are soft and flexible at normal temperatures whilst others are hard and brittle. When heated the hard thermoplastics will eventually soften and become flexible. The temperature at which this occurs is known as the *glass transition temperature* denoted by the symbol T_g. Further heating eventually causes the material to reach its melting temperature which has the symbol T_m. Figure 4.18 shows how the properties of a thermoplastic change with temperature.

KEY POINT

Thermoplastics can be softened by heating. They become soft and flexible at the glass transition temperature.

Some common thermoplastic materials and their uses are listed in Table 4.9.

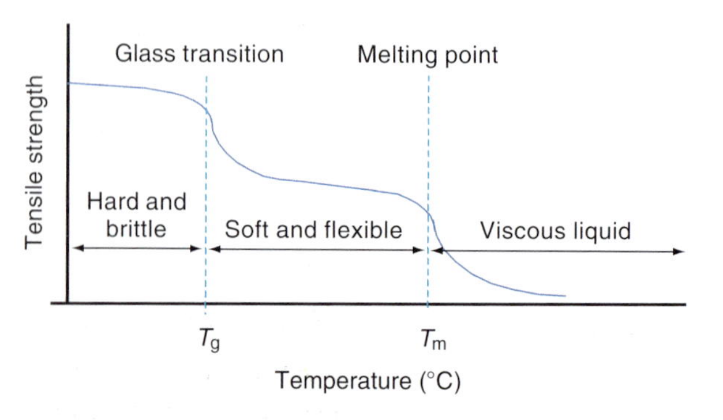

Figure 4.18 Change of thermoplastic properties with temperature

Table 4.9 Common thermoplastic materials

Polymer	Common name	Properties	Applications and uses
Polyethene	Low-density polythene	Tough, flexible, solvent resistant, degrades if exposed to light or ultraviolet radiation.	Squeeze containers, packaging, piping, cable and wire insulation.
Polyethene	High-density polythene	Harder and stiffer than low-density polythene with higher tensile strength.	Food containers, pipes, mouldings, tubs, crates, kitchen utensils, medical equipment.
Polypropene	Polypropene	High strength, hard, high melting point, can be produced as a fibre.	Tubes, pipes, fibres, ropes, electronic components, kitchen utensils, medical equipment.
Polychloroethene	PVC	Can be made tough and hard or soft and flexible, solvent resistant, soft form tends to harden with time.	When hard, window frames, piping and guttering. When soft, cable and wire insulation, upholstery.
Polyphenylethene	Polystyrene	Tough, hard, rigid but somewhat brittle, can be made into a light cellular foam, liable to be attacked by petrol based solvents.	Foam mouldings used for packaging and disposable drinks cups. Solid mouldings used for refrigerators and other appliances.
Methyl-2-methylpropenoate	Perspex	Strong, rigid, transparent but easily scratched, easily softened and moulded, can be attacked by petrol-based solvents.	Lenses, corrugated sheets for roof lights, protective shields, aircraft windows, light fittings.
Polytetra fluoroethylene	PTFE	Tough, flexible, heat-resistant, highly solvent resistant, has a waxy low-friction surface.	Bearings, seals, gaskets, non-stick coatings, tape.
Polyamide	Nylon	Tough, flexible and very strong, good solvent resistance but does absorb water and deteriorates with outdoor exposure.	Bearings, gears, cams, bristles for brushes, textiles.
Thermoplastic polyester	Terylene	Strong, flexible and solvent resistant, can be made as a fibre, tape or sheet.	Textile fibres, recording tape, electrical insulation tape.

Test your knowledge 4.5

1. What are the main constituent elements of polymers?
2. What are Van der Waal forces?
3. What change takes place in thermoplastic materials at the glass transition temperature?
4. Which thermoplastic material is widely used in moulded and granular form in packaging?
5. What is the common name for polychloroethene?

Thermosetting plastics

Thermosetting plastics, which are also called *thermosets*, undergo a chemical change during the moulding process. Strong covalent bonds are formed between the polymers that are known as *cross-links*. The cross-links may consist of a single atom or a small chain of atoms that form in the mould during what is known as the *curing process*. As a result, the bonding forces are much stronger than the Van der Waal forces and once formed, the cross-links cannot be broken. Thermosetting plastics are generally harder and more rigid than thermoplastics and cannot be softened or melted by heating.

KEY POINT

Cross-linking of the polymers in thermosetting plastics makes them harder and more rigid than thermoplastics. They cannot be softened by heating.

Filler materials are very often mixed in with the raw materials for thermosetting plastics prior to moulding. These improve the mechanical properties and include wood flour, shredded textiles, paper and various other fibres and powdered materials. Some common thermosetting plastics and their uses are listed in Table 4.10.

Table 4.10 Common thermosetting plastics

Polymer	Common name	Properties	Applications and uses
Phenolic resins	Bakelite	Hard, resistant to heat and solvents, good electrical insulator and machinable, colours limited to brown and black.	Electrical components, vehicle distributor caps saucepan handles, glues, laminates
Urea-methanal resins	Formica	As above but naturally transparent and can be produced in a variety of colours.	Electrical fittings, toilet seats, kitchen ware, trays, laminates.
Methanal-melamine resins	Melamine	As above but harder and with better resistance to heat. Very smooth surface finish.	Electrical equipment, tableware, control knobs, handles, laminates.
Epoxy resins	Epoxy resins	Strong, tough, good chemical and thermal stability, good electrical insulator, good adhesive.	Container linings, panels, flooring material, laminates, adhesives.
Polyester resins	Polyester resins	Strong, tough, good wear resistance, and resistance to heat and water.	Boat hulls, motor panels, aircraft parts, fishing rods, skis, laminates.

Rubbers

The polymers in rubbers are known as *elastomers*. They are longer and tend to be more complex than other polymers. Elastomers can be made up of more than 40 000 atoms and when the material is unloaded, they are intertwined and folded over each other in a random fashion (Table 4.11).

When loads are applied to rubber, the elastomers behave rather like coiled springs which become aligned as the material is stretched. This gives the material its elastic properties (Figure 4.19).

If the loading is excessive, the forces between the polymers may be insufficient to prevent them from sliding over each other. This will ultimately lead to failure or produce permanent deformation. In this condition the rubber is said to be both elastomeric and thermoplastic.

When sulphur or certain metal oxides are added to the rubber during manufacture, cross-links are formed between the polymers. These are

Table 4.11 Common types of rubber

Types of rubber	Properties/origins	Applications and uses
Natural rubber	Obtained as latex from the rubber tree. Perishes with time and is readily attacked by solvents.	Not used for engineering purposes except when mixed with synthetic rubbers.
Styrene rubber	Synthetic rubber developed in United States during Second World War as a substitute for natural rubber. Also known as GR-S rubber. Resistant to oils and petrol.	Blended with natural rubber and used for vehicle tyres and footwear.
Neoprene	Synthetic rubber with close resemblance to natural rubber. Resistant to mineral and vegetable oils and can withstand moderately high temperatures.	Used in engineering for oil seals, gaskets and hoses.
Butyl rubber	Synthetic rubber with good resistance to heat and temperature. Impermeable to gases.	Used in engineering for tank linings, moulded diaphragms, inner tubes and air bags.
Silicone rubber	Synthetic rubber with good solvent resistance and which retains its properties over a wider temperature range than other types. Remains flexible from −80°C to 300°C.	Widely used in chemical and process plant for seals and gaskets where a wide variation in temperature is likely. Also in aircraft where low temperatures are experienced at high altitude.

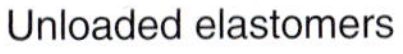

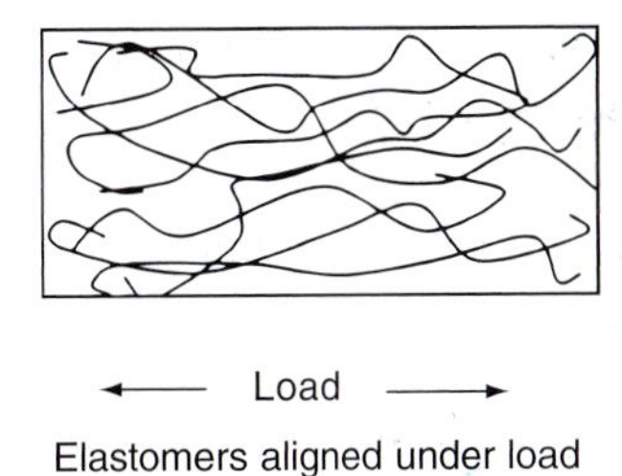

Figure 4.19 Behaviour of elastomers

similar to those formed in thermosetting plastics and the process is known as *vulcanising*. The cross-links enable the rubber to retain its elasticity and return to its original shape when loaded almost to the point of failure.The addition of larger quantities of sulphur causes the rubber to loose some of its elasticity. If sufficient is added, it becomes hard and brittle and its properties are then very similar to the hard thermosetting plastics.

Test your knowledge 4.6

1. What is the main behavioural difference between thermoplastics and thermosets?
2. What happens during the curing process in thermosetting plastic materials?
3. Name two filler materials which are used in thermosetting plastics.
4. Which kind of rubber retains its flexibility over a temperature range which extends from −80°C to 300°C?
5. What is 'vulcanising' and what effect does it have on the properties the properties of rubber?

CHAPTER 4

Activity 4.3

Take a broad rubber band and measure its unstretched length. Suspend the band from a retort stand, place a weight hanger on its lower end and add increasing weights to the hanger. (Slotted 0.5 kg masses will probably be the most suitable.)

Record the total extension after each increment of weight and continue until the length of the band is about 5 or 6 times its original length. Now remove the load and once again measure the length of the band.

Plot a graph of load against extension and examine its shape.

Explain why you think the rubber behaved as it did.

To check your understanding of the preceding section, you can solve Review questions 6–10 at the end of this chapter.

Ceramics

The term 'ceramic' comes from the Greek word for potters clay. In engineering it covers a wide range of materials such as bricks, tiles, cement, furnace linings, glass and porcelain. It also includes the abrasive grits used for grinding wheels and some cutting tools. The main ingredients of ceramics are sand and clays.

Sand contains silica which is a compound of silicon and oxygen. It also contains feldspar which is a compound of aluminium, silicon and oxygen. Clays also contain compounds of silicon and aluminium. Other elements such as potassium, calcium and magnesium might be present in combination with silicon. These compounds are called *silicates*.

Ceramics may be subdivided into amorphous ceramics, crystalline ceramics, bonded ceramics and cements. Their main properties are that they are hard wearing, good electrical insulators and strong in compression when used as structural materials. In addition, many of them can withstand very high temperatures which make them suitable for lining furnaces and kilns. Ceramics do however tend to be weak in tension, brittle and unable to withstand impact loading.

Amorphous ceramics

In amorphous ceramics the molecules are not arranged in any geometrical pattern. They include the various types of glass whose main ingredient is silica, or silicon oxide, whose chemical symbol is SiO_2. This is contained in sand which is heated until molten and then rapidly cooled. Other ingredients can include sodium carbonate, sodium borate and oxides of magnesium, calcium and boron. If the mixture is allowed to cool slowly it crystallises and the resulting solid is opaque. When it is cooled rapidly, however, there is no crystallisation and the resulting solid is amorphous and transparent. Some of the more common types of glass are listed in Table 4.12.

KEY POINT

The main constituent of glass is silicon oxide which is also called silica. When cooled quickly from its molten state the resulting solid is amorphous and transparent.

CHAPTER 4

Table 4.12 Common types of glass

Types of glass	Ingredients/properties	Applications and uses
Soda lime glass	Silica (72%) plus oxides of sodium, calcium and magnesium. Can be toughened by heat treatment.	Accounts for about 95% of all glass production. Used for windows, windscreens, bottles and jars.
Lead glass	Silica (60%), lead oxide (25%) plus oxides of sodium, potassium and aluminium. High electrical resistance.	Used for lamps and cathode ray tubes.
	Silica (40%), lead oxide (47%) plus oxides of sodium, potassium and aluminium. High refractive index.	Used for lenses and crystal glass tableware.
Borosilicate glass	Silica (70%) boron oxide (20%) plus oxides of sodium, potassium and aluminium. Low expansivity, good resistance to chemicals.	Trade name 'Pyrex'. Used for laboratory equipment, kitchen glassware and electrical insulators.

Crystalline ceramics

This group includes the hard, abrasive grits used for surface preparation and material removal. The most widely used crystals are aluminium oxide (emery), beryllium oxide, silicon carbide and boron carbide. Singly or in combinations, they are used to make emery cloth and paper. They are mixed with oil to form grinding paste and they are bonded together to make grinding wheels.

KEY POINT

Aluminium oxide (emery) is one of the most widely used crystalline ceramic grits in the manufacture of grinding paste, abrasive wheels, abrasive cloth and paper.

Also included in this group are tungsten carbide, zirconium carbide, titanium carbide and silicon nitride. These crystals are bonded together in a metal matrix to produce very hard cutting tool tips.

Bonded clay ceramics

The ingredients for this group are natural clays and mixtures of clays to which crystalline ceramics are sometimes added. After mixing and moulding, the products are fired in kilns and ovens. During the firing, a process of *vitrification* occurs in which the crystals become bonded together. Bonded clay ceramics may be subdivided into whiteware, structural clay products and refractory ceramics.

Whiteware includes china pottery, porcelain, earthenware, stoneware, decorative tiles and sanitary products. Firing takes place at temperatures between 1200°C and 1500°C after which some of the products are glazed with a thin layer of molten glass. Selected and refined clays are used for whiteware to which crystalline ceramics such as sodium borate (borax) and sodium aluminium fluoride (cryolite) have been added. These assist the bonding process by acting as a flux and they also lower the vitrification temperature.

Structural clay products are made from common clays and are fired at a higher temperature. They include bricks, drain and sewer pipes, terracotta products and flooring and roofing tiles. Natural impurities in the clay act as a flux during vitrification. Structural clay products have relatively large interstitial spaces in their crystal lattice structure. As a result they tend to absorb water and to prevent this, it is usual for drain and sewer pipes to be given a glazed finish.

KEY POINT

Vitrification is the bonding together of the crystals in bonded clay ceramics. The vitrification temperature is the firing temperature at which this occurs.

Refractory ceramics are made from fireclays which have a high content of silicon oxide (silica) or aluminium oxide (emery). They are used to make firebricks and to line the inside of furnaces and kilns. The high silica or aluminium oxide content makes them able to withstand sustained high temperatures and chemical attack from combustion gasses. They are also used to line the buckets and ladles that are used for carrying molten metal.

Cements

Cements contain metal oxides such as those of calcium, silicon and magnesium which react chemically with water. The mixture solidifies into a hard crystalline structure with good compressive strength and bonding properties. Portland cement mixed with sand is similar in colour to Portland stone and is widely used in the building industry. Stone aggregates are added to it to make concrete which has much greater compressive strength and toughness. Other types of cement for special applications are silicate cements, which resist chemical attack, and high alumina cements, which resist attack form sea water.

Test your knowledge 4.7

1. What is the main ingredient of all types of glass?
2. What is the commonly used name for borosilicate glass?
3. What is 'emery'?
4. What occurs in ceramic materials during the vitrification process?
5. What are refractory ceramics used for?

Activity 4.4

Silica, whose chemical name is silicon (IV) oxide, or SiO_2, is an ingredient in many ceramics. Find out how the atoms are arranged in its crystal lattice structure and the effect that this has on its properties.

The hob plates of many electric cookers are now made of a ceramic material. Find out what this is and why its structure makes it suitable for this purpose.

Wood

Wood is a natural sustainable product when used responsibly. Its mechanical properties and availability make it ideally suited to structural applications in both the building and engineering industries. Wood is a fibrous material and strongest when loaded in the direction of the grain. Its strength to weight ratio compared to steel is about 2:1 and it is less expensive. Wood can generally be divided into hardwoods and softwoods. The terms do not necessarily indicate hardness although in most cases the hardwoods are harder and tougher.

Hardwoods

In general hardwoods come from broadleaf trees that shed their leaves and have their seeds in a seed case. Common examples include ash, oak, elm, teak and mahogany. Woods grown in tropical climates are usually hardwoods. Hardwoods tend to have shorter fibres than softwoods and have a higher density. The main properties and typical uses of the above examples are given in Table 4.13. The properties of wood vary with moisture content and the values given are those that would be expected from a well-seasoned product. You should note that the rupture modulus is in fact the tensile stress at which the fibres farthest from the neutral axis of bending start to fail.

Table 4.13 Common hardwoods

Types of hardwood	Density ($kg\,m^{-3}$)	Moisture content (%)	Modulus of rupture (MPa)	Modulus of elasticity (GPa)	Uses
Ash	810	12	116	11.9	Vehicle bodies, tool handles, machine parts
Oak	720	12	97	10.1	Best wood for construction purposes, also used for furniture, fittings and flooring
Mahogany	720	12	78	9.0	High-quality furniture and joinery
English elm	550	12	68	7.0	Durable under water, used for piles, lock gates and external cladding
Teak	900	11	106	10.0	Indoor and outdoor furniture

Softwoods

Softwoods come from cone-bearing trees with needle-like leaves that tend to be evergreen. The larch is an exception which sheds its leaves. Other common examples are spruces, pines and firs. They have longer fibres and tend to have a lower rupture modulus. Softwoods are in general more plentiful and less expensive than hardwoods. They mature quicker and are thus more sustainable with well-planned forest management. In addition to their use as single materials, both hardwoods and softwoods are used in the production of timber composites, e.g. plywood, chipboard, etc. as will be explained (Table 4.14).

Table 4.14 Common softwoods

Types of softwood	Density ($kg\,m^{-3}$)	Moisture content (%)	Modulus of rupture (MPa)	Modulus of elasticity (GPa)	Uses
Douglas fir	530	12	93	12.7	Plywood, heavy construction work
Larch	590	13	92	9.9	General purpose, timber for outdoor use and mining
Scots pine	510	12	89	10.0	General purpose including furniture and flooring
Spruce	420	13	73	8.0	Boxes, crates, packaging, construction work

Composites

Composites are made up of two or more materials which are bonded together. They may be subdivided into laminated, particulate and fibrous composites.

Laminates

Laminates consist of two or more materials which are bonded together. Perhaps the oldest example is plywood which is made up of three or more bonded layers of wood (usually hardwood) with their grain directions at right angles to each other. Other wood laminates include blockboard and laminated chipboard. Blockboard is made up of strips of wood bonded together and sandwiched between two thin outer layers of wood. Laminated chipboard is made up of compressed wood shavings and sawdust which are bonded together and sandwiched between two thin outer layers. The bonding agent is usually a thermosetting resin. The resulting composites are resistant to warping and do not have any weakness due to grain direction.

Laminates made from a thermosetting polymer resin and a filler material are widely used for working surfaces in the home, in restaurants and in the workplace. Formica and melamine are probably the best-known trade names that come from the thermosetting resins used to make them. The filler material may be paper or cloth that is impregnated with resin and partly cured. That is to say that the cross-linking process has started, but is not complete. Alternate layers are then stacked together and placed in a hot press. The heating allows the cross-linking to proceed within and between the layers and the pressure ensures that there are no cavities present. The laminates produced are hard wearing, tough, heat resistant and resistant to staining.

Combinations of polymer materials and metals are increasingly being used to produce composites which are light in weight but extremely rigid. They have a low-density core of plastic foam or a metal honeycomb made from aluminium or titanium. This is sandwiched between two high-strength laminate skins. Composites of this kind are widely used in aircraft where low weight and high strength are essential requirements.

Particulate composites

In particulate composites, one material acts as a matrix that surrounds the particles of another. Concrete is perhaps the oldest example of a particulate composite where stone aggregate and sand are bonded in a matrix of cement. Steel reinforcing rods may also be included to give the concrete added strength. Particle-reinforced polymers, in which thermoplastics and thermosetting plastics are strengthened and toughened with silica, glass beads and rubber particles, are also particulate composites. High impact polystyrene, which contains rubber particles for added toughness, is typical of this type.

Sometimes the particles are bonded in a matrix of metal. Aluminium oxide particles contained in aluminium increase the strength of the material, particularly its tensile strength at high working temperatures.

Graphite and PTFE particles contained in phosphor–bronze improve its properties as a bearing metal by reducing frictional resistance.

In recent years a range of composite materials known as *cermets* has been developed. These are formed by mixing together powdered ceramic and metal particles which are compressed at high temperatures. The process is known as *sintering* and results in a hard and wear resistant composite in which the ceramic particles become surrounded by a metal matrix. The cemented carbides used for cutting tool tips are produced in this way. The metal is usually cobalt and the ceramic powders are tungsten, titanium, silicon and molybdenum carbides. Tungsten carbide is perhaps the oldest and best-known cermet and has been used on tipped lathe tools and masonry drills for many years.

KEY POINT

Sintering is a process used to make cermets such as tungsten carbide. The constituents are subjected to high temperatures and pressures at which they become bonded together.

An alternative type of particulate composite has oxidised metal particles contained in a matrix of ceramic material. These materials have excellent thermal insulation properties and are used in aerospace applications for heat shields.

Fibrous composites

MDF (medium-density fibreboard) is a fibrous wood composite that has seen many applications in recent years. It consists of fine wood fibres in a resin which is bonded under heat and pressure. The result is a dense, flat and stiff material which is easily machined. It can be cut and drilled without damaging the surface and it can be painted to give a high-quality finish.

The most common fibrous plastic composites are those in which glass, carbon, kevlar, silicon carbide and aluminium oxide fibres are contained in a matrix of thermosetting plastic. Epoxy and polyester resins are the usual matrix materials. The fibres increase the strength and the stiffness of the plastic materials. Glass fibre composites or GRP, which stands for 'glass-reinforced plastic', have been in use for many years and are widely used for boat hulls and motor panels.

Carbon fibre composites are not quite as strong as those reinforced with glass fibres but they are very stiff and light in weight. They are widely used in aircraft and also for fishing rods. Kevlar, silicon carbide and aluminium oxide fibres are used in composites which are required to have high impact resistance. Kevlar is an extremely strong polymer. It is used in the frames of tennis racquets, in bullet-proof body armour and to reinforce rubber in conveyer belts and tyres.

Smart materials

Smart materials have one or more properties that can be altered or controlled by an external influence. This could be external force, temperature change, electric or magnetic fields, light intensity, moisture and acidity. Smart material types include metals, crystalline and amorphous non-metals, which include some polymers, and fluids. You may be familiar with some of them, such as spectacle lenses that darken with increasing sunlight, liquid crystals that change colour with applied voltage and piezoelectric materials that produce voltage when a force is applied.

Piezoelectric materials

In certain types of crystal, such as quartz, the application of stress produces a potential difference across opposite faces. This is called the *piezoelectric effect*. Piezoelectric crystals are used in pressure sensors, vibration recorders and microphones. The same also occurs in reverse where an applied voltage produces stress within a sample. The application of this phenomenon lies in materials which can be made to bend or twist when a voltage is applied.

Shape memory alloys

Shape memory alloys or 'memory metals' are materials that remember their initial geometry. After relatively large amounts of deformation they will revert to their original crystal configuration and shape when unloaded or when heated. The three main types of shape memory alloy are copper–zinc–nickel, copper–aluminium–nickel and nickel–titanium. Of these the nickel–titanium alloys are generally more expensive but have the best properties. The ability to change shape is due to changes in the crystal lattice structure that take place in the solid material when heated. The parent shape is set by heating the material to around 500°C, and holding it in position for a period of time. When re-heated after deformation each atom 'remembers' which should be its neighbour and reverts to its parent shape. Shape memory alloys are finding applications in the petro-chemical industry for pipe couplings and in the aerospace industry for variable geometry control purposes. They have also been adopted for medical use as peripheral vascular stents and dental braces. In these applications their response to body temperature is to enlarge restricted blood vessels and exert a constant pressure on the teeth.

Shape memory alloys are being developed that change shape under the influence of magnetic fields. These are of particular interest because the response tends to be faster than for the temperature sensitive materials. Shape memory polymers are also being developed and the first ones became commercially available in the late 1990s.

Magneto-rheostatic fluids

A magneto-rheostatic fluid is a smart fluid containing microscopic magnetic particles suspended in a carrier fluid that is usually a type of oil. When subjected to a magnetic field, the viscosity of the fluid increases to the point where it virtually becomes a solid. It is then said to be in its 'on' state where the magnetic particles align themselves along the lines of magnetic flux, greatly restricting the movement of the fluid. Provided enough force is applied, the material will shear and varying the magnetic field strength can control the yield point at which this occurs. Magneto-rheostatic fluids have found applications in fast-acting fluid clutches, brakes, shock absorbers and flow control valves.

Electro-rheostatic fluids

Electro-rheostatic fluids are suspensions of microscopic dielectric particles in a non-conducting fluid. They behave in the same way as magneto-rheostatic fluids in that the application of an electric field can increase the viscosity of a fluid by a factor of 100 000. It can change from a liquid to a gel and back with response times in the order of milliseconds. The effect is again due to alignment of the particles into chains that oppose movement of the fluid. Electro-rheostatic fluids are also finding applications in fast-acting clutches, brakes, shock absorbers and hydraulic valves.

Test your knowledge 4.8

1. Describe the composition of laminated chipboard.
2. How are formica laminates produced?
3. What particulate materials are added to phosphor–bronze to improve its properties as a bearing material?
4. What is a cermet.
5. Which thermosetting resins are used to make GRP products?
6. What is a piezoelectric material?
7. What are the characteristics of magneto-rheostatic and electro-rheostatic fluids?

To check your understanding of the preceding section, you can solve Review questions 11–15 at the end of this chapter.

Material Properties and Effects of Processing

When we describe the physical properties of a material we are describing how it is likely to behave whilst it is being processed and when it is in service. We have already mentioned properties such as strength, hardness, toughness and corrosion resistance. These now need to be carefully defined so that wherever possible, they can be measured. Physical properties can be subdivided into mechanical properties, thermal properties, electrical and magnetic properties and durability.

Mechanical properties

These include density, tensile strength, hardness, toughness, ductility, malleability, elasticity and brittleness. They are defined as follows.

Density

Density is the mass in kilograms, contained in a cubic metre of a material or substance. When used in calculations, it is usually denoted by the Greek letter ρ (rho) and its units are kilograms per cubic metre ($\mathrm{kg\,m^{-3}}$). The density of steel is around $7800\,\mathrm{kg\,m^{-3}}$ and that of cast iron is around $7300\,\mathrm{kg\,m^{-3}}$. The densities of the major non-ferrous metals used in engineering are listed in Table 4.3. When designing aircraft, engineers need to be able to estimate the weight of the different components and then the total weight. Knowing the densities of the different materials helps them to do this.

Tensile strength

The tensile strength of a material is a measure of its ability to withstand tensile forces. The ultimate tensile strength (UTS) is the tensile stress that causes a material to fracture. If you have completed the unit Science for Engineers, you will know that tensile stress is the load carried per square metre of cross-sectional area. Its units are thus newtons per square metre or pascals (Pa).

The UTS of mild steel is around 500 MPa. The values of UTS for other ferrous and non-ferrous metals are given in Tables 4.1–4.3. Sometimes material suppliers quote the UTS of a material in newtons per square millimetre (N mm^{-2}). It is useful to know that the value is numerically the same in both megapascals and newtons per square millimetre, i.e. the UTS of mild steel can be written as 500 MPa or 500 N mm^{-2}.

The UTS of a material can be determined by carrying out a destructive tensile test on a prepared specimen (Figure 4.20). The specimen is gripped in the chucks of a tensile testing machine and increasing values of tensile load are applied up to the point of fracture. Depending on the type of machine, the load is applied by means of hydraulic rams, a lever system or a leadscrew. Its value is read off from an analogue or a digital display.

The extension of the specimen can be measured by means of an

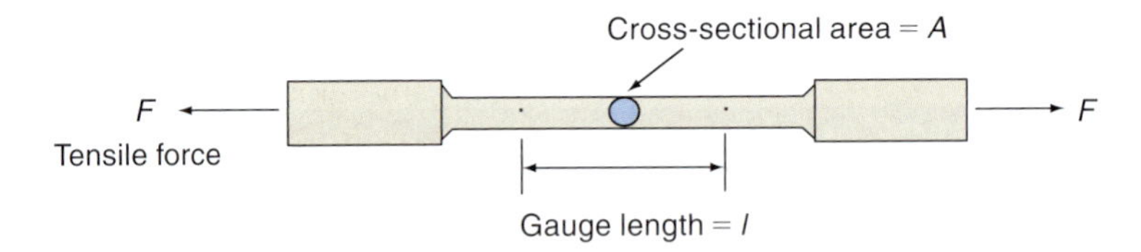

Figure 4.20 Tensile test specimen

extensometer that is attached to the specimen. This can be of the Lindley type, which records the extension on a dial test indicator, or the Monsanto type, which operates on the micrometer principle. There is also an electrical type, which is a linear variable differential transformer. This can be interfaced with an x–y plotter or a computer to produce a graph of load vs. extension.

KEY POINT

The ultimate tensile strength (UTS) of a material is the stress at which it will break. It has the same numerical value when measured in MPa and N mm^{-2}.

The extension of a specimen is usually measured over a gauge length of 50 mm. The initial diameter is checked using a micrometer and the initial cross-sectional area A is calculated. The UTS of the material can then be calculated using the formula:

$$\text{UTS} = \frac{\text{Maximum load}}{\text{Initial cross-sectional area}} \qquad (4.1)$$

Ductility

The ductility of a material is a measure of the amount by which it can be drawn out in tension before it fractures. As can be seen from the typical load vs. extension graphs in Figure 4.21, copper has a high degree of ductility and this enables it to be drawn out into long lengths of wire and tube. There are a number of different ways in which ductility can be measured. One method is to carry out a bend test in which a sample of material is bent through an angle and observed for the appearance of cracks. The angle at which cracking or breaking occurs gives a measure of ductility.

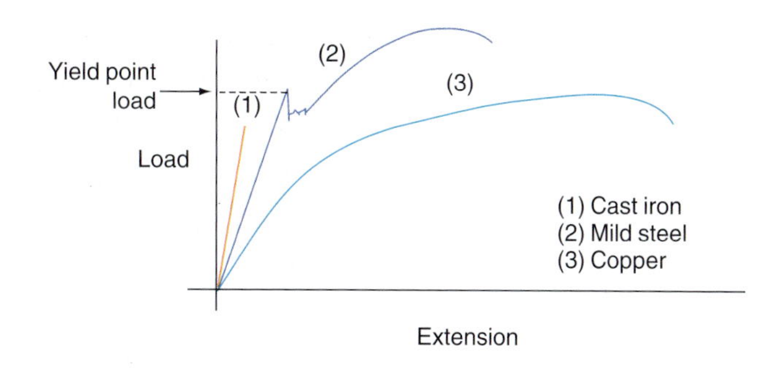

Figure 4.21 Load vs. extension graphs

Ductility measurement can also be incorporated in a tensile test. Before a tensile test, the gauge length over which extension is measured is centre punched. After the test, the two pieces of the fractured specimen are placed in contact and the elongation of the gauge length is measured. The ductility can then be calculated as the percentage increase in the gauge length.

$$\% \text{ Increase in length} = \frac{\text{Original gauge length}}{\text{Extended gauge length}} \times 100 \qquad (4.2)$$

Alternatively, the diameter at the point of fracture can be measured and the cross-sectional area calculated. The ductility can then be calculated as the percentage reduction in area.

$$\% \text{ Reduction in area} = \frac{\text{Cross-sectional area at fracture point}}{\text{Original cross-sectional area}} \times 100 \qquad (4.3)$$

When a ductile material fails, the appearance of the fracture is as shown in Figure 4.22.

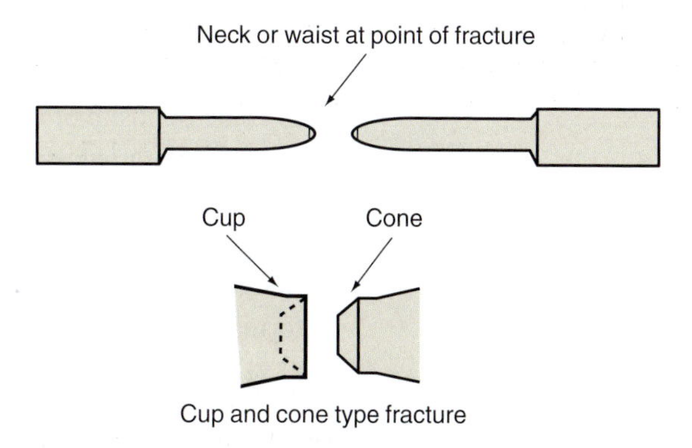

Figure 4.22 Ductile fracture characteristics

As the maximum load is approached, a neck or waist is seen to form at the point where fracture will occur. When the fracture surfaces are examined it is found that they have a characteristic cup and cone appearance which is typical of ductile materials.

CHAPTER 4

When ductile materials are loaded, a point is reached known as the *elastic limit*. If this is exceeded, elastic failure is said to have occurred. This is followed by plastic deformation up to the point of fracture. If a material is unloaded within the plastic extension range, it does not return to its original shape and there is permanent deformation (Figure 4.23).

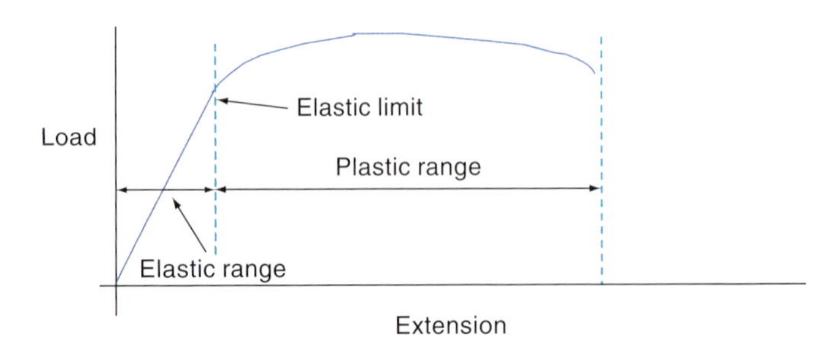

Figure 4.23 Elastic and plastic extension

Mild steel has the characteristic load vs. extension graph shown in Figure 4.21. Its elastic limit is followed closely by the yield point where the material suddenly gives way under the load. Elastic failure is then said to have occurred. If the material is loaded beyond this point there will be some permanent deformation when the load is removed.

Mild steel behaves in this way because of the interstitial carbon atoms. They are said to 'pin' the planes of iron atoms within the grains, which stops them from slipping under load. Eventually they can no longer hold the iron atoms and yielding takes place. For a short time, extension occurs at a reduced load but the material soon becomes work hardened and the load vs. extension graph then rises in a curve until the maximum load is reached. It is important to know the stress at which yielding takes place in steel components so that they can be designed to operate below the yield point load.

KEY POINT

Ductility is the ability of a material to be drawn out in tension when loaded beyond its elastic limit. Brittleness is the opposite of ductility.

Brittleness

This is the opposite of ductility. When metals are loaded in tension they first undergo elastic deformation which is proportional to the load applied. When unloaded from within the elastic range, the material returns to its original shape. Brittle materials tend to fail within the elastic range, or very early in the plastic range, before very much deformation has taken place. Cast iron is brittle, as can be seen from its load vs. extension graph in Figure 4.21.

Brittle materials very often fracture across a plane at right angles to the direction of loading. Sometimes, however, the fracture plane is at 45° to the direction of loading as shown in Figure 4.24. This indicates that the material has a low shear strength.

Elasticity

The elasticity of a material is a measure of its ability to withstand elastic deformation. If you have completed the unit Science for Engineers

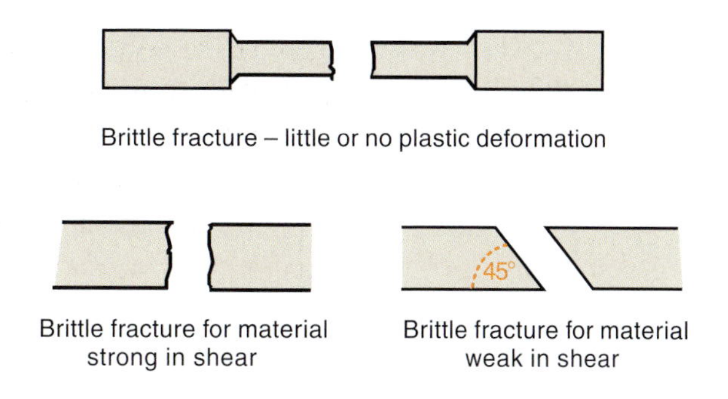

Figure 4.24 Brittle fracture characteristics

you will know that the modulus of elasticity of a material is given by the formula:

$$\text{Modulus of elasticity} = \frac{\text{Direct stress}}{\text{Direct strain}} \qquad (4.4)$$

The direct stress may be tensile or compressive and it is found that some materials have different values of the modulus when loaded in tension and compression. Aluminium is one such example, which has a slightly higher value in compression than in tension. Because strain has no units, the modulus of elasticity of a material is measured in the same units as stress, i.e. pascals (Pa) or more usually gigapascals (GPa). It should not however be confused with stress, as it is measuring something entirely different.

KEY POINT

In a perfectly elastic material, the deformation produced is proportional to the load applied and when the load is removed, the material immediately returns to its original shape.

The modulus of elasticity for mild steel is around 200 GPa and that of some aluminium alloys is around 100 GPa. This tells you that if identical specimens are subjected to the same tensile load, the elastic extension of the aluminium will be double that of the steel, i.e. the higher the modulus of elasticity of a material, the harder it will be to stretch.

Malleability

Whereas ductility is the ability of a material to be drawn out in tension, malleability is the ability of a material to be deformed or spread in different directions. This is usually caused by compressive forces during rolling, pressing and hammering operations. Copper is both ductile and malleable but the two properties do not necessarily go together. Lead is extremely malleable but not very ductile, and soon fractures when loaded in tension.

Hardness

Hardness is the ability of a material to withstand wear and abrasion. The surface hardness of a material is usually measured by carrying out a non-destructive indentation test. Depending on the material, a steel ball, or a pointed diamond indentor, is pressed into the surface of the material under a controlled load. There are three such methods of inspection in common use for engineering metals. They are the Brinell hardness test, the Vickers Pyramid hardness test and the Rockwell hardness test (Figure 4.25).

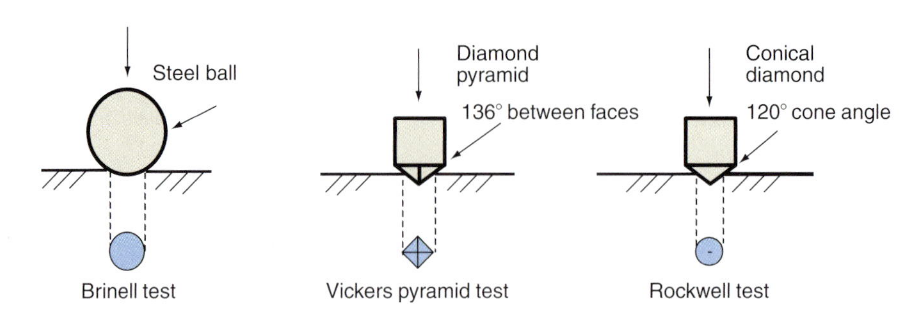

Figure 4.25 Indentation tests

The Brinell hardness tester uses a hardened steel ball indentor and is used for relatively soft materials. The Vickers hardness tester uses either a hardened steel ball or a pyramid-shaped diamond indentor and can be used for both soft and hard materials. The Rockwell hardness tester, which is American in origin, uses a hardened steel ball or a conical diamond indentor and also be used for both soft and hard materials.

Each test enables a hardness number to be calculated. For the Brinell and Vickers tests it is given by dividing the applied load by the surface area of the indentation. This requires careful measurement of the indentation, using a calibrated microscope, after which the Brinell hardness number (BHN) or the Vickers hardness number (VPN) can be calculated. More usually, however, it is read off by referring the measurements to a chart supplied with the testing machine.

For the Rockwell test, the hardness number is a function of the depth of penetration. The machine senses this automatically and the hardness number is read off directly from a display. It is important to note that these tests will give different hardness numbers for the same material. The hardness number required, and the test which should be used, is specified on the engineering drawings for components where hardness is critical.

KEY POINT

Hardness is the ability of a material to withstand wear and abrasion.

Toughness

Toughness is the ability of a material to withstand impact and shock loading. The standard tests for toughness are the Izod impact test and the Charpy test. Both use the same testing machine in which a notched specimen is subjected to a sudden blow from a swinging pendulum (Figure 4.26).

With the Izod test a notched specimen of rectangular section bar gripped in a chuck and held in a vertical position. With the Charpy test, the specimen is supported horizontally at each end, so that the pendulum strikes it at its centre (Figure 4.27(b)).

After impact with the specimen, the pendulum continues in its arc of travel and pushes the pointer around the calibrated scale. There is a separate scale for each of the two tests but both are calibrated in joules. The pointer records the amount of energy absorbed by the specimen on impact, and this gives a measure of the toughness of the material.

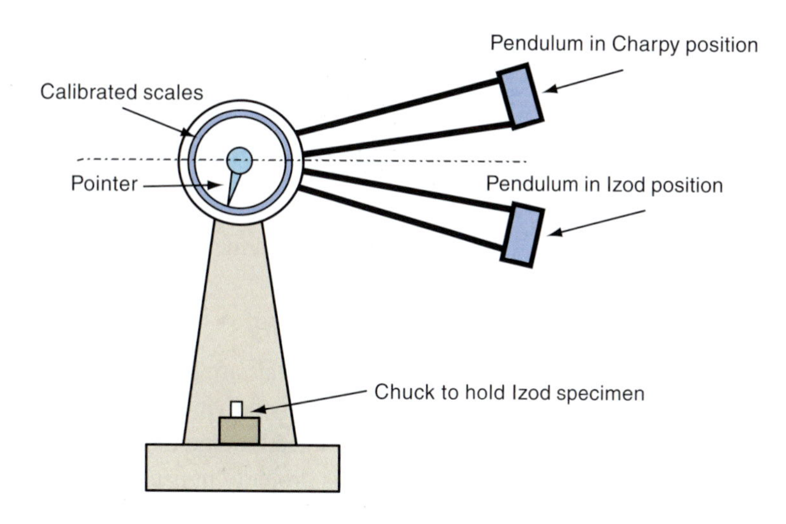

Figure 4.26 Izod/Charpy impact tester

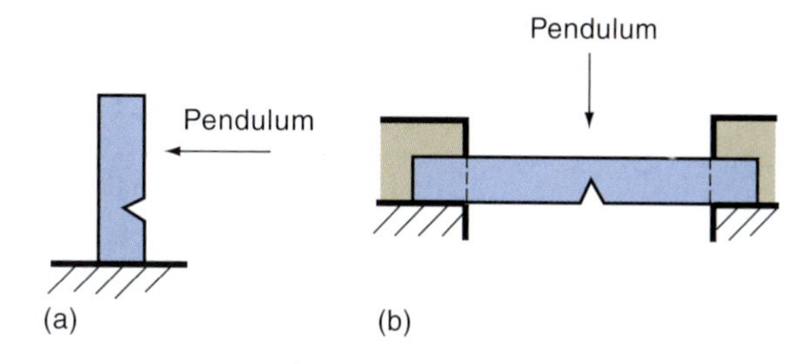

Figure 4.27 (a) Izod and (b) Charpy impact test specimens

KEY POINT

Toughness is the ability of a material to withstand impact or shock loading.

An advantage of the Charpy test, which has its origins in America and Canada, is that heated specimens from a furnace can be quickly placed in position and tested for toughness. Provided that the test is done quickly, the drop in temperature is small. The Izod test, where the specimen must be secured in a chuck, is unsuitable for heated specimens.

Test your knowledge 4.9

1. What is meant by the ultimate tensile strength (UTS) of a material?
2. What is the difference between ductility and malleability in materials?
3. How do the Brinell, Vickers Pyramid and Rockwell hardness tests differ?
4. How can the toughness of a material be measured?

Activity 4.5

You are given samples of different engineering materials that are all in the form of 10 mm diameter bar. You have at your disposal an engineer's vice, a hammer and a centre punch. Devise a series of tests which makes use of these to compare the hardness, toughness and ductility of the materials.

To check your understanding of the preceding section, you can solve Review questions 16–20 at the end of this chapter.

Thermal properties

These include the ability of a material to conduct heat energy and the changes in dimension which occur with a rise of fall in temperature.

Expansivity

When metals are heated they generally expand. The same occurs with a great many polymer and ceramic materials although the effect is not so pronounced. Some thermoplastics contract when heated and use is made of this property in packaging. Shrink insulation sleeving for electrical conductors has the same property. Gentle heating causes it to shrink and provide a firm covering for the conductor.

Thermal expansivity is a measure of the effect of temperature change on the dimensions of a material when it is heated. It is defined as the increase in length, per unit of original length, per degree of temperature rise. It is given the symbol α (alpha), and its units are $°C^{-1}$.

KEY POINT

If a material is not allowed to expand or contract freely, thermal stresses will be set up which under certain conditions can cause the material to fail.

If the original length of a component is l, and it expands by an amount x, when the temperature rises from t_1°C to t_2°C, the linear expansivity of the material is given by the formula:

$$\alpha = \frac{x}{l(t_2 - t_1)} \tag{4.5}$$

The change in length x of a component for a given change in temperature is given by transposing the formula, i.e.:

$$x = l\alpha(t_2 - t_1) \tag{4.6}$$

Thermal conductivity

Metals are mainly good conductors of heat energy whilst plastics and ceramics are generally good heat insulators. It is thought that this is because of the presence of free electrons in the crystal lattice structure of metals. When a metal is heated, the kinetic energy of the free electrons is increased in the locality of the heat source. This is passed on to other free electrons throughout the material and heat energy is transported much more quickly than in plastic and ceramic materials where the valence electrons are all employed in producing the strong covalent bonds. In these materials, it is thought that heat energy is passed on through increased thermal vibration of the atoms and molecules, which is a much slower process.

The thermal conductivity of a material is defined as the amount of heat energy per second that will pass through a 1 m cube of the material when the difference in temperature between opposite faces is 1°C. It is given the symbol k, and its units are $W\,m^{-1}\,°C^{-1}$.

Consider a piece of material whose cross-sectional area is A and thickness is l, with a temperature difference $t_2 - t_1$ across its thickness,

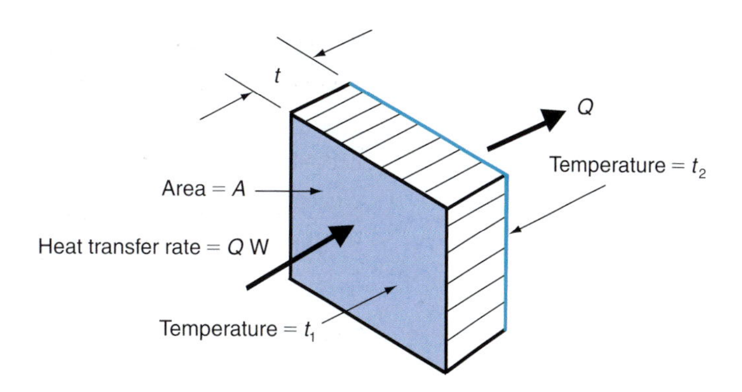

Figure 4.28 Conduction of heat energy

as shown in Figure 4.28. If the heat transfer rate through the material is Q, measured in watts, then the thermal conductivity of the material will be given by the formula:

$$k = \frac{Ql}{A(t_2 - t_1)} \tag{4.7}$$

The heat transfer rate for a given temperature difference is given by transposing the formula, i.e.:

$$Q = \frac{kA(t_2 - t_1)}{l} \tag{4.8}$$

Electrical and magnetic properties

These include the ability of a material to allow or prevent the passage of electric current and the behaviour of a material in the presence of electromagnetic fields.

Resistivity

Metals are generally good conductors of electricity. As with thermal conductivity, this is due to the presence of free electrons in the crystal lattice structure. When an electrical potential difference is applied between the ends of a metal conductor, the negatively charged free electrons drift towards the positive potential. It is this flow which constitutes an electric current.

The flow of free electrons is impeded by the fixed atoms and the resistivity of a material is defined as the resistance of a 1 m length whose cross-sectional area is 1 m. It is given the symbol ρ (rho), and its units are ohm-metres. For a conductor of length l, cross-sectional area A and resistance R, the resistivity of the material is given by the formula:

$$\rho = \frac{RA}{t} \tag{4.9}$$

Alternatively, the resistance of the conductor can be found using the formula:

$$R = \frac{\rho l}{A} \tag{4.10}$$

KEY POINT

Metals are generally good conductors of both heat and electricity due to the presence of free electrons in the material.

Temperature coefficient of resistance

The atoms and molecules of a material are thought to be in a state of vibration whose amplitude increases with temperature rise. This partly accounts for the expansion of the material and also the increase in electrical resistance since the free electron flow is further impeded by the increased vibration.

The resistance of some materials is not much affected by temperature change. Carbon is a common example of these, which are said to be *ohmic* materials. With most metals, the resistance increases uniformly with temperature and they are said to be *non-ohmic*. The temperature coefficient of resistance of a material is a measure of the effect. It is defined as the increase in resistance per unit of its resistance at 0°C, per degree of temperature rise. It is given the symbol α (alpha), and its units are $°C^{-1}$. It should not be confused with linear expansivity which has the same units and symbol.

The temperature coefficient of resistance of a material whose resistance at 0°C is R_0 and whose resistance at some higher temperature t°C is R is given by the formula:

$$\alpha = \frac{(R - R_0)}{R_0 t} \qquad (4.11)$$

Alternatively, the resistance of a material at some temperature t°C is given by the formula:

$$R = R_0(1 + \alpha t) \qquad (4.12)$$

KEY POINT

An ohmic material is one whose temperature coefficient of resistance is zero, i.e. its electrical resistance is not affected by temperature change.

With semiconductor materials such as silicon and germanium, the resistance falls with temperature rise and they are said to have a negative temperature coefficient of resistance. The fall in resistance is not uniform, however, which means that α is not a constant. As a result, the above formulae cannot be used with semiconductor materials.

Permeability

Electromagnets, transformers, motors and generators all contain current-carrying coils around which a magnetic field is produced. The core material, on which the coils are wound, can greatly affect the intensity and strength of the magnetic field. Certain metals, which are known as ferromagnetic materials, have the greatest effect. They include iron, nickel, cobalt and a number of specially developed alloys which contain these metals.

Permeability is a measure of a material's ability to intensify the magnetic field produced by a current-carrying coil. A comparison is made with the magnetic field produced by a coil which has nothing whatsoever in its core. The relative permeability μ_r of a core material indicates the increased intensity of the field produced compared to that of the same empty coil. Materials other than those listed above have little or no effect on field intensity and so have a relative permeability of $\mu_r = 1$. Soft low carbon steel on the other hand can have a relative permeability of $\mu_r = 100000$.

Ferromagnetic materials can be subdivided into 'soft' and 'hard' categories. The soft materials are the low carbon steels used for the cores of transformers and electromagnets. These have a high permeability but soon loose their magnetism when the current is switched off. Hard magnetic materials are generally made from hard alloy steels containing nickel. They are used to make permanent magnets such as those used in audio speakers. Hard magnetic materials have a lower permeability but once magnetised, they retain their magnetism for long periods of time.

KEY POINT

Permeability is a measure of the ability of a material to intensify the magnetic field produced by a current-carrying coil when it is used as a core material.

Permittivity

Permittivity is a measure of a material's ability to intensify an electric field such as that produced between the plates of a capacitor. The plates of a capacitor are separated by an insulating material known as a 'dielectric'. Depending on the use and type of capacitor, air, waxed paper, plastics and ceramic materials are used as dielectrics. When a potential difference is applied across the plates of a capacitor, electrical charge is stored on them and an electric field is set up between them.

The intensity of the electric field, and the amount of charge which can be stored in a capacitor, depends on the permittivity of the dielectric material. As with magnetism, a comparison is made with the intensity of the field produced by the same potential difference when the plates of a capacitor are separated by vacuum. The relative permittivity ε_r gives a measure of the increased field intensity when a dielectric is present. For an air space $\varepsilon_r = 1.0006$, for paper $\varepsilon_r = 2–2.5$, for mica, which is a ceramic, $\varepsilon_r = 3–7$.

KEY POINT

Permittivity is a measure of the ability of a material to intensify an electric field such as that set up between the plates of a charged capacitor.

Test your knowledge 4.10

1. Why are most metals good conductors of both heat and electricity?
2. How is the 'resistivity' of a material defined?
3. What is an 'ohmic' material?
4. Which materials have a high magnetic permeability?
5. What is meant by the relative permittivity of a dielectric material?

Durability

There is a tendency for all engineering materials to deteriorate over a period of time. This may be due to corrosion, attack by chemical solvents or degradation due to electromagnetic radiation. The durability of engineering components can be maximised by choosing materials which are best suited to their service conditions. Surface protection and shielding from the above kinds of attack can also prolong service life.

Corrosion resistance

All metals, and particularly ferrous metals, are subject to corrosion. This is the result of a chemical reaction that takes place between the metal and some other element in its service environment. The other element is very often oxygen, which is present in the atmosphere and in water. Non-ferrous metals usually have a higher resistance to corrosion than those containing iron. They react with oxygen from the atmosphere but the oxide film that forms on the surface is more dense. Once formed it protects the metal against further attack.

KEY POINT

Wet corrosion is an electro-chemical process which occurs when moisture is present. Dry corrosion is a direct chemical reaction between a material and oxygen in the atmosphere.

Ferrous metals are affected by two kinds of corrosion: wet corrosion, which occurs at normal temperatures in the presence of moisture, and dry corrosion, which occurs when the metal is heated to high temperatures. Both processes will be discussed in detail later in this chapter when considering the ways in which materials can fail in service. It is sufficient here to say that wet corrosion is an electro-chemical process where the moisture acts as an electrolyte and different areas of the metal become anodes and cathodes. Corrosion always occurs at the anodes.

The oxides that form on the surface of ferrous metals are rust, at normal temperatures, and black millscale, at high temperatures. Both are loose and flakey. They do not protect the metal from further attack but there are a number of ways in which it can be protected at normal temperatures. The most common methods are painting, plating with a corrosion resistant metal and coating with a polymer material. The choice of paint depends on the service environment but oil paints which contain bitumen and enamel paints containing polymers give good surface protection against rusting.

The most common metals used for the surface protection of steel are zinc, tin and cadmium. Plating with zinc is known as *galvanising*. It is particularly effective because, if the surface is scratched to expose the steel beneath, the steel will not rust. In the presence of moisture it is the zinc which becomes an anode and which will corrode. This is called *sacrificial* protection. The corrosion of the zinc takes place at a very slow rate and so it gives long-term protection.

EXTRACT FROM ELECTRO-CHEMICAL SERIES

Gold
Platinum
Silver
Copper
Lead
Tin
Nickel
Cadmium
Iron
Chromium
Zinc
Aluminium
Magnesium

Tin and cadmium plating give good protection provided that the coating is not damaged to expose the steel beneath. In this event, the steel becomes an anode and rusting will occur in the presence of moisture. The list of metals in the margin is part of a much longer list that is called the *electro-chemical series*. When two dissimilar metals are joined together there is a danger of electrolytic corrosion and it is the metal lower down in the series which becomes the anode and which corrodes.

You will note that zinc is below iron and that tin and cadmium are above it. This accounts for the degree of protection that they give to steel. Rubber and other polymer materials are also used to protect steel. Rubber compounds are sometimes used for tank linings. Plastic material is very often used to protect steel lamp standards and the posts which support road traffic signs.

KEY POINT

Sacrificial protection of ferrous metal occurs when they are coated with a metal whose position is below that of iron in the electro-chemical series.

Solvent resistance

There are certain chemical substances that attack plastics and rubbers. They are known generally as solvents. The liquids and gases used in many industrial processes can have such a degrading effect. Petrol, fuel oil, lubricating oils and greases can also act as solvents. Thermosetting plastics tend to have a high resistance to solvents and it is generally thermoplastics and rubbers which are most vulnerable. Checks must be made on their solvent resistance when selecting these materials for service conditions where chemical substances are present. Metals and ceramics usually have a high resistance to the solvents that attack polymers but they can be attacked by other substances, particularly acids.

Radiation resistance

The ultraviolet radiation that is present in sunlight can affect some thermoplastics and rubbers if they are subjected to it for long periods of time. This can be a particular problem with outdoor constructional materials which can become brittle and discoloured with exposure. The effect can be reduced by the addition of colouring agents, of which black is found to be the most effective.

Test your knowledge 4.11

1. How does red rust to form on the surface of iron and steel?
2. What is 'millscale' and how does it form?
3. Why is zinc able to give sacrificial protection to steel?
4. Why do non-ferrous metals usually have a high resistance to corrosion?
5. In what ways can plastics and rubbers become degraded?

To check your understanding of the preceding section, you can solve Review questions 21–25 at the end of this chapter.

Material Processing

Engineering components often pass through a number of forming and finishing processes during the production cycle. These may be subdivided into primary and secondary processes. The primary processes include moulding, casting, forging, drawing, rolling and extrusion. They are performed on the raw material to give it form and shape. Sometimes, as is the case with polymers and die-cast metals, the finished product can be made in one primary forming process. More often, a component requires secondary forming and finishing by machining, heat treatment and surface protection.

The properties of a material affect the choice of production process and conversely, a production process may affect the properties of a material. In the case of metals it is the primary forming processes and subsequent heat treatment which have the greatest effect on the final properties of a component.

Processing metals

When metal components are cast to shape, the rate at which they are allowed to cool affects the size of the grains. Slow cooling tends to produce large grains, which may leave the material weak. Fast cooling tends to produce smaller grains and a stronger material, but if the cooling is uneven there may be a variation in grain size. This can result in internal stress concentrations, which may lead to distortion and cracking.

Hot and cold forming by forging, rolling, drawing and extrusion distorts the grain structure of a metal and causes it to flow in a particular direction. This can have a strengthening effect on the material but in the case of cold forming, it can also produce work hardness and brittleness.

KEY POINT

Forming processes can leave a material with grains of uneven size, distorted grains and internal stress concentrations.

Primary forming is often followed by heat treatment, the objective of which is to refine the grain size and remove internal stress concentrations. The material can then proceed to secondary forming by machining after which it may undergo further heat treatment to give it the required degree of hardness and toughness. The most common heat treatment processes are annealing, normalising, quench hardening, tempering, case hardening and precipitation hardening. Some ceramics can also have their properties modified by heat treatment.

Annealing

The purpose of annealing is to restore the ductility and malleability of work hardened material after cold working. The material is heated in a furnace to what is known as the recrystallisation temperature. At this temperature new crystals or grains start to form and grow in the regions where the old grains are most distorted. The process continues until the new grains have completely replaced the old deformed structure. When recrystallisation is complete the material is cooled. If the annealing process is carried on for too long a time, the new grains will grow by feeding off each other to give a very course grain structure. This can make the material too soft and weak.

Carbon steels, copper brass and aluminium can all have their grain structure reformed by annealing. Figure 4.29 shows the range of temperatures at which annealing is carried out for plain carbon steels.

KEY POINT

At the recrystallisation temperature new crystals or grains start to form within the solid material at the points where the grain distortion and internal stress is greatest.

After annealing for a sufficient length of time, steel components are cooled slowly in the dying furnace. Pure aluminium is annealed between 500°C and 550°C, cold worked brass between 600°C and 650°C and copper between 650°C and 750°C. Unlike steels, these materials can be quench-cooled after annealing.

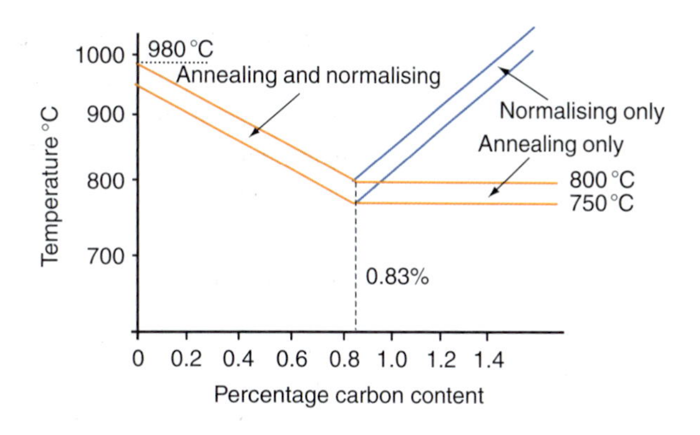

Figure 4.29 Annealing and normalising temperature range for plain carbon steels

Normalising

Normalising is a process which is mainly carried out to refine the grain structure and relieve internal stress concentrations in components which have been hot formed to shape. Components formed by hot forging and pressing are very often normalised prior to machining. The process for steel components is similar to annealing except for carbon contents above 0.83%. Here the normalising temperature is higher than that used for annealing as shown in Figure 4.29. Normalising also differs from annealing in the rate at which components are cooled after recrystallisation (Figure 4.30). The usual practice is to allow steel components to cool more quickly in still air.

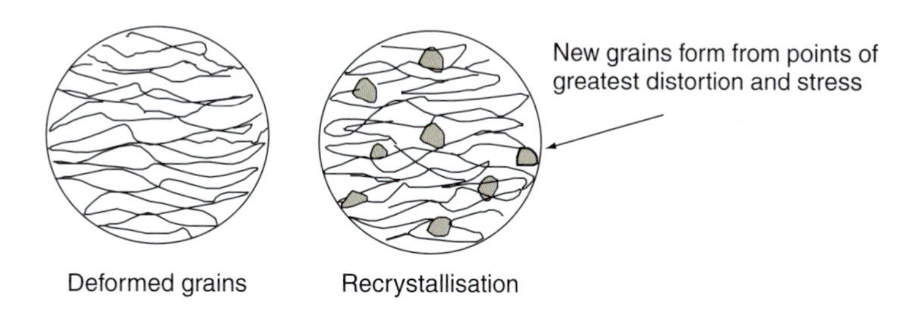

Figure 4.30 Recrystallisation

Quench hardening

Medium and high carbon steels which have a carbon content of above 0.3% can be hardened by heating them to within the same temperature band as for annealing and then quenching them in water or oil. Structural changes take place at these high temperatures. The iron atoms rearrange themselves from a BCC to an FCC structure, which is known as *austenite*, and all of the carbon atoms are taken into solution.

In this condition the steel becomes very malleable and ductile which is why hot working is done at these high temperatures. It must be remembered that the change of structure takes place with the metal still in the solid state. When the steel is quenched, it does not have time to revert to its original grain structure. The result is a new grain formation called *martensite*, which consists of hard needle-like crystals.

Quenching in water gives the fastest rate of cooling and the maximum hardness. The violence of the cooling can however cause cracking,

KEY POINT

Quench hardening can only be carried out on medium and high carbon steels.

particularly with high carbon steels. Oil quenching is slower and less violent. The steel is slightly less hard but cracking is less likely to occur. High carbon steels should always be oil quenched. Mild steel with a carbon content below 0.3% does not respond to quench hardening. It can however be case hardened on its surface, as will be described.

Tempering

Quench hardened components are generally too hard and brittle for direct use. The tempering process removes some of the hardness and toughens the steel.

Tempering is achieved by re-heating it to temperatures between 200°C and 600°C and quenching again in water or oil. The temperature to which steel components are re-heated depends on their final use as shown in Table 4.15.

Table 4.15 Tempering temperatures

Components	Tempering temperature (°C)	Oxide colour film
Craft knife blades and woodworking tools	220	Pale straw
Lathe cutting tools	230	Medium straw
Twist drills	240	Dark straw
Screwcutting taps and dies	250	Brown
Press tools	260	Brown to purple
Cold chisels and punches	280	Purple
Springs	300	Blue

KEY POINT

Quench hardened components are tempered to remove some of the hardness and increase the toughness of the material.

Large batches of components are hardened and tempered in special temperature controlled furnaces. Small single items may be hardened in the workshop by heating in a gas flame to a bright cherry red colour and quenching. The surface is then polished and they are then gently re-heated until oxide colour films start to spread over the surface. When the colour film that corresponds to the required tempering temperature starts to appear, the components are quenched.

Case hardening

Mild steel does not respond to quench hardening because of its low carbon content. Case hardening increases the surface hardness of the material whilst leaving the core in its soft and tough condition. The first part of the process is known as carburising where the components are 'soaked' for a period of time at high temperature in a carbon-bearing material. The traditional method is to pack them in cast iron boxes with a carbon rich powder. This may be purchased under a variety of trade names or made up from a mixture of charcoal and bone meal. The carbon slowly soaks into the steel to give an outer case with a high carbon content. The depth of the case depends on the time of soaking.

The second part of the process is to re-heat the components to refine the grain size in the core and then to quench harden and temper the outer case. Case hardened components have a hard and wear resistance outer case and a tough impact resistant core. This is an ideal combination of properties for many engineering components.

Case hardening is not confined to mild steels. Medium and high carbon steels in the normalised condition can be case hardened by rapidly heating and quenching the outer surface. Induction hardening is such a process, where an induction coil induces a high-frequency electric current into the component as it passes through the coil. This has a rapid heating effect after which the component passes through a water or oil jet to quench the surface.

KEY POINT

Case hardening leaves a component with a hard wear resistant surface whilst retaining a tough, impact resistant core.

Precipitation hardening

Aluminium alloys which contain copper or small amounts of magnesium and silicon among their constituents can be hardened by this process. After hot forming and slow cooling the material is quite soft and its properties are similar to those of pure aluminium. The first part of the process is to heat the material to around 500°C at which temperature the copper atoms become fully absorbed in the aluminium crystal lattice structure. The material is then quenched to retain the new formation and is then found to be harder and tougher than before.

The new structure is however unstable and over a period of time tiny particles of $CuAl_2$ start to precipitate out of the solid solution. These are evenly scattered throughout the structure and have the effect of making the material harder still. The effect is known as 'age hardening' and over a period of several days the material becomes noticeably harder and stronger.

The precipitation process can be speeded up by re-heating the material to between 120°C and 160°C for a period of around 10h. The final hardness and strength is then greater than that achieved by age hardening at room temperature. Aluminium alloys which contain silicon and magnesium also respond to precipitation hardening. Here it is the intermetallic compound Mg_2Si which precipitates. See Table 4.8 for aluminium alloys which can be heat treated.

KEY POINT

Precipitation or age hardening can only be carried out on aluminium alloys which contain copper or magnesium and silicon.

CHAPTER 4

Test your knowledge 4.12

1. What takes place at the recrystallisation temperature in a work hardened material?
2. What is the purpose of the annealing and normalising processes?
3. How should high carbon steels be quench hardened?
4. At what temperature should springs be tempered?
5. What happens in aluminium alloys during precipitation hardening?

Activity 4.6

A cold chisel that has been in use for some time has become blunted and has a heavily mushroomed head. Write an account of how you would completely recondition the chisel to make it serviceable and safe for use.

To check your understanding of the preceding section, you can solve Review questions 26–30 at the end of this chapter.

Processing polymers

The properties of polymer materials can be varied by making adjustments to their ingredients and the chemical processes that are used to produce the raw materials before forming. The tensile strength and modulus of elasticity of thermoplastics can be increased by increasing the length of the polymers and promoting side-branching. This increases the force of attraction between polymers and makes it more difficult for them to move. Increasing the amount of crystallinity can also increase the density and strength of the materials.

The rigidity and hardness of thermosetting plastics depends on the degree of cross-linking. This too can be controlled by adjusting the ingredients during the initial mixing process. They can be made more flexible by adding materials known as 'plasticisers' that have the effect of filling the spaces between the polymers. This makes it easier for the chains to move over each other.

Sometimes, different polymers are mixed together in the liquid state and allowed to solidify together. The polymers can also be encouraged to combine by a process known as 'copolymerisation' which can give improved toughness and solvent resistance. The properties of polystyrene, polyurethane and rubber can be greatly altered by injecting them with gases to form the plastic foams that are widely used for packaging, insulation and upholstery.

The most common processes used to form thermoplastic polymers are compression moulding, injection moulding, blow moulding, vacuum forming and extrusion. Compression moulding, which is also used on thermosetting polymers, involves heating and compressing a measured amount of powdered or granular material between the two parts of a mould. The combination of heat and pressure causes the material to become fluid and fill the mould cavity. In the case of thermosetting polymers, the heating also triggers cross-linking and the component quickly solidifies as the pressure is reduced. In the case of thermoplastic polymers, the mould must be cooled so that the component is allowed to solidify before being ejected. For both types of material the correct combination of pressure and temperature is essential as is the time allowed for solidification and cross-linking.

Injection moulding is carried out mainly on thermoplastic polymers and allows more complex shapes to be produced than with compression moulding. Granular or powdered raw material is fed from a hopper into a chamber where it is compressed and heated until it becomes liquid. It is then injected into a mould, the two parts of which are water-cooled. The material solidifies rapidly against the cooler mould faces after which the two parts separate and the component is ejected. The correct combination of pressure, temperature and speed of injection is essential, as is the clamping force that holds the two parts of the mould together.

KEY POINT

The temperature of the mould in compression moulding and injection moulding must be controlled to allow thermoplastics to solidify and cross-linking to occur in thermosetting plastics.

The process is generally automated with a closely controlled mould temperature and open and closed time.

Typical faults that can occur include blistering and burn marks caused by too high a material and mould temperature. Flow marks caused by the injection speed being too low and the material cooling down too quickly as it fills the mould. Sink marks, which are the localised depressions in the thicker sections, caused by too low a holding and cooling time. Flash around the mould joint line caused by a worn mould or an incorrect combination of injection pressure and holding pressure.

With extrusion the heated polymer is forced through a die to produce a variety of cross-sections. Some of these, such as those used for UPVC window frames, are quite complex. Blow moulding is an extension of the extrusion process that is used to produce a variety of plastic bottles and containers. A hot extruded tube of the material passes between the open halves of a mould. The mould is then closed to cut off and seal one end of the tube and at the same time compressed air is blown into the other end so that the material inflates to the shape of the mould. After cooling in the mould, the two parts separate and the container is ejected. With vacuum forming a heat softened sheet of thermoplastic material is placed over a die. A vacuum is then created underneath it causing it to be forced on to the die by atmospheric pressure. With all of these processes the correct combination of heating, pressure, speed and cooling rate are essential for products that have a uniform shape and surface finish.

Processing ceramics

The ingredients for ceramics have already been listed. They include naturally occurring minerals such as clay, sand, flint, quartz and also materials that have been prepared chemically. Some of the ingredients may have to be ground down into fine particles. The ingredients are generally mixed with water and an appropriate binding agent to form a malleable semi-fluid mass that can be formed to shape by machining (potters wheel), injection moulding and extrusion. The water content must be just sufficient to allow forming and for the components to maintain their shape before firing. In some cases they are heated in a partial vacuum to drive off the moisture before firing.

Fluxes such as sodium borate (borax), sodium carbonate, potassium carbonate, calcium fluoride and sodium aluminium fluoride (cryolite) are added to the clay, sand and other ingredients to lower the required firing temperature. The extent of vitrification that occurs (the bonding together of crystals) depends on the product as shown in Table 4.16.

Table 4.16 Firing of ceramic products

Ceramic	Firing temperature (°C)	Addition of fluxes	Extent of vitrification
Glasses	1500–2000	Moderate	Complete
Refractory materials	1500–2000	Small	Little
Whiteware	1200–1500	Variable	Little
Structural clay products	800–1300	Large	Little

KEY POINT

Heat treatment processes for glass can be used to remove internal stresses, give increased toughness and produce pyroceramic materials.

Glass products may be annealed to reduce internal stresses that have been set up due to uneven cooling. Glass, such as plate glass for windows, may also be toughened by heat treatment. If glass is held at its annealing temperature and very slowly cooled, it can be made to crystallise. The process is called 'devitrification' and produces a structure with crystals that are smaller and more uniform than in other bonded ceramics. Depending on their composition, glass ceramics can be made with better thermal and mechanical properties than glass itself. The 'pyroceramics' used for cooker hobs are materials of this type and they are also used for aerospace applications.

Processing composites

As we have stated, composite materials may be categorised as fibrous, laminated or particulate. With fibrous composites such as carbon fibre and glass fibre-reinforced plastics, the stiffness and tensile strength depends on the proportion of fibres to the polymer matrix material (the fibre/reinforcement ratio) and the alignment of the fibres. Some composites in which the fibres are aligned in one direction will be stiff in that direction. Carbon fibre fishing poles are such an example. Composites in which the fibres are chopped or in the form of matting will be less stiff in bending and will also have a low expansivity. Motor panels and boat hulls made from glass fibre-reinforced epoxy and polyester resins are examples of these.

The properties of laminates after processing are dependent on the type and orientation of the different layers and the bonding material. If the layers have a grain direction, as with plywood, they are generally set at right angles to each other. This gives a material with the same properties in all directions, which is then said to be 'isotropic'. A fault that sometimes occurs in laminates is failure of the bonding between the layers. It is known as 'delamination' which can seriously degrade the properties of the material. It may be due to manufacturing imperfections but can also be caused by accidental impact. The presence of cyclic loading forces that cause bending or twisting will aggravate the effect.

The final properties of the particulate type of composite depend on the nature and size of the particles and the matrix material that binds them together. These have already been described. Other factors that affect the properties of sintered cermets are the pressure and temperature that are applied during the manufacturing process. The pressure should be sufficient to cause cold welding between the matrix and reinforcement particles. The applied temperature must then be above the recrystallisation temperature of the matrix material to give uniform structure that is tough and wear resistant.

Test your knowledge 4.13

1. What effect do plasticisers have on the properties of thermosetting plastics?
2. What is blow moulding?
3. What occurs during vitrification in the production of ceramics?
4. What is a pyroceramic material?
5. What is delamination?
6. What are defines an isotropic material?

Activity 4.7

Most local authorities now have systems in place for collecting plastic containers for re-cycling. Write an account of the stages in the re-cycling process and the uses of the re-cycled material.

Selection of Engineering Materials

The success of an engineering company depends on the quality of its products. A quality product is one that is fit for its purpose and which will have an acceptably long and trouble free service life. Furthermore its material, production and maintenance costs must be kept within reasonable limits so that the company can make an acceptable profit and remain competitive. The selection of engineering materials is therefore an important part of the design process in which the following considerations have to be taken into account:

- What exactly is the function of the product and what will be its service conditions?
- What, if any, are the legal requirements of the product and are there any National or International Standards or Design Codes which must be adhered to?
- What are the properties required of the materials?
- What are the available processing facilities and labour skills?
- Which are the most suitable materials, are they readily available and what will be their cost?

Design considerations

One has only to think of the components that go into the production of a mountain bike to realise the importance of material selection. The designers must be fully aware of the mechanical properties required. The frame, forks and brake cables must have sufficient tensile strength. They must also be impact resistant as must the wheels and pedals. The chain, sprockets and bearings must be tough and wear resistant. The springs in the suspension must be tough and elastic and the tyres must be durable.

The designers must also be aware of the production processes that will be required. The frame parts will need to be formed to shape and joined by welding or hard soldering. Other components will need to be machined and heat treated. Before final assembly, all of the parts will need to be given some form of surface protection to guard against corrosion. Many of the components that go to make up a mountain bike will of course be bought-in complete from specialist suppliers. They will have followed the same material selection procedures to ensure that their products are of a high quality.

It is important to know the service conditions and loads under which engineering components will operate. This often involves complex calculations, particularly in the design of aircraft, motor vehicles and building structures. Materials with appropriate load-bearing properties can then be selected and the component dimensions calculated to ensure that the working stresses are below the strength limits of the materials by a safe margin. Where necessary, the materials must also be resistant to corrosion, solvents and other forms of degradation that might occur in service. Prototype components are often subjected to tests that simulate service conditions. Component failure or excessive wear might lead to the choice of alternative materials with better mechanical properties. New and untried materials should not be selected for their novelty value. There are many instances where unforeseen deterioration has occurred in service. New materials should be rigorously tested to confirm the properties claimed for them and information sought from other users as to their success.

Processability is an important factor that must be taken into account when selecting materials for engineered products. Materials with the required mechanical properties might be difficult and expensive to machine to the required dimensional tolerances and surface finish. The materials selected for castings must have good fluidity when molten. Steel has many good mechanical properties but it is difficult to cast into complex shapes. The materials for forging or pressing must have a sufficient degree of malleability and ductility, and some materials, such as aluminium, are difficult to weld without specialised skill and equipment. These are all factors which must be considered in material selection to make the best use of existing plant, equipment and skills.

KEY POINT

The materials selected for engineered products must be fit for their intended purpose, be suitable for the available process plant and be readily available at a reasonable cost.

Costs

Design engineers must always be aware of material and processing costs. There have been many examples of products which have been 'over engineered'. That is to say that the design has been too complex and the use of materials too extravagant. Many firms have gone out of business as a result. The material costs of a product are often a significant part of the overall manufacturing cost. The cost of a material depends on its scarcity as a natural resource, the amount of processing needed to convert it from its raw state into a usable form and the quantity and regularity with which it is purchased.

It may be that the material which appears to be the most suitable in terms of its properties is the most expensive. Cheaper alternatives may be available with similar properties but these could prove to be the more expensive in the long term. A judgement must be made of the comparative processing costs and also the cost of replacement under warrant if the alternatives prove to be unreliable in service.

Wherever possible, materials such as barstock, rolled, drawn and extruded sections should be purchased in standard sizes and with the required physical properties and surface finish. The cost of processing can then be kept to a minimum. Discount can usually be obtained by buying in bulk or placing orders for regular delivery. Bulk purchases might however require a considerable financial outlay and then take up storage space for a period of time. Overall, the cost of this might cancel out any initial savings.

KEY POINT

Discount from the bulk buying of materials can be cancelled out by the cost of storing and lack of return on the money which has been invested in them.

Many engineering firms now employ 'just in time' (JIT) systems of material supply, particularly for repeated batch and continuous production processes. This reduces the need for storage space and the carrying cost of raw material. In fact the delivery vehicles on the motorway become the raw material store. This of course needs careful co-ordination and a good working relationship with suppliers. Transport problems may interrupt the supply and there are also pitfalls in placing too much reliance on a single supplier where labour and manufacturing problems might cause interruptions.

Engineering firms have always re-cycled waste metals such as off-cuts, swarf from machining operations and sub-standard components. This gives some cost saving and the same is increasingly being done with plastics and packaging materials. Greater savings in cost can be made by adopting energy-efficient processes and procedures. The larger engineering firms often employ or engage energy management specialists to review their plant operation, processes and working procedures. This is becoming an essential part of management due to the increasing cost of energy and foreign competition.

Availability of supply

Engineering materials are available in a variety of forms. Some of the most common are shown in Table 4.17. The availability of different materials varies with material type and the form of supply. Engineering metals such as carbon steels, copper, brasses and aluminium alloys may be readily obtained in the form of bar, tube, sheet, etc. from a number of alternative suppliers in standard sizes.

Table 4.17 Forms of supply

Metals	Polymers	Ceramics	Composites
Ingots	Granules	Mouldings	Mouldings
Castings	Powders	Sheet	Sheet
Forgings	Resins	Plate	Plate
Pressings	Mouldings	Cements	
Barstock	Sheet		
Sheet	Pipe and tube		
Plate	Extrusions		
Rolled sections			
Pipe and tube			
Wire			
Extrusions			

Castings, forgings and pressings are not so easily obtained. Patterns and dies must be made before supply can commence. These are expensive to produce and the cost may only be justified if large quantities are required. Careful forward planning and close liaison with the supplier is required if supply is to commence on time. As has already been stated, problems may then arise from placing too much reliance on a single supplier. The cost of duplicate tooling and the quantities required might however dictate that this is a gamble that has to be taken.

The most common raw polymer materials in the form of granules, powders, liquid resins and hardeners are also readily available from a number of suppliers. As with metal castings and pressings, moulded polymers require the production of moulds and dies, which might take a considerable time to produce. The same applies to ceramics. Sheet glass and building materials are readily available but the supply of specially moulded components requires careful forward planning.

KEY POINT

Components that are to be cast, moulded, forged or pressed to shape will require expensive patterns, moulds and dies. The cost of these needs to be justified by the quantities that will be required.

Newly developed alloys, polymers and composites might appear to have superior properties but they may only be available in limited quantities from a single supplier. Many firms now operate on a low stock or 'JIT' system to minimise storage costs. Costly production stoppages can result from an interruption in deliveries and wherever possible it is good practice to have a back-up supplier. When selecting materials and liasing with suppliers, care should also be taken to ensure that supplies will be available to meet possible increases in demand.

Information sources

Product safety is often a concern when selecting engineering materials. This is particularly true for items such as pressure vessels, steam plant, structural supports and electrical equipment. Reference often needs to be made to British Standards Specifications (BS) and International Standards Organisation (ISO) specifications for particular sectors of engineering. Typical examples are BS 5500 for the design of pressure vessels, BSEN 60335 which lays down the durability and safety specifications for domestic products and BSEN 10002 for the tensile testing of materials. Particular grades of materials are also covered by British Standards. For example, BS 970 covers the different grades of plain carbon and alloy steels, BS 1449 for steel plate sheet and strip, BS 2875 for copper and copper alloy plate and BS 2870 which specifies the composition of brasses.

A great deal of information which is of help when selecting materials is to be found in manufacturers' and stockholders' catalogues and data sheets. They are particularly useful for selecting standard forms of supply such as barstock, sheet and rolled structural section. They may be obtained in hard copy and in many cases they may be accessed via the Internet. Trade directories which contain the names, addresses and web sites of material suppliers are often held in the reference section of public libraries and the resource centres.

The selection of materials has been made easier in recent years by the availability of computer databases that are compiled and updated by manufacturers and research bodies. Two such packages are the Cambridge Materials Selector, compiled by the Cambridge University Engineering Department, and MAT.DB which is compiled by the American Society for Metals.

They lead the user to a choice of suitable alternative materials through a structured question pathway. Access costs are involved but the time and money saved often make it well worth the outlay. They are particularly useful in introducing designers to new materials. The following free access web sites contain much useful information:

www.matweb.com
www.plaspec.com
www.steelforge.com
www.copper.org
www.avestapolarit.com
www.sandvik.com
www.structural-engineering.fsnet.co.uk

Trade association brochures and journals published by the engineering institutes often contain information on new materials. Much information can also be obtained from specialist data books such as the following:

Metals Reference Book – R.J. Smithells (Butterworth)
Newnes Engineering Materials Pocket Book – W. Bolton (Heinemann-Newnes)
Metals Data Book – C. Robb (Institute of Metals)
Handbook of Plastic and Elastomers – edited by C.A. Harper (McGraw-Hill)

Text books with titles such as *Metallurgy*, *Polymer Science*, *Engineering Materials*, etc. are useful for providing the underpinning knowledge and theory of materials engineering. These are to be found under the Dewey Codes 620 and 621 in college and public libraries. Some, from the many available, are as follows:

Properties of Engineering Materials – R.A. Higgins (Butterworth-Heinemann)
Materials for Engineering – W. Bolton (Newnes)
Materials Science – E. Ramsden (Stanley Thornes)
Engineering Materials 1 and 2 – M. Ashby and R.H. Jones (Butterworth-Heinemann)

Test your knowledge 4.14

1. Why do design engineers have to consider processability when selecting materials for engineered products?
2. What might be the advantages and disadvantages of purchasing material from a single supplier?
3. What form does the raw material take which is used for producing moulded polymers?
4. Why must the purchase orders for castings be placed well in advance of the time when they will be required?
5. What might be the most appropriate information source on the availability of engineering materials in standard sizes?

Activity 4.8

Select any of the following products and identify the materials from which it is made. Explain why you think they were selected and what, if any, are the possible alternatives. Describe the form in which the materials might have been supplied and how they might have been processed. Identify any standards which are relevant to their design and manufacture:

(i) A bicycle pump
(ii) A woodworking plane
(iii) A mains extension lead
(iv) A 250 mm (10 in) hacksaw
(v) A motor cycling crash helmet

Modes of failure

When the load on a ductile material exceeds the elastic limit, it becomes permanently deformed and *elastic failure* is said to have occurred. The material may still be intact but it is likely that the component from which it is made will no longer be fit for its intended purpose. Brittle materials, such as cast iron, very often fail in the elastic range with the brittle types of fracture shown in Figure 4.31. Brittle fracture, which is also known as cleavage fracture, is more prevalent in materials with BCC and CPH crystal lattice structures. Under certain conditions, ductile materials can also fail with a brittle type of fracture, as will be explained.

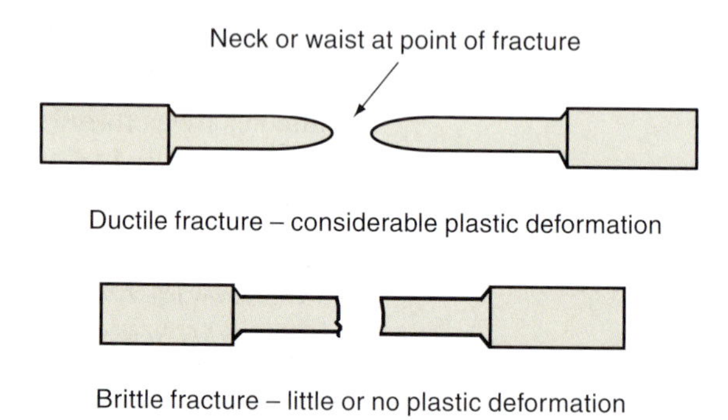

Figure 4.31 Fracture characteristics

Both kinds of failure are to be avoided by incorporating a suitable *factor of safety* into the design of engineering components. As a general rule, factor of safety of at least 2 should be employed on static structures. With this in place, the working stress in the material should always be less than half of that which will cause failure.

$$\text{i.e.} \quad \text{Factor of safety} = \frac{\text{Stress at which failure occurs}}{\text{Safe working stress}}$$

In spite of the best intentions of design engineers, components sometimes fail in service. Static loads can be hard to predict and dynamic loads on the component parts of machinery, motor vehicles and aircraft are very difficult to analyse. Combinations of direct loading, shearing, bending and twisting are very often present. A complex stress system is then said to exist, the resultant of which may exceed the predicted working stress and lead to failure.

An additional danger is the presence of stress concentrations in a component. These can occur at sharp internal corners, holes, fixing points and welds. They are known as *stress raisers*, where the stress may exceed that at which failure occurs. Under certain conditions, cracks can spread from these points, which eventually lead to failure. These kinds of failure are usually detected at the prototype stage and the design modified to prevent them occurring. Material faults such as the presence of cavities, impurities, large grain size and inappropriate heat treatment can also contribute to failure if not detected by quality control procedures.

Under certain circumstances, materials can fail at comparatively low stress levels that would normally be considered to be quite safe. The main reasons for this are changes in temperature, which can affect the properties of a material and cyclic loading. Low temperatures can cause brittleness and loss of strength. High temperatures can cause the material to *creep*, and eventually fail, under loads that are well below the normal elastic limit. A material is subjected to cyclic loading when it is repeatedly being loaded and unloaded. The loads may be well below that which would be expected to cause failure but over a period of time, failure can occur due to *metal fatigue*. Some of these failure modes will now be described in detail.

KEY POINT

Materials can fail due to metal fatigue, creep and brittle fracture at stress levels which would normally be considered safe.

The plastic deformation which precedes a ductile fracture takes a finite amount of time to take place. If a load in excess of that which will cause fracture is suddenly applied, as with an impact load, there will be insufficient time for plastic deformation to take place and a brittle form of fracture may occur. This can be observed during an Izod or Charpy impact test where an otherwise ductile material is suddenly fractured by an impact load.

Brittle or cleavage fractures usually have a granular appearance due to the reflection of light from the individual grains. Too large a grain size can affect the strength of a material and make it brittle. Grain growth can occur when materials are operating at high temperatures for long periods of time. Here the grains feed off each other in cannibal fashion, reducing the strength of the material and increasing the likelihood of brittle fracture.

Some metals which exhibit ductile behaviour under normal conditions become very brittle at low temperatures. The temperature at which the change occurs is called the *transition temperature*. Mild steel becomes brittle at around 0°C.

As can be seen from Figure 4.32 the transition temperature is judged to be that at which the fracture surface of an Izod or Charpy specimen is 50% granular and brittle and 50% smooth and ductile. It is generally the metals with BCC and CPH crystal lattice structures that are affected

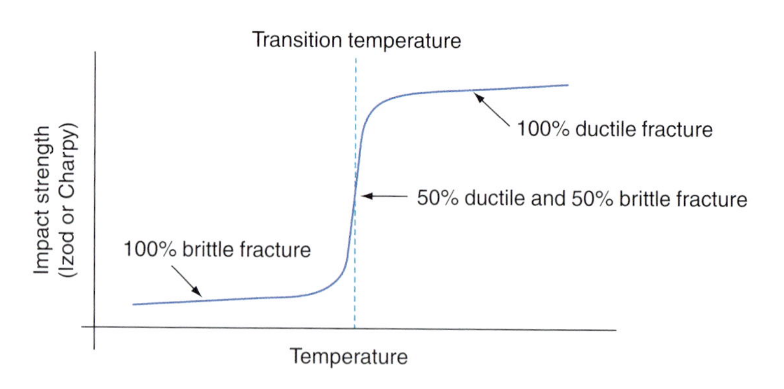

Figure 4.32 Transition from ductile to brittle material

in this way by low temperatures. The ferrite grains in steel, which are almost pure iron, have a BCC structure at normal temperatures. Chromium, tungsten and molybdenum are also BCC and suffer from low-temperature brittleness in the same way as iron.

KEY POINT

At a transition temperature of around 0°C mild steel with a high phosphorous content becomes brittle. The presence of nickel and manganese lowers the transition temperature.

The cracking and sinking of the all-welded liberty ships in the Second World War was attributed to brittle fracture. The cold north Atlantic temperatures produced brittleness in the steel hulls. Cracks appeared in places of stress concentration and these quickly spread with disastrous effects. The transition temperature in mild steel is raised by the presence of phosphorous and lowered by the addition of manganese and nickel. A relatively high level of phosphorous was present in the steel of the liberty ships and all ships hulls are now made from steel which is low in phosphorous.

Creep

Creep is a form of plastic deformation which takes place over a period of time at stress levels which may be well below the yield stress of a material. It is temperature related and as a general rule, there will be little or no creep at temperatures below $0.4 \times T$, where T is the melting point of the material measured on the kelvin scale. For mild steel, $T = 1500°C$ which is 1773 K and so there should be very little creep below $0.4 \times 1773 = 709$ K which is 436°C. It should be stressed that this is only a general rule and that some of the softer low melting point metals such as lead will creep under load at normal temperatures.

With the more common engineering metals, creep is a problem encountered at sustained high temperatures such as those found in steam and gas turbine plant. Under extreme conditions it can eventually lead to failure. A typical graph of creep deformation against time is shown in Figure 4.33.

Figure 4.33 shows the behaviour of a material which is above the threshold temperature at which creep is likely to occur. When it is initially loaded, the elastic extension OA is produced. If the stress in the material is below a level called the *limiting creep stress* or *creep limit* at that temperature, there will be no further extension. If however the stress is above this level, primary creep AB commences. This begins at a rapid rate, as indicated by the slope of the graph, and then decreases as work hardening sets in. It is followed by secondary creep BC, which takes

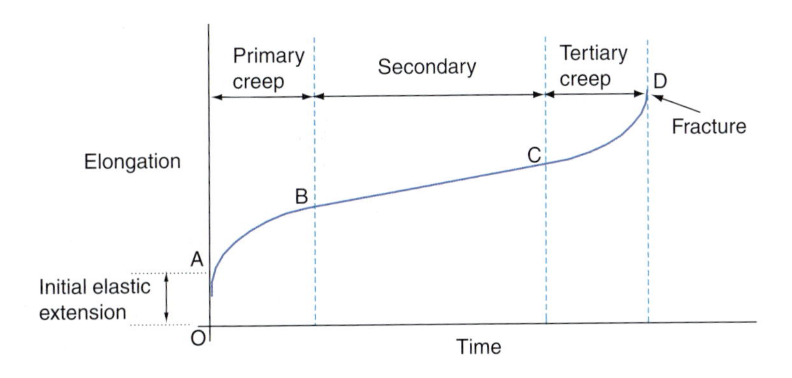

Figure 4.33 Creep vs. time graph

place over a comparatively long period at a steady rate. The final stage is tertiary creep CD, where the deformation rate increases. Necking becomes apparent at this stage, leading finally to fracture at D. Creep in polymer materials below the glass transition temperature is found to proceed in the same way.

Increases in temperature and/or increases in stress have the effect shown in Figure 4.34. The rate at which the three stages of creep take place is increased and failure occurs in a shorter time.

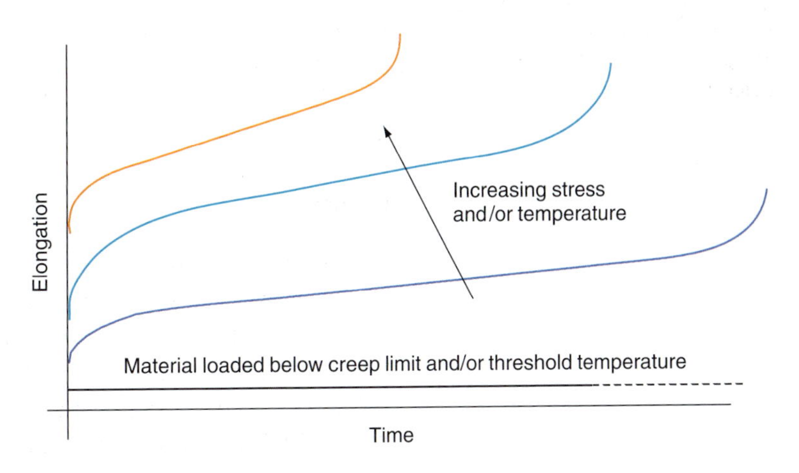

Figure 4.34 Effect of increasing the load and temperature

Study of the nature of creep suggests that the plastic deformation is partly due to slipping of the planes of atoms in the grains and partly due to viscous flow at the grain boundaries. The atoms tend to pile up in an irregular fashion at the grain boundaries which would normally lead to work hardening. The high temperatures however have a relieving effect, and the smaller the grains the greater is the viscous flow at the grain boundaries.

KEY POINT

Failure due to material creep is likely to occur if a material is loaded for sustained periods above the creep limit stress and above the creep threshold temperature.

Creep resistance can be increased in two ways. The first is to introduce alloying elements which reduce slipping within the grains. The second is to have as course a grain structure as possible, bearing in mind that this can lead to increased brittleness at normal temperatures. Many creep resistant materials have been developed over the last fifty years, in particular the nimonic series of alloys which have been widely used in gas turbines.

Fatigue

Fatigue failure is a phenomenon which can occur in components which are subjected to cyclic loading. That is to say that they are repeatedly subjected to fluctuating or alternating stresses. Typical examples are the suspension units on motor vehicles and the connecting rods and crankshafts in internal combustion engines. The forces and vibrations set up by out-of-balance rotating parts can also produce cyclic loading. The alternating stresses may be well below the elastic limit stress, and the material would be able to carry a static load of the same magnitude indefinately. Failure usually starts with a small crack which grows steadily with time. Eventually the remaining cross-sectional area of the component becomes too small to carry the repeated loads and the material fractures.

It is found that in ferrous metals there is a certain stress level below which fatigue failure will not occur no matter how many stress reversals take place. This is called the *fatigue limit* and is given the symbol S_D. As a general rule for steels, the fatigue limit is about one-half of the UTS of the material. The higher the stress above this value, the fewer will be the number of reversals or stress cycles before failure occurs. A typical graph of stress level against the number of cycles leading to failure for a material such as steel is shown in Figure 4.35. The graph is often referred to as an *S–N* curve.

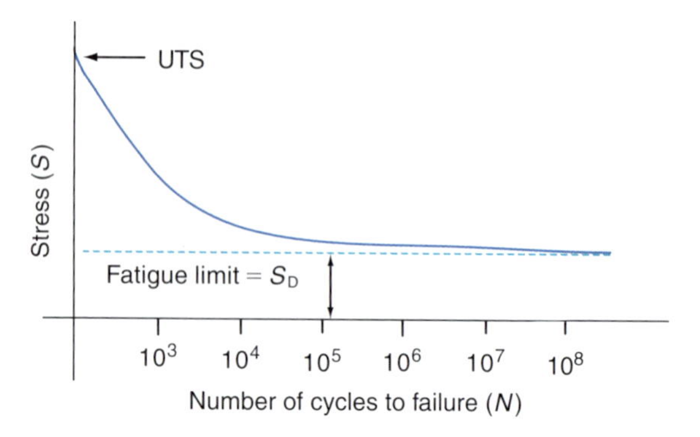

Figure 4.35 *S–N* graph

Fatigue cracks are observed to spread from points of stress concentration. Cyclic loading at stress levels above S_D produces slip in the planes of atoms in the grains of a material. This results in the appearance of small extrusions and intrusions on the surface of an otherwise smooth material, as shown in Figure 4.36. Although the intrusions are very small, they act as stress raisers from which a fatigue crack can spread.

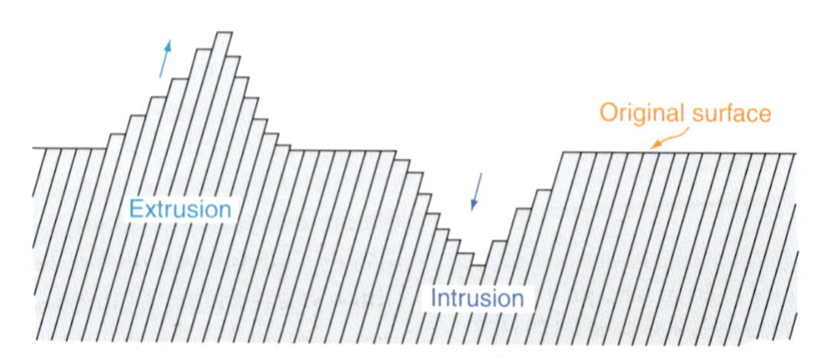

Figure 4.36 Intrusions and extrusions due to local slip (Redrawn from *Properties of Engineering Materials*, Second edition, R Higgins, Butterworth-Heinemann, p. 349, Fig. 16.11, 1994, with permission from Elsevier.)

If other stress raisers are present such as sharp internal corners, tool marks and quench cracks from heat treatment, the process can be accelerated and these should be guarded against.

The fracture surfaces of a fatigue failure have a characteristic appearance as shown in Figure 4.37. As the fatigue crack spreads, its two sides rub together under the action of the cyclic loading. This gives them a burnished, mother-of-pearl appearance. Eventually the material can no longer carry the load and fractures. The remainder of the surface, where fracture has occurred, has a crystalline or granular appearance.

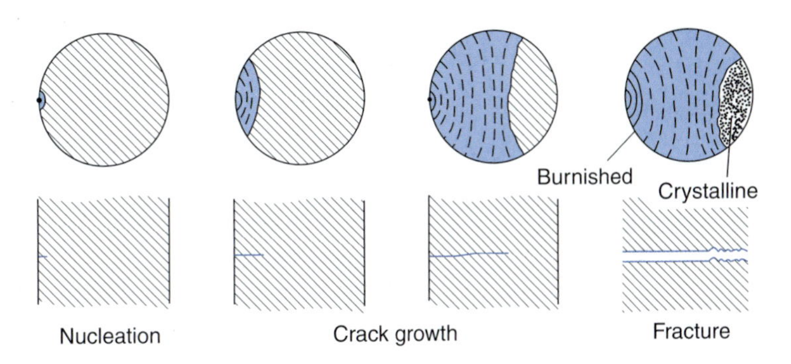

Figure 4.37 Fatigue fracture appearance (Redrawn from *Properties of Engineering Materials*, Second edition, R Higgins, Butterworth-Heinemann, p. 349, Fig. 16.12, 1994, with permission from Elsevier.)

There are many non-ferrous metals which do not have a fatigue limit and which will eventually fail even at very low levels of cyclic loading. Some steels, when operating in corrosive conditions, exhibit these characteristics. The *S–N* graph for such materials is shown in Figure 4.38.

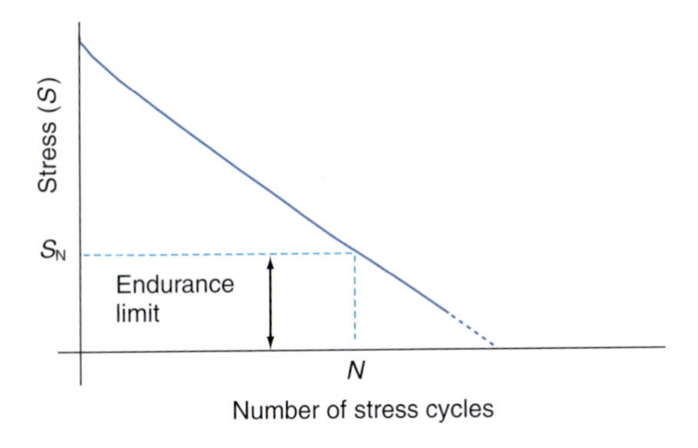

Figure 4.38 Endurance limit

KEY POINT

Fatigue failure will eventually occur when components are subjected to cyclic loading at stress levels above the fatigue limit. The presence of stress concentrations accelerates the effect.

Instead of a fatigue limit, they are quoted as having an *endurance limit* which is given the symbol S_N. The endurance limit is defined as the stress which can be sustained for a given number of loading cycles. Components made from these materials should be closely monitored, especially when used in aircraft, and replaced at a safe time before the specified number of cycles N has been reached.

CHAPTER 4

Degradation

Ferrous metals are affected by two kinds of corrosion. Low temperature or 'wet' corrosion is due to the presence of moisture and results in the formation of red rust. This, as you well know, is very loose and porous. Red rust is an iron oxide formed by electro-chemical action, in which the moisture acts as an electrolyte. Adjacent areas of the metal, which have a different composition, such as the alternate layers of ferrite and cementite in the pearlite grains, become the anodes and cathodes. Corrosion occurs at the anode areas resulting in rust formation.

KEY POINT

Wet corrosion of ferrous metals is an electro-chemical process where moisture acts as an electrolyte and different regions of the material become anodes and cathodes.

Figure 4.39 shows that the ferrite layers, which are almost pure iron, become anodes and corrode to form $FeOH_3$ which is red rust. The same kind of electrolytic action can occur between adjacent areas which have been cold worked to a different extent. Figure 4.40 show a fold in a sheet of metal which is more highly stressed than the surrounding areas. In the presence of moisture, the region in the fold becomes an anode and corrodes. This kind of electrolytic action is called *stress corrosion*. It is the form from which motor vehicle panels can suffer if they are not properly protected.

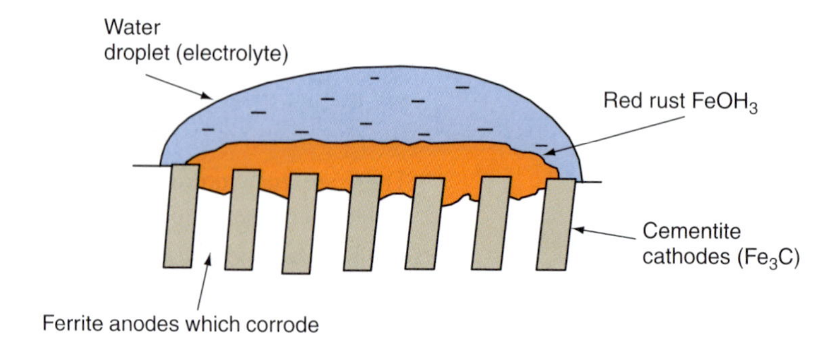

Figure 4.39 Wet corrosion of pearlite grain

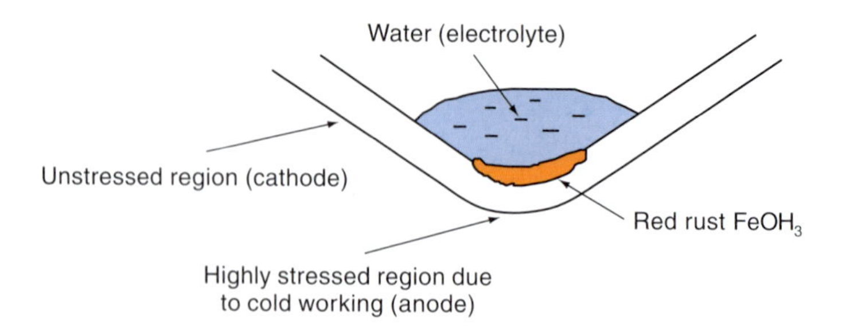

Figure 4.40 Stress corrosion

High temperature or 'dry' corrosion occurs due to a direct chemical reaction between the metal and oxygen in the atmosphere (Figure 4.41). It results in the formation of black millscale when the metal is heated for forging or for heat treatment. Millscale is another form of loose and porous iron oxide whose chemical formula is FeO.

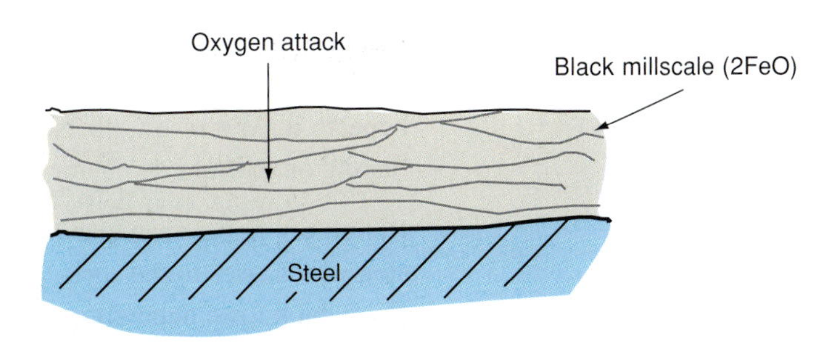

Figure 4.41 High-temperature corrosion

As has been mentioned, the oxide films that form on the surface of non-ferrous metals and alloys are generally quite dense. Polished copper, brass and silver very soon become tarnished but once a thin oxide film has formed, it protects the metal from further attack. Sometimes the oxide film is artificially thickened by an electrolytic process known as *anodising*. Aluminium alloys for outside uses such as door and window frames are treated in this way.

KEY POINT

Dry corrosion is a direct chemical reaction between a material and oxygen in the atmosphere.

Solvent attack

Thermosetting plastics tend to have a high resistance to solvents and it is generally thermoplastics and rubbers which are most vulnerable. The action of the solvent is to break down the Van der Waal forces and take the polymers into solution. Industrial solvents used for degreasing and for paint thinners, petrol, fuel oil, lubricating oils and greases can have this effect on some polymers.

Radiation damage and ageing

The ultraviolet radiation present in sunlight can have a degrading effect on some thermoplastics and rubbers. It progressively causes oxygen atom cross-links to form between the polymers. These cause the material to become brittle and can also lead to discolouration. Ultraviolet lamps and X-rays used in industrial processes can also cause this kind of degradation. Colouring pigments are often added during the polymer forming process, and this reduces the effect. The darker colours are the most effective, black being the best of all.

KEY POINT

The degradation of thermoplastics due to ultraviolet radiation can be slowed down by the introduction of colouring pigments.

Deterioration of ceramics

The ceramic tiles, bricks, cements and natural stone used for building degrade with time due to moisture and pollutant gases in the atmosphere. The absorption of rain water into the surface pores can cause deterioration in winter. When the moisture freezes, it expands and over a period of time it can cause cracking and flaking. Sulphur from flue and exhaust gases combines with moisture in the atmosphere to form sulphurous acid which falls as acid rain. This attacks many types of ceramic building material and in particular natural stone.

The refractory ceramics used to line furnaces, and the ladles for carrying molten metal can suffer from thermal shock if heated too quickly.

Because they are poor conductors of heat, there can be a very large temperature difference between the heated surface and the material beneath it. As a result, the expansion of the surface layer can cause flaking or *spalling*.

Refractory linings can also be attacked at high temperatures by the slag which rises to the surface of molten metal. There are two types of slag which form depending on the impurities present in the metal. One is acidic and the other is known as basic slag. Linings of silica brick are resistant to acidic slag whilst linings of *dolomite*, which contains calcium and magnesium carbonates, and *magnesite*, which contains magnesium oxide, are resistant to basic slag.

Test your knowledge 4.15

1. Define what is meant by a *factor of safety.*
2. How might a brittle type of fracture occur in what is normally a ductile material?
3. How does the rate of material creep vary with stress and working temperature?
4. How can fatigue failure be identified from the appearance of the fracture surfaces?
5. What are the mechanisms which lead to the formation of red rust on steel components?

Activity 4.9

You are required to design a test rig to investigate the creep characteristics of lead specimens at normal temperature when subjected to different loads. A requirement of the apparatus is that it must be able to supply an electrical output signal which is proportional to the extension to obtain graphs of creep vs. time from a *y–t* plotter.

or

You are required to design a fatigue testing machine to investigate the behaviour of mild steel test specimens when subjected to different values of cyclic stress. A requirement of the apparatus is that it must run unattended, record the number of cycles to failure and automatically switch off the power supply when failure occurs.

To check your understanding of the preceding section, you can solve Review questions 31–35 at the end of this chapter.

Review questions

1. Medium carbon steel has a carbon content of:
   (a) 0.1–0.15%
   (b) 0.15–0.3%
   (c) 0.3–0.8%
   (d) 0.8–1.4%
2. A property of grey cast iron is that:
   (a) it is strong in tension
   (b) it is easy to machine
   (c) it is ductile and malleable
   (d) it is weak in compression
3. The main constituents of brass are:
   (a) copper and tin
   (b) copper and nickel
   (c) copper and lead
   (d) copper and zinc
4. The tin–bronze used for bearings and gears contains a small amount of:
   (a) phosphorous
   (b) iron
   (c) nickel
   (d) manganese
5. Precipitation hardening is a heat treatment process which is carried out on:
   (a) medium and high carbon steels
   (b) certain aluminium alloys
   (c) aluminium–bronzes
   (d) grey cast iron
6. Polychloroethene is better known as:
   (a) perspex
   (b) polythene
   (c) PVC
   (d) polystyrene
7. At the glass transition temperature:
   (a) cross-linking occurs in thermosetting plastics
   (b) thermoplastics become liquid
   (c) thermosetting plastics become hard and brittle
   (d) thermoplastics become soft and flexible
8. The polyamide group of thermoplastics contains:
   (a) nylon
   (b) PTFE
   (c) terylene
   (d) perspex
9. Bakelite is used for:
   (a) insulation for electrical wiring
   (b) drinks cups
   (c) boat hulls
   (d) vehicle distributor caps
10. The synthetic rubber which best retains its flexibility over a wide temperature range is:
    (a) natural rubber
    (b) silicone rubber
    (c) butyl rubber
    (d) styrene rubber
11. The main constituent of glass is:
    (a) aluminium oxide
    (b) silicon oxide
    (c) beryllium oxide
    (d) calcium oxide
12. Amorphous ceramics are a group of materials which include the different types of:
    (a) cement
    (b) whiteware
    (c) abrasive
    (d) glass
13. Bonding of the crystals in clay ceramics occurs at the:
    (a) glass transition temperature
    (b) vitrification temperature
    (c) melting point
    (d) tempering temperature
14. A cermet such as tungsten carbide may be classified as a:
    (a) fibrous composite
    (b) laminate
    (c) amorphous ceramic
    (d) particulate composite
15. A piezoelectric crystal produces a potential difference across opposite faces when subjected to:
    (a) compressive force
    (b) a magnetic field
    (c) temperature change
    (d) an electric field
16. A tensile test specimen of initial diameter 10 mm which fractures at a tensile load of 31 kN will have a UTS value of:
    (a) 550 MPa
    (b) 395 MPa
    (c) 210 MPa
    (d) 430 Mpa

17. A bend test may be carried out on a material specimen to investigate its:

(a) toughness

(b) malleability

(c) elasticity

(d) ductility

18. Malleability is a property of a material which enables it to:

(a) withstand impact loading

(b) be drawn out by tensile forces

(c) be deformed by compressive forces

(d) return to its original shape when unloaded

19. The toughness of a material may be investigated by carrying out a:

(a) tensile test

(b) Brinell test

(c) Charpy test

(d) bend test

20. At the point of fracture the diameter of a tensile test specimen is 9.5 mm. If the initial diameter was 12 mm, its ductility measured as percentage reduction in area will be:

(a) 20.8%

(b) 27.6%

(c) 32.9%

(d) 37.3%

21. A bar of material is 500 mm in length at a temperature of 20°C. If its length at 150°C is 501.04 mm, the linear expansivity of the material will be:

(a) $12 \times 10^{-6}\,°C^{-1}$

(b) $13 \times 10^{-6}\,°C^{-1}$

(c) $16 \times 10^{-6}\,°C^{-1}$

(d) $19 \times 10^{-6}\,°C^{-1}$

22. The temperature difference across a tile of surface area $22.5 \times 10\,2^{3}\,m^2$ and thickness 10 mm is 100°C. If the rate of heat transfer through the plate is 200 W, the thermal conductivity of the material will be:

(a) $1.51\,W\,m^{-1}\,°C^{-1}$

(b) $0.48\,W\,m^{-1}\,°C^{-1}$

(c) $2.33\,W\,m^{-1}\,°C^{-1}$

(d) $0.89\,W\,m^{-1}\,°C^{-1}$

23. If 10 m length of resistance wire of diameter 2 mm has a resistance of 1.55 Ω, its resistivity will be:

(a) $48.7 \times 10^{-9}\,\Omega m$

(b) $36.6 \times 10^{-9}\,\Omega m$

(c) $53.6 \times 10^{-9}\,\Omega m$

(d) $21.8 \times 10^{-9}\,\Omega m$

24. The ability of a material to intensify an electric field is indicated by the value of its:

(a) permittivity

(b) conductivity

(c) permeability

(d) resistivity

25. Sacrificial protection occurs when steel is coated with:

(a) cadmium

(b) tin

(c) nickel

(d) zinc

26. Annealing is a heat treatment process carried out to:

(a) harden and toughen the material

(b) increase the carbon content of the material

(c) restore ductility and malleability after cold working

(d) induction harden the material

27. In the normalising process, steel components are cooled:

(a) by quenching in water

(b) by exposing them in still air

(c) by leaving them in the dying furnace

(d) by quenching in oil

28. Springs are tempered at a temperature of around:

(a) 220°C

(b) 240°C

(c) 260°C

(d) 300°C

29. Carburising is a process carried out on components made from:

(a) mild steel

(b) aluminium alloys

(c) cast iron

(d) high carbon steel

30. Age hardening in certain aluminium alloys results from the precipitation of:

(a) spheroidal graphite

(b) new grains or crystals

(c) stress concentrations

(d) intermetallic compounds

31. The transition temperature, below which steels become brittle, can be lowered by:

(a) the addition of magnesium and silicon

(b) increasing the phosphorous content

(c) the addition of manganese and nickel

(d) reducing the carbon content

32. Creep in materials is more likely to occur:

(a) at high working temperatures

(b) when cyclic loading is present

(c) when there is a coarse grain structure

(d) in a moist atmosphere

33. The cyclic loading of a component can result in:

(a) brittle fracture

(b) creep failure

(c) stress corrosion

(d) material fatigue

34. The chemical compound $FeOH_3$ is better known as:

(a) millscale

(b) cementite

(c) red rust

(d) silica

35. Ultraviolet radiation can cause degradation in some:

(a) cermets

(b) thermoplastics

(c) refractory ceramics

(d) ferrous alloys

The design of CNC machines is very much a team activity requiring input from mechanical, electrical and electronic engineers, IT specialists and material technologists. Marketing and financial specialists might also be involved. Alternative designs must be considered leading to a solution that is fit for purpose, safe to operate and can be produced at a reasonable cost.

Photo courtesy of iStockphoto, Shawn Gearhart, Image # 1798391

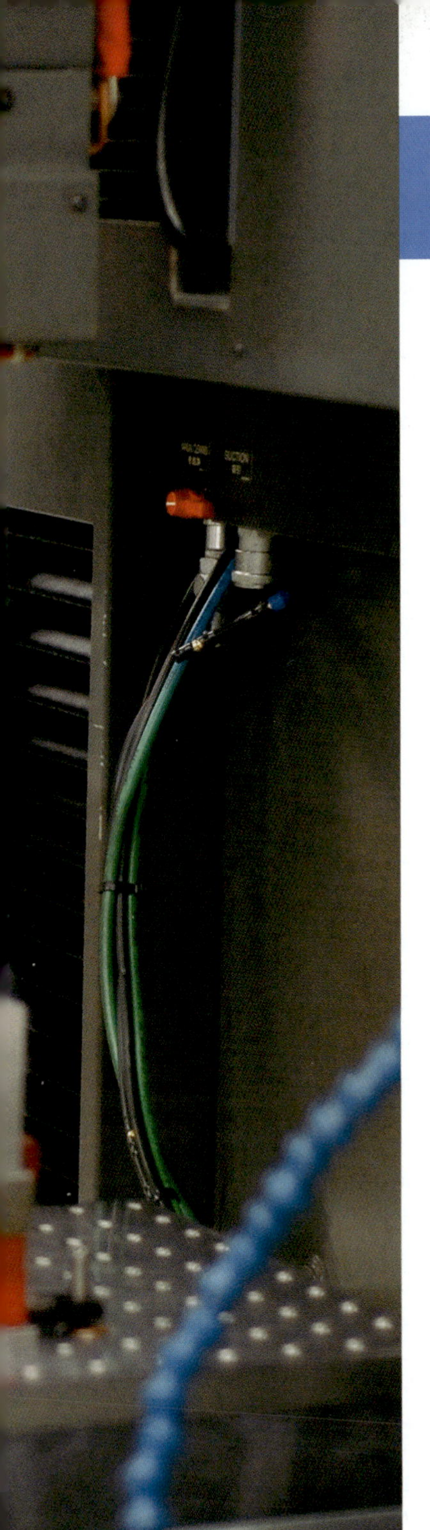

Chapter 5

Engineering Design

The aim of this chapter is to give you an understanding of how the design process operates in an engineering company. Design is a complicated activity requiring individual inputs from a whole group of people, who go to make up the *design team*. The design team must be managed efficiently bringing together expertise in marketing, customer liaison, design specification, concept design, materials technology, manufacturing methods, finance, legislation, publicity and sales. In this chapter we will be looking at the overall design process and how it is triggered. We will investigate the issues and constraints that influence product design and whether a proposal should be developed into a final solution suitable for manufacture. We will then examine the ways in which a final design solution is prepared and presented.

Knowing How the Design Process Operates When Dealing with Customers

Here you will find out about the different stages of the design process and how the needs of the customer are assessed and specified.

Marketplace pull and technology push

The design process may be triggered in different ways. One of these is known as *marketplace pull* where customers or a market research team identify the need for a new product. A product is then developed to meet that need using existing technologies. Market-driven companies need to keep a close watch on their marketplace. Their aim must be to increase their market share and maintain their profitability by developing new products and improving existing products based on customers' needs and requests. Complacency can lead to failure if their market experiences a downturn. If this happens they may have no expertise in new technologies which would enable them to move into new areas.

The design process may also be triggered by *technology push*. Technology-driven companies are research based and develop new technologies without necessarily knowing what products will flow from them. They believe that the needs of tomorrow's customers will not be met by today's technologies. This may be a high-risk strategy and companies such as these need to have big cash reserves. They also need to have a variety of projects under development as some may prove to have no commercial potential.

KEY POINT

The design process may be triggered by existing customers and market research, or by technological advance and innovation which may open up new markets.

The most successful engineering companies do not restrict themselves to one approach. In addition to satisfying the pull of the marketplace they invest in research and development. This promotes innovation, where the acquisition and development of new technologies can give rise to new product lines and marketing opportunities.

The design process

Whichever way the design process is triggered, the first task is to analyse the design parameters and draw up a *product design specification* (PDS). This document forms the essential reference point for all the activities concerned with the engineering design process. Before considering the different stages of the design process in detail, it will be useful for us to look at the overall process from the triggering of a design problem to the submission of a possible design solution. Figure 5.1 provides an overview of the process showing a possible way to reach a satisfactory conclusion.

The process may begin with a customer's brief or an idea for improving an existing product. It may also be a new concept originating from a research and development project that will fulfil an identified need. In tackling the design problem the first task is to establish the exact customer or market requirements. Problem analysis involves

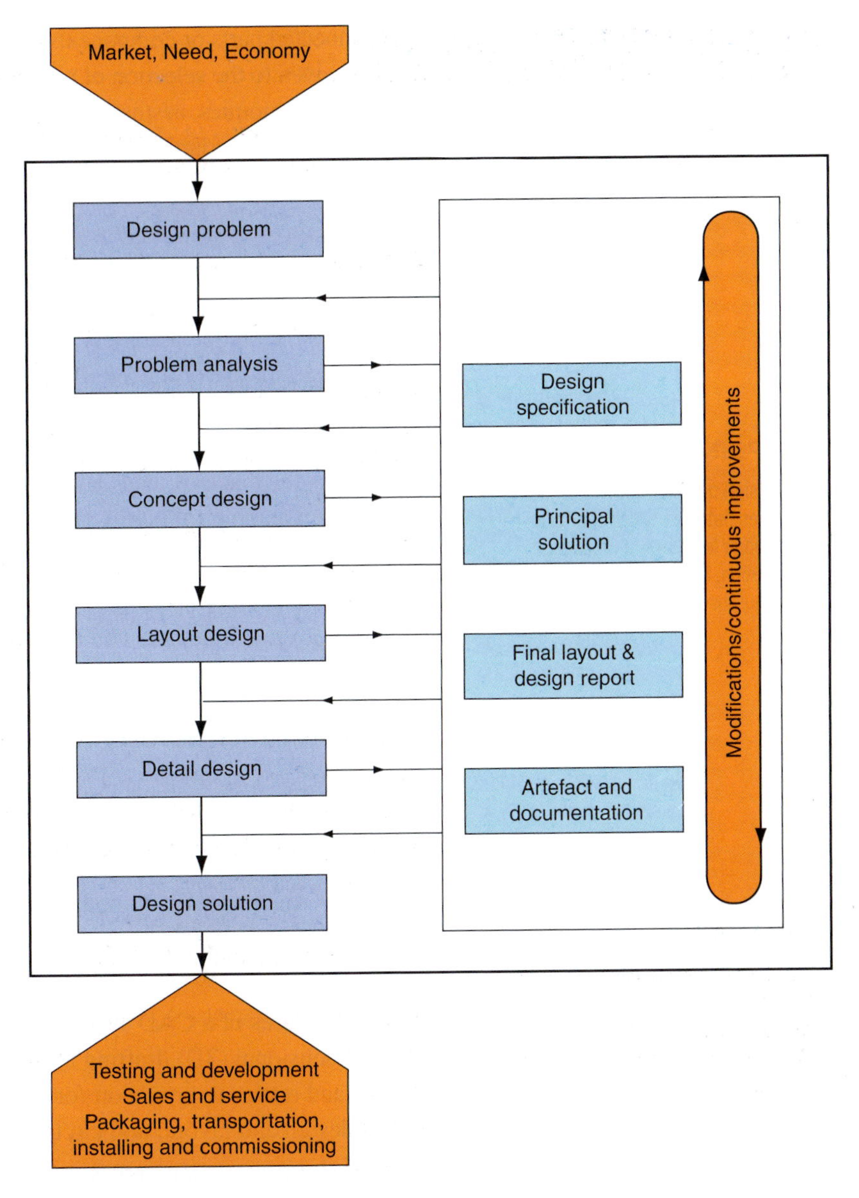

Figure 5.1 The design process.

examination of the major design parameters, researching and obtaining design information and preparing the *design specification*. This should typically contain the following information:

- Function of the product
- User requirements
- Performance
- Quality and safety issues
- Required conformance standards
- Scale of production
- Cost considerations

The PDS leads to the *concept design* stage where possible design solutions are evaluated. The principal or optimum design solution is then selected and if approved, the next stage is *layout design*. This might also be called general arrangement (GA) design of the product showing its interconnected parts or components. When producing this, consideration must be given to the selection of appropriate materials, manufacturing methods, ergonomics, aesthetics, health and safety, and conformance to regulations and standards. Care must be taken not to deviate from the PDS. The product must be fit for purpose without being 'over-engineered'.

If the final layout and design report is approved, the final stage of the process is *detail design*. Here the dimensions, tolerances, surface finish and material specifications of the product and its components are finalised. Detailed assembly drawings, component drawings and circuit diagrams are produced. Parts lists are compiled, production and purchasing costs are considered and the final design solution is presented for approval. Throughout the design process there is a continuous feedback of ideas and information between members of the design team leading to modifications and improvements to the design.

KEY POINT

The design process follows a series of stages between which there is a continuous feedback of ideas and information, leading to a solution that is fit for purpose.

The design process does not quite end with the final design solution. Prototypes generally need to be fabricated and tested before a product is approved for manufacture. The resulting test and performance data are used to refine or 'fine-tune' the design.

Computer technology in design and manufacture

The computer has become a very important and powerful tool in engineering design and manufacture. Computer-aided design (CAD) is a design process which uses sophisticated user-friendly computer graphic techniques together with computer software packages, which assist in solving the visualisation, analytical, development, economic and management problems associated with engineering design work.

The key features of a CAD system include: two-dimensional (2D) and three-dimensional (3D) drafting and modelling, parts and materials storage and retrieval, provision for engineering calculations, engineering circuit design and layout, and circuit and logic simulation and analysis.

Computers may be applied directly to the design process in a number of areas, these include: *geometric modelling*, where structured mathematics is used to describe the form or geometry of an object. The *analysis* of engineering situations where forces, motion parameters, endurance and other variables may be investigated for individual design situations. *Reviewing and evaluating* the design is made easier using computer graphics, size dimensions and tolerances may easily be checked for accuracy, and minute details can be closely scrutinised using the magnification facility of the graphics system.

Automated drafting has greatly improved the efficiency of producing hard copy drawings, which can be easily amended, as the design evolves. There is a plethora of software packages associated with the engineering design process. 2D drafting packages enable the production of single part, layout, GA, assembly, sub-assembly, installation, schematic and system drawings. 3D drafting and modelling packages, in addition to engineering drawing, enable us to visualise and model, product aesthetics, packaging, ergonomics, and the differing effects of colour change, textures and surface finish.

Design analysis packages enable us to perform calculations involving area, volume, mass, fluid flow, pressure and heat loss as well as the analysis of stress, and the modelling and analysis of static, kinematic and dynamic engineering problems. Circuit and system modelling packages enable us to simulate and modify the layout, design and operation of electronic, fluid, control and other engineering systems.

One of the most exiting developments in computer modelling has been in the introduction of rapid prototyping systems. These enable the designer to produce physical 3D models of components and assemblies which are built up layer by layer in polymer material from CAD drawing scans. The process is also known as 3D copying and can save a great deal of time when a product has reached the prototype stage. Physical models of components that are to be cast, forged or machined into intricate shapes can be quickly produced, inspected and the design fine-tuned in readiness for manufacture.

KEY POINT

Computer-aided design, rapid prototyping and computer-aided manufacturing systems may be linked together to form a fully computer-integrated manufacturing system.

In a fully *computer-integrated manufacturing system* (CIM), *computer-aided design* (CAD) is linked to *computer-aided manufacture* (CAM). Figure 5.2 shows the link and lines of communication between CAD and CAM systems. CAM systems cover activities which convert the design of a product into instructions that define how the components will be produced. CAM is any automated manufacturing process which is controlled by computer. These include CNC lathes, mills, flame cutters, process systems and robotic cells, to name but a few. In addition to NC programming, CAM systems also incorporate production planning and tool and fixture design.

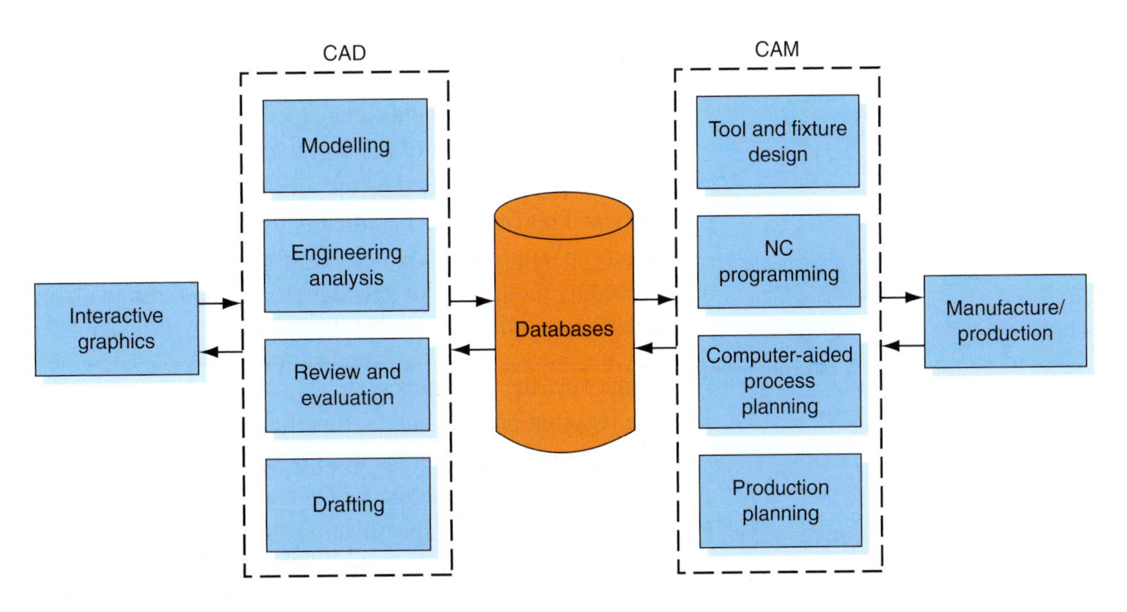

Figure 5.2 CAD/CAM system

Customer/client relationship

A customer or client is a person or organisation who receives goods or services from another individual or organisation. The word *customer* is derived from the custom or habit of someone to frequent a particular trader, rather than making a purchase elsewhere. The trader would need to maintain a good relationship with the customer to keep his or her custom, and such is the situation today.

There are two main groups of customers. *External customers* are individuals or organisations that receive products or services in return for money. An engineering company might have customers who come and place orders for its products. Alternatively it may tender to provide a product or service to a customer in competition with other firms. A company that buys-in raw materials or components for processing or assembly will itself be an external customer of its suppliers.

Internal customers are the individuals, departments or branches within an engineering company that receive a product or service from another department or branch. Each will have a budget provided for the purpose and payment will generally be made by internal computer transaction.

KEY POINT

External customers pay to receive products and services. Internal customers receive products and services from other departments or branches within the same organisation for which they may pay by computer transaction from their allocated budget.

A *relationship* may be defined as the interaction between people or organisations, how they view each other and the effect that this has on the way they do business. Good relationships are based on trust and respect. They are essential for a company to maintain or increase its market share. A company may advertise for new customers but it is often less expensive to achieve repeat business than to obtain new business. By looking after its customers a company will enhance its reputation which in turn can lead to referrals and the arrival of new customers.

Customer requirements

An engineering company must provide products and services that meet the customers' requirements and expectations. A design brief usually begins with a statement of the task that is to be achieved. The designers may also be given detailed technical specifications and the expected quality and performance parameters by the customer. This is known as a 'bottom-up' approach in which the solution is tightly defined. It may be achieved by modifying an existing product or through an arrangement of standard components.

Alternatively the designers may be provided only with general details of the required product and its functions. This is known as a 'top-down' approach in which the design team must analyse the functional needs of the customer, agree a *PDS* and liase with the customer as the design progresses. The specification should include the performance parameters, agreed physical dimensions and weight.

KEY POINT

With the bottom-up approach, designers are given detailed technical specifications and performance requirements by customers. With the top-down approach designers must liase with customers to analyse their needs and formulate an agreed product design specification.

Other factors which must be taken into account with 'top-down' design include ergonomics, aesthetics, compliance to operating standards and health and safety legislation. The design will also be affected by the quantity that the customer requires, the rate at which they are to be supplied and cost limitations. This can range from custom building, where a small quantity is produced by skilled labour, to large volume repeated batch or continuous production on automated machinery.

The customer should be assured of after sales service and wherever possible the design should enable safe disposal or re-cycling of the product at the end of its life cycle.

Product design specification

As an introduction to specification writing, consider the following modified paragraphs taken from BS 7373 *Guide to the preparation of specifications*.

A specification is essentially a means of communicating the needs or intentions of one party to another. It may be a users description, to a designer (a brief), detailing requirements for purpose or duty; or it may be a designer's detailed description to the operator; indicating manufacturing detail, materials, and manufacturing tolerances; or it may be a statement, by a sales person, describing fitness for purpose to fulfil the need of a user or possible user. It may, of course, be some or all of these in one.

The contents of the specification will, therefore, vary according to whether it is primarily from the using, designing, manufacturing or selling aspect. Specifications will also vary according to the type of material, or component being considered, ranging from a brief specification for a simple component to a comprehensive specification for a complex assembly or engineering system. We are primarily concerned here, in preparing a design specification from the point of view of the designer attempting to meet customer needs.

The requirements of the specification should be written in terms of describing the optimum quality for the job, not necessarily the highest quality. It is usually unwise to over-specify requirements beyond those for a known purpose. It is costly and restrictive to seek more refinements than those necessary for the function required. The aim should therefore be to produce a minimum statement of optimum quality in order not to increase costs unnecessarily; not to restrict processes of manufacture; and not to limit the use of possible materials.

Note that we are primarily concerned here with the PDS, i.e., the specification which conveys the designers' description of the product. As already mentioned the design specification has to take into account parameters such as function, performance, cost, aesthetics and production problems. All of these issues are concerned with *customer requirements* and this has to be the major consideration when producing a design specification. The major *design parameters* (such as layout, materials, erection methods, transportation, safety, manufacture, fabrication and legal implications) must also be considered when producing the design specification. Finally, all *design information* must be extracted from appropriate sources, such as British Standards (BS) and International Standards, and all legislative requirements concerning processes, quality assurance, and the use of new technologies must be applied.

Test your knowledge 5.1

1. What are the factors that trigger the design process?
2. What are the 'top-down' and 'bottom-up' approaches to design?
3. What is the difference between internal and external customers?
4. What is a product design specification (PDS)?
5. What constitutes a fully integrated manufacturing system (CIM)?
6. What is rapid prototyping?

Understanding the Impact of Legislation, Standards and Environmental and Manufacturing Constraints on the Design Function

Here you will find out about the some of the legislation and standards that affect the design process together with environmental, sustainable and manufacturing constraints.

Legislation and standards

The British Standards Institution (BSI) and the International Standards Organisation (ISO) specify agreed types of material, components, design codes and procedures to be used in engineering design. Examples are the BS 970 for types of steel, ISO metric screwed fastenings to BS 3692 and design management to BS 7000. The standards are agreed and periodically updated by specialists from industry, the universities and research and governmental organisations. Often the design of a product depends on the particular standards of the country in which it is to be used. Exports to America must conform to US Federal Standards which are very often enforced by federal laws. A similar situation exists with exports to other countries, which may have adopted ISO standards but also have their own national standards for particular products.

Countries in all parts of the world pass legislation to ensure that products are safe for their purpose and the environment and designers must work within this legislative framework. This also covers the expectations of the consumer and their rights with regard to product quality and length of service. A typical example is the European Electromagnetic Compatibility Directive. This states that electrical and electronic products sold in Europe must:

- Be so constructed that they do not cause excessive electromagnetic interference and are not duly affected by electromagnetic interference.
- Carry appropriate CE marking.

If a product is sold to someone, made to be used in-house or even given away, it must comply with the Essential Protection Requirements as laid down in the Directive. Ignorance of the legislation is no excuse and a punitive penalty structure is documented.

The CE mark (Conformité Européenne) is a mandatory European marking for certain product groups to indicate conformity with the essential health

KEY POINT

Designated electrical products for sale within the European Union must conform to the Electromagnetic Compatibility Directive and carry the CE mark.

and safety requirements set out in European Directives. It does not ensure quality, as a product might be perfectly safe to use but not be sufficiently robust for its service conditions and have only a short working life. Without the CE marking, however, the product may not be placed in the market or put into service in the member states of the European Union.

Activity 5.1

Making use of the Internet, find out what the CE mark looks like. Also, find out the scope and application of the Machinery Safety Directive (MSD).

Within the United Kingdom all plant, machinery and equipment that have been designed for use in factories, workshops, warehouses, offices, shops, restaurants and leisure facilities must conform to the requirements of the Health and Safety at Work Act. Although there must always be some degree of risk in the working environment, products must be designed, manufactured, installed and maintained so as to keep this to an absolute minimum. Furthermore, inspection and testing must be carried out in compliance with the legislation and a product must be supplied with clear instructions as to its installation, use and maintenance.

The fundamentals of the law regarding patents and copyright should be understood by the engineering designer. A patent, for example, does not stop anybody using your idea, it merely provides a channel for redress. The incentive gained by patenting a product normally results in the item being made and marketed, knowing that legal protection is offered. New industrial designs are protected under the provisions of the Copyright Act (1968) in the United Kingdom. This protection is offered without any form of registration and is valid for 15 years from the date of manufacture (in quantity) or when first marketed.

Some knowledge of the law of contract is also very useful for the engineering designer. A contract is a formal written agreement between two parties. Both parties agree to abide by the conditions that are laid down, in all respects. However, certain 'let-out' clauses may be included to accommodate unforeseen difficulties.

Activity 5.2

Find out how to obtain a patent and the protection that it might give to a product that you have designed.

Energy efficiency

The cost of energy, the sustainability of resources and the effect of energy use on the environment are constraints that designers of engineered products must take into account. This applies not only to the function of energy consuming products, but also to the materials used, the methods of manufacture and end of life disposal. The nineteenth century was the age of steam but steam locomotives, marine engines and stationary steam engines were notoriously inefficient. The overall thermal efficiency of these devices, measured as the percentage of useful work output to the energy supplied in the fuel, was often in single figures and seldom more than 20%.

The twentieth century might be regarded as the age of the internal combustion engine and gas turbine, which remain as the mainstay of road transport, aviation and marine propulsion. Although more efficient than steam engine, they still waste more than half of the energy available in the fuel. The same is true of the power stations that supply our electricity using coal, oil, natural gas and nuclear power. The wasted energy is given off to the atmosphere as low-grade heat in the exhaust gases from combustion and the latent heat from the condensation of steam. This in itself does not pose a direct hazard to the environment but unfortunately it is accompanied by CO_2 emissions which contribute to global warming, or nuclear waste which will need to be stored for long periods of time.

It is the aim of the government to produce 15% of our energy needs from renewable resources such as wind, wave and hydro-electric power. Of these, hydro-electric power is by far the most efficient but it is a scarce natural resource. Wave and tidal energy is plentiful but the technology to harness it has yet to be developed. Wind turbines are making an increasing contribution to our needs but they are often considered to be unsightly and are seldom able to operate at more than 30% of capacity.

KEY POINT

The onus is on designers to increase the efficiency of existing products and design new products with energy efficiency in mind.

No doubt this century will see advances in energy conservation and energy conversion technology. Fuel cells are being developed that might supersede the internal combustion engine in road transport and much work is being done in the field of energy management to increase the efficiency of industrial processes. In the long term there may be a breakthrough in the development of nuclear fusion reactors for electrical power generation.

Activity 5.3

Find out and compare the typical overall efficiencies of a diesel generating set, a hydro-electric power station and a nuclear power station. Compare also their effects on the environment.

Environmental and sustainable constraints

It has been stated that engineered products should be designed to be energy efficient and environmentally friendly in their operation. This applies to the materials used, the methods of manufacture and end of life disposal. The Environmental Protection Act 1990 introduced new controls that directly affect the manufacture of engineered products. Its aim is to provide Integrated Pollution Control (IPC) in a wide range of industries.

The worst polluters come under Part A of the Act, and include large chemical plant, power stations oil refineries, waste incinerators, etc. and are regulated by a governmental body called the Environment Agency. Smaller polluting industries such as small foundries, metal and plastics processing companies, electrical and electronic component makers, etc. come under Part B of the Act and are subject to local authority control for their waste products and emissions.

Engineering designers must take considerable care when designing plant and equipment that is to be used in these industries so that any resulting emissions to the atmosphere, effluent and noise lie within the prescribed limits. Products should be designed so that waste materials such as metal swarf, wood and plastic off-cuts, packaging, etc. is kept to a minimum. Any waste that is produced should be re-used or re-cycled into another product wherever possible. As a last resort waste materials that cannot be re-processed should be disposed of safely and in an approved manner so as not to create an environmental hazard.

In particular, the disposal of electrical and electronic products that have come to the end of their life cycle is controlled by the Waste Electronic and Electrical Equipment Directive (WEEE). Waste electronic and electrical equipment is one of the fastest growing waste categories in both the UK and EU countries. One has only to think of the number of domestic appliances, TV sets, computers, mobile phones, etc. that are discarded. Some contain lead in solder, cadmium in batteries, mercury in switches and chlorofluorocarbons in refrigerants and plastic materials. All of these must be disposed of safely.

KEY POINT

Ease of disassembly to enable the recovery of materials and the refurbishment and re-cycling of components is becoming a key requirement of engineering design.

The directive sets out to improve the way in which collection, treatment, re-cycling and recovery of WEEE is managed. It is one of a series of 'producer responsibility' directives that makes the manufacturers responsible for paying for the treatment and re-cycling of a product at the end of its useful life. This obviously places constraints of the design of electrical and electronic equipment, particularly when selecting materials and components.

Availability of labour

A major manufacturing constraint is the availability of labour. The engineering designer must take into account the people who will make the product, install it and maintain it in good working order. In third world countries there is often an abundance of unskilled labour but skilled technicians might be few in number. A product for manufacture in such a country will require a simple design which can be manufactured and assembled using basic hand and machine tools.

In technologically advanced countries, unskilled labour may be in short supply and expensive. Skilled technicians might also be expensive but available. Here the best option may be to design for manufacture on automated plant and equipment using up-to-date computer control and robotic technology.

Design solutions may have to take into account not only the availability and cost of the personnel required for manufacture, installation and maintenance, but also their training needs. This is particularly true for automated plant, aircraft and weapon systems. Training programmes, software simulation packages and service manuals often need to be designed in parallel with the product so that manufacture can proceed smoothly and the customer is equipped to make full use of the product.

Availability of material

The designers of engineered products must take account of the form, function, cost, aesthetics, safety and environmental impact when selecting materials. All of these aspects must be considered together with the ability to be processed. The form and function of a component often dictates the choice of material. Components of an intricate shape may require a material that can be die cast or injection moulded to shape. The choice will then be between a die casting alloy and an injection moulded polymer material. Service conditions, cost and appearance will then need to be considered together with the availability of the material and processing plant.

Cost is very often the prime constraint. Mild steel is perhaps the most commonly used engineering material. It is relatively cheap available of a variety of forms and can be readily forged, welded, machined or pressed into shape. Unfortunately it is liable to corrosion and needs to be adequately protected when used outdoors or in a moist atmosphere. It is widely used for motor vehicle bodies even though it is not the ideal material. Aluminium alloys and reinforced polymer materials might afford corrosion protection and are sometimes used on specialist products. They are however more expensive and do not lend themselves so well to high-volume production.

In some cases the cost of a material is of secondary importance to its performance characteristics. This is particularly true in the electronics and aerospace industries where rare materials are used, and new materials are developed for specific purposes irrespective of cost. Examples are the gold connectors used with silicon chips and the heat-resistant ceramic tiles developed for use on the space shuttle. Sometimes the choice of material is not so obvious in terms of cost, availability and ease of processing. Alternatives might fulfil the material requirements equally well. As an example, glass, plastics and metal are all used as containers for food and drinks. In such cases it is often customer preference and end of life disposal that influence the choice of material.

A material selected as being the best suited in terms of function and cost may not be readily available or supplies may be liable to interruption. This is more likely to happen if there is only one supplier. Delays in developing new materials, industrial disputes, plant breakdown, natural disasters, wars, etc. can all affect material supplies. In such cases the designer may suddenly be called upon to find an alternative material. Where there is dependency on a single source of supply it is good policy plan ahead for such contingencies. It is a well-known fact that if something can go wrong, it will go wrong.

KEY POINT

Designers must take into account function, cost, appearance, reliability of supply, safety and environmental impact when selecting materials for engineered products.

Influence of material properties

After primary processing, engineering materials are often supplied to customers in standard forms such as barstock, sheet and plate. They then undergo secondary processing such as machining, stamping and pressing to convert them into engineered components. These may then be assembled into products or sold on for use by other manufacturers. With other primary forming processes such as forging, casting and moulding, engineering materials are supplied finish-formed to shape or rough-formed prior to secondary forming by machining. At some part in the processing cycle a component may undergo heat treatment to modify its properties. If it is a part of a fabrication, it will be assembled to its mating components and may also be given a protective finish by painting it or coating it with a metal or polymer material. The manufacturing processes may be grouped into:

- Liquid forming processes
- Solid forming processes
- Machining processes
- Heat treatment processes
- Joining and assembly processes
- Finishing processes

Liquid forming processes include the sand casting and die casting of metals and the injection moulding of polymer materials. The choice of process is affected by the melting point, expansivity and fluidity of a material. High melting point metals such as cast iron are generally cast in sand moulds or ceramic moulds made by the investment or 'lost wax' technique. Die casting is generally used with lower melting point metals such as aluminium-, magnesium- and zinc-based alloys.

Some molten materials are more fluid than others are, and this affects the complexity of the components that can be produced. Steel in particular does not have good molten fluidity and can only be cast into relatively simple sections. Cast iron, certain brasses and bronzes and aluminium alloys have better fluidity and can be cast into more complex shapes. Metals generally expand when heated and contract when cooling. This must be taken into account when designing concave or hollow components for die casting as the metal might grip the die as it cools, making it difficult to eject or causing it to crack.

Melting point, fluidity and expansivity must also be considered when designing components that are to be moulded in polymer materials. These products are formed to shape by injection moulding, extrusion, blow moulding and vacuum forming, all of which involve the use of heat and pressure. It is possible to pour some thermosetting resin and hardener combinations into simple moulds but like steels, polymer materials are generally too viscous for direct casting into complex shapes.

Solid forming processes include forging, rolling, pressing, drawing and extrusion. Materials formed in this way need to be ductile and malleable. Some materials such as copper and to a lesser extent steel may be formed in the cold state and are said to have been 'cold worked'. The cold working of metals distorts the crystal lattice structure resulting in a reduced ductility and malleability and increased hardness. The phenomenon is known as 'work hardness', which may need to be removed by heat treatment before further cold working or material removal by machining.

Most metals become more malleable and ductile at elevated temperatures. This is especially true of steel which is hot formed into a variety of products by the above processes. At room temperatures, the force of attraction between the polymers of plastic material is generally too great to permit cold forming. Although they may be flexible, plastic materials lack the ductility and malleability that is required. With thermoplastic materials, however, heating loosens the interconnecting bonds and hot forming can be carried out in the semi-solid state by vacuum forming and blow moulding.

Engineering materials have different degrees of machinability. Cast irons, plain carbon steels, brasses, bronzes and aluminium alloys can be turned, drilled and milled to shape to good effect provided that the correct speeds and feeds are used and appropriate cutting lubricants employed. More difficulty is encountered with harder materials such as alloy steels and titanium alloys, requiring the use of specialist cutting tool materials. With very hard materials such as steels that have been hardened by heat treatment, grinding is often the only material removal option. Designers must be aware of these constraints together with the range of tolerance and degree of surface finish that can be obtained from the different machining processes.

Designers should be familiar with the various heat treatment processes that are used to change, modify or fix the properties of engineering materials. Metals that have been cold worked often need to have their ductility and malleability restored by the annealing process. This involves heating them to their re-crystallisation temperature at which undeformed grains grow from the points of maximum internal stress. Metals that have been cast to shape or hot worked by forging, etc. often cool unevenly. As a result, they contain internal stresses which can cause the material to deform when machined. These can be removed by the normalising process which is similar to annealing except that rate of cooling may be different. Glass products such as those for windows and windscreens are sometimes heat treated to remove internal stresses and can also be toughened by tempering.

KEY POINT

The properties of a material may affect the choice of shaping and forming process. Conversely the method of processing can determine the final properties of a material.

Designers need to know about the range of alloys that can be hardened by heat treatment. Medium and high carbon steels can be hardened and toughened by the hardening and tempering processes and certain aluminium alloys can have their hardness adjusted by a process known as precipitation hardening. The processes involve heating to a pre-determined temperature and cooling at a prescribed rate. Polymer materials cannot be heat treated in the same way as metals but certain thermosetting plastics are heated to 'cure' them. This involves heating to a prescribed temperature to promote the formation of cross-linking bonds to control the final hardness and toughness of the material.

The component parts of an engineered product may be assembled or joined together using screwed fastenings, rivets, adhesives, soft and hard soldering and welding. Screwed fastenings are used where component needs to be replaced or removed to give access for maintenance. The other methods are used where the joining is intended to be permanent or at least semi-permanent. The designer needs to take care where dissimilar metals are brought into contact as this can lead to electrolytic corrosion in the presence of moisture.

In the case of soft and hard soldering, the designer needs to be familiar with the range of metals that will accept tin–lead alloy, silver solder and brass as the joining medium. In the case of welding, the designer needs to be aware of the range of processes that have been developed to overcome the difficulty of fusing certain metals together. Fusion welding is generally associated with metals but some thermoplastics may also be welded using a hot air jet to supply the heat energy. The use of adhesives in engineered products has increased in recent years and designers should know about the different types, the way they bond to different materials and their suitability for different service conditions.

Engineering components and products are very often given a protective coating to guard against abrasion, corrosion and chemical attack. The finish may also be given for aesthetic reasons. Various types of paint and varnish have been developed for different conditions and different materials. The designer might also specify a metallic coating such as chromium plating, cadmium plating or galvanising. Alternatively it might be considered best to coat with a polymer material. Here again, the designer must know about the range of protective finishes and the materials and service conditions for which they are appropriate.

Activity 5.4

Find out how the strength, hardness, toughness and degree of machined surface finish of a material are measured and shown on design specifications.

Availability of plant and equipment

Engineered components generally pass through the following stages in the manufacture of a product manufacture:

- Material and equipment preparation
- Shaping to form
- Finishing
- Assembly
- Inspection and testing

A decision has to be made at an early stage of the design as to whether a component should be made in-house or bought-out completed. If a component is to be made in-house the designer must be familiar with

KEY POINT

Production planning and the design of jigs and tools for quantity production often take longer than the product design process. Sufficient lead time should be given for these processes when agreeing delivery dates.

the available equipment or seek the advice of the production engineer. Plant and machinery have what is known as 'process capability'. This is the tolerance range of which the equipment is capable and which varies with the age and condition. It is no use the designer specifying tight tolerances and high degrees of surface finish if they cannot be reliably achieved.

The designer needs to take account of the quantities that will be produced. Prototype and small quantity components may be produced on a CNC machining centre capable of intricate turning, boring and milling operations. If it is foreseen that large volume or continuous production will be required, the design must also be capable of manufacture on automated machines and tooling which will be dedicated to the product.

Plant capacity is another factor that must be taken into account at the design stage. If the available plant is fully loaded it may be considered more economic to buy finished components rather than purchase new equipment. Then the designer might need to modify the design to suit the supplier's process capability.

Cost-effective manufacture

As has been stated, designers and production engineers often have to decide whether a component or product should be made in-house or bought-out completed. The first consideration is the quantity required and the rate at which they will be needed. An assessment is made as to the cost of jigs, tools and possibly the purchase of new plant and machinery that will be required for in-house manufacture. To this is added the cost of material, labour and the indirect costs of production so that an estimate can be made of the likely unit cost. This can then be compared to quotations of the bought-out cost received from outside firms.

If in-house manufacture is the chosen option, the quantity requirements will dictate whether this will be a jobbing order (very small quantity), small intermittent batch production, regular repeated batch production or continuous production. In the case of batch production the most economic batch quantity must be decided. This is done by comparing the equipment set-up cost per unit, which becomes less with the number produced, to the storage or carrying cost per unit, which increases with the number produced. The following graph indicates the most economic batch size (Figure 5.3).

KEY POINT

The most economic batch size is that at which the set-up cost per unit equals the carrying cost per unit.

Health and safety

As has been stated, designers will need to be aware of the legal constraints involved with any particular product or system. With the advent of more and more EEC legislation concerning product liability, disposal of toxic waste, control of hazardous substances and general health and safety, all legal aspects must be considered during the early stages of the design process.

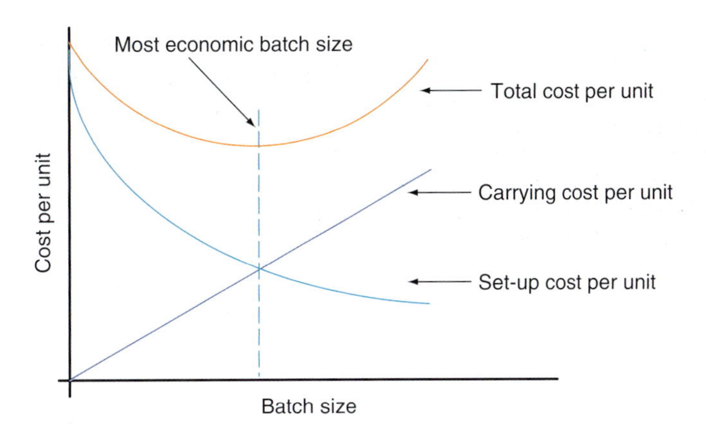

Figure 5.3 Economic batch size

Safety requirements vary according to the product or system being designed and the use to which they will be put. Many areas of engineering exist within highly regulated industries, where safety is of paramount importance, e.g., nuclear power, petro-chemical, aircraft operation, hospitals and the emergency services. For these industries and many others, adherence to the Health and Safety at Work Act (1974) and Control of Substances Hazardous to Health (COSHH) regulations and British Standards for product liability will form an essential part of the design process.

Test your knowledge 5.2

1. What does the CE mark on an electrical or electronic product indicate?
2. For what period of time does the Copyright Act (1968) give protection to industrial designs?
3. What constraints does the Waste Electrical and Electronic Equipment Directive (WEEE) place on engineering designers?
4. Who is responsible for the enforcement of Parts A and B of the Environmental Protection Act (1990)?
5. What are COSHH regulations?
6. What are the cost factors used to determine the most economic batch size for production?

Be Able to Prepare Design Proposals That Meet the Requirements of a PDS

Requirements of a PDS

To provide a successful winning design, it is essential that the requirements of the customer are met whenever possible. Because we wish to produce a specification that expresses our customer requirements it is most important that we spend time in consultation with them, to ensure that customer needs are well understood and, if necessary, to agree amendments or reach a compromise dependent on circumstances.

Remember that a specification is essentially a listing of all the parameters essential to the design. Generally a customer will list

KEY POINT

A product design specification is essentially a listing of all the parameters essential to the design. Each of the customer's requirements must be carefully examined and added to the specification only if they are achievable.

essential values as part of his/her requirements but each one must be examined before transfer to the specification. The design specification should always be formulated in this way by the designer. To illustrate the customer requirements for a particular engineering design, and the role of the designer in interpreting such requirements, consider the following example.

Example 5.1

A potential customer approaches you, as a designer, with a *brief* for the design of an electric drill. To secure the job, you need to produce a comprehensive specification, which takes into account all of your customer's requirements which are listed below:

A.B. Brown Engineering

Outline specification for electric drill

Performance:	Capable of taking drill bits up to 0.75 in. diameter. Operate from 240 V, 50 Hz power supply. Capable of two speed operation. Have hammer action. Operate continuously for long periods of time. Suitable for soft and hard drilling. Have a minimum cable reach of 5 m.
Environment:	Able to operate indoors and outdoors, within a temperature range of −20°C to +40°C. Have no adverse effects from dirt, dust, or ingress of oil or grease. Capable of operation in wet conditions. Capable of operation where combustible dusts are present.
Maintenance:	Capable of being dismantled into component parts, for ease of maintenance. Require no special tools, for dismantling and assembly operations. Component parts to last a minimum of 2 years, before requiring replacement or rectification.
Cost:	To cost a maximum of £60.00.
Quantity:	2000 required from first production run.
Aesthetics/ergonomics:	Polymer body shell with two-colour finish. Pistol grip lower body and upper body steady handle. Metal chuck assembly, with chrome finish.
Size/weight:	Maximum weight of 3 kg. Overall length not to exceed 30 cm.
Safety:	Complies with all relevant BS.

As designers we must ensure that the customer's requirements are unambiguous, complete and attainable. This is where the dialogue with the customer begins! For the purpose of this example, let us re-visit each of the requirements.

CHAPTER 5

The title for the specification needs clarification: Is *electric drill* a suitable title? We know that A.B. Brown Engineering have specified information about type of grip and body design, so perhaps a more accurate description might be: *electric hand drill (mains operated)*, since cable length and supply conditions have been mentioned.

Performance

If we consider the *performance* requirements, we note that in certain areas they are ambiguous and generally incomplete. Is the electric hand drill to be made available for export? The metric equivalent of 0.75 in. may be necessary, in any event it is a good marketing ploy to ensure that all dimensions are available in Imperial and SI units, this appeals to both the European and US markets. Thus to ensure ease of production of the 'chuck' a metric equivalent should be given.

Are the supply details and power requirements for the drill sufficient? There is, for example, no indication of the power requirements for the drill motor, this must be given together with details of the supply, i.e., alternating current. The rpm of the two-speed operation must also be given; this will depend on service loading, time in use and materials to be drilled.

Statements such as 'operate continuously for long periods of time' and 'suitable for soft and hard drilling' should be avoided. What periods of time? What types of materials are required to be drilled?

Environment

The criteria for the operating environment are quite clear, however, they do have quite serious consequences for the design. To be able to operate the drill in quite harsh external conditions, the insulation for the plug and cable assembly will need to meet stringent standards. Motor insulation and protection will be required to ensure that sparking and arcing does not occur when the drill is being used in a combustible gas atmosphere (refer to BS 4999, BS 5000, BS 5501 and BS 6467). Suitable motor caging will be required to ensure that the ingress of dirt dust, oil and grease do not adversely affect the performance of the drill motor.

Maintenance

The number and nature of component parts need to be established, prior to any detail design being carried out. There are obvious cost implications if all component parts are to last a minimum of 2 years. This needs clarification to ascertain whether we are only talking about major mechanical parts or all of the parts. The likelihood of failure dependent on service use also needs to be carefully established to determine the feasibility of this requirement.

Costs

The viability of this figure needs to be determined by taking into account the costs of component parts, tooling requirements, fabrication costs, machining costs based on required tolerances and materials finishing, and environmental protection costs.

Quantity

This will determine the type of manufacturing process, whether or not it is necessary to lay-on additional tooling, or buy-in standard parts, and concentrate only on assembly and test facilities for the production run. Future component numbers will need to be established to make predictions about the most cost-effective production process.

Aesthetics/ergonomics

Is the two-colour finish absolutely necessary? This will depend on target market and results of consumer research, which will need to be known by the designer prior to determining unit cost. Chrome finishing is an expensive process and not altogether suitable for a drill chuck which will be subject to harsh treatment in a hostile environment. Knocks, dents scratches and pollution in the work environment would quickly affect the chrome plating. Consideration needs to be given to alternative materials, which provide good protection and durability, as well as looking aesthetically pleasing.

Size and weight

The weight and size criteria are not overly restrictive and allow the designer some room for manoeuvre. Light alloys and polymers may be used to help keep the weight down, provided the performance criteria are not compromised.

Safety

There is a need to establish whether or not the product is intended for the European and/or World market. European legislation already has a major influence on safety standards, particularly relating to electrical goods. Hence, ISO standards, European legislation and other relevant quality standards may need to be followed. Is it, for instance, the intention of A.B. Brown Engineering to provide an electric hand drill which is capable of operation from the continental 220 V supply?

Since the design requirements, their associated design parameters and the requirements of the customer are inextricably linked, we need to obtain answers to the questions posed. We then need to look in more detail at the design requirements and parameters necessary to produce a comprehensive PDS. For our example there is a need to include more design requirements on the material specifications, component function, testing, prototype production and timescales. At this point it is worth remembering that it may not always be necessary to include *all* requirements since these will be dependent on the complexity (or otherwise) of the engineering components or system.

Activity 5.5

Based on the considerations provided in Example 5.1, rewrite the A.B. Brown's requirements, avoiding ambiguous information and adding design requirements for material specification and component function.

Preparing a PDS

After due consideration, the designer's description of the product is presented in the form of a PDS. A brief summary of what might be included is given below. This list is neither exclusive nor exhaustive, but it does act as a useful guide for those new to PDS writing:

- *Title*: This should provide an informative unambiguous description of the product.

- *Contents*: A list of contents, which acts as a useful introduction to the document for the reader.
- *Foreword*: This sets the scene and provides the reader with useful background information concerned with the project and the customer's brief.
- *Scope of specification*: This section provides the reader with details on the extent of the coverage and the limitations imposed on the information provided. It also gives details on the function of the product or system under consideration.
- *Consultation*: Information on any authorities who must be consulted concerning the product's design and use. These might include the health and safety executive (HSE), fire service, patent office or other interested parties.
- *Design*: Main body of text detailing the design requirements/parameters for the product or system, such as performance, ergonomics, materials, manufacture, maintenance, safety, packaging, transportation, etc.
- *Appendices*: Containing definitions that might include: complex terminology, abbreviations, symbols and units. Related information and references, such as statutory regulations, BS, ISO standards, design journals, codes of practice, etc.

Figure 5.4 shows a typical layout for the 'design' element of a PDS, detailing the customers' requirements and the parameters that might affect these requirements.

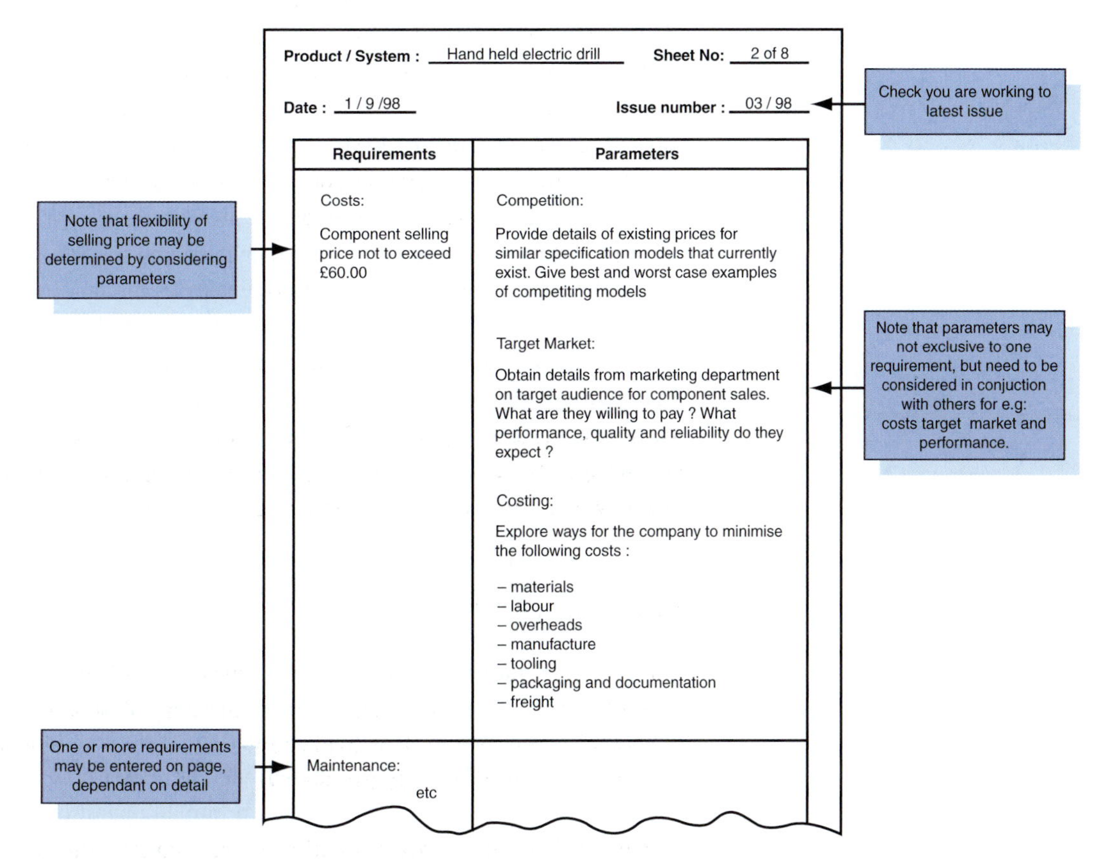

Figure 5.4 PDS layout

Information on some of the major design requirements, in the form of a list, are detailed below. You should ensure that you understand the significance of each of these and, in particular, you are aware of the important parameters associated with each.

Performance

You will already be aware of some typical performance parameters, when studying Example 5.1. The performance specified by customers must be attainable and clearly defined. The performance required by A.B. Brown Engineering in our electric hand drill example shows how easy it is to produce ambiguous performance criteria and leave out essential information.

A balance needs to be established between performance and costs, ultimate performance demanded by a customer is likely to be prohibitively expensive. Designers must be aware of the economic viability of meeting performance requirements and, where these are not feasible, dialogue must be entered into to seek a more cost-effective solution.

Over-specification of performance is more likely to occur in specialist 'one-off' products where operational information is limited. Designers should draw on the knowledge and experience gained from similar designs, or seek the advice of more experienced colleagues. In any case, the urge to over-design should be avoided; computer simulation, scale models or the use of prototypes may be a way of establishing appropriate performance data, for very large one-off products, when trying to avoid over-specification.

Ergonomics (or human factors)

The word ergonomics originates from the Greek, *ergos* – work and *nomics* – natural laws, thus ergonomics literally means *'the natural laws of work'*. It first came into prevalence during the Second World War, when aircraft pilots confused, for example, the landing gear lever with the flap lever, which on occasions, resulted in disaster. Prior to the Second World War, little interest was paid by engineers to ergonomic design. We only have to look at the arrangement of controls and displays used in old steam locomotives and heavy process equipment to realise that little consideration had been given to the needs of the operators.

Ergonomic design has risen in prominence over the past decade and the needs of the user have taken on much more importance. Consideration should now be given to one or more of the following ergonomic design parameters: controls and displays; instruments and tools; workspace arrangement; safety aspects; anthropometrics (measurement of physical characteristics of humans, in particular human dimensions); environment–visual, acoustic and thermal. No matter how complex, sophisticated or ingenious the product, if human operation is involved, the ergonomics of the situation need careful consideration by the designer.

Environment

All aspects of the product or system's environment need to be considered. In our electric drill example, the operating environment was to include dirt, dust, oil and grease, as well as combustible dust. This has a significant effect on the robustness of the design and the environmental protection required for safe and efficient use. As a result, factors such as dirt, dust, oil, grease, chemical spillage, temperature, pressure, humidity, corrosives and other pollutants, animal infestation, vibration, and noise should be considered.

Apart from the effects on the designed product or system, we as engineers should also take into account the likelihood of our design polluting the natural environment. For example, when considering the design of nuclear reactors, the vast majority of the design effort is focused on ensuring that fail-safe systems and back-up facilities exist. This is to minimise the possibility of a nuclear accident, which might result in an ecological disaster. Filtration systems to prevent the leakage of dangerous substances from plant, machinery and vehicles should be introduced into the design as a matter of routine.

Maintenance

When purchasing a domestic appliance like our electric drill, the ease with which it can be maintained and serviced is of importance to the average DIY enthusiast. Therefore, the ease with which parts can be obtained and the drill can be assembled/dismantled are important design considerations. When deciding whether or not to design-in a significant amount of maintainability, there are several factors that need to be determined. For example, we will need to know the likely market and establish their philosophy on servicing, repair and rectification. Estimates of component life will need to be found to assess the economic viability of repair, reconditioning or repair by replacement.

When designing for ease of maintenance, we need to consider the extra costs involved in using more sophisticated manufacturing processes and component parts, and balance these against customer satisfaction with the finished product or system. The service life of the product and the life of component parts also need to be established to make informed decisions about the need for maintenance.

Costs

A realistic product cost needs to be established as early as possible in the negotiating process. Estimates for products and systems are often set lower than reality dictates, because of the need to gain the competitive edge. However, it is no use accepting or giving unrealistic estimates that are likely to put a business into debt. Costing has become a science, and design engineers need to be aware of how to estimate costs accurately and evaluate the viability of new ventures. Because of the importance of finance, more information on costs and costing is given later in this chapter.

Transportation

Think of the consequences of designing and assembling a very large 100-ton transformer in Newcastle which is required in Penzance! The cost and practicalities of transporting such a monster would make the task prohibitively expensive, if not impossible. Transportation becomes an issue when products that are very large and very heavy need to be moved. Thus when faced with these problems as a design engineer, due consideration needs to be given to ease of transportation and packaging. Small items, such as our electric drill, can be packaged and transported by rail, road, sea or air, with relative ease. Size and weight restrictions must always be pre-determined, to ensure that the product will fit into the space allocated by the customer.

If, for example, we are designing, installing and commissioning an air-conditioning system for a large hotel in Cairo, then due consideration must be given, at the design stage, to ease of transportation and assembly. In fact, large structures and systems are designed in kit form and dry-assembled; this ensures that all component parts are available, that they fit together, and that the installation does not exceed required dimensions, prior to shipment. This enables the product or system to be easily installed and commissioned on sight and prevents any unnecessary transportation costs from being incurred.

Manufacture

Due consideration needs to be given at the design stage to the ease of manufacture of products and their associated parts. The cost implications of 'over-engineering' must be remembered when designing, particularly with respect to design detail. Component parts should be designed with the over-riding thought of saving costs. All non-functional features and trimmings should be omitted. For example, do not design-in radii if a square corner will do and do not waste money on extra machining operations if stock-size material is available and acceptable.

The cost and complexity of production methods also need to be considered at the design stage. One-off items are likely to require 'jobbing production', which is expensive in time and requires a high degree of skill. Batch production and mass flow production of the components for a product may need to be considered, with due consideration given to the facilities available on the manufacturing premises. Otherwise, the use of bought-in items and standard parts for product assembly may be a more economic alternative. Ultimately the decision will again depend on the customer requirements and the manufacturing facilities and tooling available.

Aesthetics

Once all the primary requirements regarding function, safety, use and economy have been fulfilled, the aim of designers is to create products that appeal to customers Consideration needs to be given to shape or form, the use of colour and surface texture. It follows that good industrial design lies somewhere between engineering and art.

The aesthetic design of an engineering product or system may be outside the remit of the engineering designer. Specialists, such as graphic artists, may be seconded to the design team to assist with product aesthetics, packaging, labelling and graphics for a particular market.

Legal implications

Designers will need to be aware of the legal constraints involved with any particular product or system. With the advent of more and more EEC legislation concerning product liability, disposal of toxic waste, COSHH and general health and safety, all legal aspects must be considered during the early stages of the design process.

The fundamentals of the law regarding patents and copyright should be understood by engineering designers, especially when competing with other firms that are designing products for the same purpose. The Copyright, Designs and Patents Act 1988 gives protection to what is called the 'intellectual property' of a person or organisation. Copyright protects materials such as technical literature, artwork and technical films and videos that a company has produced in connection with a product. The 'designs' part of the act gives protection against unauthorised copying of the shape, pattern, visual appearance or eye appeal of a product. The patents part of the act protects against unauthorised copying of the technical and functional aspects of products and processes.

Some knowledge of the law of contract is also very useful for the engineering designer. A contract is a formal written agreement between two parties. Both parties agree to abide by the conditions that are laid down, in all respects. However, certain 'let-out' clauses may be included to accommodate unforeseen difficulties. The larger engineering companies often employ a team of legal experts to draw up contracts, register new products and processes, and take action against anyone who is thought to have made unauthorised use of the company's intellectual property.

Safety

Safety requirements vary according to the products or systems being designed and the use to which they will be put. Many areas of engineering exist within highly regulated industries, where safety is of paramount importance. Typical examples are nuclear power, petro-chemicals, aircraft operation, hospitals and the emergency services. For these industries and many others, adherence to the Health and Safety at Work Act, COSHH regulations and BS/ISO standards for product liability will form an essential part of the design process.

Quality

To ensure quality in design, the design methods adopted, technical documentation, review processes, testing and close co-operation with the customer must be such that the degrees of performance, reliability, maintainability, safety, produceability, standardisation, interchangeability and cost are those required.

Many companies have achieved the Total Quality (TQ) Management System kite mark, in that they have gained BS 5750 or the ISO 9000 series equivalent. To obtain such a standard requires a company to document all of its quality assurance procedures, train personnel in these procedures, implement the procedures, verify that implementation is carried out, measure the effectiveness of the procedures, and identify and correct any problems that arise. In short it is the implementation of good business practice.

KEY POINT

A product design specification often requires input from specialists in production engineering, artistic design, costing and the legal requirements of manufacturing.

Quality control (QC) is the final part of the TQ process, if the TQ process is effective QC will be less and less needed, until eventually the ultimate *right first time* is achieved.

We have spent some time explaining the nature of the design parameters, although rather tedious, this should be treated as essential learning.

Activity 5.6

Prepare a product design specification for an electrically powered hoist suitable for raising and lowering components of up to 50 kg mass at an assembly workstation.

Preparing design proposals

Using the design specification as our guide, we next need to prepare an analysis of possible design solutions, produce concept designs, evaluate alternative concepts and select an *optimum design solution*. From the *conceptual design* phase, where engineering principles are established, we move to the *layout design* or GA design. Here we are concerned with the selection of appropriate materials, determining the preliminary form of the design, checking for errors, estimating costs and producing the *definitive layout and design report*. The final stage of the design process is the *detail design*, where we are concerned with the arrangement, dimensions, tolerances, surface finish, materials specification, detail drawings, assembly drawings and production costs of the individual parts of the product or system being designed.

In our search for an optimum solution to an engineering design problem, there are many methods available, which will help us to find such a solution. The most suitable method for a given situation will depend on the nature of the problem, the magnitude of the task, the information available, and the skill, knowledge and experience of the designers.

All solution finding methods are designed to encourage lateral thinking and foster an open-minded approach to problem solving. These methods may be conveniently divided into *general methods and problem specific* methods. The former are not linked to a specific part of the design process or to a particular product or system. They do, however, enable us to search for solutions to general problems that arise throughout the design process. Problem specific methods, as their name suggests, can only be used for specific tasks, for example to estimate costs or determine the strength of specified materials. In this section we will consider two examples of general methods: *brainstorming* and the *systematic search method*.

Brainstorming

The object of brainstorming is to generate a flood of new ideas. It is often used when there is a feeling that matters are becoming desperate. It involves using a multi-disciplinary team of people, with diverse backgrounds, who are brought together to offer differing perspectives to the generation of ideas. This method is particularly useful where no feasible solution principle has been found, where a radical departure from the conventional approach is considered necessary or where deadlock has been reached.

The group will normally consist of between 6 and 15 people drawn from a diverse variety of backgrounds, and must not be limited to specialists. They are formally asked to focus their attention on a specific problem, or group of problems, to generate ideas for a solution. This technique was originally suggested by Alex Osborn, who was an executive in the advertising industry. He devised the set of criteria given below, to ensure that all participants were given equal opportunity to express themselves freely without inhibition:

- The leader of the group should be responsible for dealing with organisational issues and outlining the problem, prior to the start of the brainstorming session. The leader should also ensure that all new ideas are encouraged and that no one criticises the ideas of other group members.
- All ideas are to be accepted by all participants, no matter how absurd, frivolous or bizarre they may seem.
- All ideas should be written down, sketched out or recorded for future reference.
- Building on the ideas of others to create a group chain reaction should be encouraged.
- The practicality of any suggestion should be ignored at first, and judged later.
- Sessions should be limited to less than 1 hour as longer sessions tend to cause participant fatigue and the repetition of ideas.
- The results should be reviewed and evaluated by experts, to find potential solutions to a design problem or problems.
- After classification and grading of the ideas, one or more suggested final solutions should be presented again to the group for interpretation, comment and feedback.

Systematic search method

This method relies on a mechanistic approach to the generation of ideas, through the systematic presentation of data. Data is often presented in the form of a *classification scheme*, which enables the designer to identify and combine criteria to aid the design solution. The choice of classifying criteria and their associated parameters requires careful thought, since they are of crucial importance. To illustrate this method, consider the following example.

Example 5.2

An engineering system is required to operate the ailerons of a light aircraft. The ailerons must be capable of being operated by the pilot from the cockpit. Use the systematic search method to produce design ideas for possible motion converters for the required aileron system. Figure 5.5 illustrates the required output motion for the system.

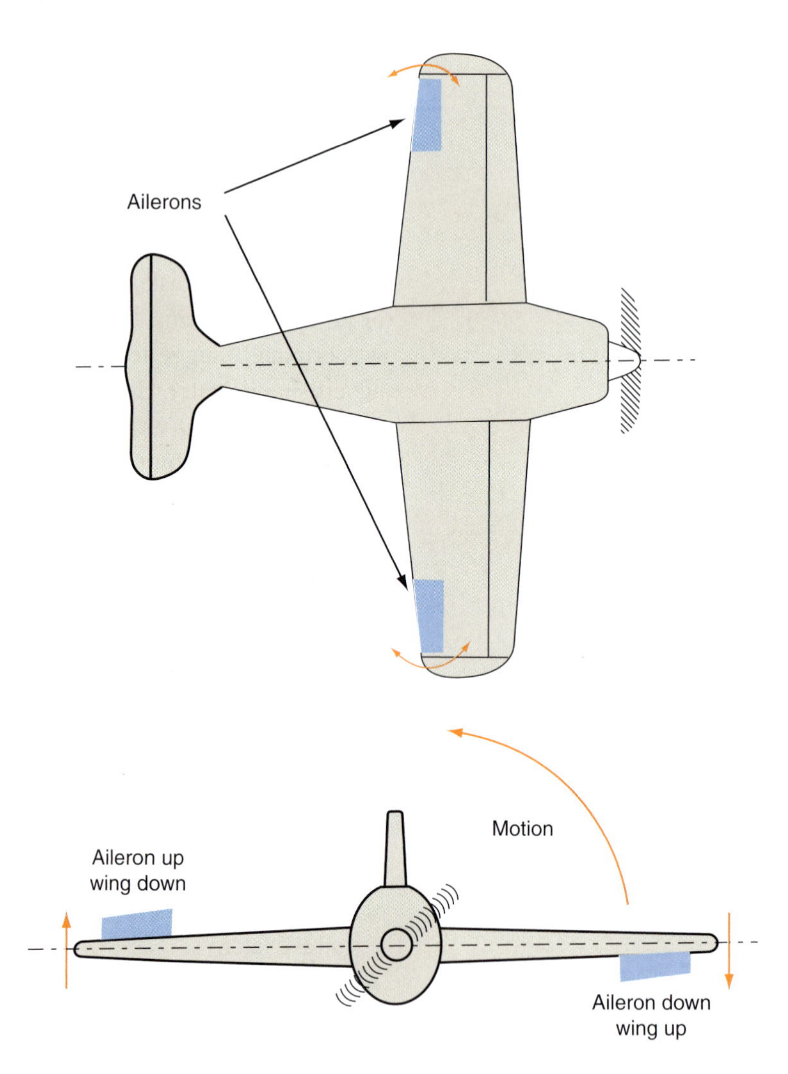

Figure 5.5 Aileron motion

Possible motion parameters might include: linear to linear, linear to rotary, rotary to linear, rotary to rotary and oscillatory, all of these forms of motion are sub-sets of translational (linear) and rotational motion.

To provide design ideas for possible motion converters, a classification scheme needs to be produced. The first attempt is shown in Figure 5.6, the column headings indicate the motion parameters and the solution proposals are entered in the rows.

In the classification scheme illustrated above, you will note that gears are included as a possible motion converter. In fact improvements to the scheme could be made by sub-dividing the solution proposals. For example, gears could be sub-divided into spur and bevel gears, helical gears, epicyclic gears and worm gear. Note that the motion parameters have already been sub-divided from translational and rotary

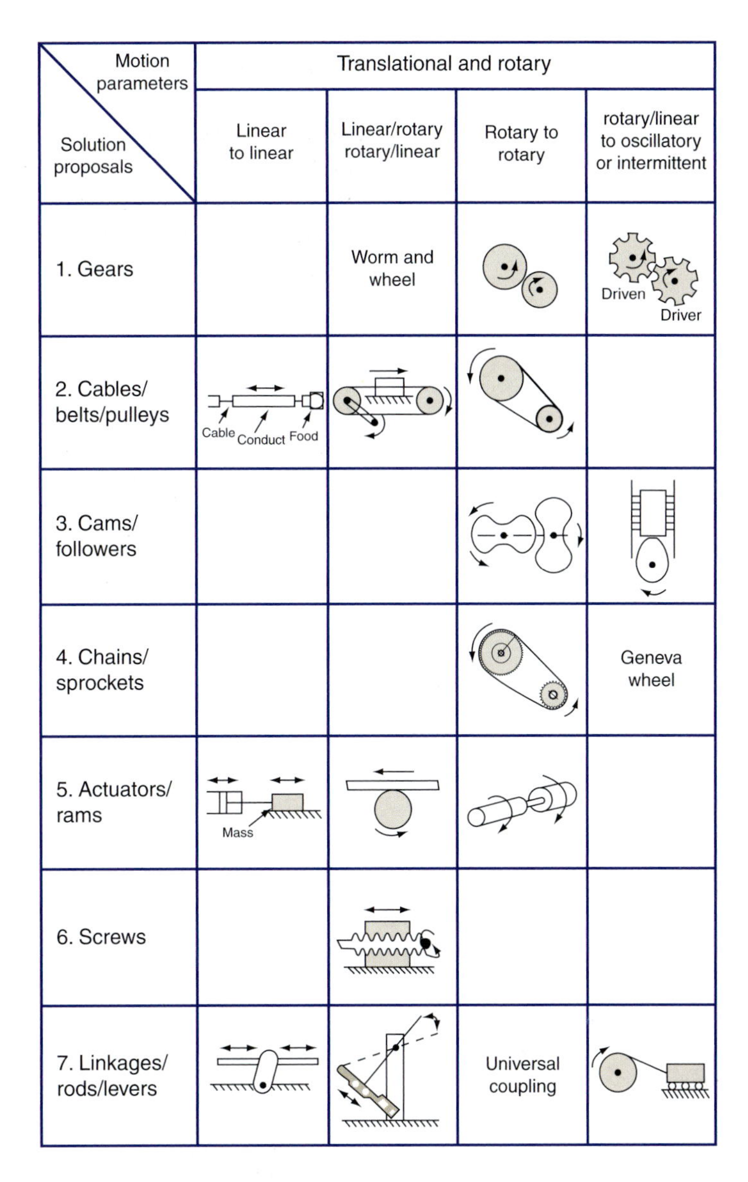

Motion parameters / Solution proposals	Translational and rotary			
	Linear to linear	Linear/rotary rotary/linear	Rotary to rotary	rotary/linear to oscillatory or intermittent
1. Gears		Worm and wheel		Driven Driver
2. Cables/ belts/pulleys	Cable Conduct Food			
3. Cams/ followers				
4. Chains/ sprockets				Geneva wheel
5. Actuators/ rams	Mass			
6. Screws				
7. Linkages/ rods/levers			Universal coupling	

Figure 5.6 Motion converters

motion. This process of layering the classification scheme enables a variety of design proposals to be considered. No mention has been made of energy sources for the system, this has been left for you as an exercise.

There are many other solution finding methods which may be used in addition to those mentioned. These include: literature search, analysis of natural and existing technical systems, model testing and many more with which design engineers need to be familiar.

KEY POINT

General design methods include brainstorming and the systematic search and evaluation of possible design solutions. Problem specific design methods include costing, testing the strength of materials and improving some particular aspect of a design solution.

Activity 5.7

List the possible power sources that might be used to provide the input power for the aileron system in Example 5.2. Also give solution proposals for the possible types of energy provider that might be used to power the system.

CHAPTER 5

Design reference material

Engineering design is essentially an exercise in gathering information, applying creative thinking, evaluating possible designs and transmitting the final design solution to the customer and manufacturing colleagues. Designers must know where to obtain information particularly for the selection of materials and components at the detail design stage. There is a variety of information sources some of which are as follows:

- *Manufacturers' catalogues*: These may be obtained in hard copy and are often also available on disc and on the Internet. Key words such as 'screwed fastenings, helical gears, roller bearings', etc. will give Internet access to a wide range of suppliers.
- *Trade associations*: Typical examples are the British Stainless Steel Association and the Aluminium Federation whose web sites contain details and addresses of specialist material suppliers.
- *Libraries/resource centres*: There are many textbooks on engineering design and materials technology from which information and information sources may be obtained. Libraries often provide Internet access and reference material such as trade directories and BS/ISO specifications.
- *Journals*: Publications such as *Engineer* and *Engineering Designer* contain articles describing advances in material technology and component design together with addresses from which further information may be obtained. Bath Information and Data Services (BIDS) has a database of world-wide journals and papers presented at conferences. It may be accessed at www.bids.ac.uk.
- *Professional and learned societies*: Two that are applicable to engineering design are *The Institution of Engineering Designers* and the *Institute of Materials Minerals and Mining* which permit student membership for a small annual fee.
- *Databases*: There are several material selection and design databases available on disc and on the Internet which are updated periodically. Most require a subscription or a one-off charge but free access can be obtained to:

  www.matweb.com
  www.plaspec.com
  www.steelforge.com
  www.copper.org
  www.avestapolarit.com
  www.sandvik.com
  www.structural-engineering.fsnet.co.uk
  www.key2study.com/matsdata/

Activity 5.8

Obtain access to one or more material databases. Report on the range of materials covered, the information available and the ease with which information may be extracted.

CHAPTER 5

Test your knowledge 5.3

1. What information might the 'Consultation' section of a product design specification contain?
2. What are 'ergonomics' and 'aesthetics' as applied to engineering design?
3. What is a contract?
4. What is 'jobbing' production?
5. What do companies have to do to gain BS 5750/ISO 9000 status?
6. What is 'brainstorming'?

Be Able to Produce and Present a Final Design Solution

Up to now we have been concerned with generating ideas for possible design proposals and we have looked in detail at one of the two methods that help to produce design concepts. We will now consider ways to evaluate and select the best of the solution variants.

The evaluation matrix

One of the most common methods is to produce an *evaluation matrix*, where each solution concept is set against a list of selection criteria. For each criterion some kind of scoring system is used to indicate how the individual design concept compares with an agreed norm. This process is illustrated in Figure 5.7.

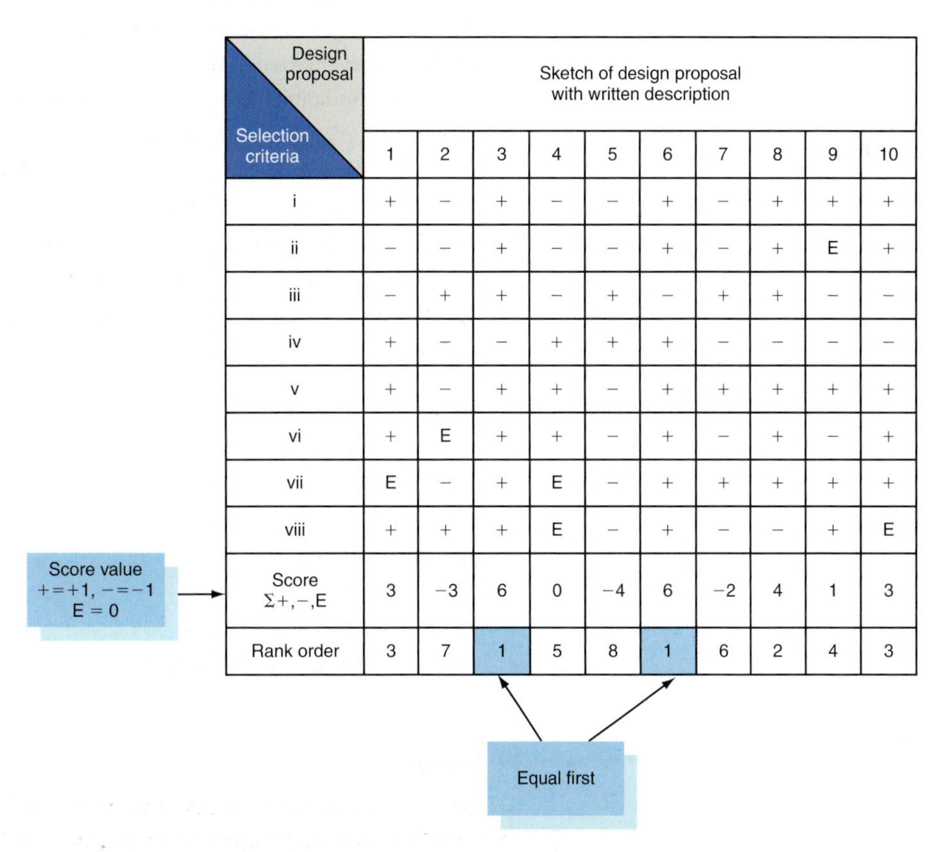

Design proposal / Selection criteria	Sketch of design proposal with written description									
	1	2	3	4	5	6	7	8	9	10
i	+	−	+	−	−	+	−	+	+	+
ii	−	−	+	−	−	+	−	+	E	+
iii	−	+	+	−	+	−	+	+	−	−
iv	+	−	−	+	+	+	−	−	−	−
v	+	−	+	+	−	+	+	+	+	+
vi	+	E	+	+	−	+	−	+	−	+
vii	E	−	+	E	−	+	+	+	+	+
viii	+	+	+	E	−	+	−	−	+	E
Score Σ+,−,E	3	−3	6	0	−4	6	−2	4	1	3
Rank order	3	7	1	5	8	1	6	2	4	3

Figure 5.7 Evaluation matrix (Redrawn from *BTEC National Engineering*, Second edition, M Tooley & L Dingle, Newnes, p. 206, Fig. 3.5, 2007, with permission from Elsevier.)

CHAPTER 5

Prior to inclusion in the evaluation matrix, if there are a large number of solution proposals, the design engineer should produce some form of pre-selection procedure, to reduce the proposals to a manageable size. The pre-selection process being based on fundamental criteria which the design proposal must meet. Such criteria might include: compatibility with required task, meets the demands of the design specification, feasibility in respect of performance, meets mandatory safety requirements, expected to be within agreed costs, etc.

The concept proposals should be in the form of sketches together with a short written explanation. Equality with the agreed norm is shown in the skeleton matrix by the letter 'E', if the design solution is considered better than the norm, in some way, then a + sign is used, conversely a − sign is used, if the design solution is worse than the norm, in some way. A score may be obtained by allocating a +1 to the positives, −1 to the negatives and 0 to the Es. More sophisticated scoring systems may be used involving 'weightings', when the selection criteria are not considered to be of equal importance. The following example illustrates the evaluation procedure.

Example 5.3

You are in charge of a small team who are designing a one-off large diameter flywheel for a heavy pressing machine. The cost is to be kept at a minimum and a number of proposed solutions have been put forward as shown in Figure 5.8.

(i) Draw up a short list of selection criteria against which the solutions can be evaluated.

(ii) Produce an evaluation matrix using the casting method of manufacture as your norm, and rank all the remaining design solutions.

First we need to produce our evaluation criteria. We could use one or more of the previous methods to generate ideas. However, for the purpose of this example, since cost is of paramount importance, we will just look closely at the manufacturing methods which minimise cost. We will assume that the requirements of the specification have been met and that all design alternatives are compatible for use with the pressing machine under all operating conditions. Then for each of the design options, we need to consider:

- Materials costs
- Degree of skill and amount of labour required
- Complexity of construction
- Tooling costs
- Machining and finishing costs
- Safety (this will be related to the integrity of the design solution assuming it is chosen)
- Amount of waste generated
- Company preference (company does not have in-house foundry facilities)

The above list of criteria is not exhaustive, but should enable us to select one or two preferred design alternatives. The matrix shown in Figure 5.9 shows a possible scoring of the proposals.

You will note that the company preference immediately skews the scores. The company does not have or does not wish to use foundry facilities; in any case, the production of the mould would be prohibitively expensive for a 'one-off' job. Obviously you, as the

1		**CAST** Hubs, spokes and rim sand cast in one operation. Wooden pattern required to produce mould. Cast iron (C1) or a steel may be used. Cast wheel requires limited final machining. Suitable finish to be applied.
2		**MACHINED** Flywheel machined from solid. Good machinability properties required of candidate material. C1 or a steel suitable.
3		**FABRICATED** Solid ring spun from standard section steel. Spokes welded to hub and outer rim. Hub machined from solid.
4		**FABRICATED** Solid ring spun from standard section. Inner disc flame-cut to accommodate hub and rim. Hub machined from standard round section bar. Welding used for assembly.
5		**LAMINATED CONSTRUCTION** Outer rim spun from standard section. Centre plate flame-cut from standard section. Bolted or welded assembly. Simple machining operation for assembly.
6		**SEGMENTED CONSTRUCTION** Number of segments varied according to stock-size materials and size of flywheel. Spokes welded to hub assembly and bolted to outer segments with the use of tie-bars.

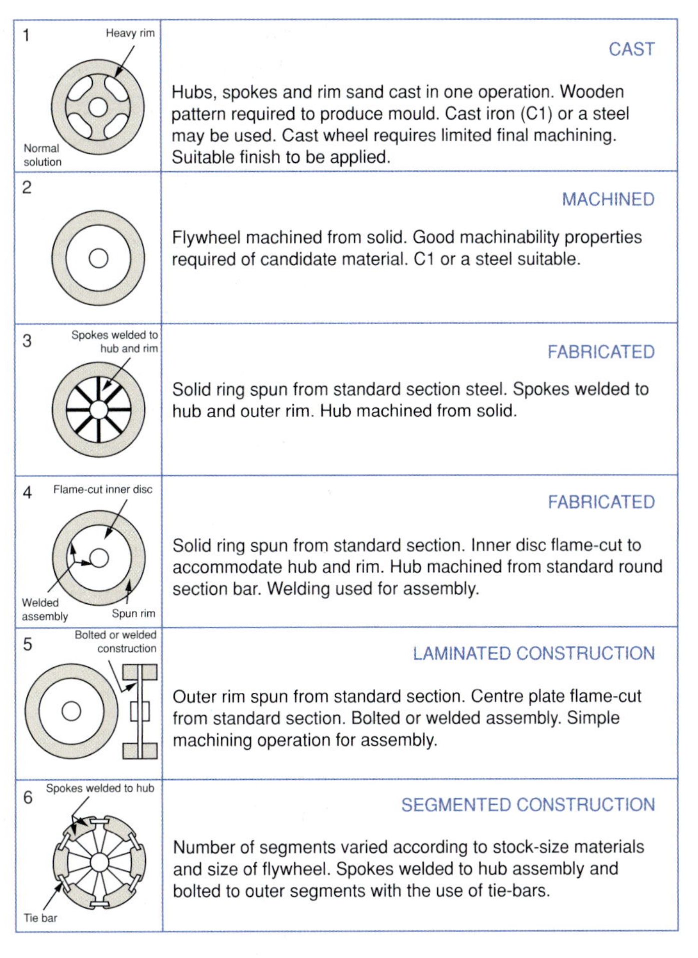

Figure 5.8 Proposed design solutions (Redrawn from *BTEC National Engineering*, Second edition, M Tooley & L Dingle, Newnes, p. 207, Fig. 3.6, 2007, with permission from Elsevier.)

Design proposal / Selection criteria	1	2	3	4	5	6
Materials	+	E	−	−	E	+
Labour	−	−	+	+	+	−
Complexity of constraint	+	+	E	E	E	−
Tooling	−	−	E	+	+	+
Machining/finish	+	−	E	−	+	+
Safety	+	+	E	E	E	−
Waste generation	+	−	+	−	E	−
Company preference	−	+	+	+	+	+
Score Σ +, −, E	2	−1	2	0	4	3
Rank order	NORM	5	3	3	1	2

Figure 5.9 Flywheel evaluation matrix (Redrawn from *BTEC National Engineering*, Second edition, M Tooley & L Dingle, Newnes, p. 208, Fig. 3.7, 2007, with permission from Elsevier.)

design engineer, would be aware of these facts before evaluating the options. Note that options 2–6 all involve some form of fabrication, assembly or machining, which we will assume is the company preference.

Proposal 2: Machining parts for a heavy flywheel will require several machining operations and the use of elaborate fixtures, not to mention operator skill for an object of such size; so labour, tooling and machining costs are relatively high. This process also involves a large amount of material waste.

Proposal 3: The major advantage of this method is that standard stock materials can be used. Difficulties include: the use of jigs and fixtures, weld decay and possibility of complicated heat treatments.

Proposal 4: Similar advantages and disadvantages to option 3.

Proposal 5: Advantages include use of standard stock materials, little machining required after assembly, relatively easy to assemble. Disadvantages include necessity for positive locking of bolts after assembly and outer rim would require skimming after spinning.

Proposal 6: Labour intensive fabrication and assembly, complex assembly and integrity of construction would raise safety issue. No specialist tooling required, finishing relatively easy and cheap, minimal waste from each machining operation, company preference.

Note that if company preference had been for casting, then option 1 would probably have been preferable, provided it met the cost requirements. Options 5 and 6 appear next to favourite, although option 3 might also be worth looking at again, dependent on the skills of the labour force. If there was insufficient evidence on which to make a decision, then more selection criteria would need to be considered. For example, do the options just meet or exceed the design specification, bursting speeds, and other safety criteria might have to be further investigated. This process would need to be adopted no matter what the artefact, an *iterative approach* being adopted, in an attempt to get ever closer to the optimum design solution.

Costing

The cost of an engineering component or system is of paramount importance. Engineering designs require the specification to be met, the artefact to be produced on time and *at the right cost*, if the design solution is to be successful. The importance of producing a successful tender cannot be over-emphasised. In fact the future of jobs within a company may depend on it. There must be an effective costing and pricing policy to ensure that commercial contracts to design, manufacture and supply on time are won.

Not only must the contract be won, against competition, but a profit margin also needs to be shown. Pricing needs careful planning. Clearly, too high a price may not result in a successful tender and too low a price may cause financial loss to the company, particularly if there are unforeseen difficulties. Thus an understanding of costs and costing procedures is something that every design engineer needs to achieve. Set out below are one or two important reasons for costing that are directly related to the production of an engineering artefact:

- Determine the viability of a proposed business venture.
- Monitor company performance.

- Forecast future prospects of a business deal.
- Price, products and/or services.
- Meet legal requirements to produce records of company viability, for public scrutiny as required.

Below are some useful definitions concerned with cost and price:

Price:	*Money paid for products or services.*
Value:	The amount of money someone is prepared to pay for products or services.
Cost:	All money spent by a supplier to produce goods and services.
Material cost:	(Volume × density × cost/kg) plus an amount for wastage.
Labour cost:	(Operational time × labour rate) plus wasted labour time that is not directly related to the task.

KEY POINT

Costing and financial control is an essential part of any organisation. Companies are legally required to maintain records and produce them as required.

Standard costing sheets

These are used to ensure that all parameters are considered when costing a product or service. Some standard costing sheet headings together with their definition are given here:

A:	Direct material cost: raw material and bought-in costs.
B:	Direct material scrap: materials subsequently scrapped (typically 3–5% of A).
C:	Direct labour cost: wages of production operations, including all incentive payments.
D:	Direct labour scrap: time spent and paid for on artefacts, which are subsequently scrapped; this would include the costs of machine breakdown or other reasons for stoppages to production (typically 3–5% of C).
E:	Prime cost: the sum of all material and labour costs, i.e. A + B + C + D.
F:	Variable overheads: cost of overheads which vary with rate of production; these might include: fuels costs, cost of power supplied, consumables, etc. (typically 75–80% of C).
G:	Manufacturing cost: this is the sum of prime costs and variable overheads (E + F).
H, I, J:	These are packaging, tooling and freight costs, respectively.
K:	Variable cost (VC): this is the sum of the previous costs, G + H + I + J.
L:	Fixed overheads (FO): overheads which do not vary with production output, these include all indirect personnel not involved with production plus marketing costs, research and development costs, equipment depreciation and premises costs (typically 30–40% of K).
M:	Total cost (TC): the sum of all direct VCs (K) plus indirect costs (L). Thus, TC = direct VCs + indirect costs (FO).

Example 5.4 illustrates the method of estimating the total cost and selling price of a product.

Example 5.4

A company has been commissioned to produce 2000 high-quality metal braided shower hoses, complete with fixtures and fittings. Assuming that:

(i) direct material cost per item is £1.50 and material scrap is estimated to be 3% of material costs;
(ii) direct labour costs total £8000 and the labour scrap rate is 4% of direct labour costs;
(iii) variable overheads are 75% of direct labour costs;
(iv) FO are 30% of VCs;
(v) packaging, tooling and freight costs are 10% of manufacturing costs.

Estimate the selling price of the shower hose if the company wish to make a 30% profit.

This problem is easily solved by laying out a costing sheet as shown below, and totalling the amounts:

Cost		Amount (£)
Direct material cost	(1.5 × 2000)	3000
Direct material scrap	(3% of 3000)	90
Direct labour cost		8000
Direct labour scrap	(4% of 8000)	320
Prime cost		11 410
Variable overheads	(75% of 8000)	6000
Manufacturing cost	(prime + variable)	17 410
Packaging, tooling and freight cost	(10% of 17 410)	1741
Variable costs (VCs)	(manufacturing + packaging, tooling and freight)	19 151
Fixed overheads (FO)	(30% of 19 151)	5745
Total cost (TC)	(VC + FO)	£24 896

Now the company is required to make 30% profit, i.e. $0.3 \times £24896 = £7469$

So the selling price per item is $\frac{£24896 + £7469}{2000} = £16.18$

Presenting the final design solution

We have looked at the design process and a number of ways in which to identify and evaluate design solutions. These evaluation methods are equally suitable for determining solutions during the concept, GA (layout) and detail phase of the design process. Information on the layout design as well as conceptual design should appear in the report. Here, we are concerned with report writing, the layout of the design report and the detail expected within each section.

During the course of the design process, a design diary and a log book should be maintained containing a record of all the meetings, communications and activities associated with the design. These will be useful for reference when compiling the final design report.

The following general information is given for guidance only. The report content and layout may differ slightly from that given, depending on the nature and requirements of the design task. More specific information on report writing may be found in BS 4811. *The presentation of research and development reports BSI* (1972).

Title page

This should include a clear and precise title for the design and contain the designer's name and company details as appropriate.

Acknowledgements

These should always appear at the front of the report. They should include individuals, companies or any associated bodies that have provided the design engineer with help and advice. This may include assistance with regard to literature, materials, information, finance or any form of resource.

Summary

This should provide a brief statement of the design problem, its solution and any further recommendations with respect to development and testing. References may be made to other areas of the report, to clarify the design description.

List of contents

This should contain a list, which provides the page number of all the main headings as they appear in the report. A separate list of all diagrams, sketches, drawings, illustrations and photographs should be provided, indicating figure numbers, page numbers, plate numbers and drawing numbers, as appropriate.

Introduction

This should provide all background details to the project and give an indication to the reader as to why the design was undertaken.

Specification

This section should include the design requirements in the form of a statement of the initial design specification.

Design parameters

A description of all the design parameters related specifically to the design in question should be given. The design parameters will include those concerned with the engineering aspects of the product or system being considered, as well as organisational factors. Any modifications to the original design specification should be given, stating all assumptions made, and giving reasons for such decisions.

Description of design

This is the most important section within the report. It should contain an explicit, succinct description of the final design solution, indicating clearly its function and operation. Sketches should be provided to clarify specific areas of the design solution and references to formal drawings; in particular, the GA drawing should be made.

Design evaluation

This section should contain a critical appraisal and appreciation of the final design solution. Recommendations for further development and testing should also be given, to enable improvements to be made to specific features of the design, as required.

References

The reference list should contain only those references that are mentioned in the text. They are normally numbered in the same order in which they appear in the text.

Appendices

These contain all supporting material necessary for the report which is not essential for inclusion or appropriate for inclusion into the main body of the report. The material contained in the appendices should be referred to in the text of the main report.

Appendices are often identified using a Roman numeral. The following list gives a typical selection of appendix material for a design report:

- Evaluation of alternative design solutions, including sketches and description of alternatives.
- GA and detail drawings for the design solution that are referred to in the report.
- Details of decision-making processes, such as evaluation matrices, decision trees, etc.
- Theoretical calculations, mathematical derivations, formulae and repetitive calculations.
- Evaluation of materials selection, for all phases of the design.
- Evaluation of appropriate manufacturing processes.
- Consideration of human factors.
- Costing considerations and pricing policy.
- Details of correspondence, associated with the design.

Engineering drawings

The freehand sketches, layout, assembly and detail drawings referred to in the description of the design plus any circuit diagrams or flow charts should conform to the relevant British Standards, e.g. BS 308, BS 8888, BS 7307, BS 3939, BS 2197. We will not go into the detail of these standards as you should already be familiar with them from the core unit Communications for Technicians. To remind you of the techniques involved, here are some typical examples.

Block diagrams

Block diagrams show the relationship between the various elements of a system. They can be used to simplify a complex design by dividing it into a number of much smaller functional elements. Figure 5.10 shows a typical example in which the links and dependencies between various elements (inputs and outputs) can be clearly seen.

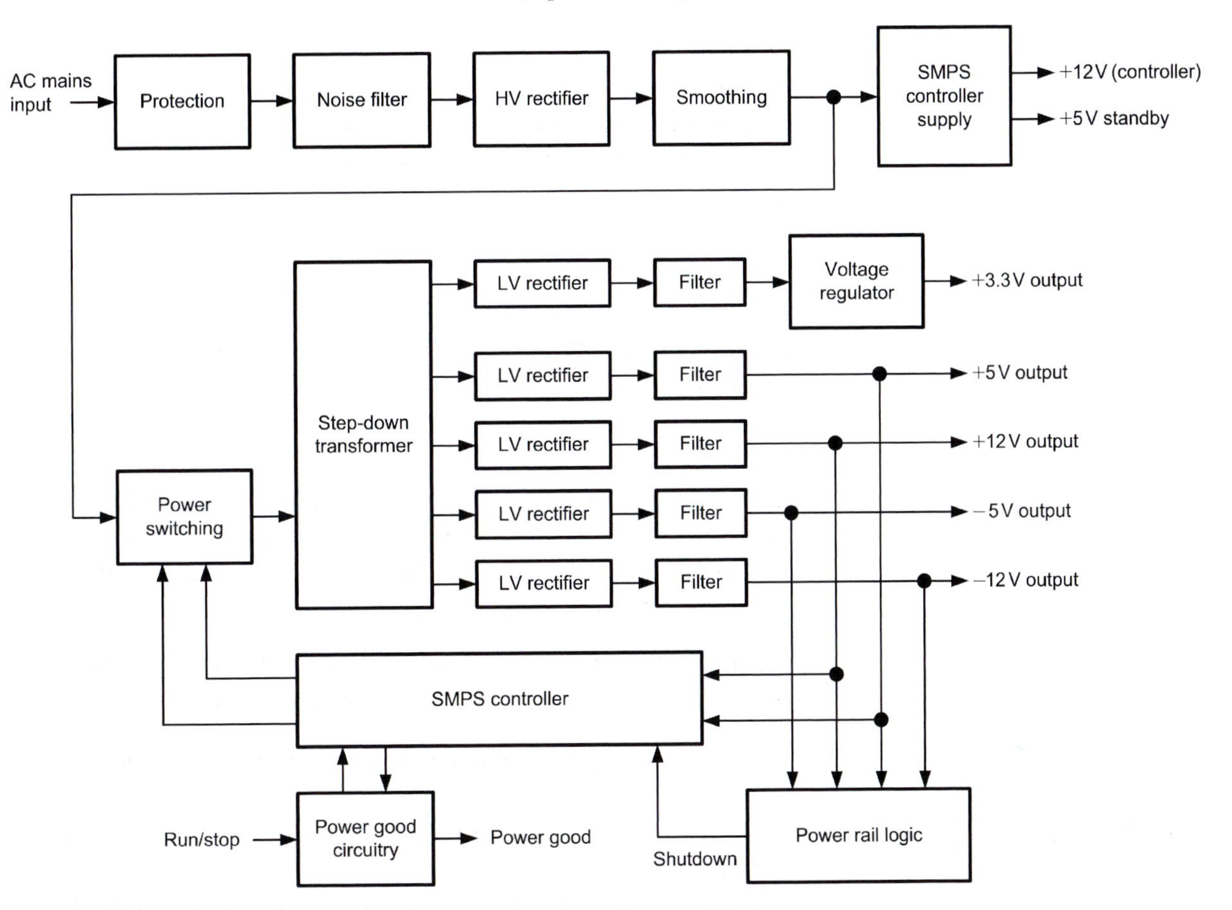

Figure 5.10 Block diagram of a computer power supply (Reprinted from *BTEC National Engineering*, Second edition, M Tooley & L Dingle, Newnes, p. 247, Fig. 3.42, 2007, with permission from Elsevier.)

Flow diagrams

Flow diagrams (or *flowcharts*) are used to illustrate the logic of a sequence of events. They are frequently used in fault-finding, computer programming (software engineering) and in process control. They are also used by production engineers when working out the best sequence of operations in which to manufacture a product or component.

Figure 5.11shows a flow chart for fault location on the computer power supply shown in Figure 5.10.

Circuit diagrams

These are used to show the functional relationships between the components in a circuit. The components are represented by symbols but their position in the circuit diagram does not represent their actual position in the final assembly. Circuit diagrams are also referred to as schematic diagrams or even schematic circuit diagrams.

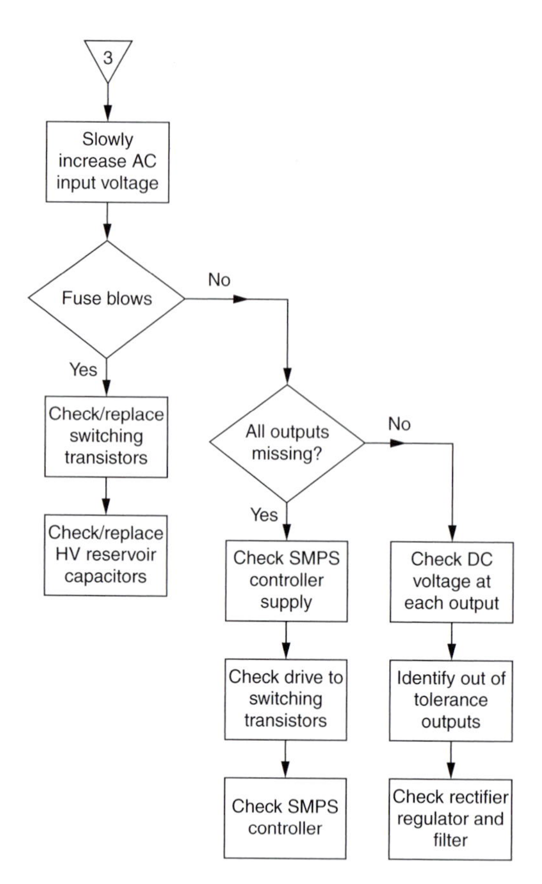

Figure 5.11 Flow diagram for fault-finding (Reprinted from *BTEC National Engineering*, Second edition, M Tooley & L Dingle, Newnes, p. 247, Fig. 3.43, 2007, with permission from Elsevier.)

Figure 5.12(a) shows the circuit for an electronic filler unit using standard component symbols. Figure 5.12(b) shows a layout diagram with the components correctly positioned.

Building services engineers use circuit diagrams to show the electrical installation in buildings. They also provide installation drawings to show where the components are to be sited. They may also provide a wiring diagram to show how the cables are to be routed to and between the components. The symbols used in electrical installation drawings and wiring diagrams are not the same as those used in circuit diagrams. Examples of architectural and topographical electrical component symbols are shown in BS 8888.

Schematic circuit diagrams are also used to represent pneumatic (compressed air) circuits and hydraulic circuits. Pneumatic circuits and hydraulic circuits share the same symbols except that pneumatic circuits should have open arrowheads, whilst hydraulic circuits should have solid arrowheads. Figure 5.13 shows a typical hydraulic circuit.

GA drawings

GA drawings are widely used in engineering to show the overall arrangement of an engineering assembly such as a pump, gearbox, motor drive, or clutch. GA drawings are often supported by a number of detail drawings that provide more detailed information on the individual parts.

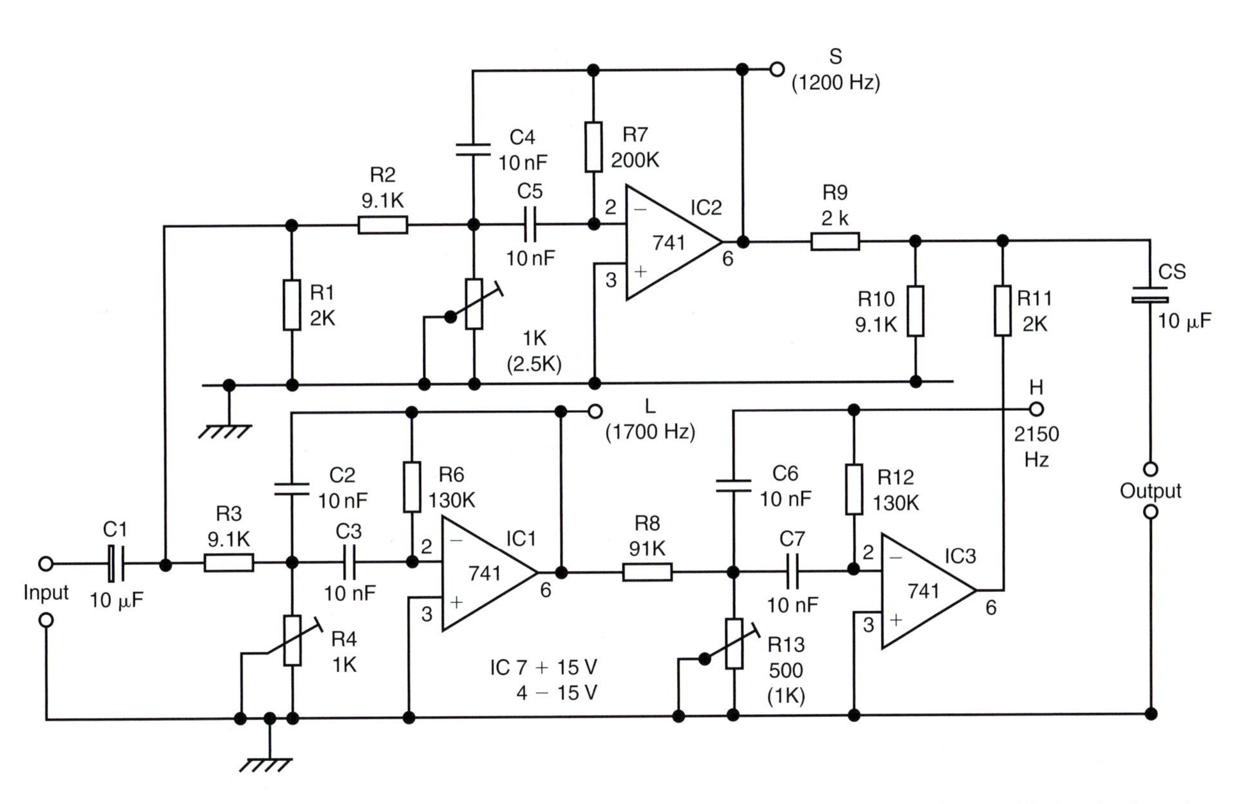

Figure 5.12(a) Electronic circuit diagram (Redrawn from *BTEC National Engineering*, Second edition, M Tooley & L Dingle, Newnes, p. 199, Fig. 3.4a, 2007, with permission from Elsevier.)

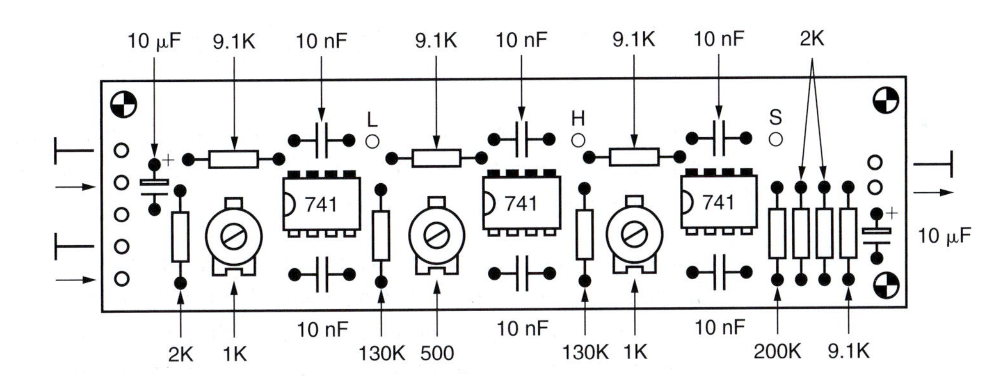

Figure 5.12(b) Component layout diagram (Redrawn from *BTEC National Engineering*, Second edition, M Tooley & L Dingle, Newnes, p. 199, Fig. 3.4b, 2007, with permission from Elsevier.)

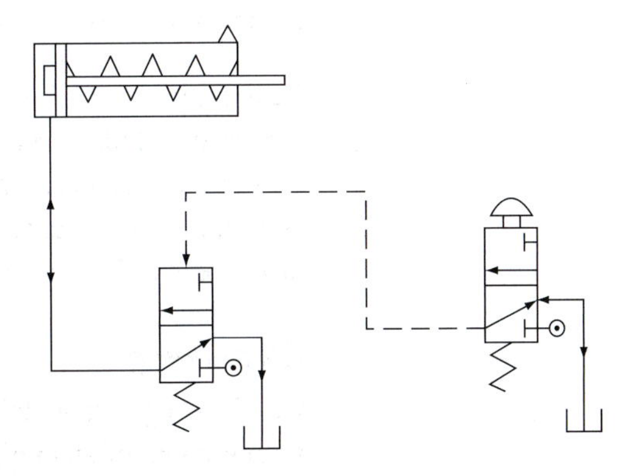

Figure 5.13 Hydraulic circuit diagram (Redrawn from *BTEC National Engineering*, Second edition, M Tooley & L Dingle, Newnes, p. 128, Fig. 2.29, 2007, with permission from Elsevier.)

CHAPTER 5

Figure 5.14 shows a typical GA drawing with all the components correctly assembled together. These are listed in a table together with the quantities required. Manufactures' catalogue references are also given for bought-in components. The detail drawing numbers are also included for components that have to be manufactured as special items.

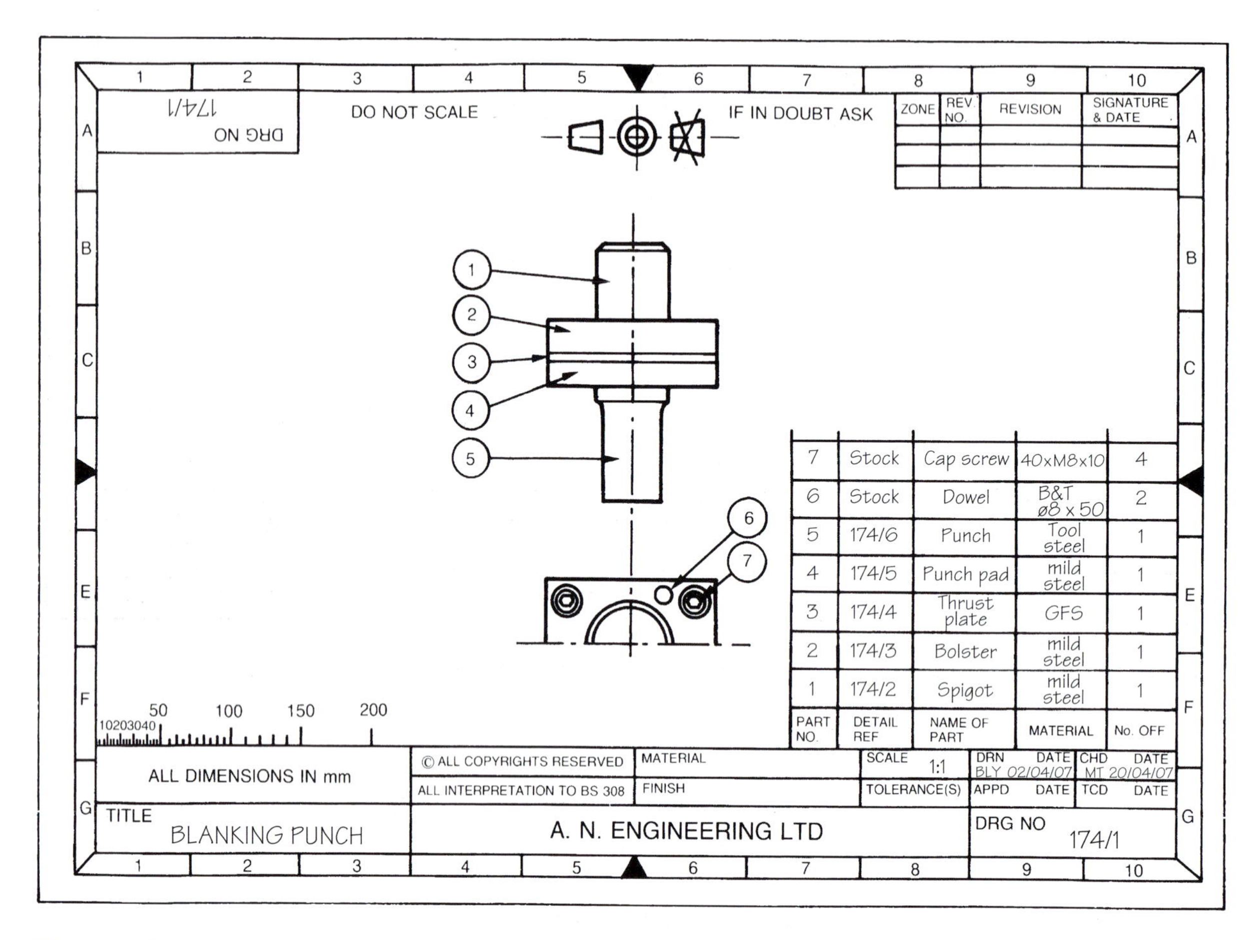

Figure 5.14 GA drawing (Reprinted from *BTEC National Engineering*, Second edition, M Tooley & L Dingle, Newnes, p. 130, Fig. 2.32, 2007, with permission from Elsevier.)

Dimensions are not usually given on GA drawings although sometimes overall dimensions will be given for reference if the GA is for a large assembly drawn to a reduced scale.

Detail drawings

As the name implies, detail drawings provide all the details required to make the components shown on the GA drawing. Referring to Figure 5.14, we see from the table that the detail drawing for the punch has the reference number 174/6.

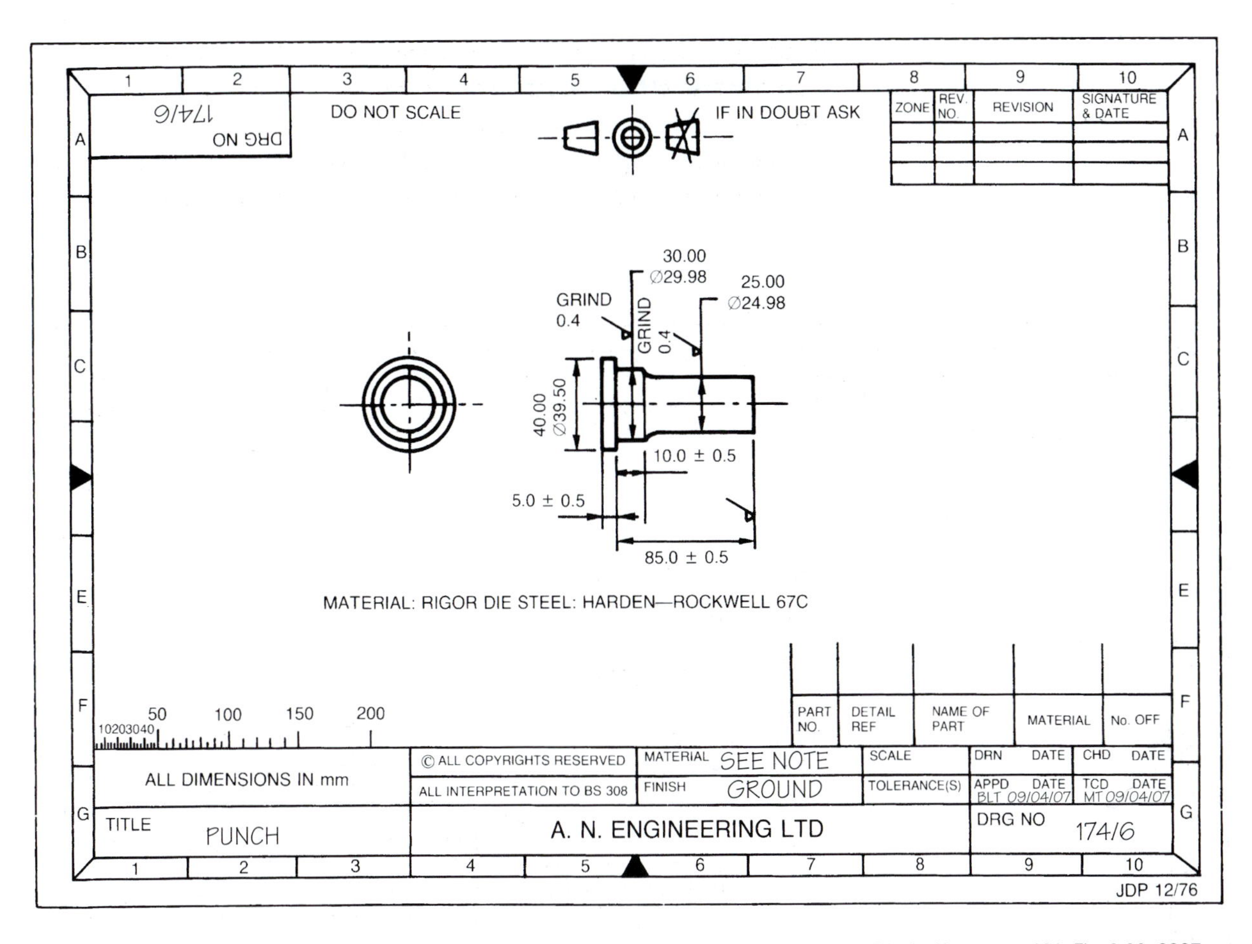

Figure 5.15 Detail drawing (Reprinted from *BTEC National Engineering*, Second edition, M Tooley & L Dingle, Newnes, p. 131, Fig. 2.33, 2007, with permission from Elsevier.)

Figure 5.15 shows this detail drawing. In this instance, the drawing provides the following information:

- The shape of the punch.
- The dimensions of the punch and the manufacturing tolerances.
- The material from which the punch is to be made and its subsequent heat treatment.
- The unit of measurement (millimetre).
- The projection (first angle).
- The finish.
- The guidance note 'do not scale drawing'.
- The name of the company.
- The name of the designer.

Test your knowledge 5.4

1. What is the purpose of an evaluation matrix?
2. What information should a design report summary contain?
3. What constitutes the prime cost of a product or service?
4. What constitutes the variable and fixed overhead costs of a product or service?
5. What kind of diagram might be used to illustrate a fault-finding procedure?
6. What information would you expect to find on a general arrangement drawing?

Review questions

1. James Dyson is the designer of one of the best selling types of vacuum cleaner. Find out how the design was triggered, the design methods he used and why the design was considered to be innovative.

2. Together with a group of colleagues brainstorm as many ideas as you can for anti-theft devices for the home.

3. Your company intends to produce an electric soldering iron suitable for use with printed circuit boards.

(a) Produce a design specification which includes at least 10 requirements and constraints.

(b) Produce a design solution in the form of a concept sketch with an accompanying explanation.

4. A device to test the strength of bicycle brake cables is required by a manufacturer.

(a) Identify three possible sources of energy that may be used for the testing device.

(b) Produce as many possible design alternatives as you can using the different power sources, giving details of the design principles and proposed operation of each design solution.

(c) Using a suitable method, rank your possible design solutions.

(d) Select your optimum solution and produce a GA drawing.

Robots are playing an increasing role in mechanical handing, component assembly and machining operations. Robotic systems contain a variety of pneumatic, hydraulic and electrical components. Pneumatic systems are primarily used where lightness of touch, speed and flexibility are required. Hydraulic systems are employed where larger forces are required for gripping and manipulating components.

Photo courtesy of iStockphoto, Paul Mckeown, Image # 5129181

Chapter 6

Electro, Pneumatic and Hydraulic Systems and Devices

Fluid power systems are widely used in manufacturing engineering. Such systems can be *pneumatic systems* (i.e. pressurised air or gas) or *hydraulic systems* (i.e. pressurised liquid), and use electrical control devices to make them work. Pneumatic systems are used to operate equipment such as automated assembly machines, packaging machines and devices used for clamping and lifting equipment. Hydraulic systems are used where greater amounts of power are required, e.g. actuators used to move arms on earth moving equipment such as excavators. The aims of this unit are to give an understanding of the design and safe operation of such systems, the components used, the basic calculations required for the design and setting up of systems, and their maintenance.

Industrial Applications

> **KEY POINT**
>
> Pneumatic systems use pressurised air or gas, hydraulic systems use pressurised liquid.

Both hydraulic and pneumatic systems have particular merits and suit some applications better than purely electrical systems would. In particular, for linear actuation, i.e. moving some object in a linear direction, electric systems are not easily used for anything other than short distances. Both hydraulic and pneumatic systems can however be used with cylinders to easily accomplish such motion and hydraulic systems are particularly useful where high forces are needed. Figure 6.1 shows the basic principle of a cylinder and how it can be used to produce linear actuation. Table 6.1 gives a general comparison of electrical, pneumatic and hydraulic systems.

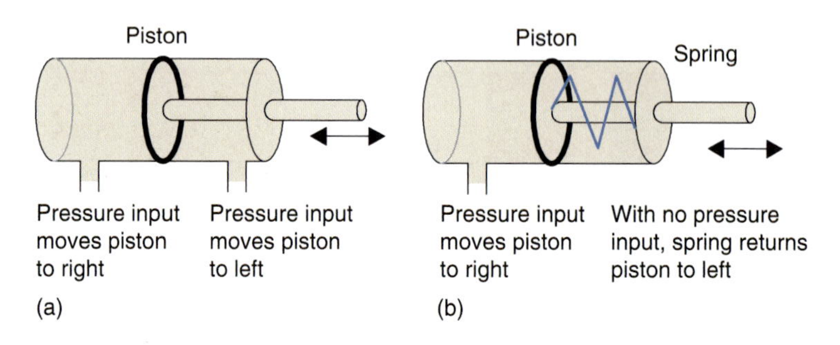

Figure 6.1 Cylinders for linear actuation: (a) double acting and (b) single acting

Table 6.1 Comparison of electrical, pneumatic and hydraulic system

	Electrical	Pneumatic	Hydraulic
Energy source	Usually external mains supply	Electric or diesel motor	Electric or diesel motor
Distribution	Easily distributed	Plant-wide possible	Limited to immediate locality
Linear actuation	Short distance via solenoid, otherwise needs a mechanical linkage; not so easy as rotation	Medium forces can be produced very simply by cylinders	High force can be produced very simply by cylinders
Rotary actuation	Via motors	Can be used but accurate control not easy	Can be used for low speed rotation
Problem issues	Hazards from electrical shock	Noise	Leakage: dangerous and fire hazard

> **KEY POINT**
>
> Fluid power systems have the great advantage over electrical systems of being able give linear actuation over significant distances.

Fluid power systems are widely used for materials handling operations such as clamping, shifting, positioning and orientating. Such operations may be in packaging, feeding and positioning components in machines into suitable positions for drilling, milling, etc., sorting and stacking parts, positioning components for stamping and embossing, and transferring materials between machines. Fluid power systems are also used to sense the position of components and feed such information to a controller which can then determine which actuator needs to be actuated.

Legislation, Regulations and Safety Precautions

Health and Safety at Work

With all pneumatic and hydraulic systems, due care and attention must be given to safety and the relevant regulations followed. In Britain, all industrial processes are governed by the *Health and Safety at Work Act 1974*. This places duties on employers to, among other matters:

- Provide and maintain safe plant and systems of work.
- Make arrangements for ensuring safe use and handling of articles and substances.
- Provide adequate information, instruction, training and supervision.
- Maintain a safe place of work.
- Provide and maintain a safe working environment.

Duties are placed on employees to, among other matters:

- Safeguard himself/herself and others by using equipment in a safe manner and follow safe working practices.
- Cooperate with the employer in respect of health and safety matters.
- Use equipment and facilities provided to ensure safety and health at work.

Regulations

Of particular relevance to those working with fluid power systems are the *Pressure Systems Safety Regulation 2000*. Useful outlines of the regulations are given in The Health and Safety Executive leaflets *Pressure Systems – Safety and You* and *Written Schemes of Examination: Pressure Systems Safety Regulations 2000* (both leaflets can be downloaded from the Health and Safety Executive sites on the Internet). The first of the leaflets advises on how to minimise the risks when working with systems containing a liquid or gas under pressure. Reducing the risks of failure of pressure systems involves:

- Providing safe and suitable equipment.
- Knowing the operating conditions.
- Fitting suitable devices and ensuring that they function properly.
- Carrying out suitable maintenance.
- Making provision for appropriate training of those operating, installing, maintaining, repairing, inspecting and testing pressure equipment.
- Having a written scheme of examination for the equipment. This typically will include an identification of the items in the system, the parts which are to be examined, the nature of the examination, the preparatory work needed for an item to be safely examined, the nature of any examination needed before an item is first used, the maximum interval between examinations, the critical parts of the system which is modified or repaired will need to be examined by a competent person before the system is used again, the name of the competent person certifying the written scheme of examination and the date of certification.

Also of relevance to working with fluid power systems are the Employment Act 2002, the Factories Act 1961, the Fire Precautions Act 1971, the Deposit of Poisonous Waste Act 1972, and regulations such as

Management of Health and Safety at Work Regulations 1999, Provision and use of Work Equipment Regulations 1998, Control of Substances Hazardous to Health (COSH) Regulations 2002, Lifting Operations and Lifting Equipment Regulations 1998, Manual Handling Operations Regulations 1992, Personal Protective Equipment at Work Regulations 1992, Confined Space Regulations 1997, Electricity at Work Regulations 1989, Control of Noise at Work Regulations 2005, Reporting of Injuries, Diseases and Dangerous Occurrences Regulations 1995, Workplace (Health, Safety and Welfare) Regulations 1992, Health and Safety (First Aid) Regulations 1981, Supply of Machinery (Safety) Regulations 2005 (SI 2005/831), the Carriage of Dangerous Goods and Uses of Transportable Pressure Equipment Regulations 2004, Dangerous Goods and Uses of Transportable Pressure Equipment Regulations 2004, and the Simple Pressure Vessels (Safety) Regulations 1991 (SI 1994/3098).

KEY POINT

With fluid power systems, due care and attention must be given to safety and the relevant regulations followed.

Safety precautions

Hydraulic and pneumatic systems, as well as the associated electrical systems, need to be approached with due consideration for safety. High pressure oil or air can be very dangerous if released suddenly and the movements of components such as cylinders can easily trap or crush limbs. Thus pressurised lines or components should never be disconnected but isolated and de-pressurised before disconnection. Gas charged hydraulic accumulators (see the next section for details) must only be pressurised with dry nitrogen since oxygen could cause an explosive mixture with hydraulic oil. Also, accumulators must be de-pressurised before any work occurs on the related hydraulic system.

Where there are electrical interfaces to pneumatic or hydraulic components, the control circuits should be electrically isolated to reduce the chance of electrical shock and also the possibility of fire. Many components contain springs under pressure and these should be released with care since if released in an uncontrolled manner they can fly out at high speed and cause injury. With hydraulic systems care should be taken to catch and mop up any oil spillages. With any plant, an assessment needs to be made of the risks involved so that safe procedures can be established.

Risk assessment

There is always some risk associated with the use of a fluid power, or indeed any, system and modern safety legislation recognises this and is concerned with minimising the risk. The term *risk* is defined as being a function of the likelihood of the hazard occurring and the severity, the term *hazard* being used to describe the potential to cause harm. In carrying out a risk assessment, the hazards of a particular system have to be identified, e.g. electrical hazards. The next stage is to assess the risk for each hazard, e.g. could it be fatal, requiring medical attention, or minor such as a small bruise, and how often people are exposed to the risk. Risks can then be classified as perhaps high, medium or low. The final stage in risk assessment is to devise methods for reducing the risks to an acceptable level.

Test your knowledge 6.1

1. What advantage do fluid power systems have over electrical systems?
2. Which fluid power system is able to exert the most force in an industrial application?
3. What are the duties of an employee with regard to workplace safety?
4. What is the major act of parliament that lays down the duties of employers and employees with regard to workplace safety?
5. What does the term 'hazard' mean in an industrial context?
6. What is the objective of a risk assessment?

Activity 6.1

Access the Internet, download and read the Health and Safety Executive leaflet *Pressure Systems – Safety and You*.

Fluid Power Devices

Production and distribution of fluid power

The basic symbol used for a compressor or motor is a circle, the figure showing some variations on this basic symbol. The basic symbol used for conditioning apparatus is a diamond. The term conditioning is used for functions such as filtering, water traps, coolers, dryers and lubricators.

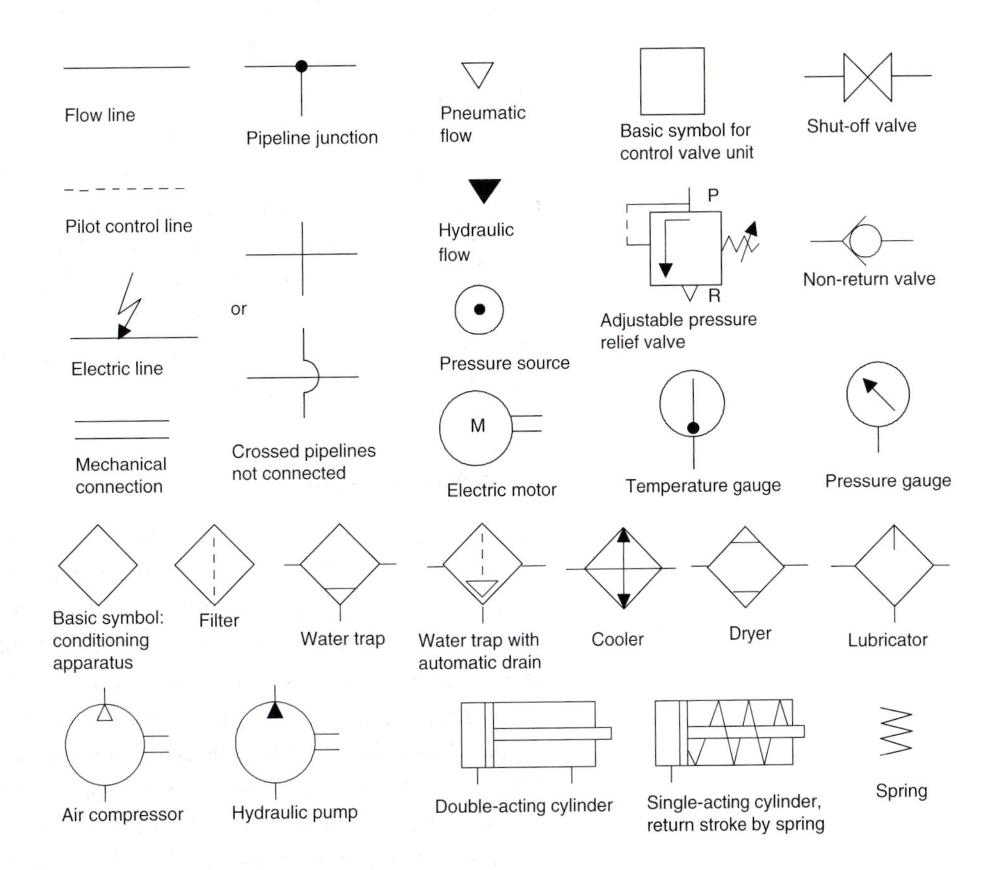

Figure 6.2 Fluid power plant symbols

The basic symbol for a control valve is a square. Valves and their control symbols are discussed in more detail later in this chapter (Figure 6.2).

Pneumatic plant

Figure 6.3 shows the typical components and layout of a typical compressed air system.

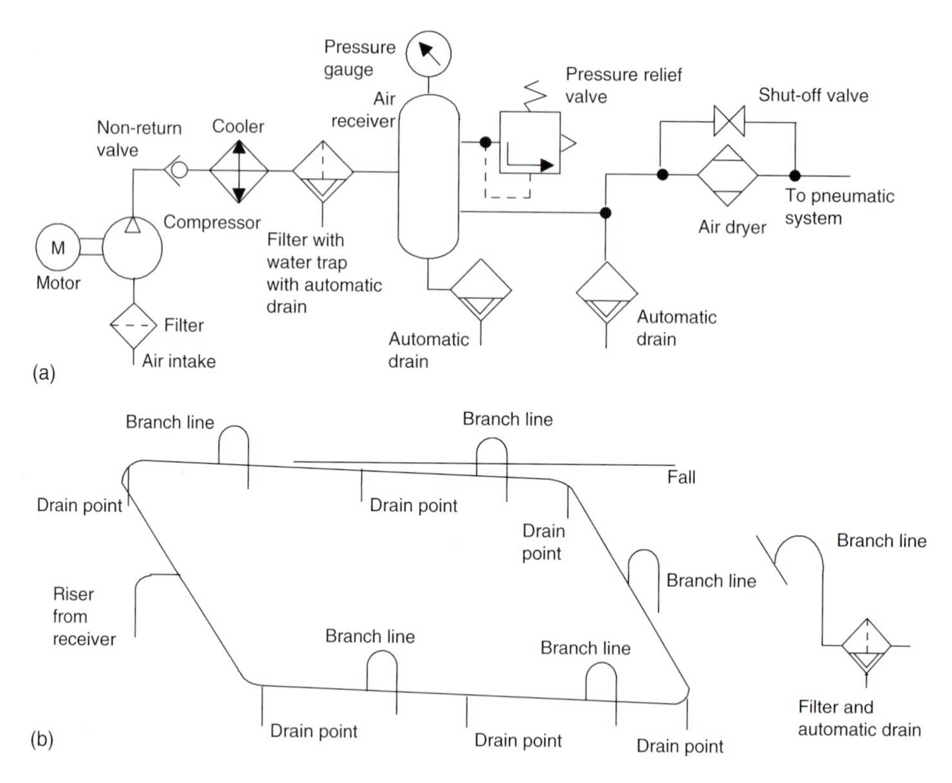

Figure 6.3 (a) Typical air production plant and (b) ring main system supplied by plant

The production of compressed air as shown in Figure 6.3(a) is likely to involve the following basic elements:

1. An *inlet filter* to remove dirt particles from the air entering the compressor. Such filters can be dry filters with replaceable cartridges or wet filters where the air is bubbled through an oil bath and then passed through a wire mesh filter, the dirt particles becoming attached to the oil drops and removed by the wire mesh.
2. Then the *compressor* to pressurise the air. The most commonly used form of compressor is the piston compressor (Figure 6.4(a)). On the air intake stroke, the descending piston causes air to be sucked into the chamber through the inlet valve. When the piston starts to rise again, the trapped air forces the inlet valve to close and so becomes compressed. When the air pressure has risen sufficiently, the outlet valve opens and the trapped air flows into the compressed air system. The cycle then repeats itself. This pressurisation causes the air temperature to rise. Such a compressor goes directly from atmospheric pressure to the required pressure in a single operation and is termed a single stage compressor. For pressures more than a few bars a multistage compressor is likely to be used with cooling

CHAPTER 6

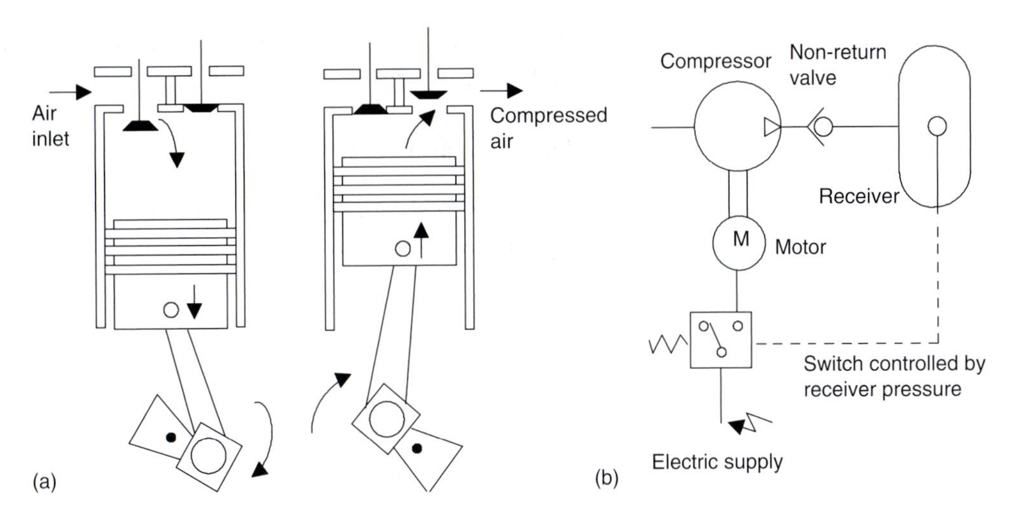

Figure 6.4 (a) Single-acting, vertical, reciprocating compressor and (b) start–stop compressor control

having to occur between stages. Control of the compressor is necessary to maintain the pressure in the air receiver. The simplest method of control is *start/stop control* where the compressor is started when the receiver pressure falls to some minimum pressure and stopped when the pressure in the receiver has risen to the required value. An electrical pressure switch can be used to monitor the pressure and provide the signals to start and stop the compressor (Figure 6.4(b)).

3. This is followed by a *cooler* and a *filter* with water trap. These are because the air leaving the compressor can be hot and contain contaminants such as oil from the compressor, moisture and dirt particles.
4. Then an *air receiver* followed by a *dryer*. Air receivers are essentially just cylindrical containers. The purpose of the air receiver is to store compressed air and so eliminate the need for the compressor to run continuously, to smooth out the pulsing of the air from the compressor and to act as an emergency supply in the event of power failure.
5. To guard against the pressure in a system becoming too high, a *pressure relief valve* is generally included with the air receiver. This vents air to the atmosphere if the pressure becomes too high. Figure 6.5 shows the basic elements of a pressure relief valve. With normal pressures, the

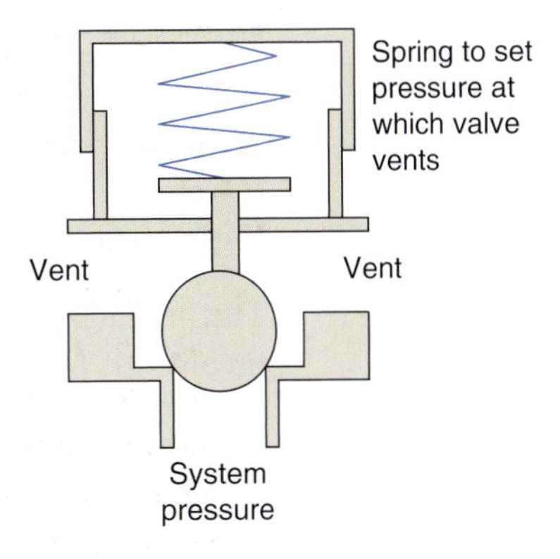

Figure 6.5 Pressure relief valve

valve is held closed by the spring exerting a greater force on the ball than the air pressure and so keeping the valve closed. However, if the pressure rises too high, it exceeds the force exerted on the ball by the spring and the valve opens and vents air from the system.

Ideally all water and oil should have separated out before the air enters the pneumatic system. However, this seldom happens and thus further separation is likely to occur in the distribution pipe work. To assist drainage, the pipe work should be given a fall of about 1 or 2 in 100 in the direction of the flow and provided with suitable drainage points. To allow part of a system to be isolated for perhaps maintenance without affecting the rest of the system, pneumatic systems are often arranged as a ring, shown in Figure 6.3(b). This is rather like the ring main supply arrangement used in households for electricity. Main pipe runs generally use black steel piping with elbow connections where bends are required; smaller brass, copper or aluminium tubing being used for smaller diameter lines. Connections are made by welding, threaded connections, flanges or compression tube connectors where one tube is flared and then clamped over the other. Where flexibility is required, at pressures less than about 6 bar, plastic tubing can be used with connections made with barbed push-on connectors while at higher pressures hoses made of three layers of material are used, an inner tube of synthetic rubber surrounded by metal braiding and then an outer layer of plastic.

Test your knowledge 6.2

(a) What is meant by a pressure of 5 bar?
(b) Why are filters necessary with pneumatic plant?
(c) Why, with pneumatic plant, is a compressor usually followed by a cooler?
(d) What is the function of a receiver in pneumatic plant?
(e) Why is it necessary to have a 'fall' in an airline?

Activity 6.2

Reciprocating compressors may be single acting as shown in Figure 6.4 or double acting. Find out how the double-acting types operate and what their advantages might be.

Hydraulic plant

Figure 6.6 shows the basic elements of a hydraulic circuit.

The production of hydraulic power as shown in Figure 6.6 is likely to involve the following basic elements:

1. The *pump* is driven by a motor which takes hydraulic fluid, via a filter, from a *supply tank* and pressurises it. The supply tank must contain enough hydraulic fluid to meet the circuit demand and still leave some reserve in the tank. Typically this means a volume

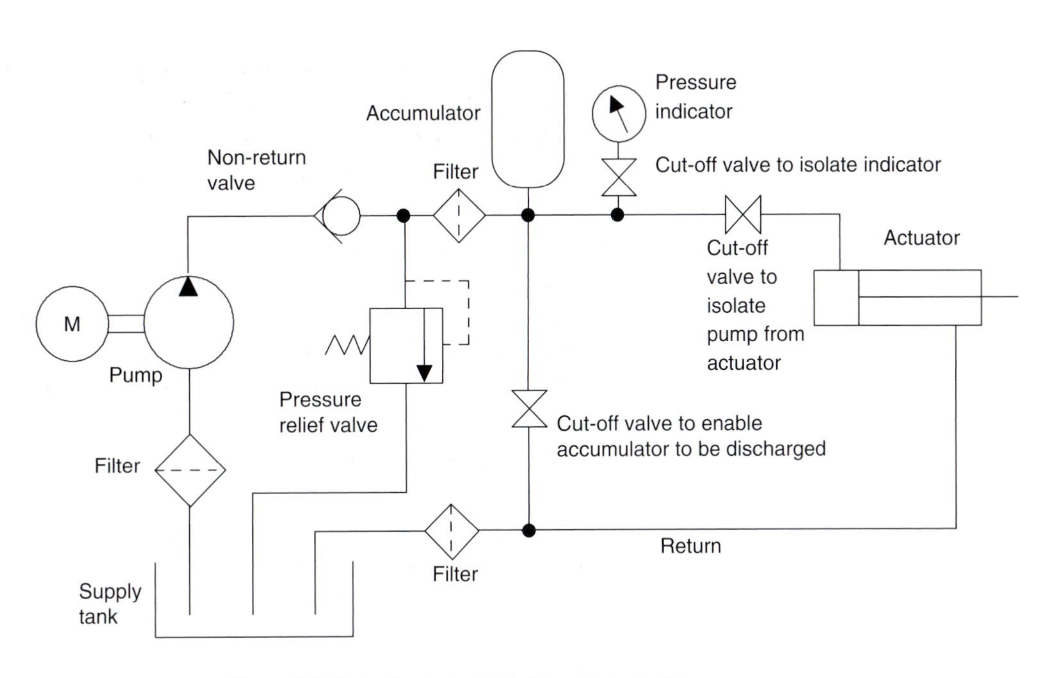

Figure 6.6 Hydraulic plant with double-acting actuator

at least 3 times the volume of fluid delivered in 1 minute. The tank also serves as a heat exchanger to cool the fluid and allow time for contaminants to settle out. A commonly used low cost pump is the gear pump as shown in Figure 6.7. This consists of two close-meshing gear wheels which rotate in opposite directions. Fluid is forced through the pump as it becomes trapped between the rotating gear teeth and the housing, and is thus transferred from the inlet port to be discharged at the outlet port. Hydraulic pumps are specified by the flow rate they can deliver and the maximum pressure they can withstand.

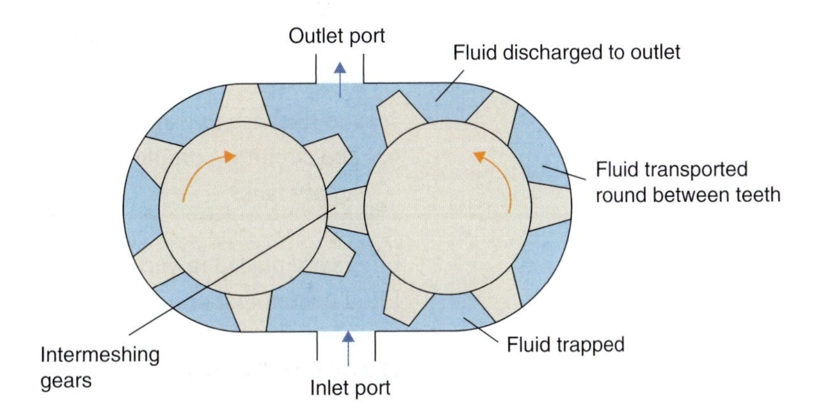

Figure 6.7 Gear pump

2. The fluid is then pumped into the system via a *non-return valve*, this being to prevent the fluid being forced back into the pump as a result of pressure pulses in the system.
3. A *pressure relief valve* is used to keep the system at a safe pressure level.

4. An *accumulator* is included primarily to smooth pressure pulses developed in the pump, cope with high transient flow demands by providing additional flow for short periods, control shock pressure loading due to such events as rapidly closing valves, and provide standby power for situations such as pump failure to allow equipment to return to a safe condition. Most accumulators are the gas-pressurised type shown in Figure 6.8. This has gas within a bladder in the chamber containing the hydraulic fluid. Smaller and older types may be of a spring-loaded design.

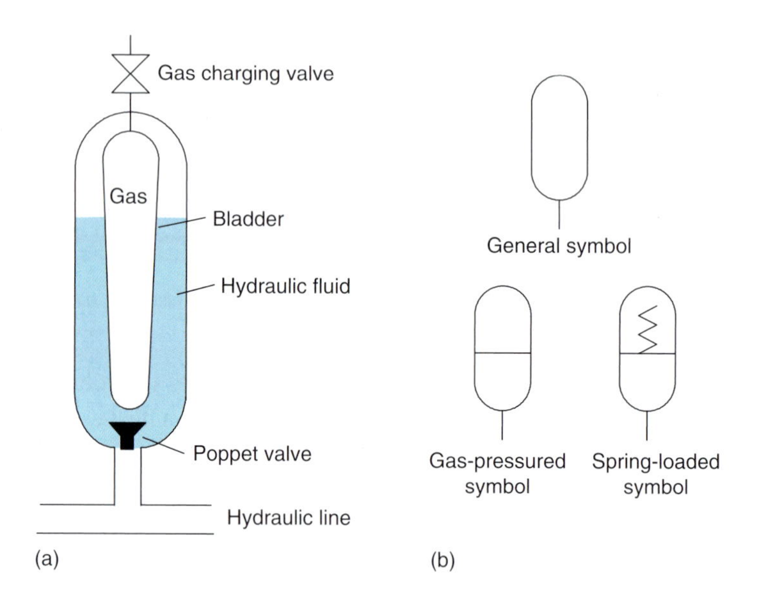

Figure 6.8 Accumulator: (a) gas pressurised, (b) symbols

5. A *shut-off valve* is included to enable the accumulator to be safely discharged, while further shut-off valves are included to enable the load part of the system, in the figure just the simple actuator shown, to be isolated from the pump part in case it has to be disconnected and likewise to isolate the pressure indicator.
6. The circuit is completed by the hydraulic fluid being returned to the supply tank.

Filters are placed in various parts of the circuit to control contamination. Fluid contamination in a hydraulic circuit can cause valves to stick, seals to fail and wear of components. Sources of contamination include debris in the hydraulic fluid as a result of storage prior to use, dirt particles in the hydraulic circuit when it is assembled, particles picked up from the working environment, sludge arising from a high running temperature, etc. Filters are commonly positioned in the following lines of a hydraulic circuit:

- In the inlet line for the fluid from the tank to the pump to protect the pump.
- In the pressure line after the pump to protect actuators and valves.
- In the return line to limit the particles being returned to the tank.

Hydraulic pipelines have to withstand higher pressures than pneumatic ones with the choice being determined by the pressures used. Joints can be made by welding, compression fittings, or threaded connections and flanges. Where flexibility is required, hoses made of three layers of material are used, an inner tube of a material compatible with the hydraulic fluid used surrounded by metal braiding and then an outer layer of material to resist abrasion.

Test your knowledge 6.3

1. What is the function of an accumulator in a hydraulic plant?
2. What dictates the size of the supply tank with hydraulic plant?
3. What are the functions of a non-return valve in a hydraulic plant?
4. What is the function of a pressure relief valve in a hydraulic plant?
5. How are hydraulic pumps specified?
6. Where are filters commonly positioned in hydraulic circuits?

Activity 6.3

In addition to gear pumps, vane pumps and swash plate pumps are used in hydraulic plant. Find out how these kinds of pumps operate.

Control valves

In addition to those already mentioned, the valves used in pneumatic and hydraulic circuits can be divided into the following groups according to what they control:

1. *Directional control valves*
   A directional control valve changes the direction of, or stops, or starts the flow of fluid in some part of the pneumatic/hydraulic circuit on the receipt of an external signal, which might be mechanical, electrical or a fluid pressure pilot signal.
2. *Pressure control valves*
   These are used to control the pressure in parts of a pneumatic/hydraulic circuit.
3. *Flow control valves*
   These are used to control the rate of flow of a fluid in the line of a circuit.

Directional control valves

A directional control valve on the receipt of some external signal, which might be mechanical, electrical or a fluid pilot signal, changes the direction of, or stops, or starts the flow of fluid in some part of the pneumatic/hydraulic circuit.

The basic symbol for a control valve is a square. With a directional control valve two or more squares are used, with each square representing the switching positions provided by the valve. Thus Figure 6.9(a) represents

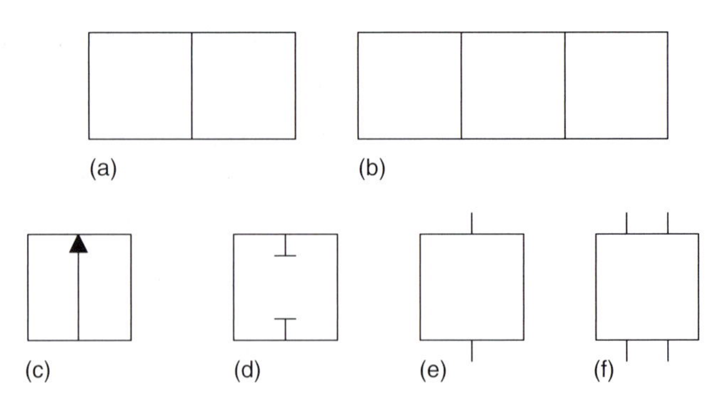

Figure 6.9 Control valve symbols: (a) valve with two positions, (b) valve with three positions, (c) flow path, (d) shut-off connections, (e) valve with two connections and (f) valve with four connections

a valve with two switching positions and Figure 6.9(b) a valve with three switching positions. Lines in the boxes are used to show the flow paths for the switching position with arrows indicating the direction of flow and shut-off positions indicated by lines drawn at right angles as shown in Figures 6.9(c) and (d). The pipe connections, i.e. the inlet and outlet ports to the valve, are indicated by lines drawn on the outside of the box and are drawn for just the 'rest/initial/neutral position', i.e. when the valve is not actuated in Figures 6.9(e) and (f). With a two-position valve, this is generally the right-hand position, with a three-position valve the central position. You can imagine each of the position boxes to be moved by the switching action of some actuator so that it connects up with the pipe positions to give the different connections between the ports.

KEY POINT

The basic symbol for a control valve is a square or a series of squares, with each one representing a switching position.

The ports of a valve are shown on the outside of the box for the initial position and labelled by a number or letter according to its function. Table 6.2 gives the standard numbers and symbols used.

Table 6.2 Port labels

Port	Numbering system	Lettering system
Pressure supply port	1	P
Exhaust port(s)	3 5, 3	R, one exhaust, 3/2 valve R, S, two exhausts, 5/2 valve
Signal outputs	2, 4	B, A
Pilot line opens flows 1–2	12 12	Z, single pilot 3/2 way valve Y, 5/2 way valve
Pilot line opens flow 1–4	14	Z, 5/2 valve
Pilot line flow closed	10	Z, Y
Auxiliary pilot air	81, 91	Pz

KEY POINT

Directional control valves are described by the number of ports and the number of positions.

Directional control valves are described by the number of ports and the number of positions. Thus a 2/2 valve has 2 ports and 2 positions, a 3/2 valve 3 ports and 2 positions, a 4/2 valve 4 ports and 2 positions, a 5/3 valve 5 ports and 3 positions.

Figure 6.10 shows some commonly used examples and their switching options.

Figure 6.10(a) shows a 2/2 valve that is normally closed with no connection between the pressure and output 2. When activated it connects the pressure port to the output port. It is thus an off–on switch.

Figure 6.10(b) shows a 2/2 valve that is normally open with the pressure port connected to the output port 2. When activated it closes both the ports and so switches the pressure to the output off. It is thus an on–off switch.

Figure 6.10(c) shows a 3/2 valve that normally has the pressure to the output off and the output exhausting via the exhaust port. When activated, pressure is applied to the output port and the exhaust port closed.

Figure 6.10(d) shows a 3/2 valve that normally has the pressure to the output and the exhaust port closed. When activated, the pressure is switched off from the output and the output exhausts via the exhaust port.

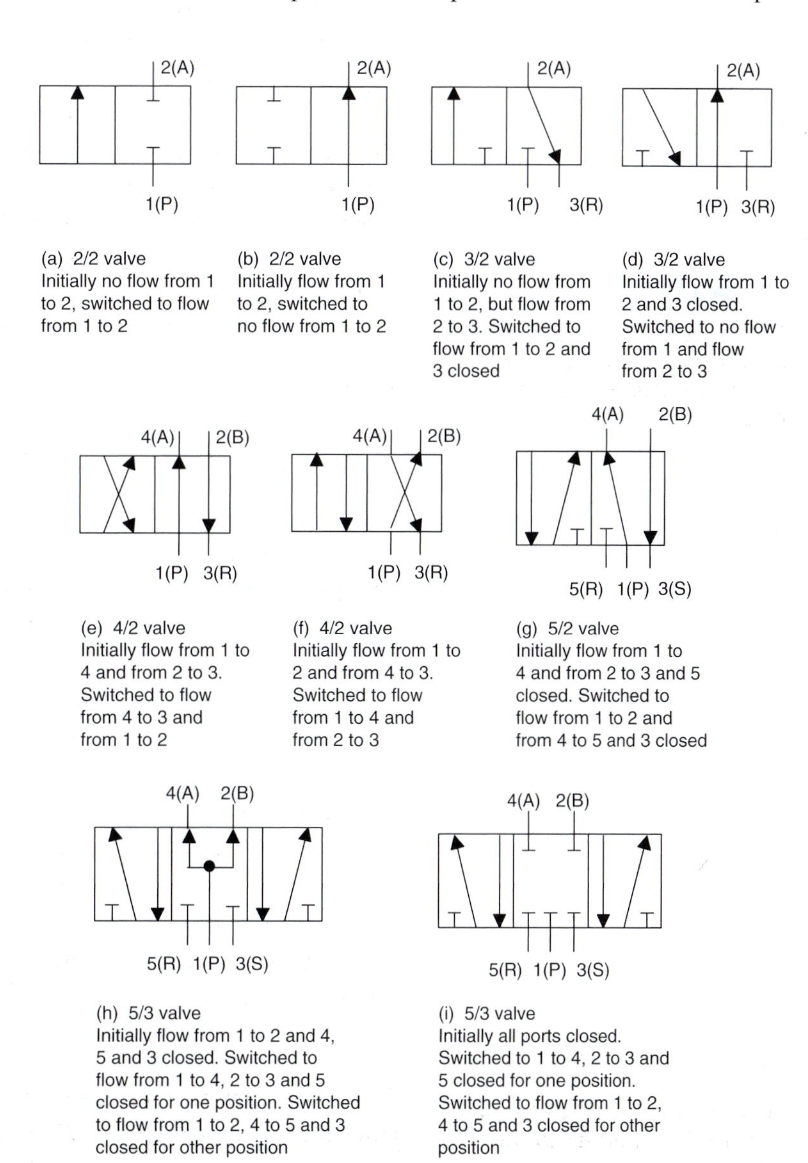

Figure 6.10 Commonly used direction valves

Figure 6.10(e) shows a 4/2 valve that normally has the pressure applied to output 4 while output 3 exhausts through the exhaust port. When activated, the pressure switches to output 3 while output 4 exhausts through the exhaust port.

Figure 6.10(f) shows a 4/2 valve that normally has the pressure to output 2 and output 4 exhausts through the exhaust port. When activated, the pressure switches to output 4 and output 2 exhausts through the exhaust port.

Figure 6.10(g) shows a 5/2 valve that normally has the pressure applied to output 4 while output 2 exhausts through exhaust port 3. When activated, the pressure switches to output 2 while output 4 exhausts through exhaust port 5.

Figure 6.10(h) shows a 5/3 valve that normally has the pressure applied to both outputs 2 and 4. When switched to one position, the pressure is applied to just output port 4 with output 2 exhausting through exhaust port 3. When switched to the other position, the pressure is applied to just output port 2 with output port 4 exhausting through exhaust port 5.

Figure 6.10(i) shows a 5/3 valve with the same switching connections but that is normally switched off rather than on as in Figure 6.10(g).

Figure 6.11 shows the symbols for commonly used valve actuation methods. Thus a complete valve symbol might be as shown in Figure 6.12.

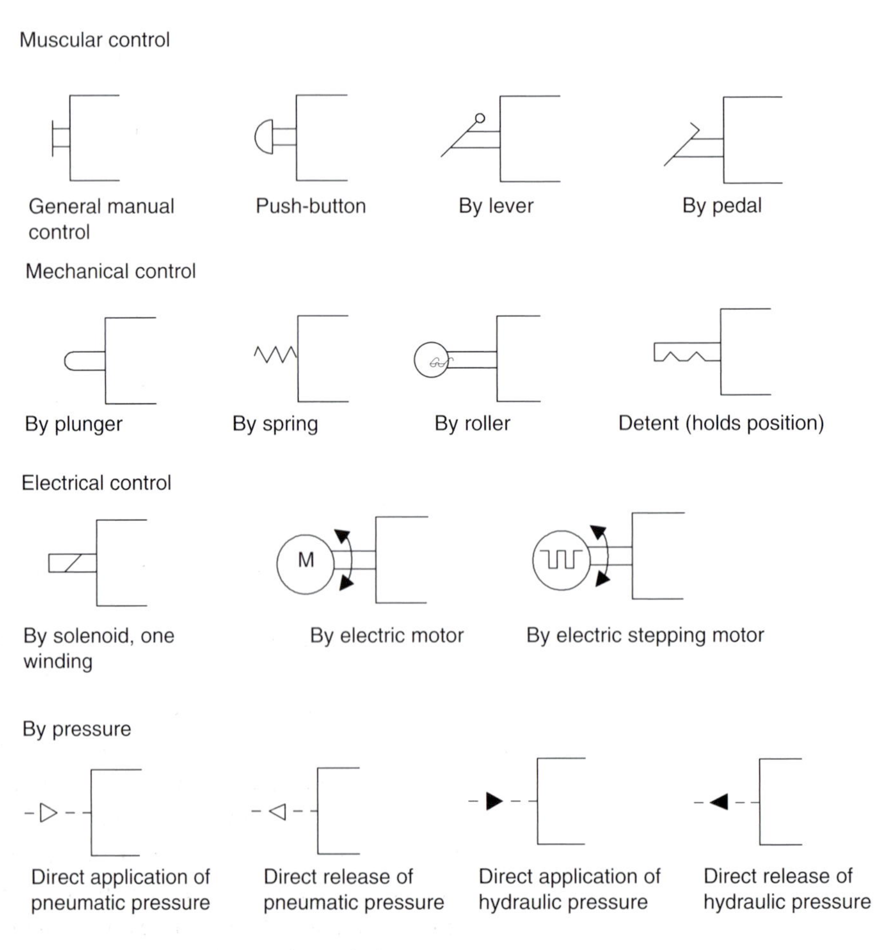

Figure 6.11 Valve actuation methods

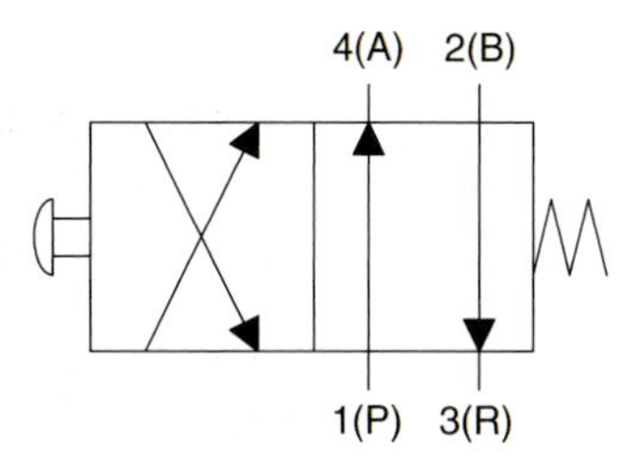

Figure 6.12 Example of a 4/2 valve with push-button and spring actuation

This represents a 4/2 valve which normally has the pressure applied to output port 4 and the other output port 2 is connected to the exhaust port. When the push-button is pressed, the valve switches to its other position and the pressure is now applied to the output port 2 and the output port 4 is connected to the exhaust port. When the push-button is released, the spring causes the initial position to be re-obtained with the pressure again applied to output port 4 and the other output port 2 connected to the exhaust port.

Figure 6.13 shows an example of a 4/3 valve with its actuation. The position with no actuation by either solenoid X or Y is the 'rest' position and has both outputs 2 and 4 switched off and not connected to either the pressure port 1 or exhaust port 3. When solenoid X is activated, the valve switches to the right-hand position and so the pressure is applied to output 4 and output 2 is connected to the exhaust port. When solenoid X is switched off, the springs cause the valve to resume the 'rest' position. When solenoid Y is activated, the valve switches to the left-hand position and so the pressure is applied to output 2 and output 4 is connected to the exhaust port. When solenoid Y is switched off, the springs cause the valve to resume the 'rest' position. Such a valve is said to be spring-centred.

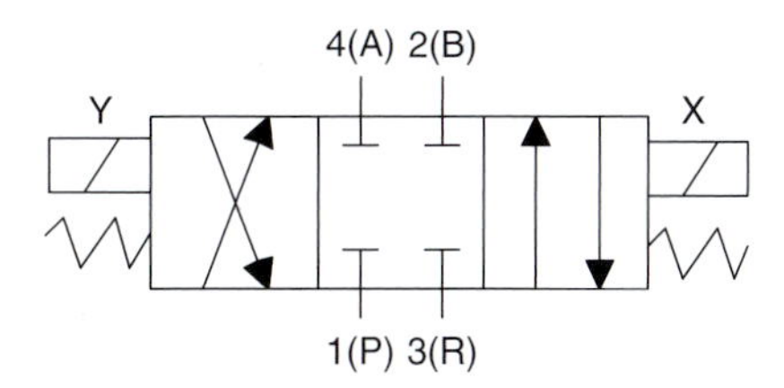

Figure 6.13 Example of a spring-centred 4/3 valve with actuation solenoids

If there is a long distance between a pneumatic valve and the cylinder it is controlling, then the air consumption per cycle can be large due to all the air in the connecting pipe work that has to be moved to carry out the required actuation, e.g. movement of a piston in a cylinder. To obtain a large air consumption means a large bore valve has to be used. With large capacity pneumatic valves, and most hydraulic valves, the operating forces required to operate the valve can be too large for actuation by push-buttons or solenoids. To overcome this, pilot operation is used. A second valve, termed the *pilot valve*, is used to provide pressure signals which can then be used to operate the main valve instead of a push-button or solenoid. In circuit diagrams, pilot pressure lines are shown as dotted.

Figure 6.14 illustrates this use of a pilot valve with a simple pneumatic circuit involving push-button control of a valve producing a pilot signal. When the push-button of the pilot valve is pressed, pilot pressure is produced which then actuates the directional control valve and switches pressure into the cylinder to cause it to extend. When the push-button is released, the pilot pressure signal ceases and the directional control valve reverts to its 'at rest' position, so causing the cylinder to retract.

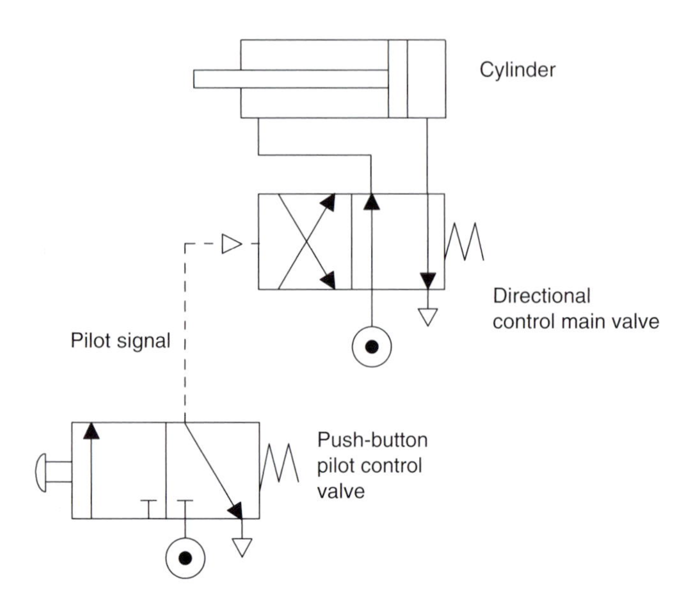

Figure 6.14 Pilot control of the main valve which is used to control the movement of the piston in a double-acting cylinder. Note, to aid interpretation of the figure, the symbols for pressure and exhaust have been used with the valves to indicate which are the pressure and exhaust ports

Figure 6.15 shows a pneumatic circuit for double pilot control of a cylinder, the extension and the retraction being controlled by push-button operated valves producing pilot signals for extension and retraction. The directional control valve here remains in the last state it

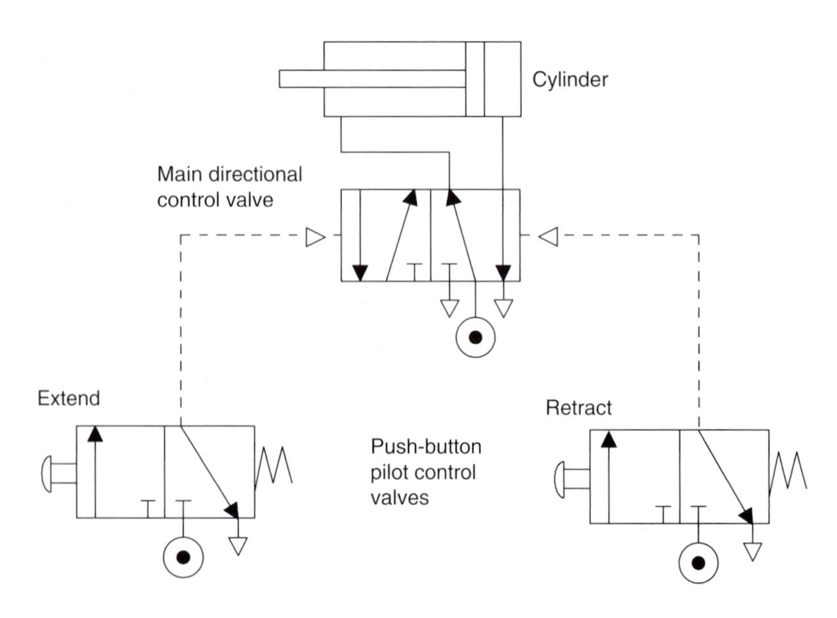

Figure 6.15 Double pilot control

has been put by a pilot signal. Although pilot operation can be achieved by the use of two separate valves, manufacturers supply combined pilot/main valve assemblies.

The most common form of directional control valve is the *spool valve*. This has a spool which can be moved within the valve body and raised areas on the spool, termed lands, block or open ports to give the required valve operation. Figure 6.16 shows the basic principle.

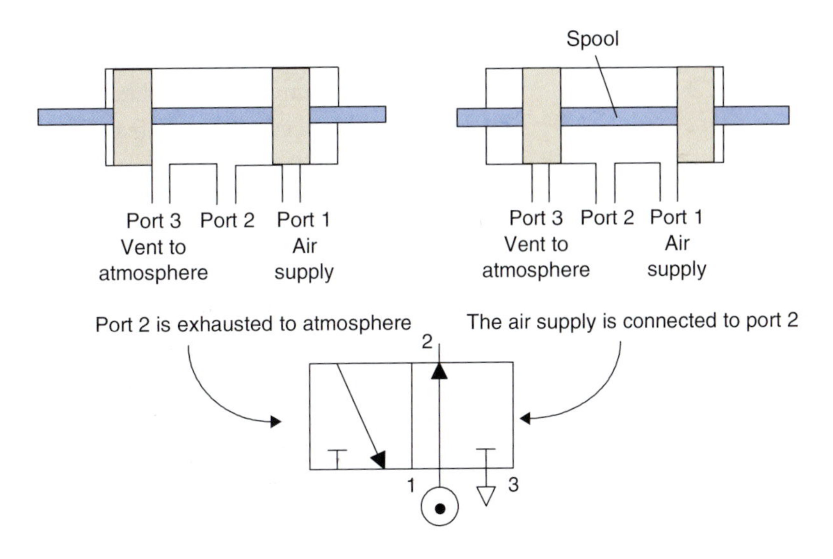

Figure 6.16 A spool valve

Figure 6.17 shows how this principle is applied in a single solenoid 3/2 valve. Passing a current through the solenoid moved the spool in one direction and when there is no such current the spring returns the spool to its 'rest' position.

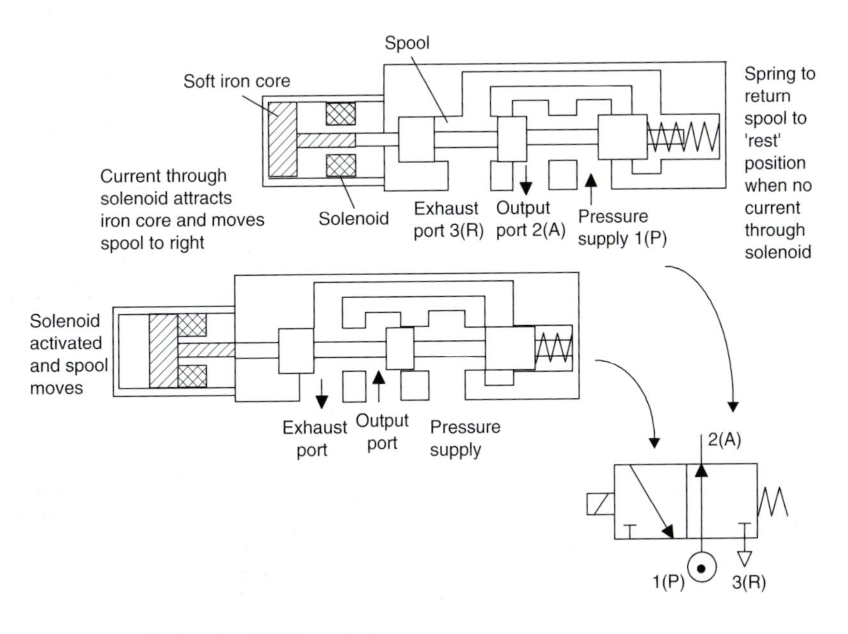

Figure 6.17 Single solenoid 3/2 spool valve

Pressure control valves

There are three main types of pressure control valves. Figure 6.18 shows their symbols:

1. *Pressure regulating valves*
   These are used to control the operating pressure in a circuit and maintain it at a constant value.
2. *Pressure limiting valves*
   These are used as safety devices, often being termed pressure relief valves, and are used to limit the pressure in a circuit to below some safe value. The valve opens and vents to the atmosphere, or back to the sump, if the pressure rises above the set safe value. Such a valve is shown in Figure 6.6 and has one orifice which is normally closed. When the inlet pressure overcomes the force exerted by the spring, the valve opens and vents to the atmosphere, or back to the sump.
3. *Pressure sequence valves*
   These valves are used to sense the pressure of an external line and give a signal when it reaches some preset value. With the pressure relief valve of Figure 6.6, the limiting pressure is set by the pressure at the inlet to the valve. We can adapt such a valve to give a sequence valve and allow flow to occur to some part of the system when the pressure has risen to the required level. For example, in an automatic machine we might require some operation to start when the clamping pressure applied to a work piece is at some particular value.

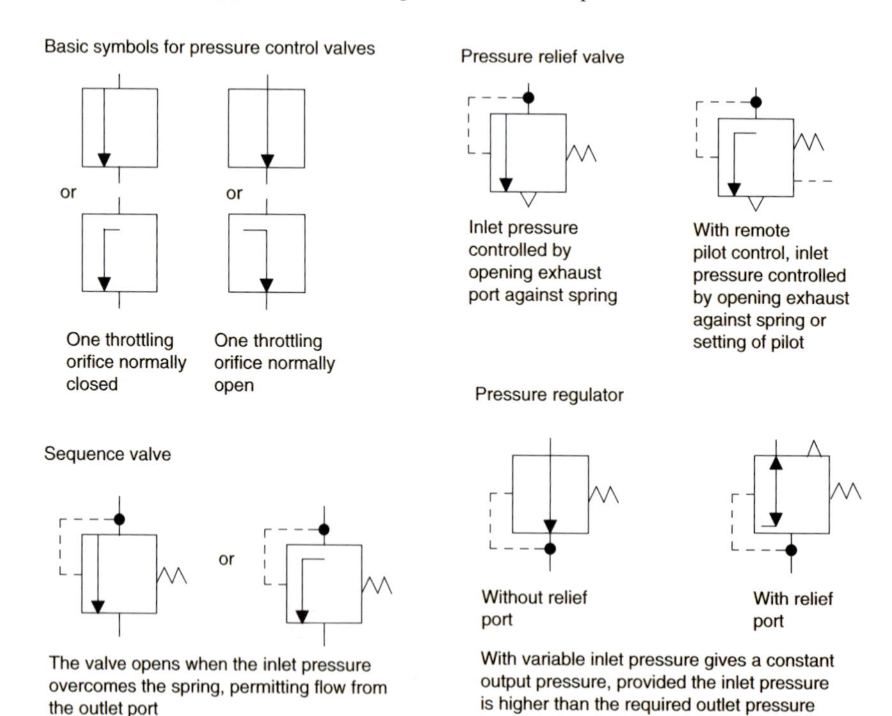

Figure 6.18 Symbols for pressure control valves

Figure 6.19 shows a system where such a valve is used. When the 4/3 valve first operates, the pressure is applied to cylinder 1 and its piston moves to the right. While this is happening the pressure is too low to operate the sequence valve and so no pressure is applied to cylinder 2.

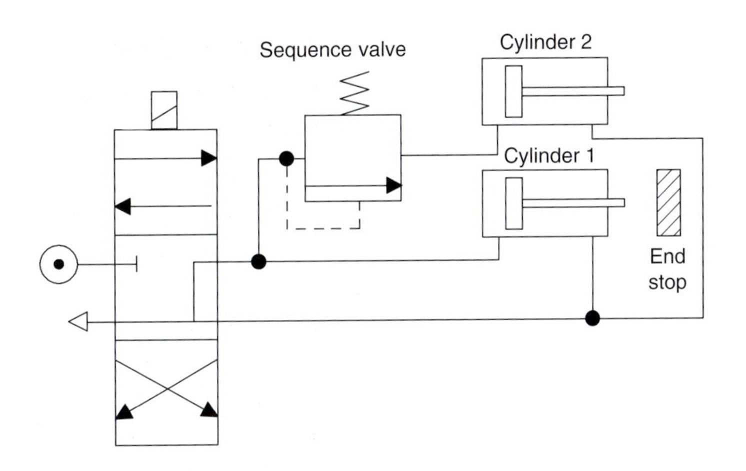

Figure 6.19 Sequential system

When the piston of cylinder 1 reaches the end stop, then the pressure in the system rises and, at an appropriate level, triggers the sequence valve to open and so apply pressure to cylinder 2 to start its piston in motion.

Flow control valves

The *flow control valve* is used to restrict the flow of fluid in a particular direction to control the rate of flow. They work by placing a variable restriction in the flow path. Figure 6.20 shows the basic principle and the symbols used. The valve has a throttling screw which can be used to adjust the size of the restriction to give the required flow. The shape of the plug determines how the movement of the throttle screw affects the rate of flow.

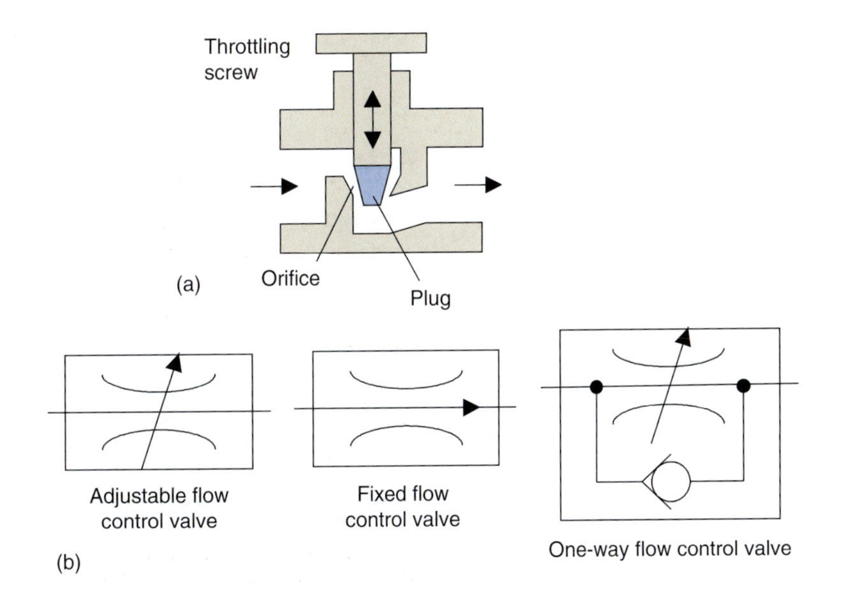

Figure 6.20 Flow control valve: (a) basic principle and (b) symbols

The *one-way flow control valve* permits flow adjustment in one direction only, the other valves permitting flow adjustment for the flow of air in both directions through the valve. It is essentially a flow control valve combined with a non-return valve. The non-return valve is used to allow fluid to flow through the valve in one direction only, blocking the flow in the reverse direction. The simplest form of non-return valve is shown in Figure 6.21. The pressure flow from left to right causes the ball to move and allow the fluid through the valve. The pressure flow from right to left forces the ball to close the valve and so prevent flow in that direction.

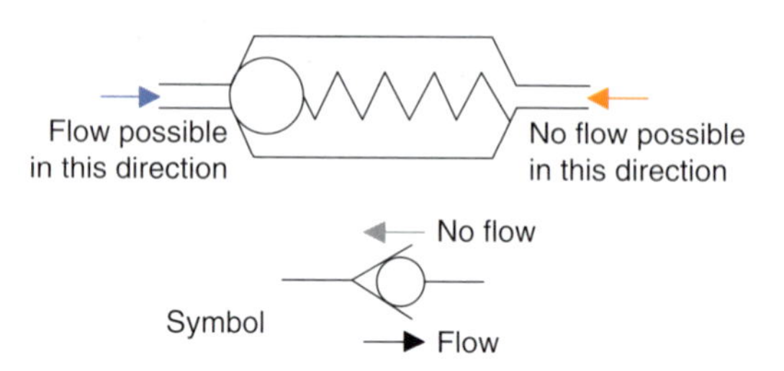

Figure 6.21 Non-return valve

Test your knowledge 6.4

1. What is the function of a direction control valve?
2. Sketch the symbol for a 3/2 control valve that normally has the pressure to the output off and the output exhausting via the exhaust port and which when activated, pressure is applied to the output port and the exhaust port closed.
3. In what different ways can a flow control valve be actuated?
4. What are the three main types of pressure control valve?
5. How does a one-way flow control valve operate?

Actuators

KEY POINT

An actuator is a device that is used to apply a force to an object to move it or hold it securely in position.

Fluid power actuators can be classified into two groups:

1. *Linear actuators* are used to move an object or apply a force in a straight line. The basic linear actuator is a cylinder of the form shown in Figure 6.1 Linear actuators can be divided into two types, single-acting cylinders and double-acting cylinders.
2. *Rotary actuators* are used to move an object in a circular path. Rotary actuators are the fluid power equivalent of an electric motor.

Linear single-acting cylinder

KEY POINT

A single-acting cylinder is powered by fluid being applied to one side of the piston to move it in one direction, it being returned in the other direction by an internal spring or some external force.

A *single-acting cylinder* is one powered by fluid being applied to one side of the piston to give movement of the piston in one direction, it being returned in the other direction by an internal spring or some external force. The other side of the piston is open to the atmosphere.

Figure 6.22 shows one with a spring return and Figure 6.23 shows how such a single-acting cylinder can be directly controlled by using a normally closed 3/2 valve. When the push-button is pressed, compressed air is admitted to the left-hand end of the cylinder and forces the piston to the right. When the button is released, the air to the left of the piston is exhausted and so the piston, under the action of the return spring, returns to its initial position.

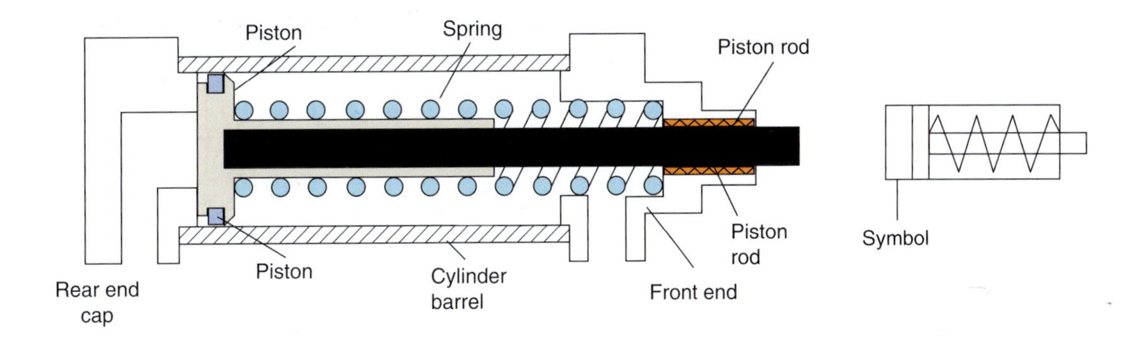

Figure 6.22 Single-acting cylinder

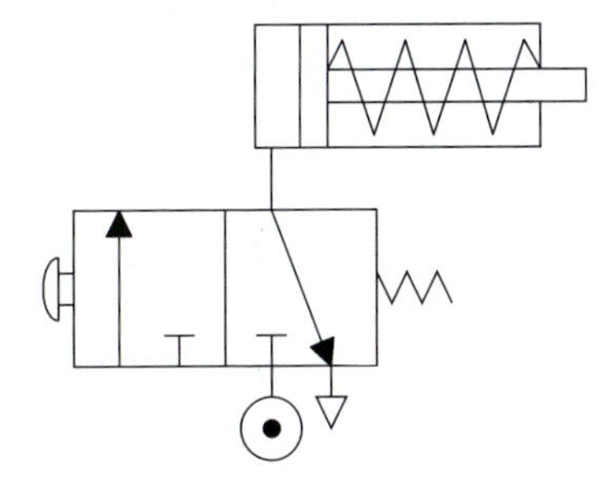

Figure 6.23 Direction control of a single-acting cylinder

Linear double-acting cylinder

A *double-acting cylinder* is one powered by fluid for the movement of the piston in both extend and return directions. With such a cylinder, two ports are used alternatively as supply and exhaust ports, so that pressure is used to give both extend and return strokes. Thus, unlike the single-acting cylinder, no spring is required for the return stroke. Because of this, the double-acting cylinder is able to do work on both extend and return strokes, the single-acting cylinder only being able to do work on the extend stroke. Figure 6.24 shows the basic form of such a cylinder and Figure 6.25 shows how such a double-acting cylinder could be controlled by the use of a lever set, spring reset valve.

KEY POINT

A double-acting cylinder is one powered by fluid for the movement of the piston in both extend and return directions.

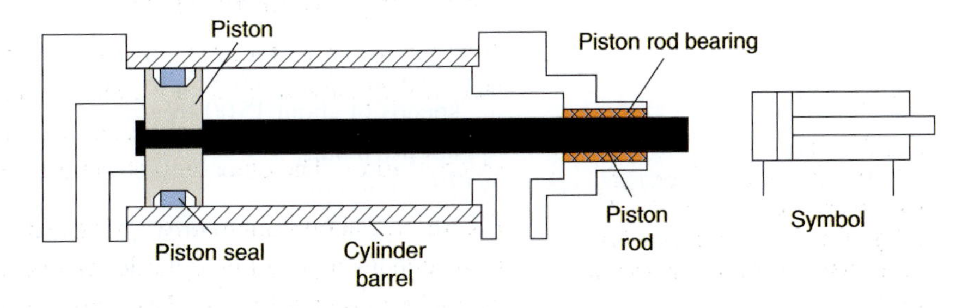

Figure 6.24 Double-acting cylinder with single piston rod

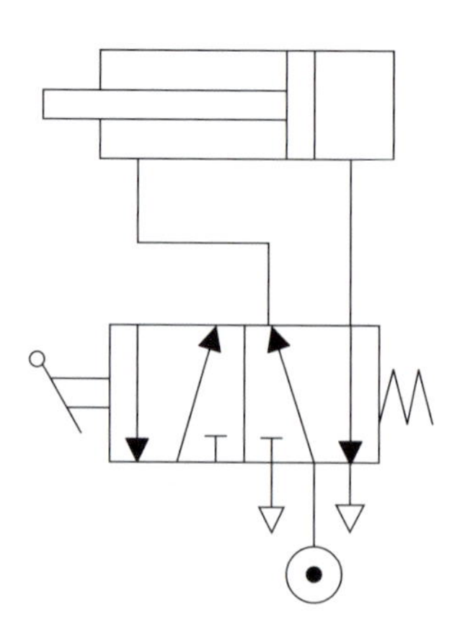

Figure 6.25 Direct control

Rotary motion

While the use of rack and pinion drives or levers can be used to obtain some rotary movement from a linear actuator, they are only able to give rotation through an angle which is less than 360°. Air motors, however, can give a continuous rotation and so unlimited angular rotation. Commonly used air motors are vane motors, gear motors, piston motors and turbine motors.

Figure 6.26(a) shows symbols for air motors. Compressors and motors are very similar, the graphical symbols used for the two reflecting this. Commonly used motors are:

1. The *vane motor* (Figure 6.26(b))
   An eccentric rotor has slots in which vanes are forced outwards against the walls of the cylinder by the rotation. The vanes divide the chamber into separate compartments, each a different size. Compressed air enters one such compartment as its volume is getting bigger. The air pressure on the vanes creates a force and hence a torque which causes the rotor to rotate, the compressed air being released from the exhaust port. Such a motor can be made to reverse its direction of rotation by using a different inlet port for the compressed air, one on the opposite side of the cylinder to the other inlet port. Vane motors are available to supply powers from less than 1 kW up to about 20 kW. Maximum speed tends to be about 30 000 rev/min. The maximum operating pressure is about 8 bar.
2. The *gear motor* (Figure 6.26(c))
   The fluid entering the upper chamber produces a pressure difference between the upper and lower chambers which results in the gears rotating. They typically have powers from about 0.5–5 kW, maximum speeds of about 15 000 rev/min and a maximum operating pressure of about 10 bar.
3. *Piston motors*
   These can be used to provide more power than vane or gear motors, typically in the range 1.5–30 kW, and are more efficient since they suffer from less leakage. They have speeds of up to about 5000 rev/min and maximum operating pressures of 10 bar. They can easily

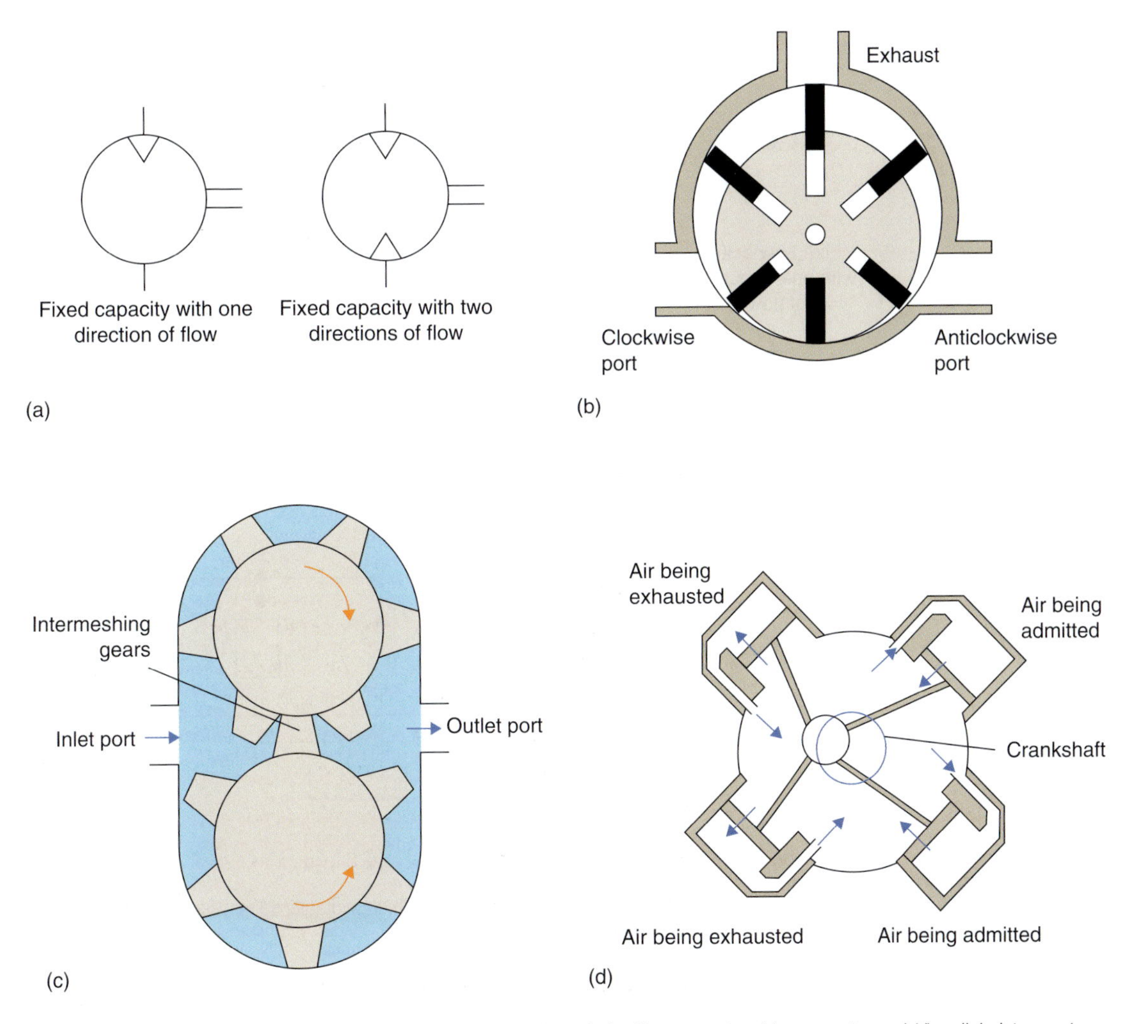

Figure 6.26 Air motors: (a) symbols, (b) vane motor, (c) gear motor and (d) radial piston motor

be reversed by reversing the direction of airflow. However, they are much more bulky and have a more complex form of construction than vane or gear motors. Figure 6.26(d) shows the basic form of a radial piston motor. It has a rotary valve which applies compressed air to above two of the pistons and exhausts air from above the other two pistons. The result is movement of the pistons which causes the crankshaft to rotate. In doing so the rotary valve rotates and the air pressure and exhausts are switched round the pistons in sequence, so maintaining the rotation.

Figure 6.27(a) shows the symbols used for hydraulic motors. The principles of hydraulic and pneumatic motors are very similar. Commonly used hydraulic motors are the gear motor and the axial piston motor. Piston motors are generally the most efficient and give the highest torques, speeds and powers. Hydraulic transmission consists of a pump, driven by a petrol engine, a diesel engine, an electric motor or some other prime mover, and one or more hydraulic motors and is used to transmit power from one rotating shaft to another at a distance from the first.

Such a transmission is used in motor vehicles. Figure 6.27(b) shows the basic form of such a transmission. The shaft of the pump is driven by

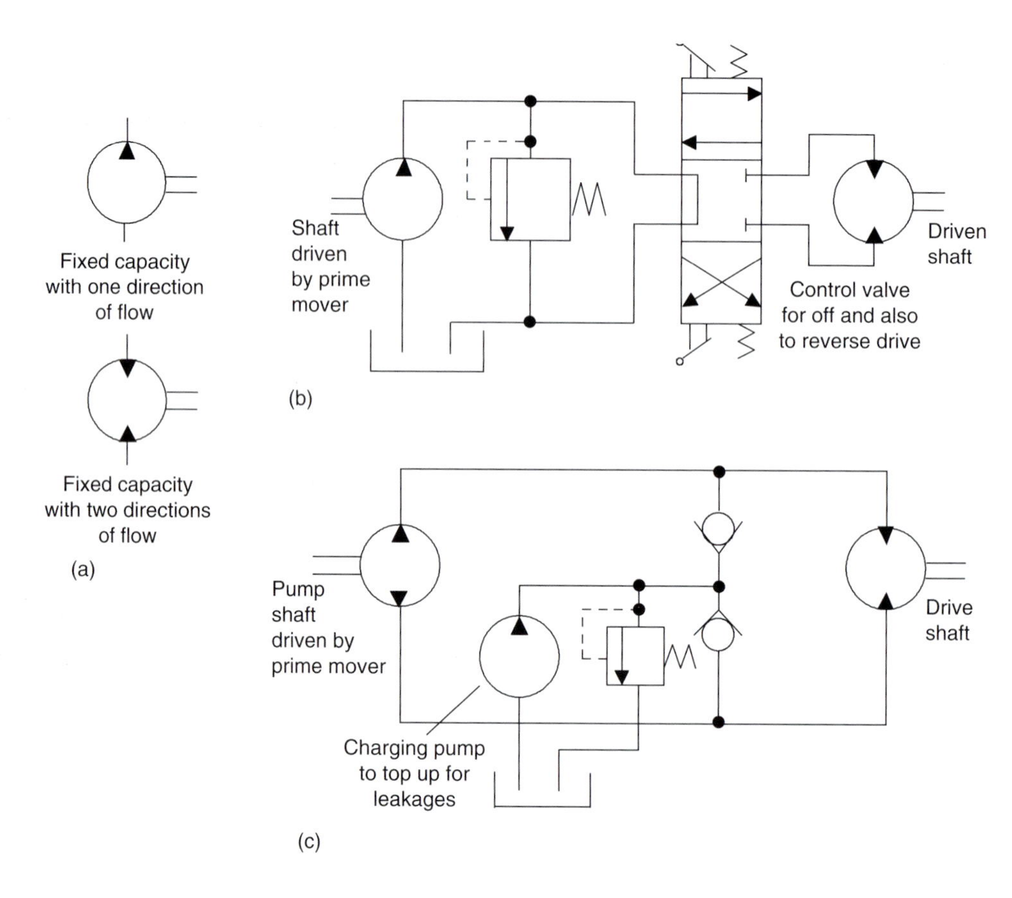

Figure 6.27 (a) Symbols, (b) open circuit hydraulic transmission and (c) closed circuit hydraulic transmission

the prime mover and pumps oil from the reservoir to the control valve. A pressure relief valve is included. When the upper lever of the control valve is used, the flow passes one way through the motor, when the other lever is used the flow passes in the opposite direction through the motor. The result is rotation of the drive shaft in one direction or the reverse, depending on which lever is used. Reversal is thus effected by a directional control valve. With this transmission system the exhausting fluid from the motor is discharged directly into the reservoir tank. Such a system is termed *open circuit*. Figure 6.27(c) shows an alternative where the discharge from the motor is fed directly back to the inlet side of the pump, no passage via the reservoir being required. This is termed a *closed circuit* system. The drive shaft can be reversed by reversing the pump, or by the use of a directional control valve.

Test your knowledge 6.5

1. What are the two basic classes of actuator?
2. How is the piston returned in a single-acting linear actuator cylinder?
3. If work is required to be done on both extend and return strokes of a linear actuator cylinder, should a single-acting or a double-acting cylinder be chosen?
4. What are the graphical symbols used to represent compressors and air motors?
5. Which kind of air and hydraulic motor is able to deliver the most power?

Activity 6.4

Look up the component symbols used in Figure 6.28 and determine the mode of operation of the circuit.

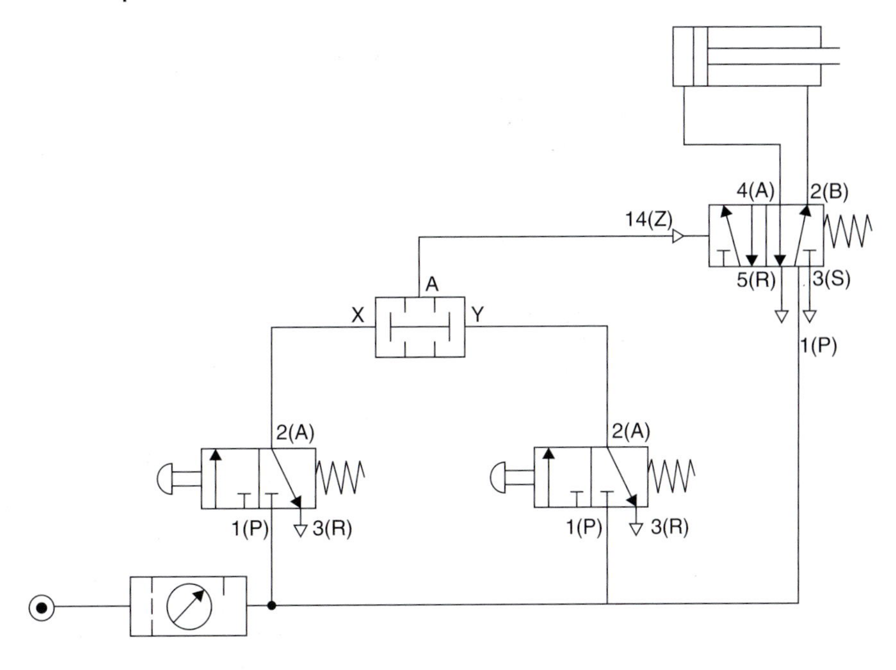

Figure 6.28

Sensors

The term *sensor* is used with instrumentation and control systems to describe the element which takes information about the variable being measured or monitored and changes it into some form which enables the rest of the measurement or control system to utilise it. Suppose a pneumatic machine is required to push a printing head onto each package produced from a packaging machine and print the name and address of the company. The packages may be of different sizes and not spaced regularly on the delivery belt. What is required is a means of determining when a package is in the correct position for a cylinder actuator to be actuated and push the printing head onto the face of the package. A position sensor is required. This might be some device which senses the package by coming into contact with it, e.g. the package might move a roller or plunger or operate an electrical switch, or it might be some device which is non-contact and detects the presence of the package by operating a reed switch or other electrical non-contact sensor. The sensor might then be used to mechanically, or by a pneumatic pilot signal, or electrically, actuate a valve.

KEY POINT

Sensors take information about the variable being measured or monitored and change it into some form which enables the control system to utilise it.

Mechanical contact sensors are very widely used for position sensing, common forms involving roller and plunger valves as in Figure 6.29. The object being detected impinges on the end of the roller or plunger and results in its movement, so switching the valve from one position to another. Precise end-position control is possible with the valve remaining switched into its new position as long as the object is in contact with the roller or plunger.

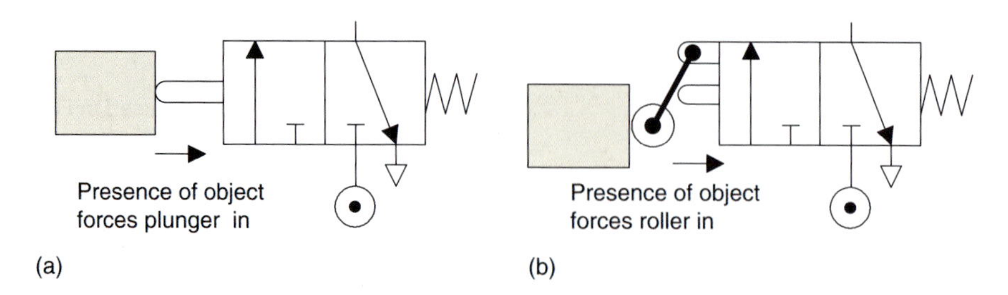

Figure 6.29 Roller and plunger position sensors

Figure 6.30 gives an example of a one-way trip valve. Such a valve uses a roller but the roller is so pivoted that it is only actuated when the projection from the object moves in just one direction past the roller. The valve is actuated and switches position for only as long as the projection depresses the roller by the right amount. Thus the trip valve can be used to give a pulse output for a period of time as the object moves past the roller. There are a number of different types of non-contacting switches, some of which are only suitable for switching as a result of the proximity of metal objects while others work with both metallic and non-metallic objects.

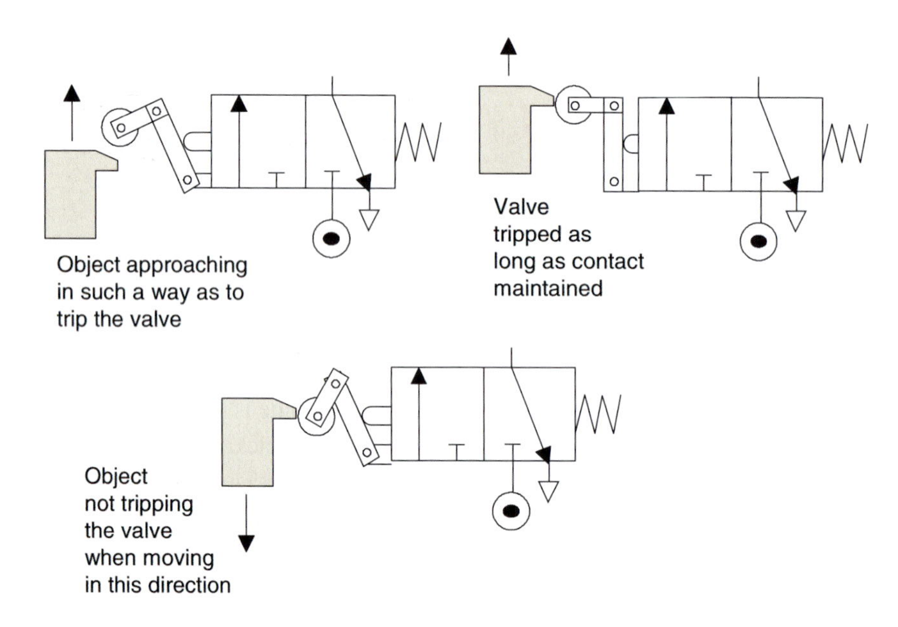

Figure 6.30 One-way trip valve

1. The *eddy current* type of proximity switch shown in Figure 6.31(a) has a coil which is energised by a constant alternating current. This current produces an alternating magnetic field. When a metallic object is close to it, eddy currents are induced in it. The magnetic field due to these eddy currents induces an e.m.f. back in the coil with the result that the voltage amplitude needed to maintain the constant current changes. The voltage amplitude is thus a measure of the proximity of metallic objects. The voltage can be used to actuate an electronic switch circuit and so give an on–off device. The range over which this sensor can operate is typically about 0.5–20 mm.

CHAPTER 6

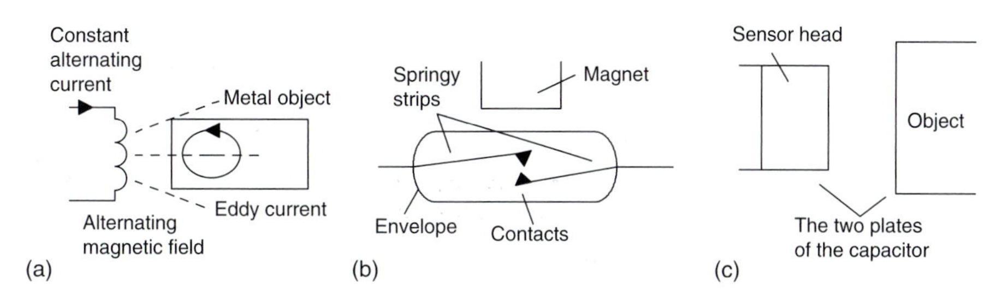

Figure 6.31 Proximity switches: (a) eddy current, (b) reed switch and (c) capacitive

2. A very commonly used proximity sensor is the *reed switch*. This consists of two overlapping, but not touching, strips of springy ferromagnetic material sealed in a glass or plastic envelope as shown in Figure 6.31(b). When a magnet or current carrying coil is brought close to the switch, the strips become magnetised and attract each other, so closing the contacts. Typically the contacts close when the magnet is about 1 mm from the switch.
3. A proximity switch that can be used with metallic and non-metallic objects is the *capacitive proximity switch* shown in Figure 6.31(c). For a parallel plate capacitor, the capacitance depends on the separation of the plates and the dielectric between them. Thus the approach of a metallic object to the plate of the capacitive sensor changes its capacitance and also the approach of a non-metallic object changes its capacitance because of a change of dielectric. The change in capacitance can be used to activate an electronic switch circuit and so give an on–off switch. Capacitive proximity switches can be used to detect objects between about 4 and 60 mm from the sensor head.
4. Another form of non-contact proximity sensor involves *photoelectric devices*, i.e. light-emitting diodes (LED) and photodetectors. With the transmission form, the object being detected prevents the beam of light from the LED reaching the photodiode and so changes its output current. With the reflective form the presence of the object reflects light onto the detector and so changes its output current.

Position sensors can be used to initiate operation of the control valves in automated processes. Figure 6.32 shows a simple circuit for the automatic return of a cylinder when it reaches the end of its stroke. The extend valve is switched by pressing the push-button on the start valve and extension occurs. When the piston rod reaches the end of its stroke, it trips a roller limit switch which then switches the retract valve so that the piston then retracts.

Situations often occur where it is necessary to activate a number of cylinders in some sequence of events. Event 2 might have to start when event 1 is completed, event 3 when event 2 has been completed, etc. For example, we might have: only when cylinder A is fully extended (event 1) can cylinder B start extending (event 2), and cylinder A can only start retracting (event 3) when cylinder B has fully extended (event 2). In discussions of sequential control it is common practice to give each cylinder a reference letter A, B, C, D, etc., and to indicate the state of each cylinder by using a + sign if it is extended or a − sign if retracted.

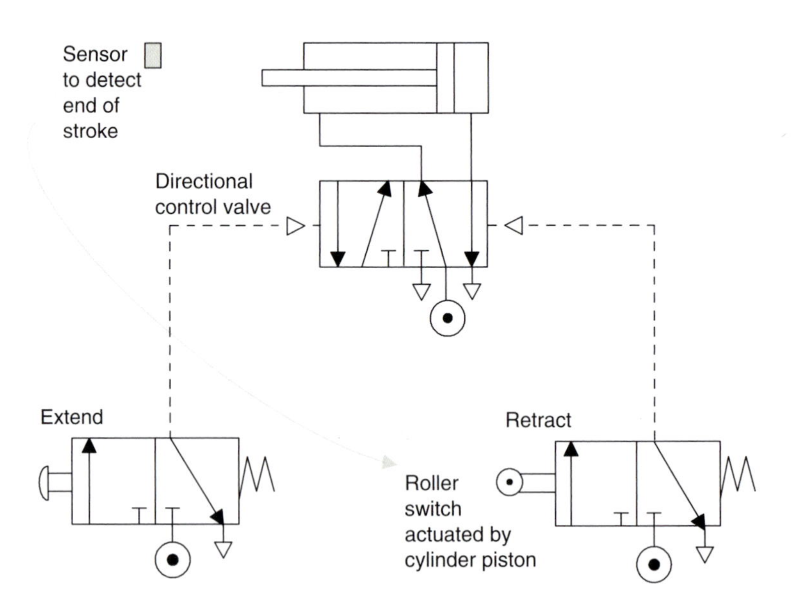

Figure 6.32 Automatic return

KEY POINT

In discussions of sequential control it is common practice to give each cylinder a reference letter A, B, C, D, etc., and to indicate the state of each cylinder by using a + sign if it is extended or a − sign if retracted.

Thus a sequence of operations might be shown as A+, B−, A−, B+. This indicates that the sequence of events is cylinder A extends, followed by cylinder B being extended, followed by cylinder A retracting, followed by cylinder B retracting.

Figure 6.33 shows a circuit that could be used to generate this sequence for two cylinders A and B and this is illustrated with a *displacement-step diagram*, such a diagram showing the displacements of the cylinders at each step in the operation.

To generate this displacement-event diagram the sequence of operations is:

Event 1

1. Start push-button pressed.
2. Cylinder A extends, releasing limit switch a−.

Event 2

3. Cylinder A fully extended, limit switch a+ operated to start B extending.
4. Cylinder B extends, limit switch b− released.

Event 3

5. Cylinder B fully extended, limit switch b+ operated to start cylinder A retracting.
6. Cylinder A retracts, limit switch a+ released.

Event 4

7. Cylinder A fully retracted, limit switch a− operated to start cylinder B retracting.
8. Cylinder B retracts, limit switch b− released.

Event 5

9. Cylinder B fully retracted, limit switch b− operated to complete the cycle.

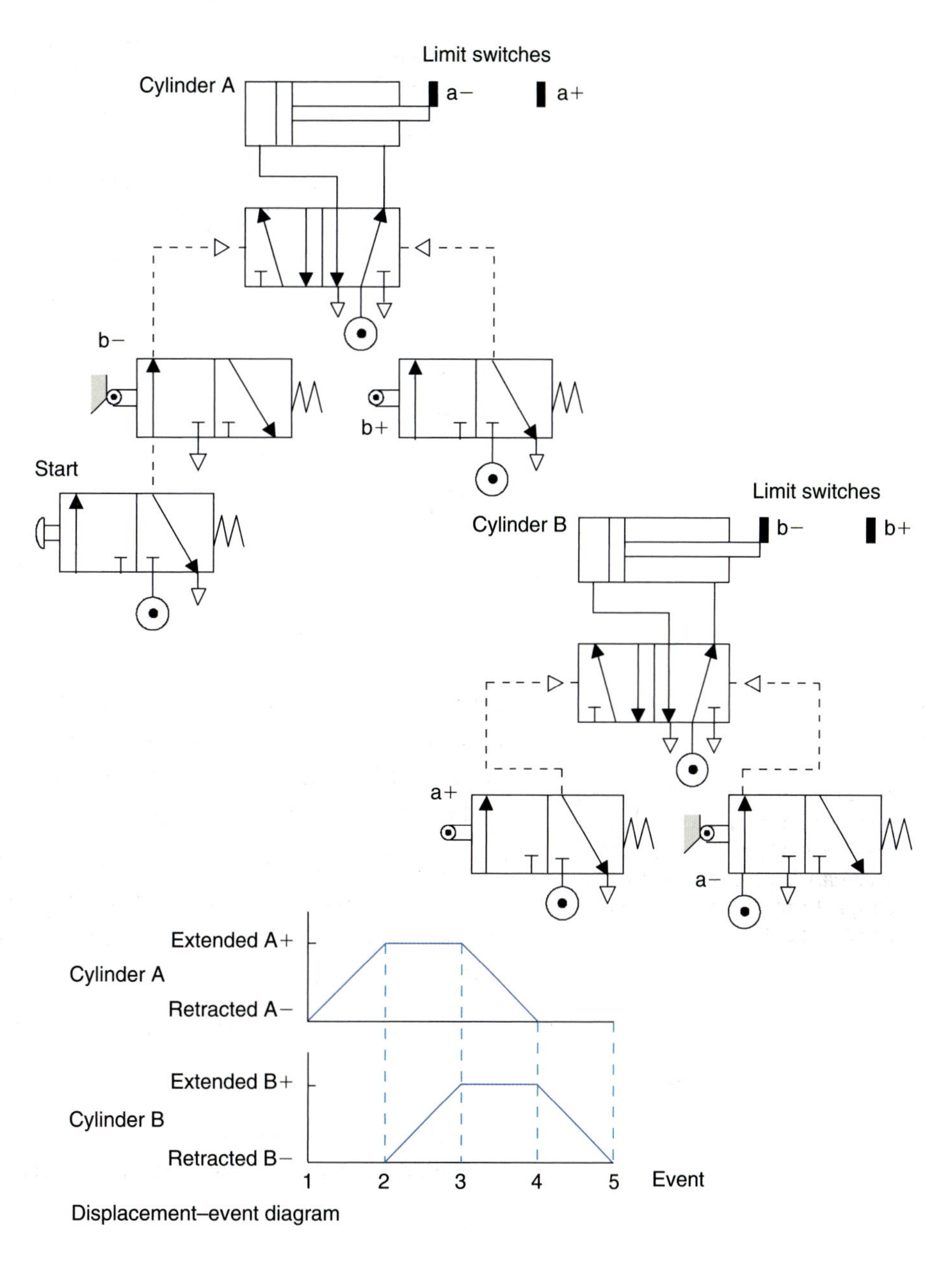

Figure 6.33 Two-actuator sequential operation

The cycle can be started again by pushing the start button. If we wanted the system to run continuously then the last movement in the sequence would have to trigger the first movement.

Test your knowledge 6.6

1. How might a mechanical contact sensor operate?
2. How does an eddy current proximity sensor operate?
3. How are light-emitting diode and photodiode combinations used to detect proximity?
4. What is a displacement-step diagram?
5. Describe the cylinder actions that will occur for the sequence A+, B+, B−, A−.

Programmable logic controllers

Solenoid directional control valves are controlled by switching the current to them on or off. This might be done by the use of electrical relays, the circuit being wired up to give the required sequence of signals to the valves. Such circuits are wired up for the sequence required. If a different sequence is required then the system has to be rewired. An alternative is to use *programmable logic controllers* (PLCs). These are microprocessor-based controllers that use a programmable memory to store instructions and implement functions to control machines and processes. Thus with inputs to the PLC of signals from sensors such as limit switches and start/stop switches, output signals are sent to the solenoids of the valves, the sequence of such signals being determined by the programme of instructions inputted to the PLC. Figure 6.33 shows the arrangement.

The PLC has the start and stop switches and the sensors connected as inputs and the solenoid coils as outputs. When the start switch is closed, the PLC starts its programme. It monitors the inputs from the limit switches and, on the results of that monitoring, gives outputs to energise the relevant solenoid coils. To change the switching sequence all that has to be done is to change the programme. With the elements shown in Figure 6.34 we might want to generate sequence A+, B+, A−, B−. The programme would thus have to be:

When start switch closed and input from limit switch b−
Output to solenoid A+

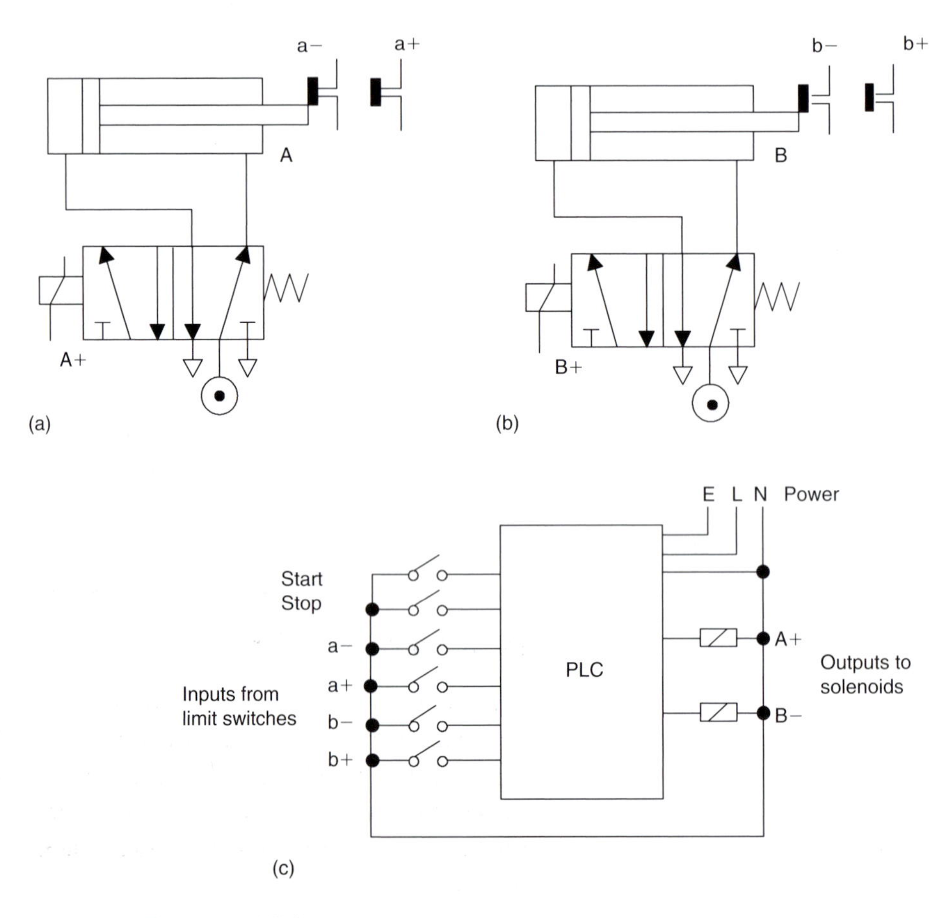

Figure 6.34 A PLC system

KEY POINT

Programmable logic controllers (PLCs) are microprocessor-based controllers that use a programmable memory to store instructions and implement functions to control machines and processes.

Latch output A+ until input from limit switch b+
When input to limit switch a+
Output to solenoid B+
Latch output B+ until input from limit switch a −
The programme then repeats itself

If we wanted a system to clamp a work piece and then drill a hole in it, then we might use the arrangement shown in Figure 6.35 and programme it to give the displacement-event diagram shown. Cylinder A is responsible for the clamping and cylinder B for lowering the drill into the work piece.

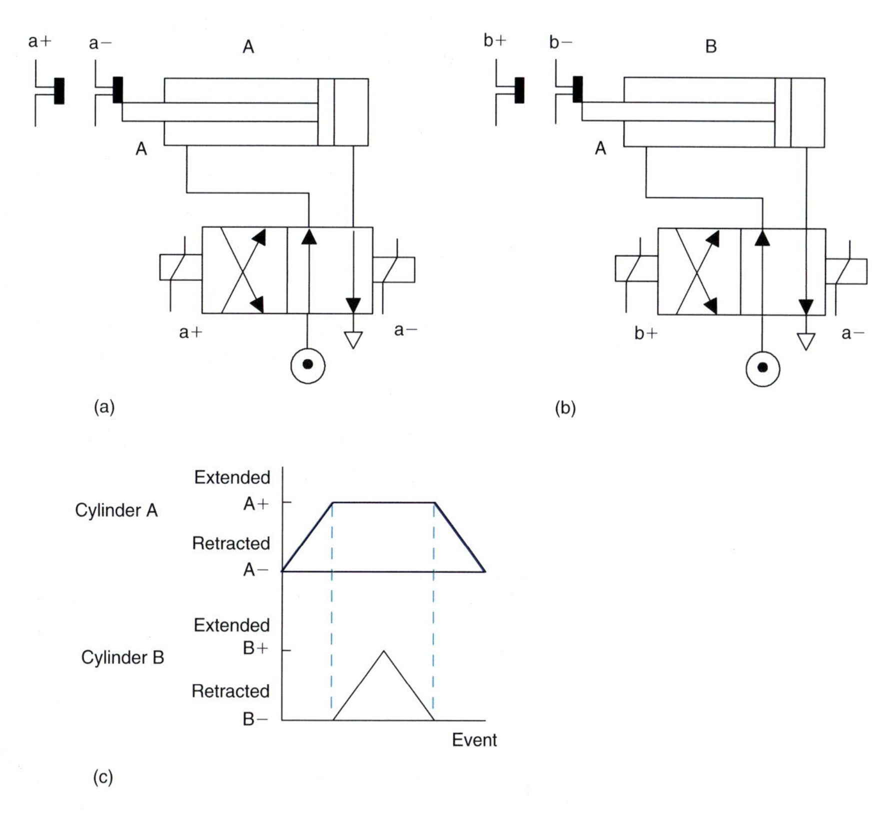

Figure 6.35 Clamp and drill

To check your understanding of the preceding section, you can solve Review questions 1–7 at the end of this chapter.

Fluid Power Principles

Temperature and pressure

Temperature is normally measured on the Celsius scale (°C) which has the freezing point of water, 0°C as its origin. This is fine for day to day measurement of temperature but gas calculations often require the measurement to be *absolute temperature* which is measured in kelvin (K). The kelvin scale has the absolute zero of temperature (−273°C) as its origin and so:

$$\text{Absolute temperature (K)} = \text{Degrees celsius (°C)} + 273$$

Pressure is defined as the force per unit area. When pressure is measured relative to the atmospheric pressure it is termed the *gauge pressure*, a pressure measured from absolute zero pressure being termed the *absolute pressure*:

$$\text{Absolute pressure} = \text{Gauge pressure} + \text{Atmospheric pressure}$$

The SI unit of pressure is the pascal (Pa) with 1 Pa being a force of 1 N acting on an area of 1 m^2. The pascal is a small unit and the prefixes kilo and mega are generally used for the types of pressures encountered in pneumatic and hydraulic systems. The atmospheric pressure varies with both location and time but for most calculations involving fluid power systems can be taken as being constant at 100 kPa or 10^5 Pa. Pressures are often expressed in terms of multiples of this pressure with 10^5 Pa being termed a pressure of 1 bar.

Pascal's laws

Blaise Pascal, in about the year 1650, determined laws governing how fluids transmit power. These are:

1. Provided the effect of the weight of a fluid can be neglected, the pressure is the same throughout an enclosed volume of fluid at rest.
2. The static pressure acts equally in all directions at the same time.
3. The static pressure always acts at right angles to any surface in contact with the fluid.

As a consequence of these laws, when a pressure is applied to one end of an enclosed volume of fluid the pressure is transmitted equally and undiminished to every other part of the fluid.

Gas laws

Air can be considered to be a reasonable approximation to an ideal gas for the range of pressures and temperatures occurring with pneumatic systems and thus obey the *ideal gas laws*. These are:

1. *Boyle's law*
   In an ideal gas in which the mass and temperature remain constant, the volume V varies inversely as the absolute pressure p, i.e.:

$$pV = a\ constant \tag{6.1}$$

2. *Charles' law*
   In an ideal gas in which the mass and the pressure remain constant, the volume V varies directly as the absolute temperature T, i.e.:

$$\frac{V}{T} = a\ constant \tag{6.2}$$

   Absolute temperatures are measured on the kelvin scale. To convert from temperatures on the Celsius scale, a reasonable approximation is to add 273. Thus 0°C = 273 K.

CHAPTER 6

3. *The pressure law*
   In an ideal gas in which the mass and volume remain constant, the pressure p varies directly as the absolute temperature T, i.e.:

$$\frac{p}{T} = \text{a constant} \qquad (6.3)$$

The combination of the above three gas laws results in the *general gas equation*, i.e.:

$$\frac{pV}{T} = \text{a constant} \qquad (6.4)$$

The constant is for a particular mass of a particular gas and thus the equation can be written in terms of the m kg of gas and a constant R which is termed the *characteristic gas constant* for the gas, i.e.:

KEY POINT

An ideal gas is one which obeys the gas laws at all temperatures and pressures.

$$\frac{pV}{T} = mR$$

$$\text{or} \quad pV = mRT \qquad (6.5)$$

Test your knowledge 6.7

1. What is the SI unit of pressure?
2. What is the difference between absolute pressure and gauge pressure?
3. What defines 1 bar of pressure?
4. Which ideal gas law will be needed to determine the effect on the pressure of a gas when its volume is changed at constant temperature?
5. Which ideal gas law will be needed to determine the effect on the pressure of a gas when, with its volume kept constant, its temperature is changed?

Example 6.1

A container has a volume of 0.10 m³ and is filled with compressed air at a gauge pressure of 600 kPa and a temperature of 40°C. If the atmospheric pressure is 101 kPa, determine the gauge pressure in the container when the air cools to 20°C. Neglect any change in dimensions of the container as a result of the temperature change.

The volume is assumed to be constant, thus using equation (6.3):

$$\frac{p_1}{T_1} = \frac{p_2}{T_2}$$

where p_1 and T_1 are the initial pressure and temperature and p_2 and T_2 the final pressure and temperature. Transposing to make p_2 the subject gives:

$$p_2 = \frac{p_1 T_2}{T_1} = \frac{(600 + 101) \times 10^3 \times (273 + 20)}{273 + 40}$$

$$p_2 = 656 \times 10^3 \text{ Pa}$$

This is an absolute pressure of 656 kPa so the gauge pressure will be 656 − 101 = 555 kPa.

CHAPTER 6

Free air

In discussing compressors and air distribution, the term *free air* is often used. Free air is defined as air at normal atmospheric temperature and pressure (n.t.p.), i.e. at a temperature of 15°C and a pressure of 101.3 kPa. Compressors, for example, are specified in terms of the *free air delivered* (f.a.d.), this being the volume a given quantity of compressed air would occupy at normal atmospheric pressure and temperature. This enables comparisons between compressors to be more easily made. The gas laws can be used to convert 'free air' to other pressures and/or temperatures or vice versa.

Example 6.2

(a) A compressor has a rated output of 3 $m^3 min^{-1}$ free air delivery at n.t.p. What will be the output at an absolute pressure of 700 kPa and 20°C?

(b) A compressor is required to deliver 0.2 $m^3 min^{-1}$ of compressed air at an absolute pressure of 500 kPa and temperature 25°C. What compressor output is required in terms of free air at n.t.p.?

Let n.t.p. conditions be $p_1 = 101.3$ kPa and $T_1 = 15$°C, and the free air delivery rate be V_1. Let compressor output conditions be p_2, V_2 and T_2.

(a) Finding compressor output flow rate

$$\frac{p_1V_1}{T_1} = \frac{p_2V_2}{T_2}$$

$$V_2 = \frac{p_1V_1T_2}{p_2T_1} = \frac{101.3 \times 10^3 \times 3 \times (20+273)}{700 \times 10^3 \times (15+273)}$$

$$V_2 = 0.44\,m^3\,min^{-1}$$

(b) Finding required compressor rating at n.t.p.

$$\frac{p_1V_1}{T_1} = \frac{p_2V_2}{T_2}$$

$$V_1 = \frac{p_2V_2T_1}{p_1T_2} = \frac{500 \times 10^3 \times 0.2 \times (15+273)}{101.3 \times 10^3 \times (25+273)}$$

$$V_1 = 0.95\,m^3\,min^{-1}$$

Air receiver sizing

Charts and formulae are available for the determination of the size of an air receiver. Alternatively the size can be determined from first principles by the use of the gas laws, the following example illustrates this.

Example 6.3

A pneumatic system requires an average delivery volume of 20 $m^3 min^{-1}$ free air delivery. The air compressor has a rated free air delivery of 25 $m^3 min^{-1}$ and a working gauge pressure of 7 bar, being controlled to switch off load when the receiver gauge pressure rises to 7 bar and back on load when the receiver gauge pressure has dropped to 5 bar. Determine a suitable receiver capacity if the maximum allowable number of starts per hour for the compressor is 20. The air temperature can be assumed to be constant throughout.

Taking atmospheric pressure as 1 bar the absolute pressure in the receiver will fall from $p_1 = 8$ bar to $p_2 = 6$ bar. Twenty starts per hour means that the average time between starts is 3 minutes. In 3 minutes the system demand is $3 \times 20 = 60\,\text{m}^3$ free air. To supply this, the compressor must run for $60/25 = 2.4$ minutes. Thus the compressor is off load for 0.6 minute. During this time the receiver has to supply the system while the gauge pressure in the receiver falls from 7 to 6 bar. Let the volume of free air supplied from the receiver in this time be $V_o = 0.6 \times 20 = 12\,\text{m}^3$ at a pressure $p_o = 1$ bar.

Consider the mass m_o of the free air supplied during the 0.6 minute period:

$$p_o V_o = m_o RT$$

$$m_o = \frac{p_o V_o}{RT}$$

Consider the mass m of the air in the receiver at a pressure of $p_1 = 8$ bar and volume V:

$$p_1 V = mRT$$

$$m = \frac{p_1 V}{RT}$$

Consider the air in the receiver at a pressure of $p_2 = 6$ bar with the same volume V:

$$p_2 V = (m - m_o)RT$$

Substituting for m and m_o gives:

$$p_2 V = \left[\frac{p_1 V}{RT} - \frac{p_o V_o}{RT}\right] RT$$

$$p_2 V = p_1 V - p_o V_o$$

$$p_o V_o = p_1 V - p_2 V = V(p_1 - p_2)$$

$$V = \frac{p_o V_o}{(p_1 - p_2)} = \frac{1 \times 12}{(8 - 6)}$$

$$V = 6\,\text{m}^3$$

If we assume that the temperature is constant at a normal 15°C, Boyle's law gives for the volume of free air V_1, i.e. at pressure $p_1 = 1$ bar, in a receiver of volume V_2 containing air at a gauge pressure of 7 bar, i.e. absolute pressure $p_2 = 7 + 1 = 8$ bar:

$$p_1 V_1 = p_2 V_2$$

$$\text{Volume of free air} = 8V/1 = 8V\,\text{m}^3$$

The volume of free air in the receiver containing air at a gauge pressure of 6 bar is:

$$\text{Volume of free air} = 7V/1 = 7V\,\text{m}^3$$

Thus the volume of free air delivered from the receiver as the pressure falls from 7 to 6 bar is $8V - 7V = 1V\,\text{m}^3$. Thus $1V = 12\,\text{m}^3$ and so the receiver volume required is $12\,\text{m}^3$.

Moisture in air

Dryers are generally used to remove water vapour from the air leaving compressors. In discussing the quantity of water vapour in air, the following terms are used:

1. *Saturation*
   Air is said to be saturated when it contains the maximum amount of water vapour which it can hold at a particular temperature.
2. *Relative humidity*
   The relative humidity is a measure of the amount of water vapour present in a sample of air at a particular temperature, being defined as:

$$\text{Relative humidity} = \frac{\text{Mass of water present in air sample}}{\text{Mass of water present in air sample when saturated}}$$

   If we have air at 20°C containing 0.005 kg m^{-3} of water then, since saturation requires about 0.0015 k m^{-3} at that temperature, the relative humidity is:

$$RH = \frac{0.005}{0.0015} \times 100 = 33\%$$

KEY POINT

Relative humidity of air at a particular temperature is the moisture content given as a percentage of moisture content when saturated at that temperature.

3. *Dew point*
   The dew point is the temperature at which a sample of air becomes saturated. For example if we have air containing 0.001 kg m^{-3} of water it will have to be cooled to −20°C before it becomes saturated and liquid starts to condense out. Thus −20°C is the dew point of that sample. Table 6.3 gives some dew point values.

Table 6.3 Dew point values

Water content (kg m^{-3})	Dew point (°C)
0.0049	0
0.0068	5
0.0094	10
0.0128	15
0.0173	20
0.0231	25
0.0304	30
0.0396	35

Flow through pipes

Consider the flow of a fluid through a pipe (Figure 6.36) and the rate at which fluid flows through the section AA. If the fluid has an average velocity v then by definition it will have moved through a distance v metres in a time of 1 second. Thus the volume-flow rate Q of fluid passing through section AA where A is the cross-sectional area of the pipe is given by:

$$Q = Av \text{ m}^3 \text{ s}^{-1} \quad (6.6)$$

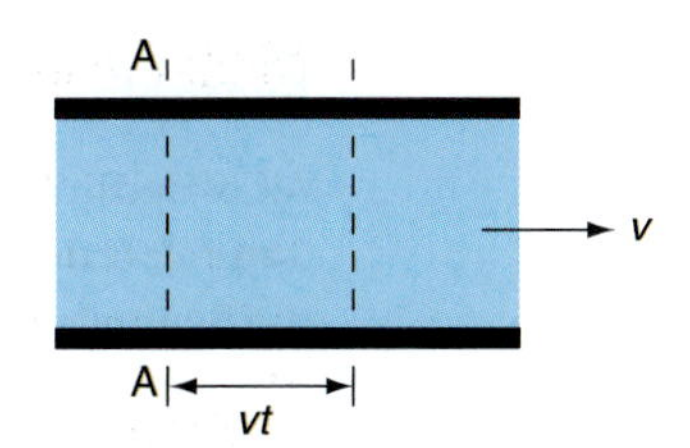

Figure 6.36 Flow through a pipe

For a fluid of density $\rho \text{kg m}^{-3}$, the mass-flow rate m of the will be:

$$m = \rho Q \text{ kg s}^{-1}$$

$$\text{or} \quad m = \rho A v \text{ kg s}^{-1} \quad (6.7)$$

Bernoulli effect

The Bernoulli effect is sometimes encountered in pipelines where there is a change in diameter or a restriction to the flow. If the cross-section of the flow passage is reduced, the flow velocity increases in this locality. This causes the kinetic energy of the fluid to increase. You might think that the pressure in the fluid would also increase but the opposite happens and there is a drop in pressure. This is called the *Bernoulli effect* which is put to good use in some flow applications. The Bernoulli effect contributes to the lift force on aircraft wings. As the airflow speeds up over the upper surface it causes a pressure drop which is greater than the pressure increase on the lower surface, and this produces the greater part of the lift force. You can demonstrate the effect by blowing over the top surface of a sheet of paper held at one end. The paper rises due to the pressure difference that you have created.

The effect is also put to use in flow measuring devices such as venture meters and orifice meters which are known as *differential pressure* flow metering devices. The pressure drop at a restriction increases with flow velocity and although the two are not directly proportional, the pressure drop can be measured and used to calculate the flow rate or display it on a calibrated scale.

Bernoulli's equation, which is used to calculate the pressure change, is based on the principle of conservation of energy. You will find out how it is derived if you study the BTEC Fluid Mechanics and Applications unit. We will not state it here in full but we will use the formula for pressure change that is obtained from it. If v_1 and v_2 are the velocities at a point where there is a change in cross-section and $\rho \text{ kg m}^{-3}$ is the fluid density, the pressure change Δp can be calculated using:

$$\Delta p = \frac{\rho}{2}(v_1^2 - v_2^2) \quad (6.8)$$

If the flow velocity increases, Δp will be negative, denoting a pressure drop and it will be positive if there is a fall in velocity, denoting a pressure increase.

Example 6.4

Air of density 1.24 kg m^{-3} flows through a pipe at a mass-flow rate of flow of 3 kg s^{-1}. If the pipe diameter gradually changes from 300 to 200 mm calculate the flow velocity at the two sections and the pressure change that occurs.

Finding cross-sectional area of pipe where $d_1 = 300$ mm:

$$A_1 = \frac{\pi d_1^2}{4} = \frac{\pi \times 0.3^2}{4}$$

$$\mathbf{A_1 = 0.0707\,m^2}$$

Finding flow velocity at this section

$$m = \rho A_1 v_1$$

$$v_1 = \frac{m}{\rho A_1} = \frac{3.0}{1.24 \times 0.0707}$$

$$\mathbf{v_1 = 34.2\,m\,s^{-1}}$$

Finding cross-sectional area of pipe where $d_2 = 200$ mm

$$A_2 = \frac{\pi d_2^2}{4} = \frac{\pi \times 0.2^2}{4}$$

$$\mathbf{A_2 = 0.0314\,m^2}$$

Finding flow velocity at this section

$$m = \rho A_2 v_2$$

$$v_2 = \frac{m}{\rho A_2} = \frac{3.0}{1.24 \times 0.0314}$$

$$\mathbf{v_2 = 77.0\,m\,s^{-1}}$$

Finding pressure change

$$\Delta p = \frac{\rho}{2}(v_1^2 - v_2^2) = \frac{1.24}{2}(34.2^2 - 77.0^2)$$

$$\mathbf{\Delta p = -2.95 \times 10^3\,Pa \quad or \quad -2.95\,kPa}$$

i.e. there is a pressure drop at the change in diameter.

Hydraulic Pump Power

If a pump forces fluid along a pipe of cross-sectional area A against a pressure p and moves it a distance x in a time t then the work done in that time is:

$$\text{Work done} = \text{Force} \times \text{Distance moved}$$
$$= \text{Pressure} \times \text{Area} \times \text{Distance moved}$$
$$\text{or} \quad W = pAx$$

The power developed is the work done per second, i.e. the rate of working:

$$\text{Power} = \frac{\text{Work done}}{\text{Time taken}} = \frac{W}{t}$$

$$\mathbf{Power} = \frac{\boldsymbol{pAx}}{\boldsymbol{t}} \qquad (6.9)$$

CHAPTER 6

But Ax is the volume moved in time t and Ax/t is the volume-flow rate Q. Hence:

$$\textbf{Power} = pQ \tag{6.10}$$

This is the power required to deliver fluid at this pressure and rate of flow. The units of power are of course watts or kilowatts.

Example 6.5

(a) Determine the power required to pump oil at a pressure of 5 MPa with a speed of 2.5 m s^{-1} in a 20 mm diameter pipe.

(b) Determine the flow velocity in a pipe 25 mm in diameter when the delivery pressure is 6 MPa and the power input is 10 kW.

(a) Finding cross-sectional area of pipe

$$A = \frac{\pi d^2}{4} = \frac{\pi \times 0.02^2}{4}$$

$$\mathbf{A = 314 \times 10^{-6}\ m^2}$$

Finding volume-flow rate

$$Q = Av = 314 \times 10^{-6} \times 2.5$$

$$\mathbf{Q = 785 \times 10^{-6}\ m^3\ s^{-1}}$$

Finding power required

$$\text{Power} = pQ = 5 \times 10^6 \times 785 \times 10^{-6}$$

$$\mathbf{Power = 3.93 \times 10^3\ W \quad or \quad 3.93\ kW}$$

(b) Finding cross-sectional area of pipe

$$A = \frac{\pi d^2}{4} = \frac{\pi \times 0.025^2}{4}$$

$$\mathbf{A = 491 \times 10^{-6}\ m^2}$$

Finding volume-flow rate

$$\text{Power} = pQ$$

$$Q = \frac{\text{Power}}{p} = \frac{10 \times 10^3}{6 \times 10^6}$$

$$\mathbf{Q = 1.67 \times 10^{-3}\ m^3\ s^{-1}}$$

Finding flow velocity

$$Q = Av$$

$$v = \frac{Q}{A} = \frac{1.67 \times 10^{-3}}{491 \times 10^{-6}}$$

$$\mathbf{v = 3.4\ m\ s^{-1}}$$

Linear actuators

For a single-acting linear actuator as shown originally in Figure 6.1(b), the fluid is applied to one side of the piston at a gauge pressure p with the other side being at atmospheric pressure. This produces a force F on the piston of area A, given by:

$$F = pA \tag{6.11}$$

To determine the actual force F' acting on the piston rod we also have to take account of friction and also the return spring force if the cylinder is single acting. The output force is thus less than the input force, i.e. $F' < F$. The *efficiency* η of a cylinder is defined by:

$$\text{Efficiency} = \frac{\text{Actual force in piston rod}}{\text{Force on piston}}$$

$$\text{or} \quad \eta = \frac{F'}{F} \tag{6.12}$$

If the piston moves at speed v the volume rate of flow of fluid into the cylinder is given by $Q = Av$. The operational speed of an actuator is thus determined by the fluid flow rate and the actuator area. The speed with which a piston moves in a cylinder depends on the rate at which fluid can enter and leave the cylinder. Thus we can control the speed by controlling the flow of fluid entering the cylinder, the so-termed *meter-in* situation as shown in Figure 6.37(a), or by controlling the rate at which fluid is allowed to leave the cylinder, this so-termed *meter-out* situation is shown in Figure 6.37(b). Meter-out control can be achieved by using a flow control valve and non-return valve in the line between the cylinder and the direction control valve or by using a flow control valve fitted to the exhaust port of the direction control valve.

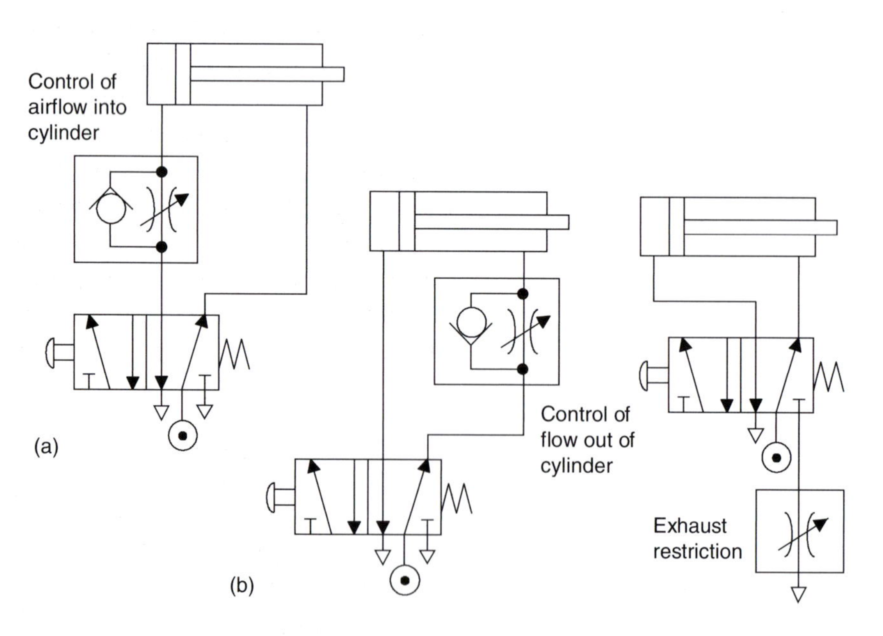

Figure 6.37 (a) Meter-in control and (b) two alternatives for meter-out control

With hydraulic systems, meter-in and meter-out control can be used in the same way. In addition, speed control can be exercised by *bleed-off*. A bleed-off valve allows a variable amount of the incoming fluid to be extracted and so control the amount entering the cylinder as shown in Figure 6.38.

CHAPTER 6

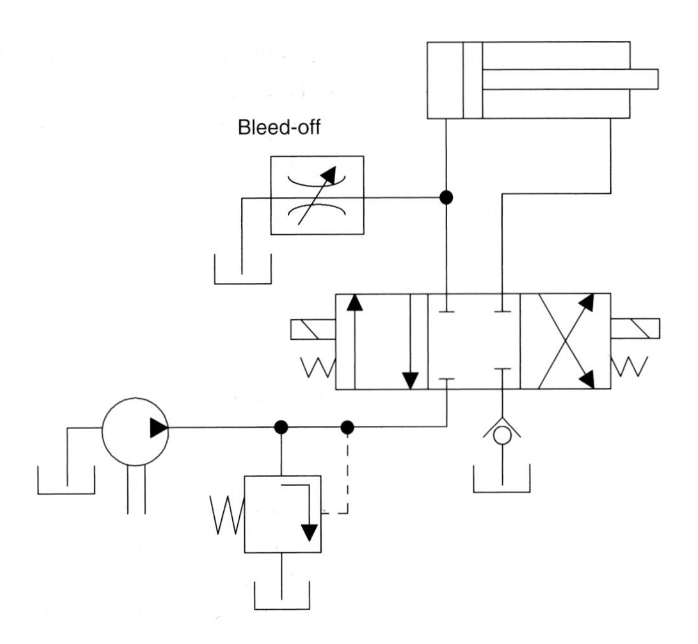

Figure 6.38 Bleed-off speed control

With the double-acting cylinder shown in Figure 6.39(a), the force produced by the pressure on one side of the piston to extend it is different to when the pressure is applied to the other side of the piston to retract it. This is because on one side we have the full surface area of the piston exposed while on the other it is reduced by the area occupied by the piston rod. Thus for a piston of area A the extend force for a pressure p is pA while the retract force is $p(A - \pi r^2)$, where r is the radius of the piston rod. The arrangement shown in Figure 6.38(b) can give equal extend and retract forces when the ratio of the area of the piston to that of the rod is 2:1. This is because on the retraction stroke the pressure p is applied to the annulus of area $A/2$ and so gives a force of $pA/2$. On the extension stroke the pressure is applied to both sides of the piston and so on one side is applied to area A and on the other side $A/2$ with the result that the resultant force is $pA - pA/2 = pA/2$.

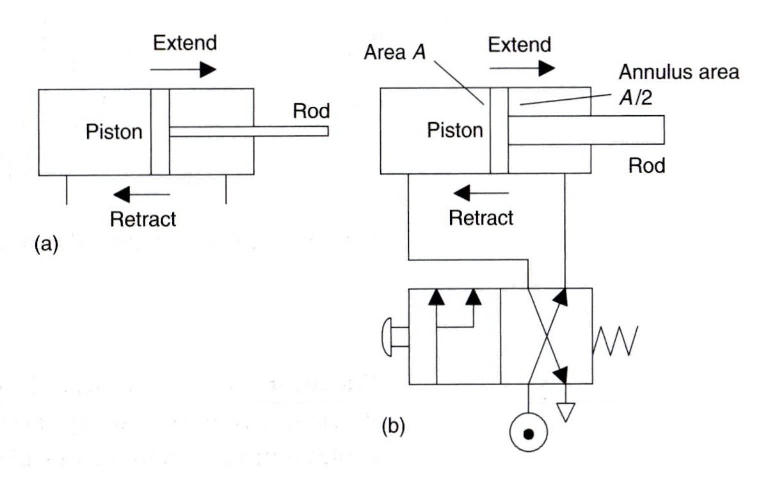

Figure 6.39 Double-acting cylinder: (a) unequal extend and retract forces and (b) equal extend and retract forces

Test your knowledge 6.8

1. What does the term 'free air' mean?
2. What are regarded as being normal temperature and pressure?
3. What is relative humidity?
4. How can pumping power be calculated from delivery pressure and volume-flow rate?
5. What form of double-acting cylinder is needed if the applied extension force is to be equal to the applied retraction force?
6. How can the speed of a linear actuator be controlled?

Example 6.6

(a) A single-acting pneumatic cylinder has an internal diameter of 50 mm and a piston which is required to exert a force of 1.5 kN. What gauge pressure is required in the system if friction is neglected?

(b) A pneumatic single-acting cylinder is to be used to clamp work pieces in a machine tool. The piston area is 0.01 m² and the system pressure is 3.5 bar. Determine the clamping force taking frictional losses to be 5%.

(a) Assuming that there is only atmospheric pressure acting on one side of the piston, the system gauge pressure will be the pressure difference across the piston. Finding area of piston

$$A = \frac{\pi d^2}{4} = \frac{\pi \times 0.05^2}{4}$$

$$\mathbf{A = 1.96 \times 10^{-3}\ m^2}$$

Finding system pressure required

$$p = \frac{F}{A} = \frac{1.5 \times 10^3}{1.96 \times 10^{-3}}$$

$$\mathbf{p = 765 \times 10^3\ Pa \quad or \quad 7.65\ bar}$$

(b) Finding efficiency of the cylinder

$$\eta = 100 - 5 = 95\% \quad \text{or} \quad 0.95$$

Finding clamping force

$$F = \frac{pA}{\eta} = \frac{3.5 \times 10^5 \times 0.01}{0.95}$$

$$\mathbf{F = 3.68 \times 10^3\ N \quad or \quad 3.68\ kN}$$

Example 6.7

(a) Determine the free air volume rate of fluid flow required for a single-acting pneumatic cylinder if the system gauge pressure is 5 bar, the piston has a diameter of 80 mm, the stroke length is 400 mm and the cylinder completes 10 strokes and returns per minute.

(b) Calculate the extension force and retraction force for a double-acting cylinder with a piston of diameter 50 mm and a rod of diameter 25 mm when the applied pressure is 4 MPa. Assume 95% efficiency.

(a) Finding volume required per minute supplied at 5 bar pressure

$$V = \text{Piston area} \times \text{Length of stroke} \times \text{Strokes per minute}$$

$$V = \frac{\pi d^2 ln}{4} = \frac{\pi \times 0.08^2}{4} \times 0.4 \times 10$$

$$\mathbf{V = 0.021\,m^3}$$

Converting this to free air conditions at n.t.p. assuming normal temperature of 15°C throughout, i.e. using Boyle's law equation and normal free pressure of $p_n = 101.3\,\text{kPa}$.

$$p_n V_n = pV$$

$$V_n = \frac{pV}{p_n} = \frac{(5+1) \times 10^5 \times 0.021}{101.3 \times 10^3}$$

$$\mathbf{V_n = 0.124\,m^3\,min^{-1}}$$

(b) Finding areas A_1 and A_2 on either side of piston

$$A_1 = \frac{\pi d_1^2}{4} = \frac{\pi \times 0.05^2}{4} = 1.96 \times 10^{-3}\,\text{m}^2$$

$$A_2 = \frac{\pi(d_1^2 - d_2^2)}{4} = \frac{\pi(0.05^2 - 0.025^2)}{4} = \mathbf{1.47 \times 10^{-3}\,m^2}$$

Finding extension force

$$F_1 = \eta p A_1 = 0.95 \times 4 \times 10^6 \times 1.96 \times 10^{-3}$$

$$\mathbf{F_1 = 7.45 \times 10^3\,N \text{ or } 7.45\,kN}$$

Finding retraction force

$$F_2 = \eta p A_2 = 0.95 \times 4 \times 10^6 \times 1.47 \times 10^{-3}$$

$$\mathbf{F_2 = 5.59 \times 10^3\,N \text{ or } 5.59\,kN}$$

To check your understanding of the preceding section, you can solve Review questions 8–15 at the end of this chapter.

Maintenance of Fluid Power Systems

The term *maintenance* is used for the combinations of actions carried out to return an item to an acceptable condition. The term *preventative maintenance* is used for maintenance carried out at predetermined intervals, or to other prescribed criteria, and intended to reduce the likelihood of an item not meeting an acceptable condition. The term *corrective maintenance* is used for maintenance carried out to restore an item which has ceased to meet an acceptable condition. Thus, for example, preventative maintenance is used for a car with annual servicing or servicing at specified mileages, while corrective maintenance is used when it has to be repaired because it has broken down. Preventative maintenance is a key method of controlling the level of corrective maintenance that might be required. If a car is not

KEY POINT

Maintenance is the actions carried out to return an item to an acceptable condition. *Preventative maintenance* is maintenance carried out at predetermined intervals and *corrective maintenance* is maintenance carried out to restore an item which has ceased to meet an acceptable condition.

regularly serviced, then it is more likely to break down. Thus preventative maintenance can reduce the amount of corrective maintenance.

In considering the best way to maintain an item, the following factors should be taken into account:

1. *Maintenance characteristics*
   There are two key factors, the deterioration characteristic and the repair characteristic. With regard to deterioration, the mean life of an item before it fails is a good indicator of the need for maintenance. If the mean life of an item is less than that of the system in which it is used, then maintenance is likely to be required. Items can be classified as: replaceable items that are likely to have to be replaced during the life of the system and permanent items that are unlikely to have to be replaced. The repair characteristic is indicated by the mean time to repair or replace.
2. *Economic factors*
   What is the cost of replacement prior to failure? What is the cost of unexpected failure? Failure might mean a system is out of action for some time and this could have economic implications. What is the repair cost?
3. *Safety factors*
   What are the safety implications of an item failing? If an item is critical and failure would result in an unsafe situation, preventative maintenance of that item is indicated rather than waiting for it to fail and use corrective maintenance.

The *maintenance plan* consists of a schedule of preventative maintenance work and guidelines for the implementation of corrective maintenance work. The plan determines the level and nature of the maintenance workload. It can involve:

1. *Fixed-time maintenance*
   The maintenance is carried out at regular intervals or after a fixed number of cycles of operation, etc. This procedure can be effective if the failure mechanism is time dependent.
2. *Condition-based maintenance*
   If the approach of failure is detectable, then condition-based maintenance is effective. Thus, for example, wear of a component can be used as an indicator of the approach of failure and maintenance based on inspection for wear. Condition-based monitoring can be simple inspection involving qualitative checks based on observation, condition checking where routine measurements are made of some quantity, or trend monitoring where measurements are made and the trend of those measurements used to forecast when failure is likely to occur.

A weekly maintenance schedule for a compressor might read:

1. Check temperature of motor. It should not be above 75°C.
2. Check noise level.
3. Clean intake filter.
4. Check V-belt tension.
5. Check compressor oil level.
6. Test safety valve.

A record of corrective maintenance for a week might include:

1. Cylinder XXX, replaced broken rod.
2. Solenoid XXX faulty and replaced.
3. Filter on airline blocked and so cleaned.
4. Faulty pressure gauge replaced.
5. Solenoid valve XXXX sticking, replaced.
6. Dirty airline cleaned out.

Inspection and testing

In hydraulic systems, oil problems are responsible for a high percentage of failures. These problems may be due to the level of oil in the reservoir or contamination by dirt, air or water. Preventative maintenance might thus involve inspection of the oil level, visual checks for oil leakage, and checking of the differential pressure across a filter element, a large drop indicating the need to change a filter before it becomes blocked. Replacement of filters might be an aspect of fixed-time preventative maintenance.

With pneumatic systems, problems can arise from contamination by dirt or water and, as with hydraulic systems, filters need to be checked and maintained. This might thus be part of preventative maintenance for such systems.

Cylinders, seals, linkages and bearings require regular inspection as part of preventative maintenance with replacement if damaged. The cylinder end caps should be removed at regular intervals and the piston assembly cleaned and inspected. Score marks can indicate dirt ingress. For long cylinder life it is essential that all dirt and moisture should be excluded from the cylinder and, in the case of pneumatic systems, the air should be adequately lubricated.

With valves, it is possible for a valve to stick due to a build-up of contamination. Cleaning might thus be required and 'O' ring seals replaced. With solenoid valves, care must be taken when replacing failed solenoids to ensure that the operating voltage of the replacement is correct.

With pumps, a noisy pump might be an indicator of air leaking into the system, a clogged or restricted intake line, a plugged air vent in the reservoir, loose or worn pump parts. Thus just listening to a pump can be a useful indicator of the need for maintenance. The oil temperature is another useful indicator. A high temperature can arise because of a clogged oil cooler, low oil, the oil used having too high a viscosity, the relief valve allowing the flow discharging with too high a pressure drop, or perhaps high internal leakages from perhaps wear. Compressors are often belt driven and the belt condition and tension requires regular checking as part of preventative maintenance.

Fault-finding

Systematic fault-finding with fluid power systems basically involves a *six-point* approach:

1. *Collect evidence of the fault.* This can include observing of the system in operation if it is safe to do so and questioning the operator. Also, refer to maintenance records, i.e. past breakdowns, planned maintenance, modifications, etc.
2. *Analyse the evidence.* Consider all possible causes of the malfunction.
3. *Locate the fault.* If this is not immediately apparent make use aids such as the system manual, functional charts and test equipment. Use a logical approach working from system input to output or use the *half-split* approach, which will be explained later.
4. *Determine and remove the cause.* This might involve the removal of damaged or faulty components and piping.
5. *Rectify the fault.* Repair or replace the components or piping in question.
6. *Check the system.* Test the safe working of the plant before returning it to operation and update the service records.

Safety is paramount in maintenance activities and appropriate personal protective equipment should be used at all times. System shut-down, isolation and permit to work procedures should be strictly observed before any repair work or unit replacement is carried out.

Fault-finding aids include *functional charts*. These are charts of what the state of each cylinder, valve and sensor should be for each step of the sequence of operations. Figure 6.40 gives an illustrative example (it is the chart for the circuit shown in Figure 6.35). The first column

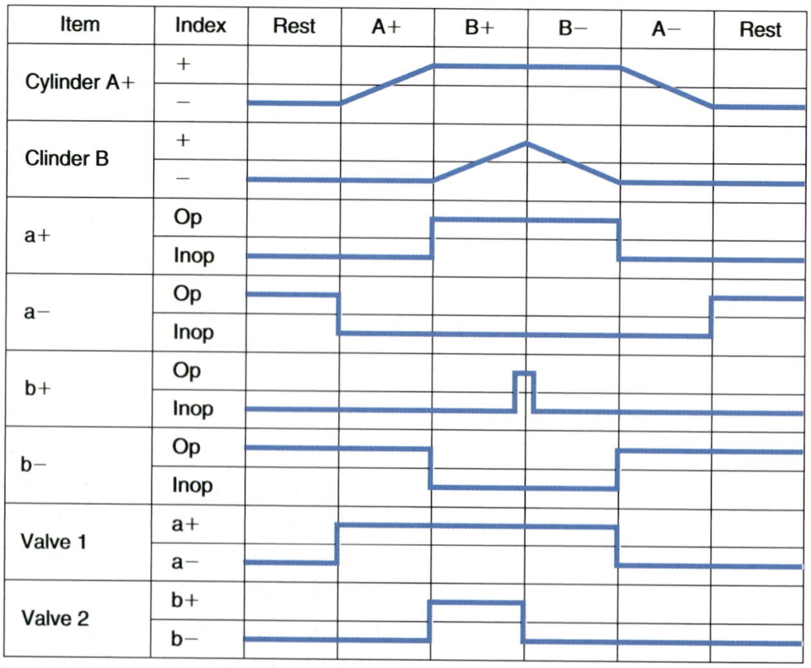

Item	Index	Rest	A+	B+	B−	A−	Rest
Cylinder A+	+						
	−						
Clinder B	+						
	−						
a+	Op						
	Inop						
a−	Op						
	Inop						
b+	Op						
	Inop						
b−	Op						
	Inop						
Valve 1	a+						
	a−						
Valve 2	b+						
	b−						

Figure 6.40 Illustrative example of a functional chart

lists each item of equipment in the system. The second column, labelled index, indicates the conditions that item can have, e.g. extended (+) or retracted (−), on or off, operative or inoperative. The next column indicates the state of each item in the rest condition. The following columns then indicate the states of each item for each step in the sequence of operations. The final column indicates the stop states.

Flow charts can also be developed to aid in a systematic search for faults in a system. Thus, for a hydraulic circuit we might have the following block items:

1. Check the level of oil in the supply tank. Is the level correct? If not, refill.
2. Check whether the pump motor is running. If not there is a motor fault.
3. Is the cut-off valve to the cylinder open? If not, open it. etc.

In fault-finding it is necessary to have the full documentation relating to the system. This should have been supplied by the manufacturers of the system and include:

1. System layout diagram with labelled valves and lines.
2. Circuit diagrams.
3. List of components, and spare parts, with component data sheets.
4. Displacement-step diagrams.
5. Operating instructions.
6. Installation and maintenance manuals.

There are different approaches to locating system faults. If a *visual examination* reveals the possible cause of the malfunction the maintenance technician can remove and replace suspect components. This is termed *unit substitution*. If the cause is not apparent, a logical approach is to work through the system from *input to output* carrying out pressure and function checks on each component in turn.

Depending on the kind of system, an alternative approach is the *half-split method*. This can save a considerable amount of time in locating a fault. Suppose a system has eight distinct stages or components, numbered 1 to 8. The method is to start at the output from stage 4 to check that there is a correct pressure or electrical signal passing to stage 5. If no such signal is present, the fault must be between stages 1 and 4. The process can then be repeated at the output from stage 2. If no signal is present, the fault must be in stage 1 or 2. Checking the output from stage 1 will then identify the source of the fault and this will have been done in three operations.

In fully automated plant such as robotic systems and food processing it is common practice to have condition sensors integral with the major components which send warning signals to a control centre should malfunction occur. Maintenance technicians are then able to home-in immediately on the location of the fault. With modern process and condition monitoring systems that incorporate hard-wire or telemetry links and programmable electronic equipment, the move is towards *self-diagnosis*, so that the time spent on fault-finding and system restoration is kept to a minimum.

Test your knowledge 6.9

1. What is condition-based maintenance?
2. What is fixed-time maintenance?
3. What is the six-point technique of fault-finding?
4. What is a functional chart?
5. What is the half-split method of fault-finding?

Activity 6.5

Draw a full functional chart for the circuit shown in Figure 6.33.

Review questions

1. For the circuit shown in Figure 6.41 state which valves need to be operated to (a) extend the piston, (b) retract the piston.

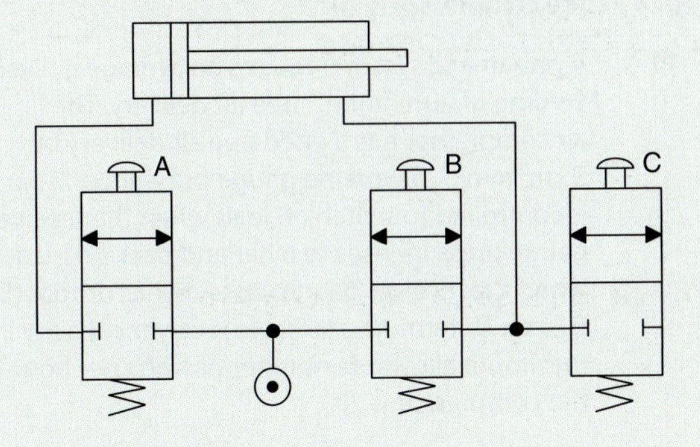

Figure 6.41

2. A pneumatically operated machine is required to have cylinder A push a component from a magazine into position, this action also ejecting the previous component, cylinder B clamps it, cylinder A retracts and then B unclamps so that the cycle can be repeated. Specify the actuator sequence required.

3. Determine the cylinder sequence for the circuit shown in Figure 6.42.

4. What sequence of cylinder actions is indicated by the displacement-event diagram in Figure 6.43?

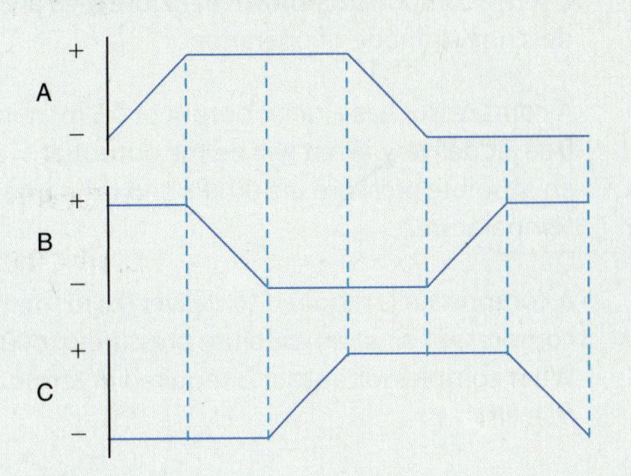

Figure 6.43

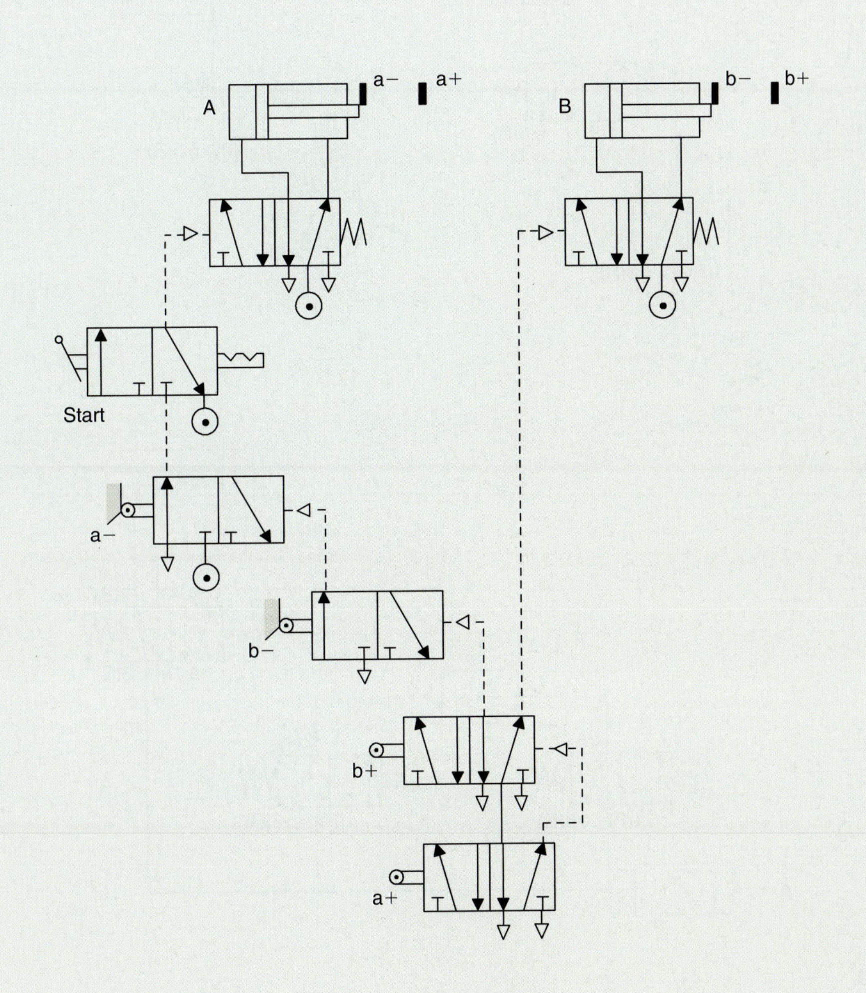

Figure 6.42

5. List the components shown in Figures 6.44(a) and (b) and compare the modes of operation of the two circuits.

6. List the components shown in Figure 6.45 and describe its mode of operation stating the logic function that it represents.

7. List the components shown in Figure 6.46 and describe its mode of operation.

8. A compressor has a rated output of $0.5\,m^3\,min^{-1}$ free air delivery. What will be the output at an absolute pressure of 500 kPa and the same temperature?

9. A compressor is required to deliver $0.6\,m^3\,min^{-1}$ of compressed air at an absolute pressure of 600 kPa. What compressor output is required in terms of free air?

10. A pneumatic system requires an average delivery volume of $20\,m^3\,min^{-1}$ free air delivery. The air compressor has a rated free air delivery of $35\,m^3\,min^{-1}$, a working gauge pressure of 6 bar and is controlled to switch off load when the receiver gauge pressure rises to 6 bar and back on load when the receiver gauge pressure has dropped to 5 bar. Determine a suitable receiver capacity if the maximum allowable number of starts per hour for the compressor is 20.

11. A pneumatic system requires an average delivery volume of $20\,m^3\,min^{-1}$ free air delivery. The air compressor has a rated free air delivery of $35\,m^3\,min^{-1}$, a working gauge pressure of 7 bar and is controlled to switch off load when the receiver gauge pressure rises to 6 bar and back on load when the receiver gauge pressure has dropped to 5.2 bar. Determine a suitable receiver capacity if the maximum allowable number of starts per hour for the compressor is 20.

12. Determine the power required to pump oil at a pressure of 6 MPa and rate of $0.2\,m^3\,min^{-1}$.

13. A pneumatic single-acting cylinder is to be used to clamp work pieces in a machine tool. The piston

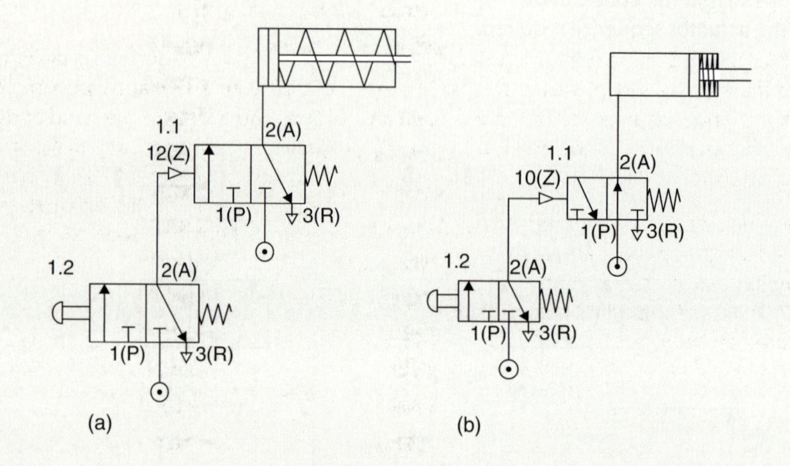

Figure 6.44

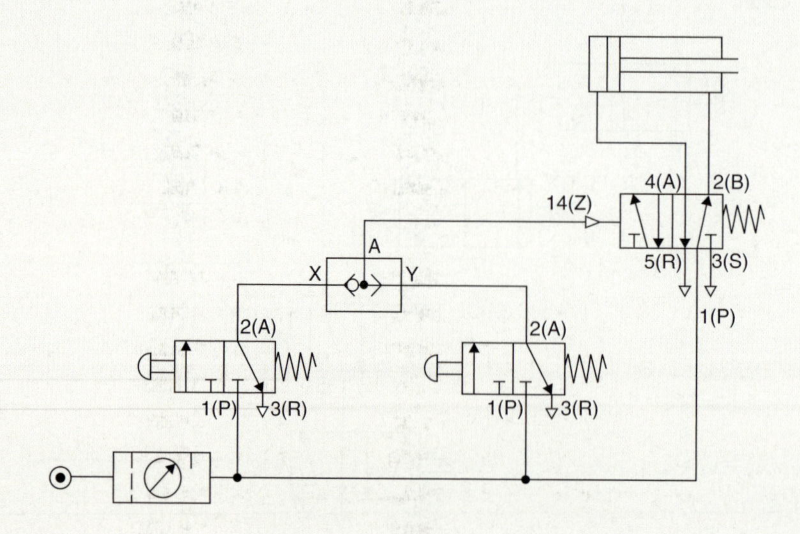

Figure 6.45

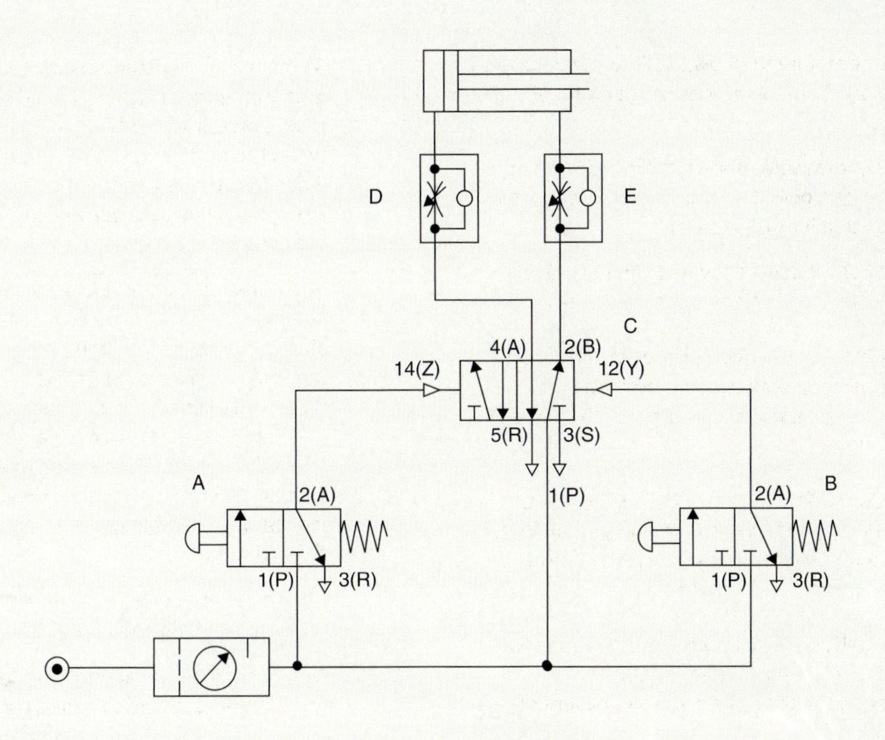

Figure 6.46

has a diameter of 80 mm and the required clamping force is 4 kN. Determine the system pressure that has to be applied to the cylinder to achieve this force. Neglect any forces due an in-built spring and take frictional losses to be 10%.

14. Determine the volume rate of flow required for a single-acting cylinder if the system gauge pressure is 6 bar, the piston has a diameter of 100 mm, the stroke length is 400 mm and the cylinder completes 5 strokes and returns per minute.

15. A single-rod, double-acting cylinder has a piston of diameter 40 mm and a piston rod of diameter 16 mm. Determine the extend and return forces if a gauge pressure of 7 bar is first applied to each side of the piston in turn with the other side in each case being connected to the exhaust line. Neglect any consideration of frictional forces.

Answers

Review questions

Chapter 1

1. 1.79 kN, 15° to horizontal, upward to the left.
2. 4.54 kN, 50.2° to horizontal, downward to the left.
3. 6.15 kN, 20.8° to horizontal, downward to the left.
4. (a) $R_X = 6.31\,\text{kN}$ $R_Y = 7.47\,\text{kN}$

Member	Force (kN)	Nature
AD	4.0	Tie
BD	5.66	Strut
BC	6.31	Strut
CD	4.63	Tie

(b) $R_X = 9\,\text{kN}$ $R_Y = 7.93\,\text{kN}$

Member	Force (kN)	Nature
AD	12.0	Tie
BD	10.4	Strut
BC	9.0	Strut
CD	5.21	Tie

(c) $R_X = R_Y = 5\,\text{kN}$

Member	Force (kN)	Nature
AD	9.66	Strut
BE	9.66	Strut
CD	7.07	Tie
CE	7.07	Tie
DE	3.66	Tie

5. (a) Maximum SF = 8 kN
Maximum BM = 21 kNm

(b) Maximum SF = 7 kN
Maximum BM = 7 kN

(c) Maximum SF = 5 kN

Maximum BM = 4.5 kNm

Point of contraflexure 2.4 m from left-hand end of beam

(d) Maximum SF = 12 kN

Maximum BM = 23 kNm

(e) Maximum SF = 7 kN

Maximum BM = 3 kNm

Point of contraflexure 1.58 m from left-hand end of beam

6. 0.18 mm, 72 MPa
7. (a) 66 MPa, (b) 131 MPa
8. 255 MPa, 1.27 mm, 111 MPa
9. (a) 10 MPa, (b) 73.6°C
10. 72 MPa
11. 143 MPa, 327 MPa, 1.24 mm
12. 42.4 MPa, 30.9 MPa, 0.6 mm
13. 103 MPa, 17.2 MPa
14. 92.9 MPa, 43.3 MPa, 0.068 mm
15. 53.6 MPa, 3.56 MPa, 0.77 mm
16. (a) 5 MPa, (b) 100×10^{-6}, (c) 50 GPa
17. (a) 648 kN, (b) 330 MPa
18. 6.86 mm, i.e use 7 mm dia rivets
19. 6.9 kN
20. (a) 10 rad s^{-2}, (b) 31.9 rad
21. 9.77 s
22. (a) 1.91 rad s^{-2}, (b) 3.27 s, (c) 679 J, (d) 208 W, (e) 314 W
23. (a) 0.21 rad s^{-2}, (b) 1 min, (c) 5.65 J, (d) 188 W, (e) 226 m
24. (a) 33.5 rad s^{-2}, (b) 0.56 rad s^{-2}
25. (a) 0.73 rad s^{-2}, (b) 35 revolutions, (c) 5.5 kJ, (d) 275 W, (e) 419 W
26. 54.5 Nm
27. (a) 5.24 rad s^{-2}, (b) 10 Nm
28. (a) 7.94 Nm, (b) 36.4 kJ, (c) 2.5 kW
29. (a) 0.0175 rad s^{-2}, (b) 913 mm
30. (a) 446 Nm, (b) 16.9 min
31. (a) 8.6 kgm^2, (b) 5.81 rad s^{-2}, (c) 64.9 s, (d) 673 kJ, (e) 10.4 kW
32. (a) 8 rad s^{-1}, (b) 2 rad s^{-2}, (c) 600 kJ, (d) 37.5 kNm, (e) 300 kW
33. (a) 31.8 s, (b) 169 kJ, (c) 5.3 kW
34. (a) 4.38 m, (b) 3.5 rad s^{-1}, (c) 4.57 kJ, (d) 522 Nm, (e) 914 W
35. 0.00669 kg m^2, 36.6 mm
36. 178 rpm
37. 49.1 mm
38. (a) 191 rpm, (b) 2.29 MW
39. (a) 622 N, (b) 124 Nm, (c) 13.1 kW
40. 6.4 kN, 961 N

41. 47.8 km h^{-1}
42. 84.4 m
43. 53.8 km h^{-1}, 1541 N, 4149 N
44. (a) 88.3 km h^{-1}, (b) 17.5°
45. (a) 118, (b) 377, (c) 31.3%
46. 200N, 6.83 kJ
47. (a) 24, (b) E = 0.085W + 13.3, (c) 42.3%
48. (a) 198, (b) 20%, (c) 100, 50.5%

Chapter 2

1. +2.04 mm, −0.0084 mm
2. +0.571 mm, −0.0366 mm
3. +0.0125 mm, +0.00162 mm
4. 160.024 mm, 69.985 mm
5. (a) +0.032 mm, (b) −0.055 mm, (c) +0.0493 mm, (d) 266 mm^3
6. −0.171 mm, −0.137 mm, −0.085 mm, 2.05 cm^3
7. 136.5 mm, 32.1%
8. 89.4 MPa, 112 m
9. 12.5 MPa, 360 m
10. 2 mm
11. 100 MPa, 4, 182 m
12. 28.8 × 57.6 mm, 76.8 m
13. (a) 25%, (b) 6.25%
14. (a) 38.5 MPa, (b) 1.1°
15. (a) 7.87, (b) 0.32° per metre
16. (a) 70 MPa, (b) 170 kW
17. 85.5 mm
18. (a) 80.3 mm, (b) 21 MPa
19. 304 km h^{-1}, 9.5° east of south
20. 1.7 m s^{-1}, 61.9° to axis of shaft
21. 57.25° east of north, 306 km h^{-1}
22. 36.1 km h^{-1} 26.3° west of north
23. (a) 58.3 km h^{-1}, (b) 3.65 km, (c) 15 min
24. 4.48 km, 6 min
25. 17° north of east, 73° north of east
26. (a) 40.5 m s^{-1}, (b) 48.6 m s^{-1}, (c) 252 rad s^{-1}
27. (a) 66 ms^{-1}, (b) 61.8 m s^{-1}, (c) 66 rad s^{-1}
28. (a) 0.22 m s^{-1}, (b) 2.13 rad s^{-1}, (c) 3.72 rad s^{-1}
29. (a) 0.524 m s^{-1}, (b) 1.1 m s^{-2}, (c) 0.291 m s^{-2}, 2.2 N
30. (a) 0.439 m, (b) 29.2 m s^{-1}, 1941 m s^{-2}, (c) 0.0946 s, 10.6 Hz
31. (a) 0.25 s, (b) 12.6 m s^{-1}, 316 m s^{-2}, (c) 11.5 m s^{-1}, 126.4 m s^{-2}
32. (a) 25.7 N, (b) 0.795 m s^{-1} (c) 1.1 s
33. (a) 0.477 m, (b) 0.274 m, (c) 0.1 s
34. (a) 0.45 s, (b) 0.35 m s^{-1}, (c) 4.9 m s^{-2}
35. (a) 2.49 Hz, (b) 0.469 m s^{-1}, (c) 7.34 m s^{-1}
36. (a) 3.7 Hz, (b) 0.233 m s^{-1}, 5.42 m s^{-2}, (c) 247N
37. (a) 6.3 kg, (b) 1.58 m s^{-2}, (c) 71.8 N
38. 1.9 s, 24.5 N m^{-1}

Chapter 3

1. (c)
2. (c)
3. (d)
4. (a)
5. (b)
6. (d)
7. (b)
8. (a)
9. (b)
10. (c)
11. (a)
12. (c)
13. (d)
14. (b)
15. (b)
16. (b)
17. (c)
18. (a)
19. (d)
20. (c)
21. (c)
22. (d)
23. (a)
24. (d)
25. (a)
26. (c)
27. (b)
28. (d)
29. (a)
30. (c)
31. (c)
32. (a)
33. (c)
34. (b)
35. (a)
36. (c)
37. (c)
38. (b)
39. (d)
40. (a)
41. (c)
42. (a)
43. (c)
44. (d)
45. (c)

Chapter 4

1. (c)
2. (b)
3. (d)
4. (a)
5. (b)
6. (c)
7. (d)
8. (a)
9. (d)
10. (b)
11. (d)
12. (d)
13. (b)
14. (d)
15. (a)
16. (b)
17. (b)
18. (c)
19. (c)
20. (d)
21. (c)
22. (d)
23. (a)
24. (a)
25. (d)
26. (c)
27. (b)
28. (d)
29. (a)
30. (d)
31. (c)
32. (a)
33. (d)
34. (c)
35. (b)

Chapter 5

No answers have been provided for the review questions in Chapter 5 as these are design problems and have no numerical answers.

Chapter 6

1. (a) A, (b) B or C
2. A+, B+, A−, B−
3. A+, B+, B−, A−
4. A+, B−, C−, C+, B+, A−
8. 0.18 $m^3 min^{-1}$
9. 3.55 $m^3 min^{-1}$
10. 12.9 m^3
11. 16 m^3
12. 20 kW
13. 884.2 kPa
14. 0.11 $m^3 min^{-1}$
15. 880 N, 740 N

Index